The Nude Mouse in Oncology Research

Editors
Epie Boven
Department of Medical Oncology
Free University Hospital
Amsterdam, The Netherlands

Benjamin Winograd
Bristol Myers-Squibb
Brussels, Belgium

CRC

CRC Press
Boca Raton Ann Arbor Boston London

Library of Congress Cataloging-in-Publication Data

The nude mouse in oncology research / editors, Epie Boven, Benjamin Winograd.
p. cm.
Includes bibliographical references and index.
ISBN 0-8493-6531-7
1. Cancer — Animal models. 2. Nude mouse. I. Boven, Epie. II. Winograd, B. (Benjamin), 1955- .
[DNLM: 1. Medical Oncology. 2. Mice, Nude. 3. Neoplasms, Experimental. QZ 206 N964]
RC267.N84 1991
616.99'4027—dc20
DNLM/DLC
for Library of Congress 91-29192
CIP

Direct all inquiries to CRC Press, Inc., 2000 Corporate Blvd., N.W., Boca Raton, Florida, 33431.

International Standard Book Number 0-8493-6531-7

Library of Congress Card Number 91-29192
Printed in the United States 1 2 3 4 5 6 7 8 9 0

PREFACE

Since it was shown in 1969 that human malignant tissue can be transplanted successfully in athymic (nu/nu) mice, these animals have become generally ac cepted as valuable tools to study the biology and therapy of human cancers. This volume in the Uniscience series will critically review the scientific work completed on the use of the nude mouse in oncology research. It will also be a reference book for those handling this animal in daily practice.

For the proper use of the nude mouse, extensive knowledge is required on the biology and immunology, the conditions for breeding and maintenance, and the measures for genetic and microbiological control. Desription of the transplantability of tumor tissue will include the influence of several host factors, the residual immune function, the site of transplantation, and hormanoal dependenc, as well as the quality of the transplant (viability, tumor tissue type, and histological appearance). A section will contain information on the characterization and the monitoring of tumor lines upon serial transfer. Aspects to be covered in the use of the nude mouse-human tumor model include studies in tumor biology (hormones, growth factors, drug resistance, tumor markers, and heterogeneity) and experimental therapy (radiotherapy, chemotherapy, hormones, monoclonal antibodies, biological response modifiers, differentiation inducers). The disadvantages of the model in respect of tumor biology or experimental therapy will be outlined. Differences between the nuse mouse model and man will be described with regard to the growth characteristics of tumors, the developement of metastases, the design of experimental therapy, pharmacology and metabolism, criteria for response to treatment, etc. The final section of the volume will cover the role of the nude mouse-human tumor model in new drug development.

THE EDITORS

Epie Boven, M.D., Ph.D., is a medical oncologist and Chief, Laboratory Section Experimental Therapy and Immunology, Department of Oncology, Free University Hospital, Amsterdam, The Netherlands.

Dr. Boven obtained her M.D. degree (with honors) in 1972 from the State University, Groningen, The Netherlands. In 1980, she was registered as a Doctor in Internal Medicine and started her training in medical oncology at the Free University, Amsterdam. In 1983, she received an NIH-visiting fellowship at the National Cancer Institute, national Institutes of Health, Bethesda. In 1986, she completed her Ph.D. thesis entitled "Human Ovarian Cancer Xenografts in Nude Mice. Application of a New Model in the Selection of Novel Agents for Ovarian Cancer".

Dr. Boven is a member if the American Association of Cancer Research, the American Society of Clinical Oncology, the European collaborative Group "Preclinical Phase II Studies", the Dutch Society of Oncology, and the Dutch Society of Pharmacy.

Dr. Boven has been the recipient of research grants from the Dutch Cancer Foundation and the Praeventiefonds. She has published more than 50 papers. Her current research interests are in the application of monoclonal antibodies in the treatment of cancer and in anticancer drug development. Her laboratory activities are closely connected to the design of clinical trials.

Benjamin Winograd, M.D., Director Clinical Cancer Research Europe, Bristol-Myers Squibb, Pharmaceutical Research Institute, Brussels, Belgium.

Dr. Winograd obtained his medical degrees in 1980 and 1981 from the Technical University, Medical School, Munich. His training in clinical pharmacology/oncology was funded by a Postgraduate Scholarship by the Deutsche Forschungs Gesellschaft, Germany and a research grant of the Dutch Cancer Socity. Dr. Winograd is a member of the American Association for Cancer Research, the European Society for Medical Oncology, the Dutch Society for Oncology and the Dutch Society for Clinical Pharmacology. In 1984 he was the first scientist to join the newly founded New Drug Development Office (NDDO) of the European Organization for Research on Treatment of Cancer (EORTC) in Amsterdam. Which he left as Associate Director in 1990.

Dr. Winograd has presented several invited lectures at International Meetings, has published more than 20 research papers and has organized the Fifth and Sixth NCI-EORTC Symposium on New Drugs in Cancer Therapy in amsterdam. During his time at the NDDO he has developed a special interest in the predictability of preclinical antitumor screening models and has focused on the validation of the Human Tumor Xenograft as a potential screening model.

CONTRIBUTORS

Ghanem Atassi
Director
Division Cancérologie Expérimentale
Institut de Recherches Servier
Suresnes, France

Frances R. Balkwill
Head of Biological Therapies Laboratory
Imperial Cancer Research Fund
Lincoln's Inn's Field
London, England

Dietmar P. Berger
Department of Internal Medicine
University of Freiburg
Freiburg, Germany

Epie Boven
Department of Medical Oncology
Free University Hospital
Amsterdam, Netherlands

Boudewijn J. M. Braakuis
Department of Otolaryngology
Free University Hospital
Amsterdam. Netherlands

Guus A. M. S. van Dongen
Department of Otolaryngology
Free University Hospital
Amsterdam, Netherlands

H. H. Fiebig
Department of Internal Medicine I,
Hematology/Oncology
University of Freiburg
Freiburg, Germany

Øystein Fodstad
Department of Tumor Biology
Norwegian Radium Hospital
Oslo, Norway

Raffaella Giavazzi
Instituto Mario Negri
Bergamo, Italy

Hidde J. Haisma
Department of Oncology
Free University Hospital
Amsterdam, Netherlands

David P. Houchens
Director of Laboratory and Clinical Sciences
Neoprobe Corporation
Columbus, Ohio

Simon Langdon
ICRF Medical Oncology Unit
Wester General Hospital
Edinburgh, Scotland

S. T. A. Malik
Biological Therapies Laboratory
Imperial Cancer Research Fund
London, England

Tatsuji Nomura
Central Inst./Exp. Animals
Kawasaki, Japan

Artemio A. Ovejera
Frederick, Maryland

Einar K. Rofstad
Department of Biophysics
Norwegian Radium Hospital

J. C. Romijn
Lab/Experimental Surgery
Erasmus University Rotterdam
Rotterdam, Netherlands

Jørgen Rygaard
Bartholin Institute Kommunehospitalet
Copenhagen, Denmark

John F. Smyth
Professor of Medical Oncology,
Director ICRF Laboratories
University Department of Clinical Oncology
Western General Hospital
Edinburgh, Scotland

G. J. van Steenbrugge
Lab/Experimental Surgery
Erasmus University Rotterdam
Rotterdam, Netherlands

Benjamin Winograd
Bristol Myers-Squibb
Brussels, Belgium

Bernd R. Winterhalter
Department of Internal Medicine I,
Hematology/Oncology
University of Freiburg
Freiburg, Germany

TABLE OF CONTENTS

Chapter 1

HISTORY AND PATHOLOGY

Jørgen Rygaard

TABLE OF CONTENTS

I. INTRODUCTION

History, according to my *Oxford English Dictionary,* can be defined as "A written narrative, constituting a continuous methodological record, in order of time, of important or public events, especially those connected with a particular country, people, individual, etc." Obviously, the individual of the present history is the nude mouse. However, thanks to modern technology, a comprehensive record of all published work on the nude mouse could be created by pressing a single — or at least very few — button(s). Since I have no ambition to compete with technology, this chapter will describe only that part of the story that relates to basic studies in immunology and oncology, with a slight personal hue that could not be retrieved from the computerized files.

When should an account of the history of the nude mouse actually begin? — When the mutation first arose, maybe unnoticed? — When the genetics of the mutation without hair was first described? — Or when the mutant was found to have no thymus?

For this introductory chapter, I will choose the latter viewpoint. Accordingly, the history of the nude mouse — in the mind of the scientific world — starts on January 27, 1968 with Dr E. M. Pantelouris' report[1] in *Nature* on the "Absence of Thymus in a Mouse Mutant".

II. A NEW MOUSE MUTANT

A. A HAIRLESS ALBINO MOUSE IN GLASGOW

The nude mouse had already been known for its lack of hair since 1962, when it was found in the Virus Laboratory, Ruchill Hospital, Glasgow, Scotland. The mutation occurred in a closed but not inbred stock of albino mice. Dr. N. R. Grist first saw the mutant, and one male without hair, a characteristic of the mutant, together with two phenotypically normal littermates, male and female, were sent to The Institute of Animal Genetics in Edinburgh, Scotland. The first reference to the mutant was a brief note by Isaacson and Cattanach.[2] A breeding colony was established at the Institute, and from here it was first described in detail by Flanagan[3] in 1966, following years of intensive genetic analysis. Flanagan studied the genetics of the hairless condition, locating it to chromosome linkage group VII; he also introduced the marker genes used for the analysis and later for breeding of the first nude mice for experimental purposes. The marker chromosomes used were Trembler (Tr) in the female and Rex in the male breeders.

Flanagan found that the nude animals had a very short life span. He suspected that a *Toxoplasma gondii* infection was the reason for this, and sent some nude mice with homozygous littermates to Dr. W. Hutchison in Glasgow, a well-known expert on toxoplasmosis. Here, the suspicion of a Toxoplasma infection could not be confirmed: no Toxoplasma organisms were found in the animals and no antibodies could be demonstrated in serological tests. Dr. E. M. Pantelouris, who lived next door at the University of Strathclyde, obtained some nude mice for infection experiments, and it was in his laboratory in 1968 that the secret of this animal was first detected.[1]

B. AN ATHYMIC MOUSE IN GLASGOW

As quoted from Pantelouris' report,[1] "A re-investigation of the mutation has yielded the following results. No thymus can be found in the homozygotes, while in all normal phenotypes (homozygote wild type or heterozygotes) the organ is normally developed.... In homozygote nu/nu, (leucocyte) counts are extremely low. In normal phenotypes, counts are either about 4 times greater, or as low as in the nude phenotypes..... It is possible that the lack, or agenesia, of thymus may be the primary defect of development in this mutation. The exact chain of causation leading to other defects remains to be investigated......" Pantelouris' report was supplemented the following year by the first publication describing some morphologic features of the lymphoid system in nude mice.[4]

It was soon realized that the athymic-born nude mice had immune deficiencies even more profound than neonatally thymectomized mice. Anatomic studies on the lymphoid system described the morphology of the lymph nodes in the spleen, where T cell-dependent areas, as we know them today, were found to be absent. Also, the number of circulating lymphocytes was very low, as was the concentration of immunoglobulins.

C. WHY WAS THE NUDE MUTANT SO INTERESTING?

In the mid-1960s, a growing awareness of the dichotomy of the immune system was spreading. A decade previously, Glick et al.[5] had described the Bursa of Fabricius in the chicken, responsible for what was later called the B lymphocytic system. In 1961, Miller[6] described the effect of neonatal thymectomy in mice with its profound consequences for cell-mediated immunity, later referred to as the T cell system. Also during the same period, the first known and well-described examples of immunodeficiency states in man were recognized.[7] Therefore, when the nude mouse became known as an athymic animal in 1968, the basis for widespread interest in this mutant was well prepared. In this context, it should be remembered that at the time, the concept of T cells and B cells had not yet been coined. They were first suggested the following year by Roitt et al.[8]

In the wake of the report on the athymic condition of the nude mouse, an intense interest in this experimental animal arose, and breeding colonies were distributed from the Institute for Animal Genetics, Edinburgh, Scotland in late 1968.

At the time I was working at The State Serum Institute in Copenhagen, the report of Cooper et al.[9] in the summer of 1966 of a mammalian (rabbit) equivalent of the avian bursa of Fabricius made me start an ambitious and heroic series of experiments in rabbits in order to make a rabbit without a thymus. Fortunately, after many unsuccessful experiments, I heard about the athymic mouse from a colleague returning from Glasgow...

D. THE COPENHAGEN COLONY

At the Patologisk-Anatomisk Institut, Kommunehospitalet, Copenhagen, we started a colony in December 1968 on the basis of two pairs of breeders with the described marker genes, kindly given to us by The Institute for Animal Genetics, Edinburgh. Unfortunately, both of the dames died after giving birth to their first litter. Left with two widowers and a few feeble young mice, we decided to make a gene transfer to another genetic background. We chose the NMRI outbred strain as the background and soon had a very productive breeding system.[10]

The athymic condition is due to a recessive autosomal mutation, nude, the symbol for which is *nu*. Based on the genetics of the mutation, various breeding schemes can be set up. Nude females are fertile in proper microbiological conditions, but they have difficulties weaning their young. Nude males are generally fertile, although infections in the genital tract may cause infertility. Breeding of nude mice can be based on mating of heterozygous male and female, or — preferably — on the mating of homozygous male with heterozygous haired female, which will give a 50% nude offspring in contrast to heterozygous/heterozygous matings where only an average of 25% are athymic.[3]

III. THE FIRST IMMUNOLOGICAL AND ONCOLOGICAL STUDIES

Thanks to the breeding of nude mice on another genetic background (NMRI), it was possible to produce enough animals to initiate studies of immune functions and transplantation studies. At the end of 1969, my late friend and colleague Carl O. Povlsen and I felt that we had enough evidence to publish preliminary reports on immune functions in nude mice[11] and to communicate the first successful heterotransplantation of a human malignant tumor to nude mice.[12]

In the summer of 1970, during the meeting of the Scandinavian Society for Immunology in Aarhus, Denmark, there were extensive discussions of the new laboratory animal. Some people's reactions were: no thymus, and so what? Others were extremely enthusiastic; for example, Cesar Milstein, who was standing in front of the podium, and continued the discussion for more than half an hour in spite of the chairman's vague protests. In retrospect, it is interesting to speculate that 10 years earlier, nobody would probably have paid any heed to a small pink mouse without a proper coat of hair. It could have suffered the same fate as the nude rat[13] that was known for nearly 20 years before it was allowed into the immunology laboratory.

The general knowledge of basic immunology in the early 1970s and the availability of an athymic experimental animal represented a very lucky coincidence. People wanted to work with nude mice. Nude mouse colonies all over the world were founded with breeders from Edinburgh. In turn, we received several requests for breeding colonies that we had no capacity to fulfill. We asked for the help of Carl W. Friis, D.V.M., Director of the Bomholtgaard Laboratory Animal Breeding and Research Center, who was soon able to supply the Basel Institute for Immunology, the National Cancer Institute in Bethesda, MD, and the Central Laboratory for Experimental Animals (CLEA) in Tokyo with breeding colonies.

Some of the most promising nude mouse research has been the study of heterotransplanted tumors in the nude mouse host, the topic of this volume. The first comprehensive review in this field was published by Povlsen[14] in 1977. The following chapters describe the state of the art in this area of cancer research.

IV. MORPHOLOGIC STUDIES IN NUDE MICE

On gross examination, the thymus is entirely absent in nude animals. By serial sectioning and histological examination it was shown, however, that some gland-like structures may occur in the thymic region, occasionally also containing mononuclear cells. These structures have been called *thymic rudiments,* and are not supposed to have a real thymic function. There have been many arguments concerning this point, particularly because of the finding of lymphoid cells with T cell markers in nude mice (see below). The developmental basis for the athymic condition is not fully known.[15,16]

Other lymphoid organs show changes, as could be expected from our knowledge of the effects of neonatal thymectomy: empty T cell compartments of both lymph nodes and spleen. Also, Peyer's patches in nude mice are poorly developed.

V. CELL POPULATIONS IN NUDE MICE

A. B CELLS AND MACROPHAGES

There are several reviews on the immune functions/dysfunctions in the nude mouse. One of the pioneers in the study of both B and T cell functions in nude mice, Berenice Kindred, authored an extensive review[17] in 1979, to which interested readers are referred.

The immune defect of the nude mice is, of course, primarily a defect in T cell functions. Examination of thoracic duct lymph also reveals a very low percentage of lymphocytes that can be shown by marker studies to be B lymphocytes.[18]

The complement of B lymphocytes seems to be normal. This can be demonstrated in Jerne's plaque assay where nude mouse spleen cells perform normally when complemented with T cells from euthymic mice.[19] Immunoglobulin production is sparse, the only immunoglobulin that is produced *in vivo* in high amounts being IgM.[20]

The macrophage system in the nude mouse is highly active. It has been shown in several studies that macrophages are activated, possibly due to a constant load of antigen that cannot be abolished by the nude organism due to its lack of responsive T cells and relevant circulating antibodies.

B. T CELL MARKER-POSITIVE CELLS

In the athymic animal, special interest has naturally focused on the appearance of mononuclear cells showing T cell markers. The first evidence came from experiments showing that some 2 to 4% of cells in adult nude mice on a Balb/c genetic background had the theta surface marker.[21,22] Much speculation was kindled by this finding. It was postulated that nude mice were actually not T cell deficient. Many were of the opinion that the demonstration of the T cell marker invalidated the use of the nude mouse as a T cell-deficient model — despite numerous reports describing its inability to reject grafts of foreign tissue.

The number of T cell marker-positive cells increases with age. This is in accordance with the observation that tumor grafts are more readily accepted in young animals than in older animals. Marincola et al.[23] found that subcutaneous tumors, derived from human pancreatic cancer cell lines, were accepted significantly more often in young than in adult animals (97.9% vs. 69%). An average of 14.7 ± 2.8 d was required for the subcutaneous tumors to become macroscopically apparent in the adult population compared with 3.1 ± 0.8 d in the young mice. Also, the maximum tumor size in young mice was found to be twice that of adult mice. It is not known in detail why there is such an age difference in responses, but the difference should be kept in mind when working with tumor transplants in this model.

C. T CELL RECEPTOR GENES

Interesting new findings[24] have been made recently in the study of the T cell receptor (TCR): lymphocytes in nude mice did actually exhibit functional gamma and delta chain gene transcripts, indicating that the rearrangement and expression of the TCR genes do occur outside the thymus. The number of TCRg/d bearing T cells gradually increases in mesenteric lymph nodes in conventional mice with age, but not in germfree mice. It was therefore supposed that an environmental factor such as intestinal microflora may be important in an age-associated increase in the number of g/d T cells in the regional lymph nodes of mice. Yoshikai et al.,[24] investigating this theory, found that nude mice treated subcutaneously with CFA (Complete Freund's Adjuvants) showed an increasing number of functional T cells bearing TCR g/d in the draining lymph nodes. Increased levels of proliferative responses against syngeneic stimulator cells were noted in the lymph nodes of CFA-injected nude mice in association with increases in the expression of g- and d-chain gene messages. They suggest that local inflammation by CFA may induce functional g/d T cells in nude mice, which may proliferate rapidly at the inflamed sites and represent a first line of defense in such animals against the invasion of various pathogens.

D. THE NATURAL KILLER (NK) CELL

The NK cell was first described in the nude mouse. In the early 1970s, we had a very active collaboration with Drs. Eva and Georg Klein at the Institute for Tumor Biology at Karolinska Institutet in Stockholm, Sweden. We obtained tumor tissue from Burkitt's lymphoma patients in Nairobi, Kenya. A Swedish surgeon performed the operations there, and fresh tumor tissue was sent by air to Stockholm, from where it was distributed to several laboratories all over the the world. Carl O. Povlsen and I were on the mailing list and obtained tissue from many lymphoma patients. One of them was Margareth, whose tumor tissue was accepted in our nude mice as the first heterotransplanted Burkitt's lymphoma.[25] I went to Stockholm to perform tests for circulating antibodies in sera from nude mice. Simultaneously, a pregraduate medical student, Rolf Kiessling, was examining tumor-bearing nude mice for the presence of cell-mediated immunity. Much to his — and our — astonishment, he could demonstrate a strong *in vitro* cytotoxicity, not only against cultured Margareth cells but also — in preliminary experiments — against other cultured cell lines in the laboratory. Eva Klein wrote us an enthusiastic letter in July 1972, even including a draft for a preliminary communication on the finding. A fortnight later, in early August 1972, Rolf Kiessling came to our laboratory, bringing with him P815 and YAC cells for further cytotoxicity studies that confirmed the previous findings. In 1975, Kiessling et al.[26,27] and Herberman et al.[28,29] gave simultaneous descriptions of a new cell type which appeared in various percentages in nude mice depending on their genetic background.

This cell type, the NK cell, was able to kill a range of cultured cell lines, first of all the P815 and YAC-cells. In analogy with the so-called natural antibodies, the cell was called the natural killer cell, because it appeared to be present without any previous priming of the animal. Also, it appeared that it did not obey the rules of the immune system cells, in that the NK cell activity was not increased after priming. Many researchers expected that the NK cell system somehow substituted for the defective cell-mediated immune function in the nude mouse. Therefore, extensive studies have been conducted particularly in the area of tumor transplantation to show a possible correlation between NK cell activity and the ability of nude mice to accept/reject transplanted human tumors. Readers are referred to the relevant chapters in this volume.

VI. MORTALITY AND SPONTANEOUS DISEASES

From the first descriptions of neonatally thymectomized mice, it was known that such animals have a very short life span.[6] The same observation was made concerning the nude athymic mice. It soon appeared, however, that the life span was very closely correlated to environmental stress. In specified-pathogen-free (SPF) conditions, the life span can be extended up to 18 months or more. By keeping animals in more protected conditions, such as in laminar flow cabinets, the life expectancy is even longer; and, as the ultimate condition, a germfree environment will allow a life expectancy comparable to that of normal mice.[10]

As has been pointed out, the term SPF is ill defined when it comes to nude mice. Obviously, this term has been coined from mice with a normal immune system, able to resist a number of bacterial attacks that may interfere with nude mouse life span. In many infections in nude mice, it can be demonstrated that the infective organisms are bacteria, viruses, or fungi that are non-pathogenic in normal mice. It is also well known that the worst scourge of a nude mouse colony, mouse hepatitis virus (MHV), can reside silently in a population of normal mice, but will cause a disastrous epizootic in nude mice if they are infected with the same virus.[30] It should be borne in mind that nude mice, although not as sensitive as cultured cell lines, should be kept and treated very much the same way as cell cultures in order to obtain reliable results.

VII. SPONTANEOUS TUMORS/IMMUNE SURVEILLANCE

One area which has attracted much interest in nude mice and has caused much discussion and disagreement is that of spontaneous tumors. According to the theory of immune surveillance,[31] malignant tumors are constantly generated in all mammals. It has been suggested that the main function of the immune system is actually to abolish such tumor cells in *statu nascendi*, and it has been further postulated that the absence of numerous tumors in all living creatures was due only to the cell-mediated immune system. It could therefore be expected that with the advent of a mouse with a total lack of cell-mediated immunity, such animals would be flooded with spontaneous tumors. But much to the distress of believers in the theory of immune surveillance, this was not the case.[32] It was shown in large populations of nude mice that they developed few if any spontaneous malignant tumors. Later observations, especially by Stutman[33] have extended these observations, showing that a range of carcinogens do not cause a higher incidence of spontaneous tumors in athymic mice than in normal mice. The one exception is tumors caused by papilloma virus. Papilloma virus resistance is T cell dependent, and therefore the tumorigenicity of this virus should rather be taken as evidence of its infectiousness in the nude mice.

Attempts have been made to call in the reserves. It has been suggested that the NK cells should be responsible for the immune surveillance in athymic animals.[34] However, even if the NK cells had such a function, it would not be in accordance with the original definition of immune surveillance and is therefore not acceptable.

VIII. NUDE MOUSE CONFERENCES

The history of the nude mouse and developments in its application can best be followed by studying the proceedings of the various workshops that have been devoted to this animal model. The first was held in Århus, Denmark in 1973. At the time, less than 100 publications on nude mice had appeared. Some 70 participants, representing a wide range of institutions and nationalities, were present. The presentations[35] appeared under the following subheadings: The developmental defect; Responses to mitogens and thymus-dependent and independent antigens; Reconstitution of nude mice; Breeding and husbandry; Pathology of the nude mouse; and Oncological studies in nude mice. — There is nothing new under the sun....

Since then, there have been triennial workshops: in 1976 in Tokyo, Japan;[36] in 1979 in Bozeman, Montana.[37] These first three workshops used the nude mouse name in their titles. In 1982, Dr. B. Sordat arranged the 4th workshop in Lausanne, Switzerland. He felt rightly that the nude mouse should recede and leave room for "the immunodeficient animals"[38] which has also occurred in the following workshops: the 5th in 1985 in Copenhagen, Denmark[39] and the 6th in 1988 in Beijing, PRC.[40]

There have also been important workshops, followed by published proceedings, in Germany in 1975[41] and in Columbus, OH in 1977.[42] Readers are referred to the proceedings for more details of the history.

Two volumes[43,44] of nude mouse review articles were edited by the late Jørgen Fogh and B. C. Giovanella. Thanks to the expertise of their editors, they present the state of art of nude research at the time they appeared and are still very useful sources of knowledge for scientists interested in nude mouse research.

IX. EPILOGUE

The advent of the nude mouse was a turning point in the study of heterotransplanted tumors. The mutant has been combined with other spontaneous mutations to establish interesting double mutants. Also, nude mice can be pretreated in various ways to give higher takes (see Chapter 5) and may, in this way, be better hosts for heterotransplanted tumors.

More recently, the scid mouse[45] has offered a new model for heterotransplantation. Comparative studies of nude mice and scid mice have shown a generally higher take rate in scid mice, and this mutant may in time take the place of the nude mouse as a host for heterotransplanted tumors.

However, the nude mouse still seems to be the most popular heterotransplantation host. The future will show whether this is due to its positive properties or to the general inertia of researchers.

ACKNOWLEDGMENTS

The original studies in our laboratory quoted herein were generously supported by The Danish Cancer Society and The Danish Medical Research Council.

REFERENCES

1. **Pantelouris, E. M.,** Absence of thymus in a mouse mutant, *Nature, (London),* 217, 370, 1968.
2. **Isaacson, J. H. and Cattanach, B. M.,** Report, *Mouse News Lett.,* 27, 31, 1962.
3. **Flanagan, S. P.,** 'Nude', a new hairless gene with pleiotropic effects in the mouse, *Gen. Res. (Camb.),* 8, 295, 1966.
4. **De Sousa, M. A. B., Parrott, D. V. M., and Pantelouris, E. M.,** The lymphoid tissues in mice with congenital aplasia of the thymus, *Clin. Exp. Immunol.,* 4, 637, 1969.
5. **Glick, B., Chang, T. S., and Jaap, R. G.,** The bursa of Fabricius and antibody production, *Poultry Sci.,* 35, 224, 1956.
6. **Miller, J. F. A. P.,** Immunological function of the thymus, *Lancet,* 2, 748, 1961.
7. **Good, R. A.,** Disorders in the immune system, in *Immunobiology.* Good, R. A. and Fisher, D. W., Eds., Sinauer Associates, Stamford, CT., 1971.
8. **Roitt, I. M., Torrigiani, G., Greaves, M. F., Brostoff, J., and Playfair, J. H. L.,** The cellular basis of immunological responses, *Lancet,* 2, 367, 1969.
9. **Cooper, M. D., Perry, D. Y., McKneally, M. F., Gabrielsen, A. E., Sutherland, D. E. R., and Good, R. A.,** A mammalian equivalent of the avian bursa of Fabricius, *Lancet,* 1, 1388, 1966.
10. **Rygaard, J.,** *Thymus and Self. Immunobiology of the Mouse Mutant Nude,* FADL's Publisher, Copenhagen, 1973, John Wiley & Sons, London, 1975; Tuttle-Mori, Tokyo 1979.
11. **Rygaard, J.,** Immunobiology of the mouse mutant "nude". Preliminary investigations, *Acta. Path. Microbiol. Scand.,* 77, 761, 1969.
12. **Rygaard, J. and Povlsen, C. O.,** Heterotransplantation to a human malignant tumour to "nude" mice, *Acta. Path. Microbiol. Scand.,* 77, 758, 1969.
13. **Festing, M. F., May, D., Connors, T. A., Lovell, D., and Sparrow, S.,** An athymic nude mutation in the rat, *Nature, (London),* 274, 364, 1978.
14. **Povlsen, C. O.,** *Heterotransplantation of Human Malignant Tumors into Nude Mice,* (in Danish) FADL's Publisher, København, Denmark, 1977.
15. **Holub, M., Rosman, P., Tlaskalova, H., and Vidmarova, H.,** Thymus rudiment of the athymic nude mouse. *Nature, (London),* 256, 492, 1975.
16. **Cordier, A. C.,** Le Thymus de la Souris Normale et de la Souris Nude, thesis, University of Louvain, Belgium, 1980.
17. **Kindred, B.,** Nude mice in immunology, in *Progress in Allergy,* Vol. 26, Kallos, P., et al., Eds., 1979, 137.
18. **Sprent, J.,** Migration and lifespan of circulating B lymphocytes of nude (nu/nu) mice, in *Proceedings of the First Int. Workshop Nude Mice,* Rygaard, J. and Povlsen, C. O., Eds., Gustav Fischer Verlag, Stuttgart, 1974.
19. **Lefkovits, I.,** *In vitro* complementation experiments with nude mice. I. The allogeneic effect in the antibody response to sheep red cells, *Eur. J. Immunol.,* 3, 397, 1973.
20. **Reed, N. D. and Jutila, J. W.,** Immune response of congenitally thymusless mice to heterologous erythrocytes, *Proc. Soc. Exp. Biol. Med.,* 139, 1234, 1972.
21. **Loor, F. and Roelants, G. E.,** High frequency of T lineage lymphocytes in nude mouse spleen, *Nature, (London),* 251, 229, 1974.
22. **Loor, F., Hägg, L.-B., Mayor, K. S., and Roelants, G. E.,** Θ-Positive cells in nude mice born from homozygous nu/nu mothers, *Nature, (London),* 255, 657, 1975.
23. **Marincola, F. M., Drucker, B. J., Siao, D. Y., Hough, K. L., and Holder, W. D.,** The nude mouse as a model for the study of human pancreatic cancer, *J. Surg. Res.,* 47, 520, 1989.
24. **Yoshikai, Y., Matsuzaki, G., Inoue, T., and Nomoto, K.,** An increase in the number of T-cell receptor γ/δ-bearing T cells in athymic nude mice treated with complete Freund's adjuvants, *Immunology,* 70, 61, 1990.
25. **Povlsen, C. O., Fialkow, P. J., Klein, E., Klein, G., Rygaard, J., and Wiener, F.,** Growth and antigenic properties of a biopsy-derived Burkitt's lymphoma in thymus-less (nude) mice, *Int. J. Cancer.,* 11, 30, 1973.
26. **Kiessling, R., Klein, E., and Wigzell, H.,** 'Natural' killer cells in the mouse. I. Cytotoxic cells with specificity for mouse Moloney leukemia cells. Specificity and distribution according to genotype, *Eur. J. Immunol.,* 5, 112, 1975.
27. **Kiessling, R., Klein, E., Pross, H., and Wigzell, H.,** 'Natural' killer cells in the mouse. II. Cytotoxic cells with specificity for mouse Moloney leukemia cells. Characteristics of the killer cell, *Eur. J. Immunol.,* 5, 117, 1975.
28. **Herberman, R. B., Nunn, M. E., and Lavrin, D. H.,** Natural cytotoxic reactivity of mouse lymphoid cells against syngeneic and allogeneic tumors. I. Distribution of reactivity and specificity, *Int. J. Cancer,* 16, 216, 1975.
29. **Herberman, R. B., Nunn, M. E., and Lavrin, D. H.,** Natural cytotoxic reactivity of mouse lymphoid cells against syngeneic and allogeneic tumors. II. Characterization of effector cells, *Int. J. Cancer,* 16, 230, 1975.
30. **Fujiwara, K.,** Spontaneous virus infections of nude mice, in *The Nude Mouse in Experimental and Clinical Research,* Vol. 2, Fogh, J. and Giovanella, B. C., Eds., Academic Press, New York, 1982, 1.

31. **Burnet, F. M.,** *Immunological Surveillance,* Pergamon Press, Sydney 1970.
32. **Rygaard, J. and Povlsen, C. O.,** The nude mouse vs. the hypothesis of immunological surveillance, *Transplant. Rev.,* 28, 43, 1976.
33. **Stutman, O.,** Spontaneous, viral, and chemically induced tumors in nude mice, in *The Nude Mouse in Experimental and Clinical Research,* Vol. 1, Fogh, J. and Giovanella, B. C., Eds., Academic Press, New York, 1978, 411.
34. **Herberman, R. B.,** Natural cell-mediated cytotoxicity in nude mice, in *The Nude Mouse in Experimental and Clinical Research,* Vol. 1, Fogh, J. and Giovanella, B. C., Eds., Academic Press, New York, 1978, 135.
35. **Rygaard, J. and Povlsen, C. O.,** Eds., *Proc. First Int. Workshop on Nude Mice,* Gustav Fischer Verlag, Stuttgart, 1974.
36. **Nomura, T., Ohsawa, N., Tamaoki, N., and Fujiwara, K.,** Eds., *Proc. Second Int. Workshop on Nude Mice,* University of Tokyo Press, 1977.
37. **Reed, N. D.,** Ed., *Proc. Third Int. Workshop on Nude Mice,* Vol. 1 and 2, Gustav Fischer Verlag, New York, 1979.
38. **Sordat, B.,** Ed., *Proc. Fourth Int. Workshop on Immune-Deficient Animals in Exp. Res.,* (formerly Nude Mice Workshop), Karger, Basel, 1984.
39. **Rygaard, J., Brünner, N., Græm, N., and Spang Thomsen, M.,** Eds., *Proc. Fifth Int. Workshop on Immune-Deficient Animals,* Karger, Basel, 1987.
40. **Wu, B. Q. and Zheng, J.,** Eds., *Proc. Sixth Int. Workshop on Immune-Deficient Animals,* Karger, Basel, 1989.
41. **Houchens, D. P. and Ovejera, A. A.,** Eds., *Proc. Symp. Use of Athymic (Nude) Mice in Cancer Res.,* Columbus, Ohio, 1977, Gustav Fischer Verlag, New York, 1978.
42. **Bastert, G. B., Fortmeyer, H. P., and Schmidt-Matthiesen, H.,** Eds., *Proc. Symp. Klinikum der Johann-Wolfgang-Goethe-Universität,* Frankfurt am Main, 11.06.-15.06.79, Gustav Fischer Verlag, Stuttgart, 1981.
43. **Fogh, J. and Giovanella, B. C.,** Eds., *The Nude Mouse in Experimental and Clinical Research,* Vol. 1, Academic Press, New York, 1978.
44. **Fogh, J. and Giovanella, B. C.,** Eds., *The Nude Mouse in Experimental and Clinical Research,* Vol. 2, Academic Press, New York, 1982.
45. **Bosma, G. C., Custer, R. P., and Bosma, M. J.,** A severe combined immunodeficiency mutation in the mouse, *Nature, (London),* 301, 527, 1983.

Chapter 2

TYPES OF IMMUNODEFICIENCY IN MICE

Simon P. Langdon and John F. Smyth

TABLE OF CONTENTS

I. INTRODUCTION

With the realization that the immune system is responsible for the rejection of human tumor tissue after transplantation to other species, several approaches have been adopted to circumvent this obstacle. These include implantation into immune-privileged sites, use of mutant mice with inherited immunodeficiencies, and immunosuppression and immunodeprivation of normal and mutant animals.

Initially, human tumor xenografts were grown in immune-privileged sites of normal rodents and lagomorphs. These sites include the anterior chamber of the eye,[1-3] the brain,[4-8] and the cheek pouch.[9-19] More recently, the subrenal capsule has been extensively investigated as a potential location for short-term growth of xenografts.[20-22] Apart from the inconvenience of these sites for tumor implantation and measurement, there is limited space for growth and hybridization may occur between tumor cells and normal host cells.[18,19] Furthermore, an immune response will eventually cause rejection of the xenograft which may compromise the interpretation of the experiments. Use of immature immunocompetent hosts has also been attempted with some success; but with age, the immunity develops presenting a limitation to the time period for studying tumor transplants.[23-28] Another approach has been the implantation of tumor cells into millipore diffusion chambers which are then placed into the peritoneal cavity.[29,30] In addition to the limitation of space for tumor cells to expand, assessment of tumor growth is only feasible after removing the chamber from the animal.

Short-term immunosuppression can be achieved by the use of irradiation or drugs and these have been used for immunosuppression of several species permitting xenograft growth.[15,31-36] Ultimately, long-term immunodeficiency became a possibility with the development of mutant strains of rodents with inherited immunodeficiencies, as well as with the use of combined immunodeprivation and immunosuppressive techniques.

The observation that the nude mouse with thymic hypo/aplasia could successfully act as a host for human tumor xenografts[37] led to its widespread use for this purpose (see Chapter 6). The nu gene has now been introduced into many defined inbred backgrounds of mice. A wide variety of other mutant strains of mice with specific immunodeficiencies have also been derived and characterized. These strains have proven valuable models in helping to define the role of a number of components of the immune system and the way in which they interact. Several have been investigated for their ability to host tumor xenografts. Combining strains possessing different immunodeficiencies may improve transplantability of human tumor xenografts and several doubly congenic strains have been developed for this purpose.

The discovery of the role of the thymus in T lymphocyte function and its importance in tumor rejection led to the development of the use of thymectomy and anti-thymocyte serum (ATS) as specific immunosuppressive techniques. In conjunction with whole body irradiation, combination regimens have been developed to produce effective long-term immunosuppression.

It is the purpose of this chapter to review some of the types of immunodeficient strains used in oncology research and the various techniques of immunosuppression and immunodeprivation, with particular emphasis on their usefulness in providing murine hosts for human tumor xenografts.

II. STRAINS OF NUDE MICE AND TRANSPLANTABILITY

The autosomal recessive nu gene has been introduced into at least 75 different congenic strains of nude mouse. These are predominantly from inbred sources including the following: A.SW, A/Ḥe, AKR, BALB/c, Bl0.LP, C17, C3H, C57BL, C57, CBA, DBA, DDD, I/st, NC, NFR, NFS, NSF/N, NZB, NZC, NZW, P, RIII, and SJL.[38] Nude mice congenic with outbred sources include the NMRI, NIH, and Swiss strains.[38] Many of these nude mouse strains have been derived by C. Hansen who used the following method of repeated cycles of outcrossing and

TABLE 1
Characteristics of Immunodeficient Mice

Mutant locus	Chromosome	Characteristics	Ref.
A^y (yellow agouti)	2	Semidominant; spleen cells have reduced graft vs. host reactivity	43
an (Hertwig's anemia)	4	Recessive; normochromic macrocytic anemia	44,45
Aia-1,-2 (autoimmune hemolytic anemia-1)	4	Occurs in NZB and NZC mice	46
bg (beige)	13	Recessive; predominantly NK defect	47–55
Dh (dominant hemimelia)	1	Semidominant; hemimelia and asplenia	56–58
lh (lethargic)	2	Recessive; thymic involution, diminished cell-mediated immunity and lymphocyte number	59
Lps (lipopolysaccharide response)	4	B cell deficiency in C3H/HeJ mice	60,61
me (motheaten)	6	Recessive; immature B cells, defective T cell function and low NK cell activity	62,63
nu (nude)	11	Recessive; athymic	see Chapter 1
scid (severe combined immune deficiency)	16	Recessive; loss of B and T cell immunity	64–66
sl (Steel anemia)	10	Semidominant; macrocytic anemia	67,68
W (dominant spotting)	5	Semidominant; macrocytic anemia	69
Xid (X-linked immunodeficiency)	X	Recessive; B cell defect in CBA/N mice	42,70

intercrossing to introduce the nu gene into the strain of interest.[39] The outcrossing step consisted of mating a female of the recipient strain with a homozygous nude male. The intercross then involved mating the heterozygous outcross progeny with one another to recover a homozygous male. This cycle was then repeated at least 11 times before the strain was considered congenic with the original strain. At this point, the nude mouse should be genetically identical with the strain into which it was introduced, except for the presence of the nu gene.

The influence of the strain of nude mouse on xenograft transplantability has been studied by Maruo et al.[40] The growth of a human gastric cancer was compared in nude mice from seven different genetic backgrounds [BALB/cA, CBA/N, NFS/N, NIH(s), C3H/HeN, C57BL/6N, and Lasat (i.e., BALB/cA + Dh gene — see below)]. Rapid growth was observed in the CBA/N, NSF/N, and NIH(s) nudes and, thus, these might represent preferred nude strains. However, as the authors of this study point out, in the choice of a nude mouse strain, differences in the productivity of the mice or drug pharmacology and toxicity should be considered as more important factors.

III. OTHER TYPES OF IMMUNODEFICIENT MICE

A wide variety of animals with inherited immunological defects other than the nude mouse have been bred and used to study specific aspects of the immune system (reviewed in detail by Festing[41] and Green[42]). The mutants best described are listed in Table 1.

Combining mutants has resulted in mice with multiple immunodeficiencies and provided valuable models to study interactions within the immune system. Some of the more commonly used mutant strains and combinations which act as recipients of xenografts are described in more detail below.

A. THE BEIGE MOUSE

The beige mutation (bg) occurred spontaneously in an autosomal recessive gene on chromosome 13 in C57BL/6 mice.[47,48] The presence of this gene results in a complete impairment of natural killer (NK) cell activity; the effect of this is a state analogous to the

Chediak-Higashi syndrome in man characterized by marked susceptibility to infections.[41,47,48] The defect was claimed to be selective for NK cells since various forms of cell-mediated immunity exerted by T lymphocytes, activated macrophages, promonocytes, and cells participating in antibody-dependent cell-mediated lysis appeared normal.[49,50] However, later studies demonstrated that the deficiency was not restricted to NK cells since the generation of cytotoxic T cells in response to alloimmune challenge was also markedly impaired in beige mutants.[51] The NK abnormality appears to lie in the cytolytic machinery and not at the level of population size or target cell recognition since the frequency of binding to target cells is normal.[47,50] This model has been used extensively to help study the cytotoxic and anti-metastatic mechanisms of NK cells toward syngeneic tumor cells.[47,51,52]

B. THE DH MUTANT

The Dh (dominant hemimelia) mutant possesses an autosomal dominant gene on chromosome 1 resulting in hemimelia and asplenia.[56] This is lethal for homozygotes which die soon after birth from visceral abnormalities, while heterozygotes possess skeletal abnormalities and no spleen.[56] The absence of a spleen results in normal cellular immunity, but depressed IgM synthesis and lower serum protein levels.[57] The lymph nodes of these mice contain double the number of lymphocytes compared to normal mice; however, the thymocyte count is normal. In addition, B and T cell cooperation is abnormal, unlike the situation found in splenectomized mice.[58]

C. THE XID MUTANT

The Xid (X-linked immune deficiency) recessive mutation is found in the central region of the X chromosome and was identified in the CBA/N strain in 1966.[42,70] In these mice, the size of the spleen and the number of nucleated cells per spleen is reduced. Their B cells are characterized by high surface IgM, low surface IgD, low IgG3 production and low complement receptor. The defective B cells have been shown to be abnormal and not simply immature.[42] Normal B cells which would not survive in normal non-irradiated mice will restore immune responsiveness to normal in Xid/Xid mice.[42] The B cell abnormalities result in a defective response to type 2 thymus-independent antigens, a reduced response to thymus-dependent antigens and a normal response to type 1 thymus-independent antigens.[42,70]

D. THE SCID MOUSE

The congenital syndrome, known as severe combined immunodeficiency (scid), was reported in a mutant strain of CB-17 mice in 1983.[64] This disease is characterized by loss of both T and B cell immunity. The autosomal recessive scid gene has been shown to be located on chromosome 16.[65] Both B and T cell functions are deficient, as indicated by a lack of response to mitogens such as concanavalin A and lipopolysaccharide. Levels of the serum immunoglobulin classes IgM, IgG3a, IgG2b, IgG2a, and IgA are not detectable in these animals. Lymphoid organs are 1/10 or less of their normal size, while Peyer's patches of the intestinal tract are rarely visible. The number of lymphocytes is greatly reduced, while the proportion of granulocytes is elevated. The hematocrit in these animals is normal. The deficiency of mature B and T cells can be partly corrected by engrafting bone marrow or fetal liver cells from normal BALB/c mice into adult scid mice.[65] For full reconstitution, scid mice need to be sublethally irradiated just prior to cell transfer.[65,66]

E. THE LASAT MOUSE

The "lasat" (Lozzio asplenic athymic) mouse contains both the nu and the Dh mutations, resulting in agenesis of both spleen and thymus.[71,72] While lymphocyte surface markers and serum immunoglobulin levels of lasat mice are similar to their nude and asplenic (Dh) littermates, there are also several significant differences.[72] Levels of serum IgA are higher than in nudes, but lower than in either nu/+ or Dh/+ mice. Hematocrit, hemoglobin, and reticulocyte counts of lasat mice are normal, but they are markedly leukopenic and have thrombocytosis and an increased number of bone marrow CFU-C cells. Like nude mice, they fail to respond to sheep red blood cell immunization.

F. THE NUDE-XID MOUSE

The nude-Xid mouse combines the nu mutation with the X-linked immune defect in B lymphocyte function, resulting in a mouse with both T and B cell deficiency.[73] Azar et al.[73] first introduced the nu gene into N:NIH(s) non-inbred stock to give the N:NIH(s) I mouse. This was then combined with the CBA/N mice (which possesses the Xid gene). The resulting double congenic N:NIH(s) II-nu/nu mice maintain the thymic hypoplasia and poor hair development of nu/nu mice. T and B cell zones of the lymph nodes and the spleen are depleted of lymphocytes when these mice are compared with nu/nu or CBA/N mice. The levels of NK cells in these animals are, however, relatively high.

Nude-Xid mice have also been derived by Wortis et al.[74] who demonstrated, in addition to the above, that these animals had less than 10% of normal serum immunoglobulin levels, while spleen cells studied *in vitro* did not respond to thymus-independent antigens nor to lipopolysaccharide.

G. THE BEIGE-NUDE MOUSE

The beige-nude mouse has been derived by a number of investigators.[52,75-77] The extent of NK cell activity has been studied in these mice and found to be intermediate between that of the beige and nude strains from which it was derived.[52,75] For instance, the NK cell activity in beige-nude mice derived from the NIH nude mouse was much lower than in adult or even 2-week-old NIH nude mice.[52] The response to T cell mitogens, such as phytohemagglutinin and concanavalin A, was lower relative to adult NIH mice, while the B cell response to lipopolysaccharide was similar to 2 to 3-week-old NIH mice.[52] This strain, together with the "beige-Xid" and "beige-nude-Xid" combinations, has helped provide evidence that lymphokine-activated killer (LAK) cells and NK cells are distinct.[76] Thus, while splenocytes from the beige-nude and the beige-nude-Xid mice failed to generate LAK cells but displayed significant NK cell activity, those obtained from the beige-Xid mouse generated LAK cells, but did not display NK cell activity.[76]

IV. TRANSPLANTABILITY INTO VARIOUS IMMUNODEFICIENT MICE

Several of the strains of immunodeficient mice described above have been investigated for their ability to act as hosts for human tumor xenografts in attempts to propagate tumors which are only poorly transplantable in nude mice. In general, these studies have compared the success rate of growth of the xenograft in the mutant host with that in nude mice. Transplantability into nude mice in general is described in Chapter 6.

A. THE LASAT MOUSE

Following the observation that human myeloid leukemic cells could be transplanted successfully in lasat mice,[71] a number of studies have compared growth of human tumor material in lasat and nude mice.[72,78-80] Gershwin et al.[78] demonstrated an increased growth rate of a human melanoma and a lower threshold for the critical number of cells to form a tumor in lasat compared to nude mice.[72] This was followed by a larger study comparing the latency period, success of growth, and minimum cell inoculum number required for transplantation of 11 continuously transferred human tumor cell lines into both strains of mice.[78] Of 11 cell lines studied (1 breast cancer, 1 bladder cancer, 2 osteosarcomas, 2 fibrosarcomas, 3 melanomas and 1 rhabdomyosarcoma), only 3 (1 fibrosarcoma and 2 melanomas) were successfully transplanted to nude mice, while 9 could be transplanted to lasat mice. Furthermore, the latency period was shorter and the minimum inoculum number lower for lasat mice.

In a study by Tropé et al.,[79] the growth of a nasopharyngeal cancer and a testicular germ cell tumor were compared in both nude and lasat mice. No significant differences in the take rates were observed with 13 of 16 inocula growing in nude mice compared with 11 of 12 in lasat mice. With respect to a variety of primary transplants and subsequent passages, Giovanella and Fogh[80] found no difference in take rates and growth between the two strains.

In conclusion, transplantation rates into lasat mice are at least as good as in nude mice and occasionally may be higher.

B. THE SCID MOUSE

Philips et al.[81] have compared growth of xenografts in scid and nude mice. Samples of 28 human tumors (representing 10 different tissue types) were compared for their ability to grow in either mutant strain. Retinoblastoma and osteosarcoma samples grew better in scid mice. Thus, while 69% of retinoblastoma samples were accepted in scid mice, 28% grew in nude mice; similarly, 90% of osteosarcoma samples were accepted in scid mice compared to 0% in nudes. Cancers of the bladder, prostate, pancreas, kidney, and testis, and also Wilm's tumor, were also transplantable into scid mice, while samples of melanoma and breast cancer were not. The A549 lung adenocarcinoma has also been shown to grow in these mice.[82]

C. THE NUDE-XID MOUSE

Azar et al.[73] have compared growth of six tumor types in nude-Xid or N:NIH(s)II-nu/nu (NIH-II) mice and nude mice. In NIH-II mice, 12 of 33 injections produced tumors compared to 11 of 34 in nude mice. The two transplantable tumor types of the six studied in NIH-II mice were a mesenchymal chondrosarcoma of the pelvis and a Burkitt's lymphoma cell line (Raji).

Giovanella and Fogh[80] have also compared xenograft growth in NIH-II and in Swiss nude mice. They observed no major differences in either tumor take or growth for lung, breast, or colon cancers, for melanomas or Burkitt's lymphoma.

D. THE BEIGE-NUDE MOUSE

The growth of several human tumor cell lines, CEM (a T cell line), K562 (CML), and LOX (malignant melanoma), have been compared in beige-nude mice and in NIH adult nude mice.[52] While the LOX and CEM lines grew equally well in both strains, the K562 line grew in only 2 of 10 nudes and in 0 of 10 beige-nude mice. Therefore, for these cell lines, transplantability into beige-nude mice was not more successful than into nude mice.

V. ARTIFICIAL IMMUNOSUPPRESSION OR IMMUNODEPRIVATION AND TRANSPLANTABILITY

A. ANTI-THYMOCYTE SERUM (ATS)

In 1963, Woodruff and Anderson[83] reported that the survival of skin homografts in rats was significantly prolonged by the use of a rabbit antiserum to rat lymphocytes (ALS). Philips and Gazet[84] were the first to show that human tumor cell lines (HeLa and Hep 2) would grow in mice treated with anti-thymocyte serum (ATS). The results with these human tumors were confirmed and extended to other tumor types.[85-88] ATS is the form of anti-lymphocyte serum (ALS), wherein the antigen used to raise the antiserum is the thymocyte.[89,90] It achieves its immunosuppressive effect through a selective depletion of the recirculating pool of lymphocytes.[91] Many investigators have used the protocol of Levey and Medewar[90] to prepare ATS, which involves the injection of 10^9 thymocytes from mice i.v. into New Zealand rabbits on 2 occasions 14 days apart. Some 7 d after the 2nd injection, the rabbits are bled and the serum heated to 50°C for 30 min, filtered, and stored at –20°C. While ATS alone, as an immunosuppressant, allows a percentage of tumors to grow, it is much more effective when combined with other methods (see below).

B. THYMECTOMY

In 1961, Miller[92] demonstrated that neonatal thymectomy of mice prolonged the survival of allogeneic skin grafts. Osaba and Auersperg[93] reported the growth of a human cervical cancer in mice thymectomized at birth. As with ATS treatment, thymectomy is associated with a decrease in the number of circulating pool lymphocytes.[94,95] The standard method of thymectomy, as described by Steel et al.,[96] is as follows. The animal is first anaesthetized and an incision (5 to 7 mm) is made in the skin overlying the suprasternal notch. The neck muscles are then pulled apart with forceps and the sternum split to a distance of 3 mm using sharp scissors. The two lobes of the thymus can then be seen and are removed by vacuum suction. The skin is then closed.

Like ATS, thymectomy alone as a modality produces suboptimal immunosuppression and has been combined with other methods as described below.

C. THYMECTOMY AND ALS OR ATS

ALS treatment in conjunction with thymectomy has been shown to depress lymphocyte counts and allow skin graft survival twice as long as for ALS treatment alone.[97] The combination of ATS and thymectomy for xenograft transplantability was investigated by Phillips and Gazet.[87] Only 6 of 40 biopsies grew in ATS-treated mice compared to 6 of 26 in ATS plus thymectomy treated animals. Xenografts with successful growth included samples from colon, rectal, breast, and stomach cancers.

D. THYMECTOMY AND WHOLE BODY IRRADIATION

The combination of thymectomy and whole body irradiation (WBI) rescued with syngeneic bone marrow cells (to protect against the lethal effects of irradiation) provides an improved method of effective immunosuppression compared to thymectomy alone. The standard method was first described by Miller et al.[98] in 1963. Castro,[99] Franks et al.,[100] and Cobb et al.[101,102] removed the thymus in mice at 4 weeks of age, gave 900 cGy WBI 2 to 4 weeks later, rescued with an injection of 5×10^6 or 10^7 syngeneic bone marrow cells within 24 h, and transplanted tumors either under the renal capsule[99] or s.c.[100,102] Castro[99] successfully transplanted 15 tumors representing 8 different tumor types from 23 patients, while Franks et al.[100] obtained growth in 7 of 8 tumor types (breast cancer without success). Cobb[101] implanted three colorectal cancers into an s.c. site and then demonstrated that these xenografts could grow in either s.c., i.m., i.p., or intrapleural sites. In 1974, he and co-workers[102] reported on the successful transplantation of 22 tumors from 83 patients, including cancers of the colon, rectum, ovary, kidney, and lung.

By a number of investigators (Franks et al.,[103] Detre et al.,[104,105] and Steel et al.[96,106,107]) it was recognized that the few T cells present after WBI, or from addition of the injected bone marrow cells, represented potential sources of residual immunity which could compromise the effectiveness of the immunosuppression. In an attempt to reduce the T cell population further, Franks et al.[103] tried a higher dose of 1200 cGy WBI, rather than the standard 900 cGy WBI in conjunction with thymectomy and bone marrow rescue. This approach, however, did not improve growth of HeLa cells or a human bladder cancer relative to the dose of 900 cGy WBI. ATS treatment in conjunction with thymectomy, WBI, and bone marrow rescue was also investigated and appeared to have a small but positive effect on the improved growth of these 2 xenografts. This latter approach had also been investigated by Detre and Gazet,[104] who obtained 5 xenografts from 18 samples, and by Berenbaum et al.,[108] who reported survival in 75% of 104 tumors. Steel et al.[96,106,107] looked at ploys to minimize the amount of syngeneic bone marrow cells inoculated. They demonstrated that use of the standard number of cells (5×10^6 to 10^7) represented a large excess for a protective effect and that the take rate for a human pancreatic cancer cell line increased from 30% using 5×10^6 cells to 71% using only 2×10^5 cells, without any marked changes in animal deaths. Detre[105] has also reduced the number of bone marrow cells for reconstitution and investigated the percentage of animals surviving 850 cGy WBI having been reconstituted with 5×10^6 or 10^5 bone marrow cells or, alternatively, 10^6 fetal liver cells. A slight reduction in the survival of animals receiving fetal cells was shown in comparison to those receiving 5×10^6 bone marrow cells, while 10^5 and 5×10^6 bone marrow cells produced similar survival rates. This was compensated though by a greater take rate for a variety of tumors in the fetal liver and low bone marrow treated animals, as compared to the high bone marrow group. Reduction below 10^5 bone marrow cells resulted in high mortality rates.

The requirement for bone marrow reconstitution can be avoided by the use of cytosine arabinoside (Ara-C). If injected 2 d before irradiation, Ara-C protects against an otherwise lethal effect of 1000 cGy WBI.[106,107] The mechanism of action of Ara-C is not due to reduction of damage to bone marrow stem cells, but to an enhanced recovery of the marrow and intestinal epithelium.[109,110] Use of the higher dose of WBI (1000 cGy) after Ara-C protection also extended the time period of effective immunosuppression from 6 weeks to 6 months.[107] The regimen of thymectomy, Ara-C, and WBI probably represents the most effective and convenient means of immunodeprivation/suppression in mice for the purpose of tumor transplantation.[106,111]

E. IMMUNOSUPPRESSION IN NUDE MICE

The use of immunosuppressive methods and immune-privileged sites in nude mice has been investigated in attempts to improve transplantability in situations where the take rates are poor. For example, WBI has been employed to enhance transplantion rates of leukemias and lymphomas in nude mice.[112-114] Ohsugi et al.[114] demonstrated that 10 of 10 lymphoblastoid lines would grow in irradiated mice compared to 2 of 10 in non-manipulated nude mice. The immune-privileged intracranial site in nude mice is another alternative for hemopoietic tumors (also for small cell lung cancer) and an improved transplantation rate has been shown in comparison with the s.c. site.[115,118] The anterior chamber of the eye of the nude mouse has also been explored in order to grow retinoblastomas. In a study by Gallie et al.,[119] while 12 of 13 retinoblastoma tumors were able to grow at this site, none would grow s.c. unless the immunosuppressive drug cyclophosphamide was used.

Transplantation to nude mice may also be enhanced by the use of ALS treatment. Thus, Ohsugi et al.[114] obtained growth of 6 of 10 lymphoblastoid lines in ALS-treated nude mice compared to 2 of 10 in normal nudes, while Gershwin et al.[78] obtained growth of 5 in 10 different tumor types in ALS-treated nude mice, but only 3 of 11 in untreated nude mice.

VI. TRANSPLANTABILITY INTO NUDE MICE VS. IMMUNODEPRIVED/IMMUNOSUPPRESSED MICE

Several investigators have compared growth of xenografts in nude mice with that in thymectomized/irradiated animals (Steel et al.,[96] Detre,[105] and Floersheim[120]). No major differences have been observed between these immunodeficient hosts in terms of take rates or growth.

In a review by Steel et al.[111] of 329 tumors transplanted to thymectomized/irradiated animals, the overall take rate was 36%. Initial take rates which gave rise to serially transplantable lines were as follows: for colorectal cancers, 52%; breast cancer, 10%; and lung cancer, 78%. In another large study of Berenbaum et al.,[108] again using thymectomized/irradiated animals (with ALS), progressive growth in 36 of 104 tumors (35%) with a 51% success rate for colorectal cancer was obtained. In comparison, one of the largest studies using nude mice as host is the series of 342 tumors described by Sharkey et al.,[121] where 36% of tumors were found to be transplantable. The percentages were: for colon cancer, 72%; for breast cancer, 16%; and for lung cancer, 51%. Thus, in terms of transplantability and growth, nude mice and thymectomized/irradiated animals (with added modalities) appear to be equivalent.

There are, however, distinct advantages (and disadvantages) in the use of nude mice, as opposed to immunodeprived/immmunosuppressed animals. The nude mouse possesses a much more reproducible and stable immunological background and, together with other types of immunodeficient mutants, is probably the only practical means of maintaining xenografts for prolonged periods of time (e.g., 6 months). Nude mice are, however, expensive to breed and require a germfree environment. In contrast, thymectomy is a relatively easy procedure in the hands of a skilled technician, but is time-consuming and can be labor-intensive. While satisfactory for short-term experiments (when used in conjunction with WBI), the restoration of immunity may be a problem. The latter difficulty also applies for ATS treatment. In addition, ATS treatment shows minor inconsistencies because of batch-to-batch variation and occasionally results in toxic and allergic reactions.

VII. CONCLUSIONS

A variety of murine immunodeficient hosts are now available for the purpose of maintaining human tumor xenografts. The nude mouse represents the most widely used immunodeficient mouse and the nu gene has been successfully introduced into many strains to provide animals with defined genetic backgrounds. Other genes associated with immunodeficiency, such as the bg, Xid, and Dh genes, when present with the nu gene, provide hosts with NK and B cell immunodeficiencies in addition to T cell immunodeficiency. In some cases, these represent possible alternatives when xenograft growth is poor in nude mice.

Methods of immunodeprivation and immunosuppression of immunocompetent mice have been developed and combinations of these provide short-term alternatives to nude mice as recipients for xenografts. In situations where transplantation is poor in nude mice, these techniques may also be usefully applied to these animals.

REFERENCES

1. **Greene, H. S. N.,** Heterologous transplantation of human and other mammalian tumors, *Science*, 88, 357, 1938.
2. **Greene, H. S. N.,** Heterologous transplantation of mammalian tumours. I. The transfer of human tumours to alien species, *J. Exp. Med.*, 73, 475, 1941.
3. **Greene, H. S. N.,** The significance of the heterologous transplantability of human cancer, *Cancer*, 5, 24, 1952.
4. **Greene, H. S. N.,** Transplantation of tumors to brains of heterologous species, *Cancer Res.*, 11, 529, 1951.
5. **Lumb, G.,** Transplantation of human tumours into mouse brain, *Br. J. Cancer*, 8, 434, 1954.
6. **Chesterman, F. C.,** Intracranial heterotransplantation of human tumours, *Br. J. Cancer*, 9, 51, 1955.
7. **Epstein, A. L., Herman, M. M., Kim, H., Dorfman, R. F., and Kaplan, H. S.,** Biology of human lymphomas. III. Intracranial heterotransplantation in the nude athymic mouse, *Cancer*, 37, 2158, 1976.
8. **Gazdar, A. F., Carney, D. N., Sims, H. L., and Simmons, A.,** Heterotransplantation of small cell carcinoma into nude mice: comparison of intracranial and subcutaneous routes, *Int. J. Cancer*, 28, 777, 1981.
9. **Billingham, R. E., Ferrigan, L. W., and Silvers, W. K.,** Cheek pouch of the Syrian hamster and tissue transplantation immunity, *Science*, 132, 1488, 1960.
10. **Cohen, S. N.,** Comparison of autologous, homologous and heterologous normal skin grafts in the hamster cheek pouch, *Proc. Soc. Exp. Biol. Med.*, 106, 677, 1961.
11. **Galton, M., Goldman, P. B., and Holt, S. F.,** Karyotypic and morphological characterisation of a serially transplanted human choriocarcinoma, *J. Natl. Cancer Inst.*, 31, 1019, 1963.
12. **Yohn, D. S., Hammon, W., and Aitchison, R. W.,** Influence of implant site on the immunologic response of unconditioned Syrian hamsters to heterotransplantable human tumours, *Cancer Res.*, 25, 484, 1965.
13. **Goldenberg, D. M. and Witte, S.,** Chemotherapy of two morphologically similar human tumours growing in the cheek pouch of the golden hamster, *Eur. J. Cancer Clin. Oncol.*, 3, 95, 1967.

14. **Smith, G. M. R.,** The effect of cytotoxic agents on human tumours transplanted to the hamster cheek pouch, *Br. J. Cancer,* 23, 78, 1969.
15. **Williams, D. E., Evans, D. M. D., and Blamey, R. W.,** The primary implantation of human tumours to the hamster cheek pouch, *Br. J. Cancer,* 25, 533, 1971.
16. **Adams, R. A., Flowers, A., Sundeen, R., and Melk, L. P.,** Chemotherapy and immunotherapy of 3 human tumour lymphomas serially transplantable in the neonatal Syrian hamster, *Cancer,* 29, 524, 1972.
17. **Cobb, L. M.,** The hamster as a host for the growth and study of human tumour cell populations, *Cancer Res.,* 34, 958, 1974.
18. **Goldenberg, D. M., Bhan, R. J., and Pavia, R. A.,** *In vivo* human hamster somatic cell fusion indicated by glucose-6-phosphate dehydrogenase and lactate dehydrogenase profiles, *Cancer Res.,* 31, 1148, 1971.
19. **Goldenberg, D. M., Pavia, R., and Tsoa, M. C.,** *In vivo* hybridisation of human tumour and normal hamster cells, *Nature (London),* 250, 649, 1974.
20. **Bogden, A. E., Kelton, D. E., Cobb, W. R., and Esber, H. J.,** A rapid screening method for testing chemotherapeutic agents against human xenografts, in *Proc. Symp. Use of Athymic (Nude) Mice in Cancer Res.,* Houchens, D. P. and Ovejera, A. A., Eds., Gustav Fischer, New York, 1978, 231.
21. **Bogden, A. E., Haskell, P. M., LePage, D. J., Kelton, D. E., Cobb, W. R., and Esber, H. J.,** Growth of human tumor xenografts implanted under the renal capsule of normal immunocompetent mice, *Exp. Cell Biol.,* 47, 281, 1979.
22. **Bogden, A. E., Cobb, W. R., LePage, D. J., Haskell, P. M., Gulkin, T. A., Ward, A., Kelton, D. E., and Esber, H. J.,** Chemotherapy responsiveness of human tumors as first transplant generation xenografts in the normal mouse: 6-day subrenal capsule assay, *Cancer,* 48, 10, 1981.
23. **Murphy, J.,** Transplantability of malignant tumours to the embryos of a foreign species, *JAMA,* 57, 874, 1912.
24. **Hurst, E. W., Cooke, B., and McLennan, G. C.,** A note on the survival and growth of human and rabbit tissues (normal and neoplastic) on the chorioallantis of the chicken and duck embryo, *Aust. J. Exp. Biol. Med. Sci.,* 17, 215, 1939.
25. **Sommers, S. C., Sullivan, B. A., and Warren, S.,** Heterotransplantation of tumour cancer. III. Chorioallontoic membranes of embryonated eggs, *Cancer Res.,* 12, 915, 1952.
26. **Kaufman, N., Kinney, T. D., Mason, E. J., and Prieto, L. C., Jr.,** Maintenance of human neoplasm on the chick chorioallontoic membrane, *Am. J. Path.,* 32, 271, 1956.
27. **Kutner, L. J. and Southam, C. M.,** Growth of human cancer cells (HEP2) in newborn rats, *Proc. Soc. Exp. Biol. Med.,* 104, 785, 1960.
28. **Vogel, H. B. and Berry, R. G.,** Chorioallontoic membrane heterotransplantation of human brain tumours, *Int. J. Cancer,* 15, 401, 1975.
29. **Smith, I. E., Courtenay, V. D., and Gordon, M. Y.,** A colony forming assay for human tumour xenografts using agar in diffusion chambers, *Br. J. Cancer,* 34, 476, 1976.
30. **Selby, P. J. and Thomas, J. M.,** Clonogenic survival curves for human melanoma xenografts using agar diffusion chambers and lung colony assays, *Br. J. Cancer,* 41, 150, 1980.
31. **Toulan, H. W.,** Successful subcutaneous growth and transplantation of human tumours in X-irradiated laboratory animals, *Proc. Soc. Exp. Biol. Med.,* 77, 572, 1951.
32. **Toulan, H. W.,** Growth of human tumours in the subcutaneous tissues of X-irradiated laboratory animals: their practical use for experimental purposes, *Cancer Res.,* 12, 302, 1952.
33. **Toulan, H. W.,** Growth of human tumours in cortisone treated laboratory animals: the possibility of obtaining permanently transplantable human tumours, *Cancer Res.,* 13, 389, 1953.
34. **Toulan, H. W.,** The transplantable human tumour, *Ann. N. Y. Acad. Sci.,* 76, 733, 1958.
35. **Foley, E. J.,** Effect of cortisone acetate on growth of strain specific tumours in alien strains of mice, *Proc. Soc. Exp. Biol. Med.,* 80, 669, 1952.
36. **Patterson, W. B.,** Transplantable human tumours, *Cancer,* 10, 1281, 1957.
37. **Rygaard, J. and Povlsen, C. O.,** Heterotransplantation of a human malignant tumour to "nude" mice, *Acta Path. Microbiol. Scand.,* 77, 758, 1969.
38. **Festing, M. F. W.,** *International Index of Laboratory Animals,* 5th ed., Festing, M. F. W., Ed., Medical Research Council, Surrey, UK, 1987.
39. **Hansen, C. T.,** The nude gene and its effects, in *The Nude Mouse in Experimental and Clinical Research,* Fogh, J. and Giovanella, B. C., Eds., Academic Press, New York, 1978, 1.
40. **Maruo, K., Ueyama, Y., Hioki, K., Saito, M., Nomura, T., and Tamaoki, N.,** Strain-dependent growth of a human carcinoma in nude mice with different genetic backgrounds, *Exp. Cell Biol.,* 50, 115, 1982.
41. **Festing, M. F. W.,** Inherited immunological defects in laboratory animals, in *Immunodeficient Animals for Cancer Research,* Sparrow, S., Ed., Macmillan Press, London, 1980, 5.
42. **Green, M. C.,** Catalog of mutant genes and polymorphic loci, in *Genetic Variants and Strains of the Laboratory Mouse,* Lyon, M. F. and Searle, A. G., Eds., Oxford University Press, Oxford, 1989, 12.
43. **Gasser, D. L. and Fishgrund, T.,** Genetic control of the immune response in mice. IV. Relationship between graft vs. host reactivity and possession of the high tumour genotypes $A^{y}a$ and $A^{vy}a$, *J. Immunol.,* 110, 305, 1973.

44. **Hertwig, P.,** Neue mutationem und Koppelungsgruppen bei der Hausmaus, *Z. Indukt. Abstammungs-Vererbungsl,* 80, 220, 1946.
45. **Bannerman, R. M., Edwards, J. A., and Pinkerton, P. H.,** Hereditary disorders of the red cells in animals, *Prog. Hematol.,* 8, 131, 1973.
46. **Knight, J. G. and Adams, D. D.,** Genes determining autoimmune disease in New Zealand mice, *J. Clin. Lab. Immunol.,* 5, 165, 1981.
47. **Roder, J. C. and Duwe, A.,** The beige mutation in the mouse selectively impairs natural killer cell function, *Nature (London),* 278, 451, 1979.
48. **Lane, P. W. and Murphy, E. D.,** Susceptibility to spontaneous pneumonitis in an inbred strain of beige and satin mice, *Genetics,* 72, 451, 1972.
49. **Roder, J. C.,** The beige mutation in the mouse. I. A stem cell predetermined impairment in natural killer cell function, *J. Immunol.,* 123, 2168, 1979.
50. **Roder, J. C., Lohmann-Matthes, M.-L., Domzig, W., and Wigzell, H.,** The beige mutation in the mouse. II. Selectivity of the natural killer cell defect, *J. Immunol.,* 123, 2174, 1979.
51. **Saxena, R. K., Saxena, Q. B., and Adler, W. H.,** Defective T-cell response in beige mutant mice, *Nature (London),* 295, 240, 1982.
52. **Fodstad, Ø., Hansen, C. T., Cannon, G. B., Statham, C. N., Liechtenstein, G. R., and Boyd, M. R.,** Lack of correlation between natural killer cell activity and tumor growth control in nude mice with different immune defects, *Cancer Res.,* 44, 4403, 1984.
53. **Karre, K., Klein, G. O. Kiessling, R., Klein, G., and Roder, J. C.,** Low natural *in vivo* resistance to syngeneic leukemias in natural killer-deficient mice, *Nature (London),* 284, 624, 1980.
54. **Gorelik, E., Wiltrout, R. H., Okumura, K., Habu, S., and Herberman, R. B.,** Role of NK cells in the control of metastatic spread and growth of tumor cells in mice, *Int. J. Cancer,* 30, 107, 1982.
55. **Talmadge, J. E., Meyers, K. M., Prieur, D. J., and Starkey, J. R.,** Role of natural killer cells in tumor growth and metastasis, *J. Natl. Cancer Inst.,* 65, 929, 1980.
56. **Searle, A. G.,** Hereditary absence of spleen in the mouse, *Nature (London),* 184, 1419, 1959.
57. **Fletcher, M. P., Iheda, R. M., and Gershwin, M. E.,** Splenic influence of T cell function: the immunobiology of the inbred hereditarily asplenic mouse, *J. Immunol.,* 119, 110, 1977.
58. **Batisto, J. R., Borek, F., and Bucsi, R. A.,** Splenic determinants of immunocompetence: influence on other lymphoid organs, *Cell Immunol.,* 2, 627, 1971.
59. **Dung, H. C.,** Deficiency in the thymus-dependent immunity in lethargic mutant mice, *Transplantation,* 23, 39, 1977.
60. **Glode, L. M. and Rosenstreich, D. L.,** Genetic control of B cell activation by bacterial lipopolysaccharide is mediated by multiple distinct genes or alleles, *J. Immunol.,* 117, 2061, 1976.
61. **Watson, J. K., Kelly, K., Largen, M., and Taylor, B. A.,** The genetic mapping of a defective LPS response gene in C3H/HeJ mice, *J. Immunol.,* 120, 422, 1978.
62. **Green, M. C. and Schultz, L. D.,** Motheaten, an immunodeficient mutant of the mouse. I. Genetics and pathology, *J. Hered.,* 66, 250, 1975.
63. **Schultz, L. D. and Green, M. C.,** Motheaten, an immunodeficient mutant of the mouse. II. Depressed immune response and elevated serum immunoglobulins, *J. Immunol.,* 116, 936, 1976.
64. **Bosma, G. C., Custer, R. P., and Bosma, M. J.,** A severe combined immunodeficiency mutation in the mouse, *Nature (London),* 301, 527, 1983.
65. **Bosma, M. J.,** The Scid mutation: occurrence and effect, in *Current Topics in Microbiology and Immunology. The Scid Mouse, Characterisation and Potential Uses,* Bosma, M. J., Phillips, R. A., and Schuler, W., Eds., Springer-Verlag, Berlin, 1989, 3.
66. **Fulop, G. M. and Phillips, R. A.,** Full reconstitution of the immune deficiency in scid mice with normal stem cells requires low-dose irradiation of the recipients, *J. Immunol.,* 136, 4438, 1986.
67. **Sarvella, P. A. and Russel, L. B.,** Steel, a new dominant gene in the house mouse, *J. Hered.,* 47, 123, 1956.
68. **Murphy, E. D.,** Effect of mutant steel alleles on leukemogenesis and life span in the mouse, *J. Natl. Cancer Inst.,* 58, 107, 1977.
69. **Russel, E. S.,** Abnormalities of erythropoiesis associated with the W-locus in mice, in *Regulation of Hematopoiesis,* Vol. 4, Gordon, A. L., Ed., Appleton-Century-Crofts, New York, 1970, 649.
70. **Scher, I.,** The CBA/N mouse strain: experimental model illustrating the influence of the X-chromosome on immunity, *Adv. Immunol.,* 33, 1, 1982.
71. **Lozzio, B. B., Machado, E. A., Lozzio, C. B., and Lair, S.,** Hereditary asplenic-athymic mice: transplantation of human myelogenous leukemic cells, *J. Exp. Med.,* 143, 225, 1976.
72. **Gershwin, M. E., Ahmed, A., Ikeda, R. M., Shifrine, M., and Wilson, F.,** Immunobiology of congenitally athymic-asplenic mice, *Immunology,* 34, 631, 1978.
73. **Azar, H. A., Hansen, C. T., and Costa, J.,** N:NIH(S)-nu/nu mice with combined immunodeficiency: a new model for human tumor transplantation, *J. Natl. Cancer Inst.,* 65, 421, 1980.

74. **Wortis, H. H., Burkly, L., Hughes, D., Roschelle, S., and Waneck, G.,** Lack of mature B cells in nude mice with X-linked immune deficiency, *J. Exp. Med.,* 155, 903, 1982.
75. **Karre, K., Klein, G. O., Kessling, R., Klein, G., and Roder, J. C.,** *In vitro* NK activity and *in vivo* resistance to leukemia: studies of beige/nude and wild-type hosts on C57BL background, *Int. J. Cancer,* 26, 789, 1980.
76. **Andriole, G. L., Mule, J. J., Hansen, C. T., Marston, Linehan, W., and Rosenberg, S. A.,** Evidence that lymphokine-activated killer cells and natural killer cells are distinct based on an analysis of congenitally immunodeficient mice, *J. Immunol.,* 135, 2911, 1985.
77. **Fodstad, O., Hansen, C. T., Cannon, G. B., and Boyd, M. R.,** Immune characterisation of the beige-nude mouse, a model for studying immune surveillance, *Scand. J. Immunol.,* 20, 267, 1984.
78. **Gershwin, M. E., Ikeda, R. M., Erickson, K., and Owens, R.,** Enhancement of heterotransplanted human tumor graft survival in nude mice treated with antilymphocytic serum and in congenitally athymic-asplenic (lasat) mice, *J. Natl. Cancer Inst.,* 61, 245, 1978.
79. **Tropé, C., Wennerberg, J., Alm, P., and Von Eyben, F. E.,** Heterotransplantation of two human tumours in athymic mice and asplenic-athymic mice, *Acta Path. Microbiol. Immunol. Scand., Sect. A,* 90, 75, 1982.
80. **Giovanella, B. C. and Fogh, J.,** The nude mouse in cancer research, in *Advances in Cancer Research,* Klein, G. and Weinhouse, S., Eds., Academic Press, Orlando, FL, 44, 69, 1985.
81. **Phillips, R. A., Jewett, M. A. S., and Gallie, B. L.,** Growth of human tumors in immune-deficient Scid mice and nude mice, in *Current Topics in Microbiology and Immunology. The Scid Mouse, Characterisation and Potential Uses,* Bosma, M. J., Phillips, R. A., and Schuler, W., Eds., Springer-Verlag, Berlin, 1989, 259.
82. **Reddy, S., Piccione, D., Takita, H., and Bankert, R. B.,** Human lung tumor growth established in the lung and subcutaneous tissue of mice with severe combined immunodeficiency, *Cancer Res.,* 47, 2456, 1987.
83. **Woodruff, M. A. and Anderson, N. F.,** Effect of lymphocyte depletion by thoracic duct fistula and administration of anti-lymphocytic serum on the survival of skin homograft in rats, *Nature (London),* 200, 702, 1963.
84. **Phillips, B. and Gazet, J.-C.,** Growth of two human tumour cell lines in mice treated with antilymphocytic serum, *Nature (London),* 215, 548, 1967.
85. **Phillips, B. and Gazet, J.-C.,** Effect of antilymphocytic serum on the growth of Hep 2 and HeLa cells in mice, *Nature (London),* 220, 1140, 1968.
86. **Stanbridge, E. and Perkins, F. T.,** Tumour nodule formation as an *in vivo* measure of the suppression of cellular immune response by anti-lymphocytic serum, *Nature (London),* 221, 80, 1969.
87. **Phillips, B. and Gazet, J.-C.,** Transplantation of primary explants of human tumour to mice treated with antilymphocyte serum, *Br. J. Cancer,* 24, 92, 1970.
88. **Arnstein, P., Taylor, D. O. N., Nelson-Rees, W. A., Huebner, R. J., and Lennette, E. H.,** Propagation of human tumors in antithymocyte serum-treated mice, *J. Natl. Cancer Inst.,* 52, 71, 1974.
89. **James, K.,** Anti-lymphocytic antibody — a review, *Clin. Exp. Immunol.,* 2, 615, 1967.
90. **Levey, R. H. and Medawar, P.,** Nature and mode of action of antilymphocytic serum, *Proc. Natl. Acad. Sci. U.S.A.,* 56, 1130, 1966.
91. **Lance, E. M.,** The selective action of antilymphocyte serum on recirculating lymphocytes: a review of the evidence and alternatives, *Clin. Exp. Immunol.,* 6, 789, 1970.
92. **Miller, J. F. A. P.,** Immunological function of the thymus, *Lancet,* ii, 748, 1961.
93. **Osoba, D. and Auersperg, N.,** Growth of a human carcinoma of the cervix in mice thymectomised at birth, *J. Natl. Cancer Inst.,* 523, 1966.
94. **Miller, J. F. A. P.,** The thymus: yesterday, today and tomorrow, *Lancet,* ii, 1299, 1967.
95. **Miller, J. F. A. P., Mitchell, G. F., and Weiss, N. S.,** Cellular basis of the immunological defects in thymectomized mice, *Nature (London),* 214, 992, 1967.
96. **Steel, G. G., Courtenay, V. D., and Rostom, A. Y.,** Improved immune-suppression techniques for the xenografting of human tumours, *Br. J. Cancer,* 37, 224, 1978.
97. **Monaco, A. P., Wood, M. L., and Russel, P. S.,** Adult thymectomy: effect on recovery from immunologic depression in mice, *Science,* 149, 132, 1965.
98. **Miller, J. F. A. P., Doak, S. M., and Cross, A. M.,** Role of the thymus in recovery of the immune mechanism in the irradiated adult mouse, *Proc. Soc. Exp. Biol. Med.,* 112, 785, 1963.
99. **Castro, J. E.,** Human tumours grown in mice, *Nature New Biol.,* 239, 83, 1972.
100. **Franks, C. R., Perkins, F. T., and Holmes, T. J.,** Subcutaneous growth of human tumours in mice, *Nature (London),* 243, 91, 1973.
101. **Cobb, L. M.,** The behaviour of carcinoma of the large bowel in man following transplantation into immune-deprived mice, *Br. J. Cancer,* 28, 400, 1973.
102. **Cobb, L. M. and Mitchley, B. C. V.,** Growth of human tumours in immune deprived mice, *Eur. J. Cancer Clin. Oncol.,* 10, 473, 1974.
103. **Franks, C. R., Bishop, D., and Reeson, D.,** The growth of tumour xenografts in thymectomised high dose irradiated mice reconstituted with syngeneic bone marrow cells incubated with anti-thymocyte serum, *Br. J. Cancer,* 33, 112, 1976.

104. **Detre, S. I. and Gazet, J.-C.,** Transplantation of human tumour to immune deprived mice treated with anti-thymocyte serum, *Br. J. Cancer,* 28, 412, 1973.
105. **Detre, S. I.,** Comparison of the growth of xenografts in various kinds of immunodeficient mice, in *Immunodeficient Animals for Cancer Research,* Sparrow, S., Ed., MacMillan Press, London, 1980, 81.
106. **Steel, G. G. and Peckham, M. J.,** Human tumor xenografts — a critical reappraisal, *Br. J. Cancer,* 41(Suppl. IV), 133, 1980.
107. **Steel, G. G., Courtenay, V. D., Phelps, T. A., and Peckham, M. J.,** The therapeutic response of human tumour xenografts, in *Immunodeficient Animals for Cancer Research,* Sparrow, S., Ed., MacMillan Press, London, 1980, 179.
108. **Berenbaum, M. C., Sheard, C. E., Reittie, J. R., and Bundick, R. V.,** The growth of human tumours in immunosuppressed mice and their response to chemotherapy, *Br. J. Cancer,* 30, 13, 1974.
109. **Millar, J. L., Blackett, N. M., and Hudspith, B. N.,** Enhanced post-irradiation recovery of the haemopoietic system in animals pretreated with a variety of cytotoxic drugs, *Cell Tiss. Kin.,* 11, 543, 1978.
110. **Phelps, T. A. and Blackett, N. M.,** Protection of intestinal damage by pretreatment with cytarabine (cytosine arabinoside), *Int. J. Rad. Oncol. Biol. Phys.,* 5, 1617, 1979.
111. **Steel, G. G., Courtenay, V. D., and Peckham, M. J.,** The response to chemotherapy of a variety of human tumour xenografts, *Br. J. Cancer,* 47, 1, 1983.
112. **Watanabe, S., Shomosato, Y., Kameya, T., Kuroki, M., Kitahara, T., Minato, K., and Shimoyama, M.,** Leukemic distribution of a human acute lymphocytic leukemia cell line (Ichikawa strain) in nude mice conditioned with whole body irradiation, *Cancer Res.,* 38, 3494, 1978.
113. **Ziegler, H. W. L., Frizzera, G., and Bach, F. H.,** Successful transplantation of a human acute leukemia cell line into nude mice: conditions optimising graft acceptance, *J. Natl. Cancer Inst.,* 68, 15, 1982.
114. **Ohsugi, Y., Gershwin, M. E. Owens, R. B., and Nelson-Rees, W. A.,** Tumorigenicity of human malignant lymphoblasts: comparative study with unmanipulated nude mice, antilymphocyte serum treated nude mice and X-irradiated nude mice, *J. Natl. Cancer Inst.,* 715, 1980.
115. **Epstein, A. L., Herman, M. M., Kim, H., Dorfman, R. F., and Kaplan, H. S.,** Biology of the human malignant lymphomas. III. Intracranial heterotransplantation in the nude athymic mouse, *Cancer,* 37, 2158, 1976.
116. **Schaadt, M., Kirchner, H., Fonatsch, C., and Diehl, V.,** Intracranial heterotransplantation of human hematopoietic cells in nude mice, *Int. J. Cancer,* 23, 751, 1979.
117. **Giovanella, B., Nilsson, K., Zech, L., Yim, O., Klein, G., and Stehlin, J. S.,** Growth of diploid Epstein-Barr virus carrying human lymphoblastoid cell lines heterotransplanted into nude mice under immunologically privileged conditions, *Int. J. Cancer,* 24, 103, 1979.
118. **Gazdar, A. F., Carney, D. N., Sims, H. L., and Simmons, A.,** Heterotransplantation of small cell carcinoma of the lung into nude mice: comparison of intracranial and subcutaneous routes, *Int. J. Cancer,* 28, 777, 1981.
119. **Gallie, L., Albert, D. M., Wong, J. J. Y., Buyukmihci, N., and Puliafito, C. A.,** Heterotransplantation of retinoblastoma into the athymic nude mouse, *Invest. Opthalmol., Vis. Sci.,* 16, 256, 1977.
120. **Floersheim, G. L.,** Comparative growth of human tumours in pharmacologically immunosuppressed, immune-deprived cyclosporin A and nude mice, *Eur. J. Cancer Clin. Oncol.,* 18, 589, 1982.
121. **Sharkey, F. E., Fogh, J. M., Hajdu, S. I., Fitzgerald, P. J., and Fogh, J.,** Experience in surgical pathology with human tumor growth in the nude mouse, in *The Nude Mouse in Experimental and Clinical Research,* Fogh, J. and Giovanella, B. C., Eds., Academic Press, New York, 1978, 187.

Chapter 3

BREEDING AND MICROBIOLOGICAL CONTROL

Tatsuji Nomura

TABLE OF CONTENTS

I. INTRODUCTION

For experiments in animals to be performed, it is important to carefully consider the rationale and aims of the experiment, and prepare an appropriate experimental plan to achieve these aims. It is also essential to assure the quality of the animals used in the experiments. If this is not done, reliable and reproducible results cannot be obtained. To assure the quality of experimental animals involves both genetic and microbiological control. Genetic control is achieved by genetic management, while microbiological control requires not only control of the animals themselves, but also simultaneous microbiological control of the rearing environment of the animals. Particularly in oncology research using nude mice, the animals by themselves have a deficient T cell-mediated immune response and the host defense mechanism is often further suppressed by anticancer drugs or radiotherapy, which makes them susceptible to opportunistic infections. Therefore, microbiological control is especially important for nude mice.[1,2]

Based on recent breeding results for nude mice and experience gained from experimentation using these animals, it has become possible to handle nudes in the same way as ordinary euthymic animals by means of strict microbiological control in accordance with the specific pathogen free (SPF) production technology. In this respect, this chapter will outline the breeding and care and management of nude mice based on genetic and microbiological control.

II. BREEDING OF THE NUDE MOUSE

In oncology research, nude mice are often used in quantitative experiments in the form of bioassays. For example, the results of screening of anticancer drugs in xenografts are often inferred from differences in tumor growth between the mice treated with the drug and untreated controls. In such experiments, it is necessary to have sufficient numbers of mice in each group for statistical analysis. In cases where relative evaluations among several test drugs are performed, large numbers of animals must be used in each experiment. In addition, it should be possible to make comparisons between experiments performed at different time points. For this purpose, nude mice must be genetically and phenotypically uniform (within a fixed range even in lots of mass-produced animals) and reproducible over time (no changes in the characters of the animals during continued production).

Nude mice are congenitally hairless and athymic, and it is difficult to breed them under normal animal rearing conditions. Therefore, a system which incorporates both genetic and microbiological control is especially necessary for the production of nude mice. Mass production techniques for nude mice based on uniformity and reproducibility form the backbone for high-quality nude mice in oncology research, such as screening procedures for anticancer drugs and cancer therapy experiments.

A. OUTLINE OF THE BREEDING SYSTEM

Nude mice usually survive in germfree or SPF environments and live basically as long as conventional mice. Both males and females have normal reproductive functions, making breeding possible.

Germfree rearing is performed in special equipment, and the most reliable rearing method will completely prevent infections. Such methods are unfortunately inappropriate for the rearing of large numbers of animals. SPF rearing takes place in special facilities, and large-scale rearing and production of nude mice are possible since infections are prevented in units of one building or one room. However, there is always a possibility that personnel will bring pathogens into the facilities, because animal caretakers and researchers must enter the facilities. Therefore, we have devised a system whereby microbiological control is easier than in the ordinary SPF rearing system through a combination of germfree and SPF rearing techniques, and apply this system in the production of nude mice.

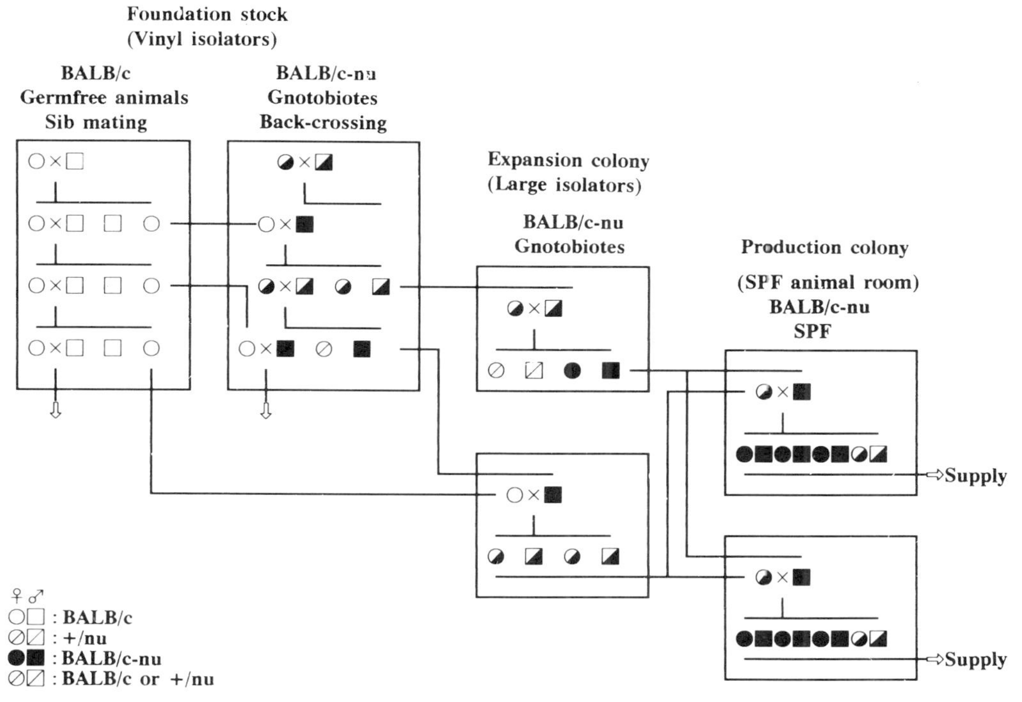

FIGURE 1. Outline of the nude mouse production system.

Nude mice are hairless, athymic mutants in which the recessive gene nu is expressed as the homozygous type (nu/nu). Among the characters of nude mice, those due to the expression of genes other than nu are defined by the genetic background of the mouse strains in which the nu gene is introduced. One example is the BALB/c-nu strain, in which the nu gene was introduced into the inbred BALB/c strain and which we produce. BALB/c mice are widely used in biomedical research as the standard albino strain. To enhance the uniformity and reproducibility of nude mice as laboratory animals, it is necessary to increase the uniformity in the gene composition other than the nu gene; that is, for BALB/c-nu mice, to make the gene composition as close as possible to that of the BALB/c strain which has provided the genetic background. For this purpose, we continuously back-cross inbred BALB/c-nu mice with ordinary BALB/c mice during nude mouse production. The production system for nude mice which we use is shown in Figure 1 and is described below.

The inbred BALB/c strain which provides the genetic background is maintained by sib mating in vinyl isolators. The BALB/c-nu foundation stock is also maintained in isolators, and they are back-crossed with BALB/c mice in every other generation.

The expansion colony used for production (the animals which are the parents for production) is maintained and bred in large isolators; that is, female and male heterozygous (+/nu) mice and male nude (nu/nu) mice from the BALB/c-nu foundation stock isolator are moved to the large-scale isolator, where they are made SPF by the introduction of specific flora. The heterozygous females and homozygous male nude mice that are bred are used for the expansion colony. For the ultimate production of nude mice, these animals are placed in SPF production facilities for mating. The nude mice produced are supplied to research institutes for screening of antitumor agents and various other types of animal experimentation.

With this breeding system, genetic control is performed by back-crossing, and microbiological control is carried out by keeping the nude mouse foundation stock in germfree isolators, whereafter the nude mice obtained are moved to the SPF production facilities.

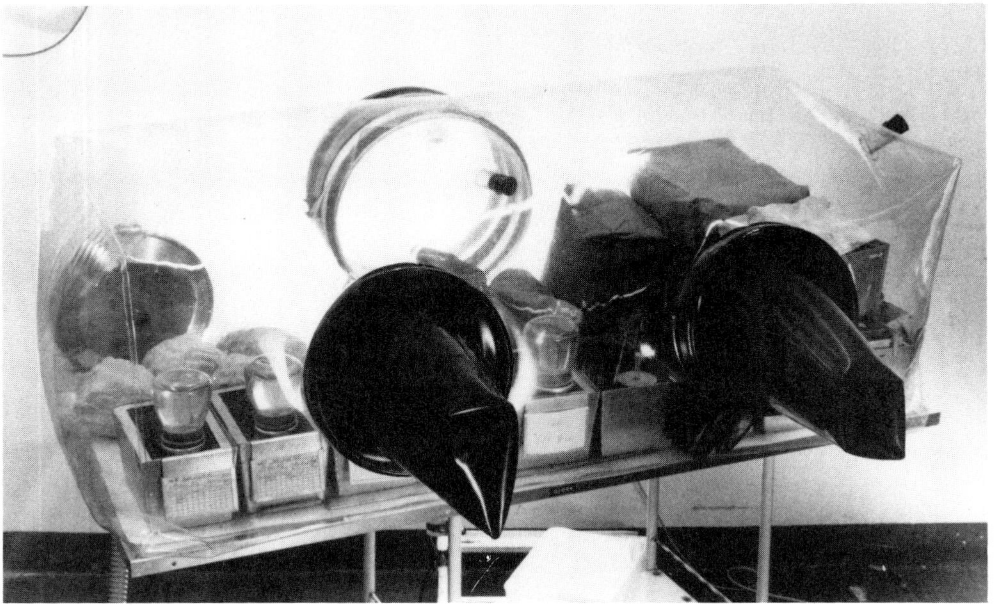

FIGURE 2. Vinyl isolator.

B. FACILITIES AND EQUIPMENT

1. Vinyl Isolators

Since vinyl isolators were first proposed by P.C. Trexler for germfree animals in 1957, they are now widely used for animal rearing throughout the world. Various types of isolators have been constructed, including those with soft vinyl frames. We use isolators with vinyl sections 115 cm wide, 60 cm deep, and 50 cm high for the foundation stock of nude mice (Figure 2). These vinyl isolators are used in accordance with the basic procedures for rearing germfree animals.

2. Large Vinyl Isolators

We have developed large vinyl isolators for germfree rearing of large numbers of animals in expansion colonies. They have a soft vinyl frame and the rearing procedures are the same as those for the vinyl isolators described above. They are 200 cm wide, 160 cm deep, and 90 cm high; they can hold approximately 60 mouse cages (Figure 3).

3. SPF Animal Production Facilities

These facilities were designed for rearing and production as well as experimentation on SPF animals. To prevent infections by pathogens from outside, personnel can only enter the clean areas after showering and changing into a special uniform. The materials brought into the facilities are all sterilized, and for air ventilation, passing through high-efficiency particulate air (HEPA) filters is ensured. Although there are various types of SPF facilities for animal experimentation, the two-corridor system shown in Figure 4 is often used, while in production facilities the small independent building system also shown in Figure 4 is more appropriate. The small, isolated type is recommended for mass production of nude mice since infections can be reliably prevented and effects on other reproduction colonies are eliminated.

4. Other Rearing Materials

Rearing materials used in the production of nude mice are essentially the same as those used in the rearing of ordinary laboratory animals.

FIGURE 3. Large vinyl isolator.

Breeding cages, water bottles, and bedding are all sterilized in an autoclave before use. Autoclavable diets are used in the isolators, and irradiation-sterilized diets are used in the SPF facilities. Ordinary tap water can be used as drinking water in cases where bottling of water is possible in the rearing site, but sterilization is required when the water is brought into the rearing rooms after bottling.

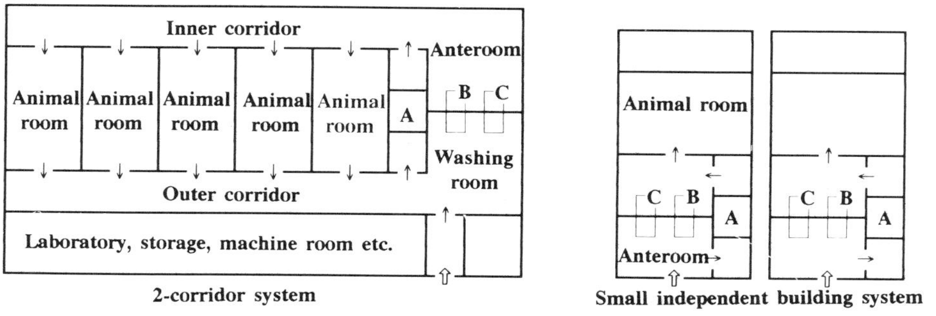

FIGURE 4. SPF animal facilities. A: Shower pass, B: Autoclave, C: Pass box.

C. REPRODUCTION METHODS

Nude mice, which are bred and maintained in germfree or SPF environments, will survive for long periods of time and both sexes are able to reproduce. However, the nursing ability of the females is poor because they are hairless. Therefore, reproduction for production purposes is performed by mating heterozygous females (+/nu with hair), and selecting the hairless nude mice from the offspring.

1. Foundation Stocks

The foundation stocks of BALB/c-nu are maintained by back-crossing of cross-intercross mice (Figure 1). Firstly, female inbred BALB/c and male nude (nu/nu) mice which have reached the reproductive age are mated by 1:1 continuous cohabitation, and heterozygous offspring (+/nu) are obtained. The next generation is obtained by mating male and female littermates from among the heterozygous mice by continuous cohabitation and selecting the nude offspring which appear at a probability of 1:4. For the third generation, inbred BALB/c mice are again mated with nude (nu/nu) mice. Several pairs are mated in each generation, and the mice are used for reproduction up to the third or fourth delivery. When female heterozygous and male nude mice are supplied to the expansion colony, they are both retired at the time reproduction of the subsequent generation is confirmed.

2. Expansion Colonies

Production of several thousands of nude mice per month is impossible, if mice have to be supplied from the foundation stock. Therefore, nude mice are produced by increasing the number of breeders for production in one more stage based on mice supplied from the foundation stock. Production in this expansion colony is performed in large isolators by one of the following two methods, depending on the number of mice required.

In the case of relatively few mice necessary, heterozygous females and nude males, obtained as offspring from mating by 1:1 continuous cohabitation of a heterozygous female and a nude male supplied from the foundation stock, are moved to the expansion colony. For efficient use of nude males from the foundation stock, rotation mating of six heterozygous females with one nude male is performed. The number of pairs required varies, depending on the size of the expansion colony needed for mass production. Heterozygous females are retired after they have produced four or five times and the nude males are retired after three rotations.

Rather large expansion colonies necessitate the following method. Nude males for the expansion colony are produced by mating one heterozygous male or one nude male supplied from the foundation stock with a heterozygous female or an inbred BALB/c female mouse. Rotation mating is used in both instances (Figure 1).

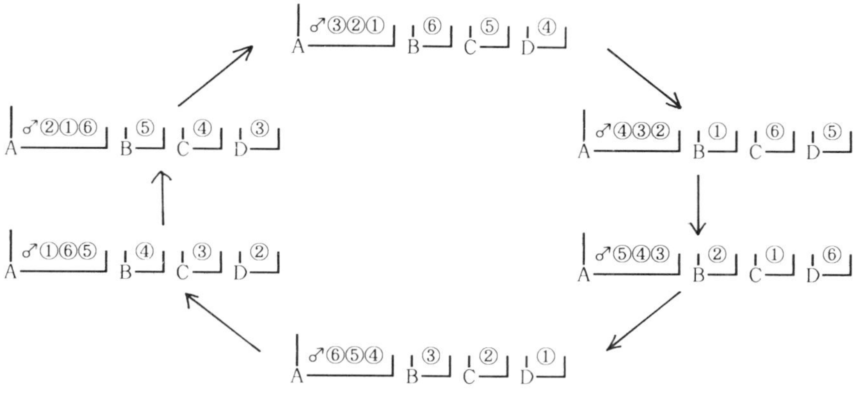

FIGURE 5. Reproduction system by rotation mating using one male and six females. A female of at least 7 weeks of age is placed each week in a mating cage containing 1 male from the foundation stock, and they are mated. If things go as planned, pregnancies occur each week in succession, and the females are then transferred to delivery cages when they are ready to deliver. After delivery, they are transferred to nursing cages. After weaning of their litters, the mothers are again cohabited with the male for the next mating.

TABLE 1
Reproductive Results of the BALB/c-nu Mouse Expansion Colony in an SPF Environment

Delivery	No. of females	No. of deliveries		Total no. of pups (average)		No. of nu/nu weaned				Production index
						Females	Males	Total	(%)	
1	248	236	(95.2)	1479	(6.3)	361	268	629	(42.5)	2.6
2	236	212	(89.8)	1463	(6.9)	328	339	687	(45.6)	2.8
3	212	176	(83.0)	1119	(6.4)	267	255	522	(46.7)	2.5
4	176	134	(76.1)	790	(5.9)	195	160	355	(45.0)	2.0
Total	872	758	(86.9)	4851	(6.4)	1151	1022	2173	(44.8)	2.5

3. Production under SPF Conditions

Production is performed by the supply of males and females from the expansion colony to the SPF production facilities. Reproduction is achieved by rotation mating of one nude male with six heterozygous females (Figure 5). A female of at least 7 weeks of age is cohabited in a cage containing 1 male breeder for 1 week. Pregnant females (those with a swollen abdomen) are isolated in succession (if things go as planned, one female becomes pregnant with each cohabitation period), and they deliver and nurse their pups separately. Mothers from which pups have been weaned are again cohabited with the same male for the next reproduction. Females are retired after four to five reproductions and males after three rotations. The reproductive results of an SPF production colony are shown in Table 1.

D. QUALITY TESTING

The quality of both the animals and the reproduction system should be tested at each stage in order to assure reliable breeding of nude mice. Since nude mice are highly susceptible to infections, microbiological monitoring is necessary on a routine basis. Genetic monitoring is

TABLE 2
Items for Genetic Monitoring of Mice

Biochemical Marker Genes

Gene symbol	Chromosome	Gene name
Idh-1	1	Isocitrate dehydrogenase-l
Pep-3	1	Peptidase-3
Akp-1	1	Alkaline phosphatase-1
Car-2	3	Carbonic anhydrase-2
Mup-1	4	Major urinary protein-1
Gpd-1	4	Glucose phosphate dehydrogenase-1
Pgm-1	5	Phosphoglucomutase-1
Ldr-1	6	Lactate dehydrogenase regulator-1
Gpi-1	7	Glucose phosphate isomerase-1
Hbb	7	Hemoglobin β-chain
Es-1	8	Esterase-1
Es-2	8	Esterase-2
Mod-l	9	Malic enzyme
Trf	9	Transferrin
Es-3	11	Esterase-3

Immunogenetic Marker Genes

Gene symbol	Chromosome	Gene name
Hc	2	Hemolytic complement
Thy-1	9	Thymus antigen-1
H-2K	17	Histocompatibility-2K
H-2D	17	Histocompatibility-2D

also required because nude mice are mutant strains with various genetic backgrounds.

In our facilities, we sample animals periodically at each production stage for microbiological control. If mice are not available from a specific colony, sentinel mice (usually C57BL/6) placed in the same isolator or room are used for testing. When accidental infections arise in isolators or SPF rooms, the infected colony is immediately destroyed and other colonies are tested to detect spreading of the infection. Countermeasures against such accidents and smooth disposal of infected animals are very important for the care and management of nude mice. The items which are subject to microbiological control, the checking methods, and the measures to prevent infection are described in detail in Section III on microbiological control.

In the case of genetic monitoring, six mice are sampled from the foundation stock in each generation or once a year, from retired breeders and the production colonies once every year. Since both the hairlessness and athymic condition are closely linked with the nu gene, it can be assumed that a mouse is athymic if it is hairless. Therefore, genetic monitoring of nude mice is performed for the genetic profile of the inbred strain, which has formed the genetic background. The items used in genetic monitoring are shown in Table 2. At present, 19 gene loci on chromosomes are used.

III. MICROBIOLOGICAL CONTROL

A. NECESSITY OF MICROBIOLOGICAL CONTROL

The athymic condition with the resulting T cell-mediated immunodeficiency renders the nude mouse highly susceptible to microorganisms. Various infectious diseases may interfere with experiments carried out in nude mice. The most prominent feature of these diseases is the wasting syndrome, characterized by weight loss, a scaly skin, and a hunched posture. Since, in 1966, Flanagan[3] described that infection by *Toxoplasma gondii* caused wasting of nude mice,

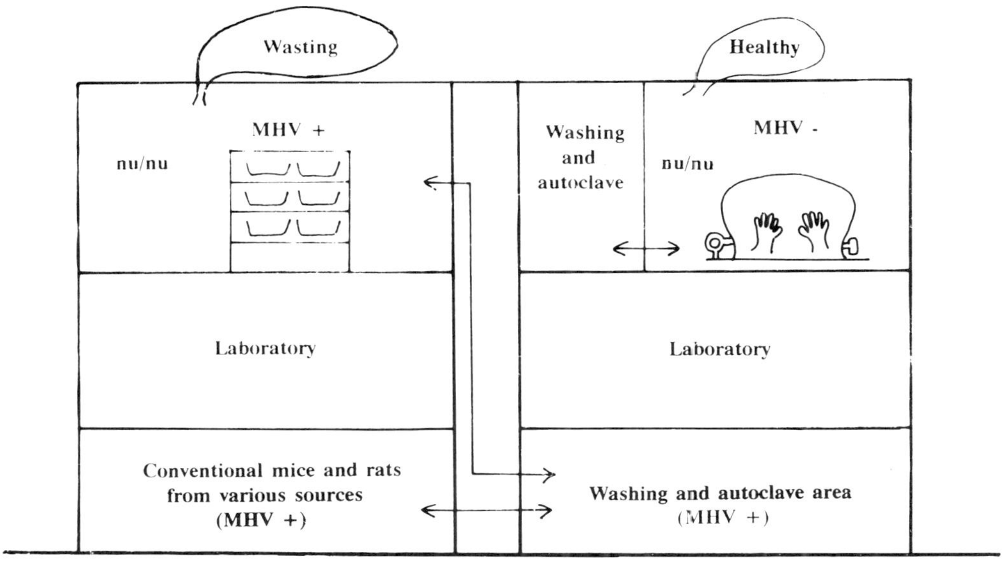

FIGURE 6A. Accidental infections of nude mice, case A. Cause: contamination of cages in washing and sterilizing area. Problem: design of facilities.

there have been many reports describing that *Giardia muris*,[4] *Hexamita muris*,[4] and *Pneumocystis carinii*[5] as well as mouse hepatitis virus (MHV),[6-9] Sendai virus,[6,10-12] and mouse adenovirus[13] can play a role in this syndrome. A weak pathogen, which can be carried by euthymic conventional animals without causing disease symptoms, can often induce the wasting of nude mice. Furthermore, certain types of bacteria, such as *Staphylococcus aureus* which exists on the surface of the human body and is carried by personnel (caretakers, researchers, etc.), may cause severe abscesses in nude mice, even though these microorganisms are usually harmless to man or conventional animals.

The quality of the care and the microbiological environment of nude mice, and the microbiological quality itself, have to be kept at a high level in order to obtain reliable experimental data. Just one nude mouse can contaminate an animal experimentation facility and all of the human tumor lines maintained in the facility. In the following sections, some examples are given of accidental infections and preventive measures are discussed.

B. ACCIDENTAL MICROBIOLOGICAL INFECTIONS

Among the various accidents due to infections where the causes were clarified by microbiological tests in the Microbiological Monitoring Center (Asia) of ICLAS (International Council for Laboratory Animal Science), in the Central Institute for Experimental Animals, Japan, six typical cases have been selected for discussion in this section. In Figures 6A to 6F, these accidental infections have been drawn in a schematic order.

Case A occurred in a nude mouse colony used for transplantation to maintain tumor lines and for screening of antitumor agents (Figure 6A). Although a vinyl isolator system was used for transplantation to maintain the tumor lines, therapeutic experiments were carried out in an open animal room. Wasting disease appeared only in nude mice placed in the open room, and MHV was subsequently isolated from such animals. Cages for these mice were supplied from the washing and sterilizing area on the first floor of the facility. This area also handled cages for conventional euthymic mice and rats, latently infected with MHV. Under such conditions, the route of MHV infection to nude mice was suspected to be via the cages supplied from the washing and sterilizing area. The vinyl isolator system was kept in a

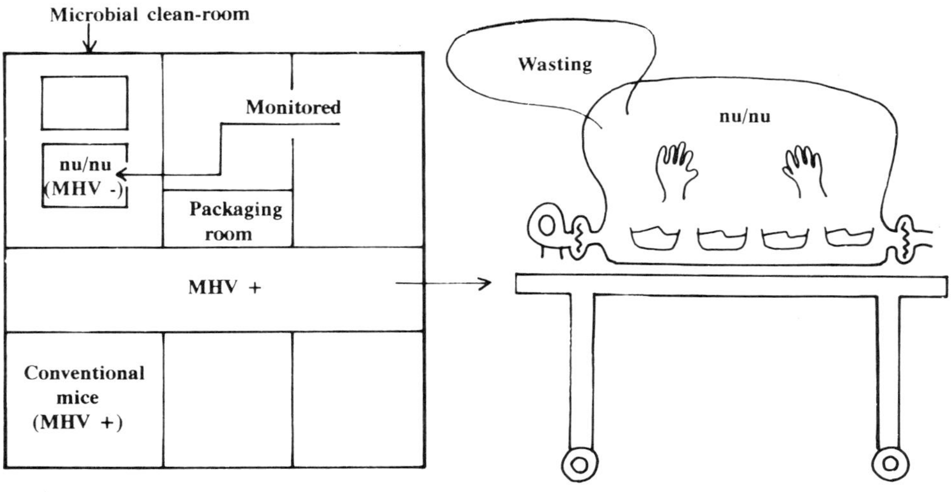

FIGURE 6B. Accidental infections of nude mice, case B. Cause: contamination during packaging. Problem: design of facilities.

room provided with an adjacent sterilizing area, and mice in cages supplied from this system were all healthy.

In the other facility shown in Figure 6B, nude mice were produced in a microbiologically clean-room system and were sent to a separate area where they were reared under strict microbiological control using a vinyl isolator system. Nevertheless, the mice in the vinyl isolator suffered from wasting disease and MHV was isolated, although the original nude mouse colony kept in the clean-room system had been proved MHV-free by periodic monitoring. To investigate the origin of this MHV infection, the supply route of these infected animals was traced and it was found that they had been transferred and packaged in a room adjacent to a conventional area, where mice latently infected with MHV were reared. Therefore, it was suggested that MHV from these conventional mice presumably had contaminated the packing room and eventually infected the nude mice.

In case C, nude mice were also kept in a vinyl isolator system and used for tumor transplantation in this facility (Figure 6C). Because of the lack of space in the facility, the vinyl isolators were placed in a conventional room where conventional rats were reared. The vinyl isolator system is able to give protection against microbial infections in such a conventional environment, because of high air filtration, tight sealing, and strict sterilization of food and water. However, the wasting syndrome appeared in nude mice kept in the vinyl isolator and MHV infection was detected upon microscopical examination of liver tissues. Inspection of the isolator revealed leakage. Microbiological tests showed that conventional rats reared in the same room had anti-MHV antibodies. This case suggests that microbiologically controlled rearing systems are useless if the technical quality is not carefully maintained.

Case D gives an example of the importance of quarantine of experimental materials. Tumor transplantation to nude mice had been carried out in a vinyl isolator system, and some of these tumor-bearing nude mice were found to suffer from arthritis in the foot joints (Figure 6D). Microbiological monitoring of +/+ or +/nu mice obtained from the isolator as sentinel animals did not show signs of microbial contamination, but *Mycoplasma sp.* was isolated from the inflamed joints as well as the tumor tissues. In addition, it was found that arthritis appeared only in nude mice that were inoculated with the contaminated tumor tissue. Thus, microbiological

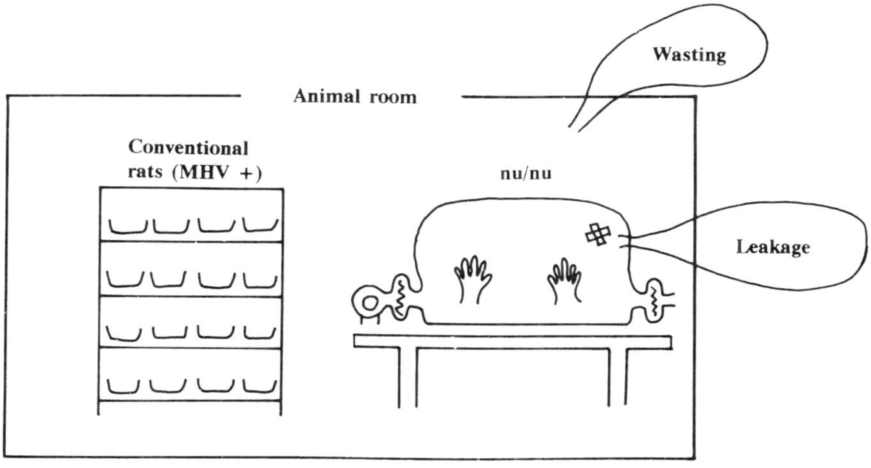

FIGURE 6C. Accidental infections of nude mice, case C. Cause: isolator leakage during experiment. Problem: technical maintenance of isolator.

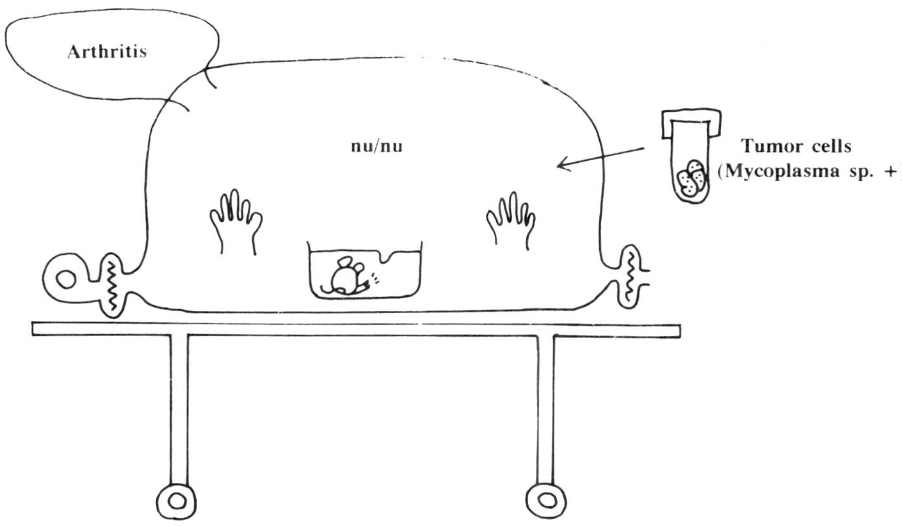

FIGURE 6D. Accidental infection of nude mice, case D. Cause: contaminated tumor cells. Problem: quarantine system.

control is necessary for experimental materials, as well as for the rearing equipment and the animals themselves.

In another facility shown in Figure 6E, nude mice were produced in a barrier system. Supplies were all autoclaved. The caretaker took a shower, wore a clean uniform, and used gloves and boots. Nevertheless, a high incidence of severe abscesses appeared in the faces of the nude mice. Examination of the abscesses, conjunctiva, cecal contents, etc., resulted in the isolation *Staphylococcus aureus*. These microorganisms were also detected on the skin of the caretaker, suggesting that he was the origin of the infection. This case is an example of the problem of sterilization limits within the barrier system.

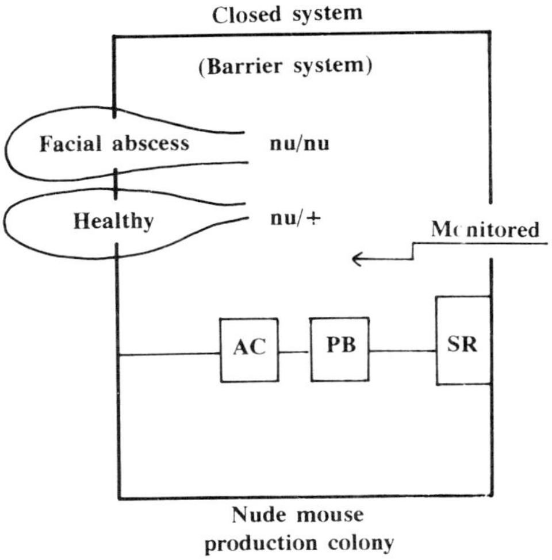

Staphyiococcus aureus	
Facial abscess	+++
Conjunctiva	+++
Cecal contents	+

FIGURE 6E. Accidental infections of nude mice, case E. Cause: contamination from caretaker (?). Problem: limitation of barrier system. AC: autoclave; PB: pass box; SR: shower room.

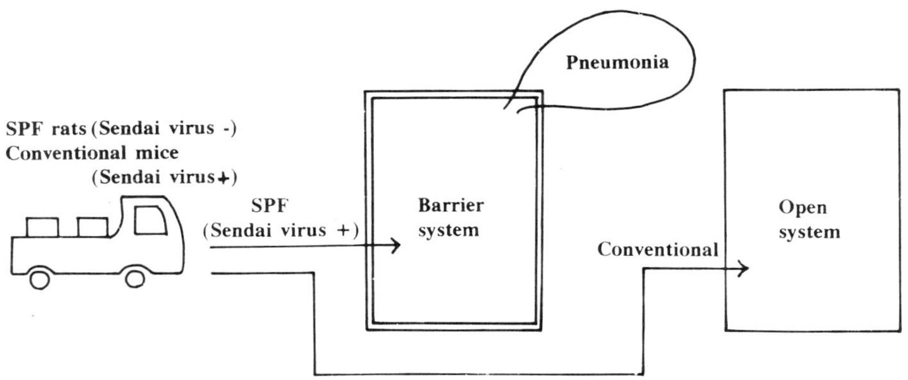

FIGURE 6F. Accidental infections of SPF rats, Case F. Cause: contamination during transportation. Problem: placing SPF rats together with conventional animals during transportation.

The final example, case F, involved SPF rats. The SPF rats were transported to an experimental facility where a barrier system was employed (Figure 6F). After a few days, most of the rats had rough coats and showed body weight loss. Random autopsies were performed, and an interstitial pneumonitis was found in the lungs. In addition, anti-Sendai virus antibodies were detected in the remaining rats after several weeks. Microbiological data of the breeding facility were rechecked, but the data showed that rats reared in the facility were free from Sendai virus. However, further investigation revealed that these rats had been placed together with conventional mice latently infected by Sendai virus, and were transported in the same truck.

In the six cases described, the wasting syndrome was caused by MHV infection in cases A to C, arthritis by *Mycoplasma sp.* in case D, facial abscesses by *Staphylococcus aureus* in case

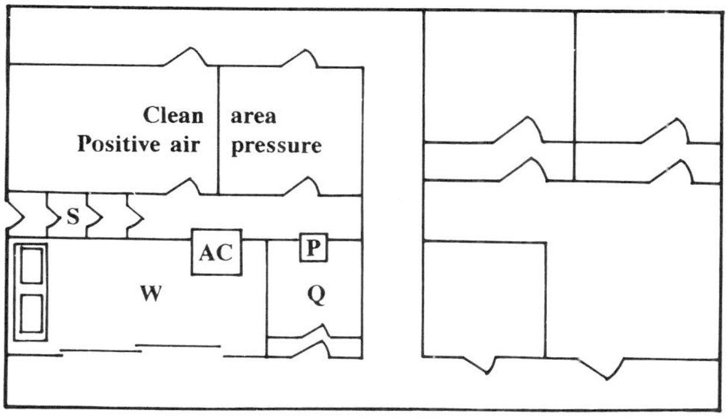

FIGURE 7. An example of desirable design of facilities. AC: autoclave; S: shower room; P: pass box; W: washing room; Q: quarantine room.

E, and pneumonia in euthymic SPF rats by Sendai virus in case F. In each case, the infection in nude mice was caused by exterior impurities, such as contaminated cages, technical failures in isolator systems, close contact with contaminated animals, infected tumor tissue used for transplantation, or distribution of infections by animal caretakers. Knowledge of microorganisms known to cause the wasting syndrome, as well as the routes of contaminations, will define the measures to prevent accidental infections.

C. MEASURES TO PREVENT CONTAMINATION

1. Care and Management

Consideration of the problems mentioned in the previous section indicates the great importance of microbiological control in the rearing systems. Infection from other animals should be avoided by establishing special areas and equipment for nude mice only within the facilities. Since immunodeficient animals, such as nude mice, are very susceptible to infections, their maintenance requires strict microbiological control. Experiments carried out in such immunodeficient animals can only offer reliable data when the rearing systems are kept completely sterilized under strict microbiological control. Personnel handling nude mice should be careful not to touch any other animals or equipment; if they must handle other animals, they should start their work with nude mice before other daily routine work. This is probably the best way to avoid infections carried by man.

The design of the animal facilities requires particular attention. Figure 7 shows an example of facilities based on the necessary principles. Even in facilities with adjacent contaminated and clean areas, it is possible to prevent contamination of the clean area by using an appropriate construction. The left part of this facility is the clean area, a barrier system which must have a positive air pressure and must be equipped with an autoclave, a pass box, and a shower room. This part should be completely isolated from the contaminated area (the right side in Figure 7), which is kept under negative air pressure. Although such a barrier system is desirable for rearing euthymic animals, it alone does not achieve perfect microbiological control for nude mice. Therefore, the barrier system should also be provided with special rearing systems or units. The following three systems can be applied according to the aim of the experiments involved.

The first example is the filter rack system shown in Figure 8, which was introduced by Lane-Petter[14] in 1970. Although this system is simple, its use in barrier facilities makes it possible to prevent cross-contamination between cages. The filter rack is suitable for performing short-term experiments and it is employed widely in many countries.

The second example is the microbial clean-room system shown in Figure 9. The principle of this system was introduced by McGarrity et al.[15] Personnel can enter this room to work. It is possible to keep the room permanently clean by means of circulating refiltered air so that reliable

FIGURE 8. Filter rack system with laminar flow racks and filter caps.

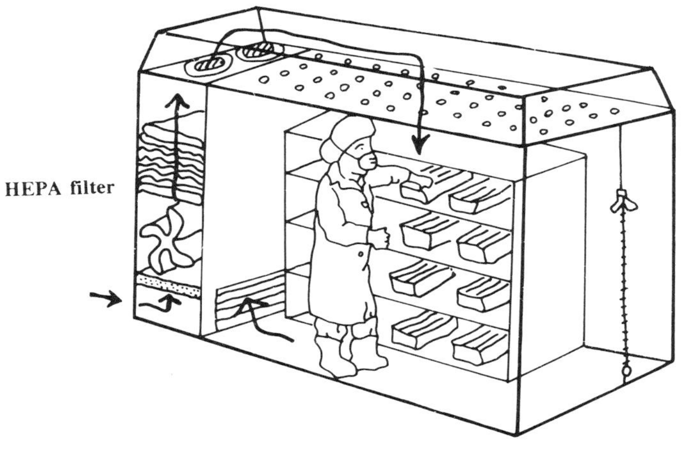

FIGURE 9. Microbial clean-room system.

sterilization of air can be performed. This system is suitable for relatively long-term experiments and can be applied to rooms of any size.

The third example is the isolator system shown in Figure 10. This is a large-scale isolator which was first introduced by Trexler[16] in 1975. The system was originally developed for the isolation of patients with infectious diseases. In this isolator, personnel can work without directly touching the animals and complete microbiological control can be achieved. Complete sterilization of the system and equipment is possible. When a negative pressure is applied, the isolator can be used for experiments on various types of biohazardous agents.

The three systems mentioned have the following common advantages:

1. Sterilized air is supplied by means of high-efficiency particulate air (HEPA) filters
2. Cross-contamination by animals is prevented

FIGURE 10. Large isolator system (P.C. Trexler).

TABLE 3
Items and Checking Methods for Microbiological Monitoring of Nude Mice

Items	Techniques[a]		nu/+	nu/nu
MHV	S		x	
Mouse adenovirus	S		x	
Sendai virus	S		x	
Ectromelia	S		x	
Mycoplasma pneumoniae	S	E	x	x
Corynebacterium kutscheri	S	E	x	x
Escherichia coli 0115 a,c:K(B)	S	E	x	x
Pasteurella pneumotropica	S	E	x	x
Pseudomonas aeruginosa	S	E		x
Salmonella sp.	S	E	x	x
Staphylococcus aureus		E		x
Tyzzer's organism	S		x	
Enteric protozoa		E		x
Syphacia obvelata		E		x

[a] S, serological; E, etiological.

3. Sterilization of equipment or units can be performed easily

2. Microbiological Tests and Monitoring

The basic concept of microbiological animal control is that nude mice used in experiments must be quarantined just before they are brought into the facilities and that they must be periodically monitored during the experiments. The rearing systems, which include the facilities, equipment, and care, must be examined often to determine whether they are clean. Once an accidental infection has occurred, its cause must be clarified by microbiological tests as soon as possible.

In 1979, the ICLAS designated the Central Institute for Experimental Animals as an ICLAS

TABLE 4
Diagnostic Methods for Clarification of Accidental Infections in Nude Mice

Example	Micro-organisms	Diagnostic methods
A	MHV	Serological testing of mice in the same facilities as nu/nu mice are kept
B	MHV	Serological testing of specially prepared sentinel mice
C	MHV	Serological testing of rats in the same room as nu/nu mice are kept
E	*Staphylococcus aureus*	Cultivation of samples from both nu/nu mice and caretaker
F[a]	Sendai virus	Serological testing of recovered rats

[a] Euthymic SPF rats.

Monitoring Center (ASIA). Table 3 indicates the items and tests used for microbiological monitoring of nude mice. These items have been selected with the following conditions in mind: (1) diseases or latent infections are commonly detected in euthymic conventional mouse colonies in Japan, and (2) effective test techniques are available. Other considerations may be added according to local needs. The standard items and techniques for nude mice given in the table should be minimal requirements in all experimental facilities.

Since nude mice are athymic animals, serological monitoring is not applicable to the animals themselves. Instead, such monitoring has to be performed on euthymic mice kept in the same room as the nude mice. A characteristic method of microbiological monitoring of nude mice is the periodic testing of specially prepared euthymic sentinel mice. Table 4 indicates the diagnostic methods for clarification of accidental infections, such as described for the six accidents. In cases A through C, serological methods were used to explain the causes of infections, as shown in the right hand column in Table 4. In cases A and C, serological monitoring was performed on euthymic mice or rats which were reared in the same facilities as nude mice. In case B, similar serological monitoring of specially prepared sentinel mice solved the problem.

In nude mice, it has been reported that Sendai virus infection can be detected by the fluorescent antibody method.[17] The method should be useful for monitoring Sendai virus in nude mice themselves, because these animals show more persistent viral infections than euthymic mice.

For successful experiments using nude mice, the application of microbiologically controlled animals is indispensable. Animals from colonies produced under strict microbiological-controlled conditions can only be supplied to other facilities or to research institutes by means of careful transportation to prevent contamination. The facilities or institutes that receive these animals should be provided with special areas for nude mice alone. All personnel should be careful to avoid needless handling prior to any other daily routine work. Such precautions should lessen or eliminate cross-contamination within the facilities. Finally, it is necessary to check the animals themselves. It is highly desirable to perform microbiological monitoring periodically, no matter whether the nude mice used for experiments are infection-free or not. When any accidental infection is detected, its cause should be detected at once and eliminated as soon as possible.

ACKNOWLEDGMENTS

The author gratefully acknowledges the assistance and advice of Dr. M. Saito concerning breeding of the nude mouse, Dr. N. Kagiyama concerning microbiological control, and also thanks Dr. Y. Ohnishi for his help with the organization of this chapter.

REFERENCES

1. **Gullino, P. M., Ediger, R. D., Giovanella, B., Merchant, B., Outzen, H. C., Reed, M. D., and Wortis H. H.,** Guide for the care and use of the nude (thymus-deficient) mouse in biomedical research, *ILAR News,* 19, 1976.
2. **Nomura, T.,** Future problems of nude mice as laboratory animals, in *Proc. 2nd Int. Workshop on Nude Mice,* Nomura T., Ogawa, N., Tamaoki, N., and Fujiwara, K., Eds., University of Tokyo, Tokyo, 1977, 569.
3. **Flanagan, S. P.,** 'Nude', a new hairless gene with pleiotropic effects in the mouse, *Genet. Res.,* 8, 295, 1966.
4. **Boorman, C. A., Lina, P. H. C., Zurcher, C., and Nieuwerkerk, H. T. M.,** *Hexamita* and *Giardia* as a cause of mortality in congenitally thymus-less (nude) mice, *Clin. Exp. Immunol.,* 15, 623, 1973.
5. **Ueda, K., Goto, Y., Yamazaki, S., and Fujiwara, K.,** Chronic fatal *pneumocystosis* in nude mice, *Jpn. J. Exp. Med.,* 47, 475, 1977.
6. **Eaton, G. J., Outzen, H. C., Custer, R. P., and Johnson, F. N.,** Husbandry of the "nude" mouse in conventional and germfree environments, *Lab. Anim. Sci.,* 25, 309, 1975.
7. **Hirano, N., Tamura, T., Taguchi, F., Ueda, K., and Fujiwara, K.,** Isolation of low-virulent mouse hepatitis virus from nude mice with wasting syndrome and hepatitis, *Jpn. J. Exp. Med.,* 45, 429, 1975.
8. **Sebestany, A. and Hill, A. C.,** Hepatitis and brain lesions due to mouse hepatitis virus accompanied by wasting in nude mice, *Lab. Anim.,* 8, 317, 1974.
9. **Ward, J. M., Collins, M. J., and Parker, J. C.,** Naturally occurring mouse hepatitis virus infection in the nude mouse, *Lab. Anim. Sci.,* 27, 372, 1977.
10. **Iwasaki, Y.,** Experimental virus infection in nude mice, in *The Nude Mouse in Experimental and Clinical Research,* Fogh, J. and Giovanella, B. C., Eds., Academic Press, New York, 1978, 457.
11. **Ueda, K., Tamura, T., Machii, H., and Fujiwara, K.,** An outbreak of Sendai virus infection in a nude mouse colony, in *Proc. 2nd Int. Workshop on Nude Mice,* Nomura, T., Ohsawa, N., Tamaoki, N., and Fujiwara, K., Eds., University of Tokyo, Tokyo, 1977, 61.
12. **Ward, J. M., Houchens, D. P., Collins, M. J., Young, D. M., and Reagan, R. L.,** Naturally-occurring Sendai virus infection in athymic nude mice, *Vet. Pathol.,* 13, 36, 1976.
13. **Cohen, B. J. and De Groot, F. C.,** Adenoviral infection in athymic (nude) mice, *Lab. Anim. Sci.,* 26, 995, 1976.
14. **Lane-Petter, W.,** A ventilation barrier to the spread of infection in laboratory animal colonies, *Lab. Anim.,* 4, 125, 1970.
15. **McGarrity, G. J., Coriell, L. L., Schaedler, R. W., Mandle, R. J., and Greene, A. E.,** Medical application of dust-free rooms. III. Use in an animal care laboratory, *Appl. Microbiol.,* 18, 142, 1969.
16. **Trexler, P. C., Spoers, A. S. D., and Gaya, H.,** Plastic isolators for treatment of acute leukaemia patients under "germ-free" conditions, *Br. Med. J.,* 6, 549, 1975.
17. **Iwai, H., Goto, Y., and Ueda, K.,** Response of athymic nude mice to Sendai virus, *Jpn. J. Exp. Med.,* 49, 123, 1979.

Chapter 4

SUBCUTANEOUS XENOGRAFTS

Jørgen Rygaard

TABLE OF CONTENTS

I. INTRODUCTION

Since the first human malignant tumor was implanted subcutaneously (s.c.) in nude mice in 1969,[1] this site for implantation seems to be the most popular, as based on a survey of the literature (Figure 1). In a variety of heterotransplantation models, other sites of inoculation have been chosen, mostly because only these sites would allow acceptance of the tumor tissue. Thus, in many studies, tumors have been transplanted to the anterior chamber of the eye,[2] intracerebrally in mice or other rodents,[3] or to the cheek pouch of the hamster.[4] In none of these places can tumor growth be readily monitored.

In a mouse without an immune system, it was tempting to develop the tumor in a place that could easily allow inspection of tumor take and growth. The obvious answer to that would be the implantation of tumors into the s.c. space. The first tumor,[1] a colon cancer derived from a 71-year-old patient, was implanted in solid tumor blocks measuring $2 \times 2 \times 2$ mm. The tumor tissue was taken from the serosal side of the surgical specimen immediately after surgery. A pouch was formed in the lateral aspect of the abdominal wall of nude mice under ether anesthesia. The tumor grew in all inoculated animals and reached a considerable size in the longest surviving animals, measuring $2 \times 2 \times 3.5$ cm after about 3 months, at which time the tumor-bearing mice were killed because of extensive ulceration of the tumors.

The mode of growth of the first s.c. human tumor xenograft was characteristic of what we have observed later.[5] The tumor was a local nodule and was encapsulated in a thin connective tissue capsule. As mentioned, the skin was ulcerated on top of the tumor. On gross examination after sacrifice of the animals, the tumor was found to be mobile, free of the underlying fascia, and covered with a mesh of vessels, both medium size and small arteries and veins.

On histological examination, the tumor appeared to be similar to the patient's tumor. It was a well-differentiated adenocarcinoma. However, the amount of tumor stroma was less in the mouse-grown tumor. Both small and larger necrotic areas were found throughout the tumor tissue. No inflammation was seen in the capsule or in the surrounding connective tissue. Vital tumor tissue from this first implanted tumor was serially transferred to other nude mice, again inoculated s.c. and developed in the same manner. The tumor was maintained over 7 years for 76 passages. It was lost because of an epizootic of mouse hepatitis. Already after the third and fourth passage, we attempted to make a cell suspension from the tumor using mild trypsin treatment. With the tumor in question, it was difficult to obtain a sufficient number of vital tumor cells for further inoculations, but we succeeded in getting a take of inoculated cell suspensions.

II. TRANSPLANTATION PROCEDURES

The s.c. space in mice, also in the nude mouse, is very suitable for tumor implantation. Anatomically, the integument of the mouse differs substantially from that of man. The skin of the mouse is freely mobile. The epidermis and the thin layer of dermal connective tissue have a structure very similar to that in man if compared with haired areas. Below the skin is the panniculus musculosus, the muscle that allows free movement of the skin in a way that is only seen on the scalp and, to a lesser degree, in the anterior part of the neck in man, where strands of the platysma muscle are located. Under the panniculus musculosus, layers of fascia connected with thin strips of connective tissue allow free movements of the skin. It is in this area between the panniculus musculosus and fascia that tumors are routinely inoculated s.c.

Tumors can be placed s.c. as solid blocks, as needle biopsies by use of a trocar, or as cell suspensions. In all instances, a space is easily prepared, either by surgical procedures or by the pressure of the trochar or the inoculated cell suspension, that creates a graft bed for the implant. The area is richly vascularized. Therefore, the blood supply and lymph drainage can be readily established. The free mobility of the skin allows an expansion of the skin over the growing tumor.

FIGURE 1. Nude mouse with a subcutaneous tumor implant in the anterior part of the left flank.

A thorough knowledge of the structure of the primary and mouse-grown tumor tissue is important for the selection of material for implantation. Of course, only vital tissue/cells will lead to subsequent tumor growth. This point seems banal, but there have been examples of large series of unsuccessful attempts at tumor transplantation that failed because the tissue was taken from the mucosal side of the colon. Tissue from this location is frequently necrotic and always heavily contaminated with bacteria. Also, when it comes to later passages from mouse to mouse, including selection of tissue for the production of cell suspensions for inoculation, it is important to distinguish vital tissue from necrotic areas. This can most often be done just by inspection based on a previous study of the histological appearance of the tumor.

For solid block or trochar implantation, tumor tissue fragments of suitable size are prepared and introduced/injected in the s.c. space. The thin skin of the nude mouse allows easy observation of the location of the implant and, by slight pressure, the implant can be brought to the desired position. If an incision was made or if the trocar caused too much injury to the skin, the defect can be closed with a suture or clamp or, more easily, with a tissue-compatible methacrylate glue. The technique for injection of cell suspensions freshly prepared from primary tumors or from cultured cell lines needs no further explanation.

III. REGIONAL GROWTH DIFFERENCES

The growth rate of tumors implanted s.c. into nude mice shows pronounced regional differences. This fact was also noticed in murine tumors grown in their syngeneic hosts by Auerbach.[6] There seems to be a significant predilection for tumor growth in the anterior region of the mouse trunk, as opposed to the posterior region. In fact, the difference may be so pronounced that anterior tumors may grow three to four times as rapidly as posterior tumors. Also, there is a growth difference between tumors transplanted in the dorsal and the ventral region, ventral tumors developing more rapidly than dorsal tumors. There are no differences between the right site and the left site of the mouse trunk. The reason for the difference in growth in the various regions is not known. Obviously, it cannot be immunological factors that account for the difference. It is important to be aware of this phenomenon, particularly when comparing tumor sizes. In order to obtain reproducible results, it is necessary to use a highly standardized procedure for tumor implantation, also with regard to the site of implantation.

Most investigators seem to prefer the lateral anterior aspect of the trunk, that is, the lateral side of the lower thoracic part. Using this site, it is also possible to inoculate tumors bilaterally and use the contralateral tumor as a control for, for example, irradiation or other local treatment experiments. For some experiments (e.g., NMR-scanning), it is to be preferred that tumors are implanted on the top of the back because they can more easily be immobilized during scanning and spectroscopy experiments.

Is the s.c. space the ideal location for all tumors? Not necessarily. It has been known since the first comparative studies of different tumors were performed that some tumor types are very hard to establish by s.c. inoculation. Some of these may grow more readily in immunologically privileged sites, for example, intracerebrally (see Chapter 7). It was also shown in early studies[7] that breast tumors and breast premalignant diseases which could not be transplanted to the subcutis of the back or flanks of nude mice would grow readily when transplanted to the mammary fat pad. It has also been shown for other tumors such as, for example, pancreatic tumors that orthotopic transplantation allows or accelerates tumor take and growth.[8-11]

There seems to be an interplay between stroma and epithelium, as demonstrated most clearly in experiments with recombination of mouse stroma and epithelium from various regions.[12] In such experiments, it is obvious that the stroma has a decisive influence on the differentiation of the epithelium. Obviously, this is not a must for tumor implants any more than it is for metastatic growth in man and other mammals where tumors metastasize to regions widely different from their origin. Conversely, it is also known that there are certain areas or organs of predeliction to which certain tumor types metastasize more regularly or even exclusively. By comparison, the s.c. space of the nude mouse may be similar to the place in which metastases from a given tumor are able to grow, but in which other tumor cells will never establish themselves.

Some changes in the immunodeficiency can result in an accelerated take and growth of tumors in nude mice (Chapters 2, 5, and 6). Despite the profound T cell deficiency, there are some host mechanisms that may be responsible for the rejection of tumors. These mechanisms are also active for s.c. implantation. In 1982, an astonishing observation was reported by Wilson et al.[13] They found that the concomitant inoculation of melanoma cells with cultured human fibroblasts would accelerate take of tumors and allow for acceptance of very small inocula even down to 10^1 tumor cells. These observations have been extended by Picard et al.[14] and Aagaard et al.[15] It seems that there is a feeder-cell effect of such cultured fibroblasts, irrespective of their species origin. It is noteworthy that also killed fibroblasts and tissue culture supernatants will accelerate growth of small inocula. With our present-day knowledge of growth factors, it can be supposed that both growth factors and angiogenesis factors are involved. In fact, it has been shown that angiogenic factors produced by genetic engineering may have a similar effect.

The phenomenon of tumorigenicity in nude mice is used as a characteristic of tumor cell lines, for example, in the American Tissue Culture Collection Catalogue. It will be obvious from what has been mentioned here that tumorigenicity in the unadulterated nude mouse is not an absolute value. Also, strain differences between the mice may account for different outcomes of tumor inoculation. For this reason, a more well-defined model for tumorigenicity may be needed. The scid mouse[16] may actually represent a better tumorigenesis model, but does not seem to be so widely accepted at the present time.

IV. THE SUBCUTANEOUS TRANSPLANT

S.c. implanted tumors will generally grow as local well-circumscribed tumors, encapsulated in a thin connective tissue capsule.[17,18] Some tumors, however, tend to be locally invasive. They may infiltrate between the muscle cells of the panniculus musculosus and may even invade the abdominal muscles if implanted in this area. In some instances, they may also give rise to metastases in local lymph nodes or to distant metastases in lungs and other organs.[19-21] This is a relatively rare event, however, and has been considered a characteristic of individual tumors. There are certain indications, however, that the genetic background of the mouse host may determine the metastasizing capacity of tumors, at any rate for many tumors investigated. Using a nude mouse strain, NCr, obtained from the National Cancer Institute, Frederick, MD, we have recently found that tumors that do not normally metastasize in our nude mice (NMRI, NC, or Balb/c background) will metastasize in these mice.[22] We are at present performing a genetic analysis in order to study this characteristic.

V. TUMOR ANGIOGENESIS

It was already noticed in early transplantations that the tumors did contain some mouse tissue, presumably from connective tissue and blood vessels. The presence of mouse stroma was first described in isoenzyme studies of transplanted Burkitt's lymphoma.[23] Recent studies have reported the development of blood vessels in the s.c. space after injection of tumor cells.[24] Following the initial death of most inoculated tumor cells, host-supplied inflammatory cells infiltrated the xenografts and built up solid strands of connective tissue which tightly contacted surviving tumor cells and next were used as guide-rails for invading blood vessels.

Konerding et al.[25] have carried out elegant studies on the vascular system of implanted human melanomas and sarcomas into nude mice. They used microcorrosion casting and found that the vascular system of xenografts can be described in terms of a simply structured system, lacking not only a hierarchy with respect to wall formation, but also to vascular distribution. The vasculature of such transplanted tumors clearly differs from that in normal tissues. They subsequently found by scanning and transmission electron microscopy that structurally complete real arteries or veins are neither seen in the periphery nor in the center of transplanted tumors.[26] Large-caliber vessels basically have a capillary wall structure. The endothelial cells show an alternating height and electron density. Different cell types and tumor cells are themselves apparently involved in the formation of the vessel wall. The endothelium is characterized by very simple immature cell contacts. In some xenografts, the number of vessels without or with incomplete endothelium seems to be higher than the number of structurally real capillaries.

These findings on angiogenesis call for two comments. In the first place, the vasculature of the mouse seems to mimic the vasculature in the primary tumor, thus apparently obeying some rules of growth that are inherent to the tumor. Secondly, knowing that in many cases tumor cells form part of the wall in the small tumor vessels, it may seem even more astonishing that metastases are relatively infrequently observed in the ordinary nude mouse-xenograft model.

VI. SUBDERMAL IMPLANTS

One particular modification of the s.c. implantation is subdermal injection of tumor cells. In this location, the inoculum is deposited between the lower part of the dermal connective tissue and the s.c. tissue. Cornil et al.[27] compared cells from an amelanotic human melanoma cell line, MeWo, originally established from a melanotic lymph node metastasis and two MeWo variants resistant to wheat germ agglutinin (WGA) which were selected for altered malignant capacities. They found that 5×10^5 MeWo cells were required to achieve a 100% tumor take with s.c. injection, but only 2×10^4 cells were required with subdermal inoculation. Subdermal injection of the MeWo cells furthermore resulted in the development of highly melanotic and nonencapsulated primary tumors which grew quickly into the epidermis and metastasized frequently to draining lymph nodes. The tumors developing after s.c. injection were found in the deepest layer of the subcutis and were predominantly amelanotic and encapsulated. They rarely metastasized to lymph nodes. The authors suggested that the subdermis may facilitate or induce the expression of phenotypic traits which are characteristic of human melanomas in their human host and that it may enhance the success rate of transplantation of such tumors in nude mice.

The transplantation described in the subdermal region can be characterized as an orthotopic transplantation. The high success rate, the specific growth pattern, and pattern of spread of the human melanoma may be caused by the nature of the primary tumor and the orthotopic transplantation. It might be worthwhile to inoculate other tumor types in the same place in order to elucidate whether the success is due to the orthotopic transplantation or can be ascribed generally to the site of inoculation.

VII. SUBCUTANEOUS TUMOR GROWTH

Traditionally, the size of s.c. implanted tumors has been described by measurement with a slide caliper in two or, in some instances, three dimensions. There are several objections to this. The measurement in only two dimensions (length × width) does, of course, not mathematically describe the volume of the tumor, but only the two measured dimensions. By using suitable formulae, the two measurements can be transformed into an approximate tumor volume.

Euhus et al.[28] have made an extensive study on the reliability of tumor measurement in the nude mouse. Diameters of s.c. tumors were expressed as length, area, and volume; tumor weights were also recorded. Estimates of area and volume derived from caliper measurements correlated well with water displacement volume. At necropsy, tumor weight was the most consistent and reproducible reflection of tumor volume. Interobserver variation averaged 15% of the mean calculated volume. Volumes were compared in order to quantify the average variation that occurs when a single investigator repeatedly measures the same mass. When large masses were measured, the error was only 7%; whereas, when small masses were measured, the error was 27%. *Sic!*

Tomayko and Reynolds[29] surveyed the literature and revealed the use of 19 different formulae for determining the size of s.c. tumors grown in nude mice. They compared the volumes, areas, and diameters predicted by each of the 19 formulae with the actual weights of 50 tumors ranging from 0.46 to 22.0 g established in nude mice as xenografts from human cell lines. In addition, they analyzed how well each formula estimated the actual tumor mass. They found that the ellipsoid volume formulae ($\pi/6 \times L \times W \times H$ and $1/2 \times L \times W \times H$) were best for estimating tumor mass ($r = 0.93$), whereas measurement of diameter correlated poorly with tumor mass (r less than 0.66). Also, although the determined tumor area correlated well with tumor mass in small tumors ($r = 0.89$), correlations of area with tumor mass for large tumors were poor ($r = 0.41$). The authors conclude that determination of the ellipsoid volume from measurements of three axes consistently yields the most accurate estimations of both relative and actual tumor mass. From tumor measurements, growth curves of individual tumors and mean values can be calculated.

Whereas such measurements may be illustrative of tumor growth, they are very difficult to handle when therapy effects are to be monitored. Such therapeutic effects are usually described as treatment-induced growth delay and specific growth delay. Also, an important variable in the description of tumor growth is the tumor volume doubling time. The calculation of these values[30] can be based on a transformation of the Gompertz function $\ln[\ln A(\max) - \ln A(t)] = \ln(\beta/\alpha) - \alpha t$. This growth function depicts the growth rectilinearly with the slope $-\alpha$ when the tumor size $y = \ln[\ln A(\max) - \ln A(t)]$ is plotted against time t. A computer-based program for the calculation and description of tumor growth has recently been reported,[31] and allows easy handling of large amounts of data.

The advantage of obtaining a mathematical expression to describe tumor growth and changes in tumor growth following therapy is obvious. However, this does not exclude some of the pitfalls in tumor measurement. Changes in growth variables are, of course, most important as indicators of treatment effect. It should be remembered that changes following therapy are actually the result of several factors: initially, tumors may become edematous, with both intra- and extracellular edema. This may cause an increase in tumor size despite a positive effect of treatment on the tumor. Following this, necrosis and inflammatory responses to necrosis will influence the gross size of the tumor. These changes can be monitored by a combination of a reliable morphometric method[32] and flowcytometric DNA-analysis.

VIII. EPILOGUE

As mentioned initially, the s.c. space is the most popular site for inoculation of human malignant tumors into nude mice. There can be several objections to this procedure. It is known that there is an interplay between mesenchymal and epithelial tissue. According to that, tumors not derived from the skin or the appendages should perhaps not be grown s.c. In most instances tumors transplanted s.c. do not exhibit one of their most important characteristics: the ability to metastasize. Some people find that they are not "real" malignant tumors,—even if they maintain a range of structural and biochemical properties, including appropriate responses to therapy.

Even if the objections are many, advantages are obvious: s.c. tumors are easily amenable for inspection and local treatment. Their responses can easily be measured — although perhaps not correctly. In case the s.c. space is not considered to be ideal, it should not be forgotten that exact measurements of tumor size in any other location would be even more difficult.

ACKNOWLEDGMENTS

The original work from our laboratory reported herein was generously supported by The Danish Cancer Society and The Danish Medical Research Council.

REFERENCES

1. **Rygaard, J. and Povlsen, C. O.,** Heterotransplantation of a human malignant tumour to "nude" mice, *Acta Path. Microbiol. Scand.*, 77, 758, 1969.
2. **Greene, H. S. N.,** Heterologous transplantation of human and other mammalian tumours, *Science*, 88, 357, 1938.
3. **Greene, H. S. N.,** The transplantation of tumours to the brains of heterologous species, *Cancer Res.*, 11, 529, 1951.
4. **Lutz, B. R., Fulton, G. P., Patt, D. I., and Handler, A. H.,** The growth rate of tumor transplants in the cheek pouch of the hamster *(Mesocritecus auratus)*, *Cancer Res.*, 10, 231, 1950.
5. **Povlsen, C. O. and Rygaard, J.,** Heterotransplantation of human adenocarcinomas of the colon and rectum to the mouse mutant nude. A study of nine consecutive transplantations, *Acta Path. Microbiol. Immunol. Scand.*, 80, 713, 1971.
6. **Auerbach, R., Morrissey, L. W., and Sidky, Y. A.,** Regional differences in the incidence and growth of mouse tumors following intradermal or subcutaneous inoculation, *Cancer Res.*, 38, 1739, 1978.
7. **Jensen, H. M. and Wellings, S. R.,** Preneoplastic lesions of the human mammary gland transplanted into the nude athymic mouse, *Cancer Res.*, 36, 2605, 1976.
8. **Tan, M. H. and Ming Chu, T.,** Characterization of the tumorigenic and metastatic properties of a human pancreatic tumor cell line implanted orthotopically into nude mice, *Tumour Biol.*, 6, 89, 1985.
9. **Naito, S., Von Eschenback, A. C., and Fidler, I. J.,** Different growth pattern and biologic behavior of human renal cell carcinoma implanted into different organs of nude mice, *J. Natl. Cancer Inst.*, 78, 377, 1987.
10. **Vezeridis, M. P., Doremus, C. M., Tibbets, L. M., Tzanakakis, G., and Jackson, B. J.,** Invasion and metastasis following orthotopic transplantation of human pancreatic cancer in the nude mouse, *J. Surg. Oncol.*, 40, 261, 1989.
11. **Rodolfo, M., Balsari, A., Clemente, C., Parmiani, G., and Fossati, G.,** Tumorigenicity and dissemination of primary and metastatic human melanomas implanted into different sites in athymic nude mice, *Invasion Metast.*, 8, 317, 1988.
12. **Mackenzie, I. C. and Hill, M. W.,** Connective tissue influences on patterns of epithelial architechture and keratinization in skin and oral mucosa of the adult mouse, *Cell Tissue Res.*, 235, 551, 1984.
13. **Wilson, E. L., Gärtner, M., Campbell, J. A. H., and Dowdle, E. B.,** Growth and behaviour of human melanomas in nude mice: effect of fibroblasts, in *Proc. 4th Int. Workshop on Immune-Deficient Animals Exp. Res. (formerly Nude Mice Workshop)*, Sordat, B., Ed., Karger, Basel, 1984, 357.
14. **Picard, O., Poupon, M. F., and Rolland, Y.,** Fibroblast cooperation for the graft of tumor cells in nude mice, in *Proc. 5th Int. Workshop on Immune-Deficient Animals*, Rygaard, J., Brünner, N., Græm, N., and Spang-Thomsen, M., Eds., Karger, Basel, 1987, 204.

15. **Aagaard, J., Roed, H., Rygaard, J., and Tromholt, V.,** Adjunctive effect of cultured human fibroblasts in establishing human tumor cell lines in nude mice, in *Proc. 5th Int. Workshop on Immune-Deficient Animals,* Rygaard, J., Brünner, N., Græm, N., and Spang-Thomsen, M., Eds., Karger, Basel, 1987, 207.
16. **Bosma, G. C., Custer, R. P., and Bosma, M. J.,** A severe combined immunodeficiency mutation in the mouse, *Nature (London),* 301, 527, 1983.
17. **Rygaard, J.,** *Thymus and Self. Immunobiology of the Mouse Mutant Nude,* FADL's Publisher, Copenhagen, 1973, John Wiley & Sons, London, 1975; Tuttle-Mori, Tokyo 1979.
18. **Povlsen, C. O.,** *Heterotransplantation of Human Malignant Tumors into Nude Mice,* (in Danish) FADL's Publisher, København, Denmark, 1977.
19. **Neulat-Duga, I., Sheppel, A., Marty, M., Lacroux, F., Pourrat, J., Caverivière, P., and Delsol, G.,** Metastases of human tumor xenografts in nude mice, *Invasion Metast.,* 4, 209, 1984.
20. **Morikava, K., Walker, S. M., Nakajima, M., Pathak, S., Jessup, J. M., and Fidler, I. J.,** Influence of organ environment on the growth, selection, and metastasis of human colon carcinoma cells in nude mice, *Cancer Res.,* 48, 6863, 1988.
21. **Sekikawa, K., Arends, J. W., Verstijnen, C. P. H. J., Van der Linden, E., Dinjens, W., Schutte, B., and Bosman, F. T.,** Influence of implantation site on growth, antigen expression and metastatic potential of human colonic cancer HT29 and 5583 xenografts in nude mice, *Invasion Metast.,* 8, 238, 1988.
22. **Brünner, N., Spang-Thomsen, M., and Rygaard, J.,** Metastatic rate of human malignant tumors in Balb/c/NCr/nu/nu mice, (in preparation).
23. **Povlsen, C. O., Fialkow, P. J., Klein, E., Klein, G., Rygaard, J., and Wiener, F.,** Growth and antigenic properties of a biopsy-derived Burkitt's lymphoma in thymusless (nude) mice, *Int. J. Cancer,* 11, 30 1973.
24. **Köpf-Maier, P. and Kestenbach, U.,** Host-supplied connective tissue as a guide for proliferating tumor cells in human tumor xenografts, *Acta Anat.,* 135, 289, 1989.
25. **Konerding, M. A., Steinberg, F., and Streffer, C.,** The vasculature of xenotransplanted human melanomas and sarcomas on nude mice. I. Vascular corrosion casting studies, *Acta Anat.,* 136, 21, 1989.
26. **Konerding, M. A., Steinberg, F., and Streffer, C.,** The vasculature of xenotransplanted human melanomas and sarcomas on nude mice. II. Scanning and transmission electron microscopic studies, *Acta Anat.,* 136, 27, 1989.
27. **Cornil, I., Man, S., Fernandez, B., and Kerbel, R. S.,** Enhanced tumorigenicity, melanogenesis, and metastases of a human malignant melanoma after subdermal implantation in nude mice, *J. Natl. Cancer Inst.,* 81, 938.
28. **Euhus, D. M., Hudd, C., LaRegina, M. C., and Johnson, F. E.,** Tumor measurement in the nude mouse, *J. Surg. Oncol.,* 31, 229, 1986.
29. **Tomayko, M. M. and Reynolds, C. P.,** Determination of subcutaneous tumor size in athymic (nude) mice, *Cancer Chemother. Pharmacol.,* 24, 148, 1989.
30. **Spang-Thomsen, M., Nielsen, A., and Visfeldt, J.,** Growth curves of three human malignant tumors transplanted to nude mice, *Exp. Cell. Biol.,* 48, 138, 1980.
31. **Rygaard, K. and Spang-Thomsen, M.,** 'GROWTH' — A computer program for determination of mean growth curves and calculation of response to therapy of solid tumor grafts, in *Proc. 6th Int. Workshop on Immune-Deficient Animals,* Wu, B. Q. and Zheng, J., Eds., Karger, Basel, 1989, 301.
32. **Albrectsen, J., and Rygaard J.,** Morphometric studies of treated and untreated heterotransplanted tumors in nude mice, (in preparation).

Chapter 5

TUMORIGENICITY

J. C. Romijn

TABLE OF CONTENTS

I. INTRODUCTION

Tumorigenicity, operationally defined as the ability to grow and produce tumors in an *in vivo* environment, is a crucial feature of the malignant behavior of tumor cells. As it has been known for many years that the lymphoid system is involved in the rejection of xenogeneic tumor grafts,[1] the discovery of a thymusless mutant (nu/nu) mouse[2] was soon followed by attempts to transplant human tumor cells to such animals. Since the first report on the successful transplantation of a human tumor into athymic hosts in 1969,[3] the nude mouse has become a major tool in oncology research, being used for tumor biology studies, preclinical drug testing, and assessment of the tumorigenic potential of cancer cell populations.

A large variety of human cancer cell lines and tumor tissue specimens could be propagated in nude mice,[4-7] often as serially transplantable tumor lines. Whereas certain types of human cancer appear to grow readily in nude mice, others, such as prostate and breast cancers, are more difficult to establish. The reasons for such a difference are not entirely clear. Tumor-related as well as host-related factors may be responsible for the lack of growth of certain types of xenografts. In this chapter, a number of such factors will be discussed and some of the methods that have been used to circumvent the supposed suppression of tumorigenicity will be reviewed.

II. FACTORS IN GROWTH CONTROL

A. SURVIVAL OF IMPLANTED TUMOR CELLS

The probability of acceptance of human tumor transplants in the nude mouse varies considerably for different tumor types. A series of factors, a number of which is listed in Table 1, may determine whether a tissue fragment will grow. In this respect, not only tumor cell properties (such as hormonal and nutritional requirements), but also the reactions in the host (e.g., the immunological defense mechanisms) and the microenvironmental conditions for the tumor cells, will have an impact on tumor establishment and growth. Many of these factors are likely to affect primarily the tumor "take" rate (the percentage of transplants that grow out and produce tumors), rather than the growth rate of already established tumors.

Contrary to what has often been assumed, histopathological studies show that the majority of tumor cells will die soon after implantation.[8] Although the fate of tumor cells after inoculation *in vivo* has not been studied extensively, a few attempts have been made to quantitate their survival by use of radioactively labeled tumor cells.[9-13] In our studies,[12,13] using ^{3}H-thymidine-labeled prostate cancer cells attached to Millipore filter discs, we calculated that less than 10% of the implanted cells survived for more than 24 h. Immunosuppression of the host increased this percentage slightly, but significantly,[12,13] demonstrating that immune as well as non-immune factors play a role in the prevention of tumor growth. Increased survival correlated with an enhanced tumor take rate, while tumor growth rate remained unchanged.[13]

B. SUPPORTIVE STROMA

Human tumors grown in nude mice are composed of malignant cells of human origin, together with stromal cells of murine origin.[14,15] Within a few days following implantation, host fibroblasts start to invade the patient's tumor tissue, replacing the original human connective tissue elements.[8,15] Penetration of host fibroblasts into the xenografts precedes the initiation of tumor cell proliferation, the first signs of which are usually observed in the vicinity of the newly formed connective tissue strands. These observations suggest that the stroma plays a critical role in the development of xenogeneic tumors in the nude mouse.

The mechanisms by which stromal cells exert their influence on tumor growth are still poorly understood. Undoubtedly, the stroma may serve as a physical support or 'ground substance', allowing the tumors to form their characteristic structures and guiding the blood vessels that supply the cells with the required nutrients. Also, the formation of the extracellular matrix, that

TABLE 1
Factors of Influence on the Growth of Implanted Human Tumors in Nude Mice

Tumor-related factors:
Hormonal requirements
Growth factor dependence
Induction of angiogenesis
Host-related factors:
Immunological defense mechanisms:
- Macrophages
- NK cells
- NC cells
- Antibody response
Supply of favorable microenvironment:
- Extracellular matrix production
- Vascular supply

may have a regulatory effect on tumor growth and differentiation, is dependent on the production of macromolecular components by the stroma, although some of these components are produced by the tumor cells as well.[16] In addition, stromal cells may secrete certain chemical signals, such as angiogenic factors and soluble growth factors, that are able to promote tumor cell growth.

Based on the importance of stromal cells for growth, a number of investigators have co-injected cultured fibroblasts along with different types of tumor cells.[17-19] In all studies, a significant, stimulating effect of the fibroblasts on the tumor take was observed. The effects were not dependent on the species or on the origin of the fibroblasts, since skin fibroblasts were as effective as tumor-derived fibroblasts.[18,19] In support of the view that the stimulating effects are mediated by secreted, diffusible factors, it was shown that fibroblast-conditioned media also increased the tumor take,[18] whereas inactivated, fixed fibroblasts greatly reduced the stimulatory effect.[18,19] It should be noticed, however, that additional mechanisms might be involved. In some models, inactivated, irradiated tumor cells may substantially enhance the tumor take rate upon mixing them with untreated tumor cells before the injection,[20-22] or even upon administration shortly therafter.[22] This phenomenon, mostly studied in immunocompetent mice and now known as the "Révész effect", was described as early as in 1956.[20] Although alternative explanations have been presented,[23] it is usually believed that interference with the immune system, for example, by a saturation effect, may increase the chances of the viable tumor cells to survive. It is also possible that tumor growth is promoted by a 'feeder layer effect' due to secretion of growth factor-like substances.

C. IMMUNOLOGICAL DEFENSE MECHANISMS

1. Macrophages

Mononuclear phagocytic cells, or macrophages, are important elements in the surveillance against foreign invaders such as microorganisms and parasites. In addition, a role of macrophages in the host defense system against malignant diseases is now well recognized.[24,25]

The mechanisms by which macrophages are able to destroy tumor cells are multiple,[26] very complex, and still incompletely understood. Interactions with other types of immune cells, including T lymphocytes and natural killer (NK) cells,[27] are known to occur. Optimal expression of the cytotoxic activity of macrophages requires their "activation", which is effectively achieved by the interaction with microorganisms or bacterial products, or by exposure to lymphokines. Also, the activation of macrophages upon interaction with tumor cells was reported,[25,28,29] but it is unclear whether[28] or not[29] T cells are required and whether activation is mediated by secretion of lymphokines.

The occurrence of activated macrophages in nude mice is well documented.[30-32] Comparative studies, using peritoneal and pulmonary macrophages of normal and of nude mice, indicated an enhanced phagocytic activity in the latter.[32-34] As the activity of macrophages from nude mice reconstituted with thymic tissue appeared to be similar to that of normal mice, it was suggested that the absence of T cell-mediated suppression could be responsible for such enhanced activity.[32] Another important observation is the lack of cytotoxic activity of macrophages from germfree nude mice; the supply of a defined intestinal flora resulted in a significant rise of macrophage activity.[32] These results emphasize the microbiological status to be considered in nude mice upon their use in experiments that may be influenced by immunological factors.

The role macrophages play in the prevention of human tumor growth in nude mice is based largely on indirect evidence. Activators of macrophage function were shown to reduce the growth of several melanoma cell lines, whereas the suppressive effects were abolished by blockade of NK cell activity.[35] Further studies revealed that the tumorigenicity of these cell lines inversely correlated with the capacity to activate nude mouse macrophages.[29,36] The cytotoxic effect of the *in vivo* activated macrophages was proven by the significant growth inhibition observed after the co-injection of tumor cells and these macrophages.[36]

2. Natural Killer Cells

The occurrence of large granular lymphocytes that are capable of mediating spontaneous cytotoxity against a variety of target cells without prior sensitization is a general phenomenon in many vertebrate species, including mice and man. Such cells, named natural killer (NK) cells, are involved in a variety of immunological functions, of which the immune surveillance against cancer cells is an important aspect. The characteristics of NK cells have been reviewed extensively.[37,38] In mice, NK cells are present in most of the lymphoid organs, especially in the spleen and the lymph nodes, and in the peripheral blood, but to a limited extent in the bone marrow and the peritoneal cavity.[39,40]

NK cell activity is clearly not thymus dependent since high levels are observed in nude and neonatally thymectomized mice.[37] As a matter of fact, it has been demonstrated that a higher NK cell activity is present in the spleen of nude mice than in euthymic controls.[39] High levels of NK cell reactivity were observed in nude mice of many backgrounds, although slight variations were noted between the various strains.[39]

Maturation of NK cells appears to be rather slow and the expression of NK cell activity is markedly age dependent. NK cell reactivity was generally undetectable in mice less than 3 weeks old, reached peak levels in mice between 6 and 8 weeks of age, and slowly declined at further increasing age.[39,40] The latter observation does not implicate that NK cell reactivity may play a role only in young adult mice of 4 to 10 weeks of age. NK cell reactivity appears to be well regulated and can be rapidly stimulated upon exposure to a variety of agents. In this respect, interferons and interferon inducers were shown to be highly potent.[41] Treatment for 24 h with Poly I:C, an inducer of interferon, resulted in stimulation of NK cell reactivity in the spleens of young nude mice to the same extent as that in 9-week-old animals.[42] Apart from environmental factors,[37] it is of relevance to note that the inoculation of tumor cells or murine viruses also resulted in augmentation of NK cell activity.[43]

NK cells have a rather wide spectrum of reactivity. Originally, it was suggested that the cytotoxicity was largely restricted to lymphoma cells,[43] but further studies clearly revealed that a variety of other tumor cells,[37,45] including human malignancies,[43,46] and even normal cells[45] were susceptible to lysis by murine NK cells. Such information was mostly obtained from short-term ^{51}Cr-release cytotoxicity assays in which labeled tumor cells, usually derived from cultured cell lines, were incubated with NK cell rich suspensions, usually isolated from the spleen. Although the growth conditions may have influenced the sensitivity to NK cell-mediated lysis,[39,44] the results have made clear that large variations exist for NK cell susceptibility of different cell lines.

Two major approaches have been used to evaluate the possible participation of NK cells in tumor rejection. Several investigators have studied the correlation between NK cell susceptibility of tumor cells and tumor growth *in vivo*. Such studies have yielded a number of conflicting results. Some investigators have noticed that tumor cells sensitive to NK cell activity grew poorly in nude mice,[47,48] while others failed to find a significant correlation between the susceptibilty to NK cells and the degree of tumorigenicity.[42,49,50] Another method to study the role of NK cells in tumor rejection is the use of hosts with modified NK cell reactivity. For that purpose, advantage of age- and strain-related variations may be taken or pharmacological manoevres applied to induce differences in NK cell activity. Such studies have provided evidence for a role of NK cells in the inhibition of metastasis formation as the number of metastatic deposits was increased in young nude mice[51] and reduced in cyclophosphamide-[52] or estradiol-pretreated[53] hosts or in animals selectively depleted from NK cells by use of anti-NK cell antibodies.[21] A promising new development is the introduction of another mouse mutant, the beige mouse, that has a selective NK cell defect[54] and is indeed shown to be less resistant to tumor growth.[55,56] Double mutant beige-nude mice, being athymic with a relatively low NK cell activity,[56] have been generated. Surprisingly, the growth and metastatic rate of human tumor cells was not enhanced in such beige-nudes, as compared to normal nude mice.[57,58] These findings, therefore, do not support the view that NK cell activity is generally a major factor in the prevention of xenografts from growth in nude mice.

D. OTHER FACTORS

A number of other factors, apart from those directly affecting the immunological defense system, were shown to be important determinants for the growth of human tumors in nude mice. Whereas some of these factors may partially be related to effects on the immune system, other aspects are believed to be involved as well. For instance, it is a common observation that tumors grow better in young than in adult nude mice.[42,59] NK cell reactivity in young animals has not fully matured, but does not necessarily correlate with improved tumor growth of the transplanted cells.[42] It was suggested that in younger mice, the stronger vascularization may also account for the observed effects. The possible influence of hormonal conditions and growth factor supply on human take in young vs. adult nudes has not yet been investigated.

A comparative study in nude mice of different genetic backgrounds demonstrated substantial strain-related variations in tumor growth.[60] These variations did not seem to correlate with the strain dependent differences in NK cell reactivity.

The site of implantation may have a considerable impact on the tumor growth. With regard to the s.c. inoculation of tumor cells into the flank, it was repeatedly shown that implantation in the anterior lateral region had a favorable effect on the tumor growth rate when compared to the posterior lateral region.[61-64] The variable transplantability may be explained by differences in the regional vascular supply.[63]

III. POSSIBILITIES TO PROMOTE TUMOR GROWTH

Since it had become clear that the residual immunological defense system may play a role in the prevention of tumor growth in nude mice, attempts were made to improve the take rate of xenografts by further immunosuppression. A variety of immunosuppressive agents, most of which are known to inhibit either macrophage or NK cell activity, have been used for that purpose (Table 2).

A. BLOCKADE OF MACROPHAGE ACTIVITY

1. Silica

In vitro studies have indicated that silica particles are selectively toxic to macrophages.[65] When injected into mice or rats, a rapid suppression of the phagocytic activity of the reticuloendothelial system was observed,[66] without any apparent effects on cells other than macrophages,[66] including NK cells.[67] In some experimental tumor systems, silica treatment

TABLE 2
Some Agents That May Be Used for Further Immunosuppression in Nude Mice

Agent	Cell system inhibited
Silica	Macrophages
Carrageenan	Macrophages, (NK cells)
Anti-asialo-GM1	NK cells
Anti-NK 1.1	NK cells
Irradiation	NK cells, macrophages
Cyclophosphamide	NK cells, macrophages[a]
Cyclosporin A	T cells, (NK cells)
Anti-interferon	NK cells[b]

[a] Dose- and time-dependent inhibition.
[b] Inhibition of augmentation, not of basal levels.

markedly shortened the time of tumor appearance after inoculation of tumor cells, resulting in improved tumor growth.[68] This effect could be prevented by the administration of polyvinyl-pyridine-*N*-oxide,[68] a potent macrophage stabilizing agent.[65] Silica treatment has been applied to nude mice only occasionally. It was reported that the suppressive effect of BCG (which activates macrophages) on the growth of rat hepatoma cells in nude mice was completely abolished after administration of silica particles.[69]

Since the suppression of macrophage activity by silica is transient, timing of the treatment is of crucial importance for its effect on tumor growth. Optimal effects were usually obtained upon administration of the drug within 2 d before the inoculation of tumor cells; treatment at 5 or more days before or a few days after tumor implantation did not influence tumor take and growth rate.[68]

2. Carrageenan

Carrageenans are high molecular weight sulfated polygalactose compounds that can be extracted from marine algae. These compounds are cytotoxic to macrophages because, following ingestion and accumulation in the lysosomes, the cells are lethally damaged by the release of hydrolytic enzymes from disrupted lysosomes.[65] Furthermore, it was suggested that the functional activity of macrophages might be impaired by changes induced at their cell surface.[70]

In vivo phagocytic activity is suppressed by carrageenan in a time-dependent fashion.[71-73] Maximal reduction of peritoneal macrophage activity in mice occurred at 5 to 7 d after treatment.[73] Growth[68,70,73-75] and metastasis[70] of certain tumor cells in mice was promoted by carrageenan treatment, presumably due to such suppression of macrophage activity. Treatment with carrageenan 1 d prior to tumor grafting increased the take rate, but not the growth rate of human colorectal cancer cells in athymic mice.[73] In this study, administration of carrageenan to mice with tumors in an advanced stage had no effect on the growth rate.[76]

The specificity of carrageenan as an anti-macrophage agent has been a matter of controversy. Direct or indirect effects on other elements of the immune system, including NK cells, were reported.[72,77] A possible limitation in the practical use of carrageenan might be formed by the toxic side effects, of which ischemic lesions in body extremities (legs and tail), hepatotoxicity, and nephrotoxicity are most pronounced.[72]

B. INHIBITION OF NK CELL ACTIVITY

1. Specific Antibodies

Specific NK cell depletion in animals may be obtained by use of antibodies specifically directed against antigens expressed by NK cells. Anti-asialo-GM1 antibodies, raised in rabbits against the glycolipid asialo-GM1 (gangliotetraosylceramide), were shown to react selectively with NK cells.[78] This finding is somewhat surprising because asialo-GM1 is also present on the cell surface of other lymphocytes. Nevertheless, it was convincingly demonstrated that anti-asialo-GM1 specifically inhibited NK cell activity *in vitro* and *in vivo*.[67,79,80] Since the suppressive effect was only transient, multiple injections were required to maintain NK cell depletion. A marked enhancement of the take rate and growth of NK cell-sensitive syngeneic and human tumors in nude mice was found to be associated with the antibody-mediated reduction of NK cell activity.[79,80] In contrast, the growth of NK cell-resistant lymphoma cells in nude mice was not affected by anti-asialo-GM1.[80] These results provided strong evidence for a role of NK cells in the control of certain types of tumors.

Experiments in euthymic mice using another NK cell-specific antiserum, anti-NK 1.1, also resulted in a decrease of the NK cell activity.[81] The reduction of NK cell effects measured by the ^{51}Cr release assay *in vitro* correlated well with a substantial improvement in the survival of radiolabeled tumor cells *in vivo*.[81]

A slight improvement of tumor transplantability has been reported earlier upon the use of "anti-lymphocyte serum" (ALS).[82,83] Such antisera, that are cytotoxic to nude mouse spleen cells,[83] are of variable quality and probably of low specificity. The efficacy in tumor growth stimulation is relatively low.

2. Whole Body Irradiation

Total body irradiation at a sublethal dose results in a rapid and drastic decrease in the number of spleen lymphocytes in mice.[37,84] A number of different cell types are affected, but NK cells appear to be relatively radioresistant, as compared to other immune cells.[37] A sharp decline in the activity of NK cells (on a per cell basis) was not observed within 2 weeks after the irradiation. However, subsequent recovery of the NK cell activity was delayed and slower than the restoration of the total number of nucleated cells in the spleen.[84] Administration of fractionated doses for an extended period of time resulted in a profound suppression of NK cell activity lasting for several weeks after the last dose of irradiation.[85]

Whole body irradiation of nude mice was reported to improve the transplantability of syngeneic tumors[86] and human cancer cells, including lymphoma and leukemia cells,[21,83,87] nasopharyngeal cancer,[88] and hepatoma.[89] Irradiation appeared to be highly effective and showed greater take rate enhancement than treatment with anti-lymphocyte serum for human lymphoblast transplants.[83]

The mechanism of irradiation-induced stimulation of tumor growth has not been studied in detail. The fact that administration of single doses given shortly before tumor cell inoculation proved to be successful[83,87,88] suggests that factors other than NK cell suppression may be involved. Interestingly, it was shown that the positive effect on tumor growth was increased upon the combination of total body irradiation with splenectomy.[87,88]

3. Cyclophosphamide

Cyclophosphamide, an alkylating agent with antitumor activity, has been shown to modulate immune reactions in a very complex manner. Interactions with various elements of the immune system may lead to a disturbance of the regulatory mechanisms and result in either a decrease or an increase of the immune response.[90] The dose and schedule of administration of the drug are critical determinants for its effects on the immune system. Relatively high doses were shown to impair the NK cell activity in euthymic[91,92] and in nude mice.[41,93] At a similar dose, the humoral immune response was also suppressed.[94] A more prolonged exposure to lower doses of cyclophosphamide was required to impair macrophage activity.[95,96] Recovery of NK cells from cyclophosphamide-induced suppression occurred within 7 d.[91]

Cyclophosphamide treatment of nude mice with the purpose to enhance human tumor transplantability has been carried out with limited success. In order to promote tumor growth, the drug should be administered prior to tumor implantation to prevent direct antitumor effects.

Using such a protocol, tumor take rates were substantially increased for a number of ovarian cancer lines.[93,97] In contrast, the transplantability of various other tumors was not improved under such conditions.[93] It is not clear whether this difference reflects the variation of NK cell susceptibility of these tumor lines.

4. Cyclosporin A

Cyclosporin A is a fungal metabolite with immunosuppressive activity that is preferentially directed against T cell-mediated responses.[98] The production of humoral antibodies in nude mice[98] and the phagocytic activity of macrophages[99] are not inhibited. *In vitro* or *in vivo* exposure to cyclosporin A resulted in a dose-dependent inhibition of NK cell activity in normal and in nude mice,[100,101] and also in the absence of other cell types such as macrophages, B cells, or T cells. The latter property would make this drug useful for application in nude mice, but such studies have yet to be reported. Attempts to grow human tumors in cyclosporin-treated normal mice have failed,[102] but whether the immunosuppression was sufficient or other causes were involved has not been established.

C. OTHER TREATMENTS

A number of other compounds with immunosuppressive properties have been studied with respect to their effects on the immune system of normal and nude mice. Hydrocortisone,[103] prostaglandins,[104] pristane,[105] and strontium-90[106] were all shown to reduce the NK cell activity in mice. However, none of these agents has been evaluated systematically for its tumor growth promoting effect.

The rationale of the treatment with anti-mouse interferon serum is based on the prevention of interferon-mediated augmentation of NK cell activity upon tumor implantation. It was demonstrated that anti-interferon administration to nude mice stimulated the growth and invasive capacity of tumor cell lines persistently infected with RNA viruses.[107] Likewise, the growth of two primary human prostate cancer specimens was stimulated by this treatment.[107] With anti-interferon administration, we observed a significant increase in tumor cell survival using a prostate cancer cell line,[13] but were unable to induce growth of primary prostate cancer specimens.

IV. CONCLUDING REMARKS

For many years, it has been recognized that human tumors, of various origin, differ widely in their transplantability into nude mice. These differences have not yet been fully explained. Since it was realized that the nude mouse has a residual immune system, attempts have been made to manipulate the immunological defense mechanisms for promotion of human tumor growth. Several of these approaches have been successful, but a general improvement of human tumor transplantability has not been achieved. Insufficient reduction of immune reactivity or involvement of non-immune factors may account for the persisting lack of tumor growth.

In view of the high macrophage and NK cell activities that may occur in nude mice, it seems logical to employ procedures in order to suppress such activities. As the macrophage and NK cell systems are complementary to each other in a number of aspects, including the kinetics of their induction and their target cell specificity,[108] it may not be sufficient to inhibit only one of these systems. Relatively non-specific immunosuppressive treatments, such as total body irradiation, were remarkably effective in tumor growth stimulation in several cases.[21,83,87-89] For that reason, strategies may be considered for a more complete blockade of immune reactions, using a combination of different immunopharmacological agents.

Attempts to reduce immune reactivity against human tumors in nude mice have mostly focused on the suppression of the NK cell activity. However, it may well be that the role of NK cells has been overemphasized and that the mechanisms involved in host defense are more complicated than previously thought. In contrast to (murine) leukemia and lymphoma-type cells, human solid tumor cells often have relatively low NK cell susceptibility.[42] As a matter of fact, transplantability of human tumors was not increased in beige-nude mice in comparison to normal nude mice.[57] Therefore, if immune resistance plays a role in the prevention of human tumor growth in nude mice, other cell types must be involved. The development of athymic mice that are hereditary asplenic (lasat mice) and therefore produce significantly less antibodies[109] has provided the possibility to study tumor growth in T and B cell deficient mice. Although it was reported that the growth of a number of human tumors was improved,[82] these results could not be reproduced by others.[60,110,111] The precise role of another type of immune cell, the natural cytotoxic (NC) cell, is still poorly understood. Whereas NC cells share many properties with NK cells, differences with respect to target specificity (solid tumor cells) and sensitivity to immunopharmacological treatment (cyclophosphamide) may exist.[112] Further characterization of NC cell properties may contribute to the understanding of their role in tumor growth prevention and may perhaps provide other ways to circumvent tumor resistance in nude mice.

The partial responsibility of non-immune factors offers another explanation for the observed lack of tumor growth. It has been noticed that only a small percentage of inoculated tumor cells succeed to survive *in vivo,* even when the host animals are immunosuppressed.[42,113] Insufficient access for nutrients (due to the absence of vascularization) or lack of physical support may account for such poor survival rates. Co-injection of irradiated tumor cells or fibroblasts (that may secrete angiogenesis factors) have been shown to stimulate tumor growth.[18,19,21] Other investigators have used gelatine sponges to provide anchorage for the implanted tumor cells.[113,114] This procedure significantly reduced the latency period for tumor appearance.[113] Implantation of sponge material prior to tumor cell inoculation resulted in a diminished immune response and a more rapid vascularization.[114] This technique may be modified to allow supplementation of certain essential substances (e.g., extracellular matrix components) with a stimulatory effect on tumor growth. When combined with extensive immunosuppression, this approach may ultimately provide optimal conditions for growth of human tumors in nude mice.

REFERENCES

1. **Miller, J. F. A. P.,** Immunological significance of the thymus of the adult mouse, *Nature (London),* 195, 1318, 1962.
2. **Pantelouris, E. M.,** Absence of a thymus in a mouse mutant, *Nature (London),* 217, 370, 1968.
3. **Rygaard, J. and Povlsen, C. O.,** Heterotransplantation of a human malignant tumor to "nude" mice, *Acta Pathol. Microbiol. Scand.,* 80, 713, 1972.
4. **Giovanella, B. C., Stehlin, J. S., and Williams, L. J.,** Development of invasive tumors in the "nude" mouse after injection of cultured human melanoma cells, *J. Natl. Cancer Inst.,* 48, 1531, 1972.
5. **Shimosato, Y., Kameya, T., Nagai, K., Hirohashi, S., Koide, T., Hayashi, H., and Nomura, T.,** Transplantation of human tumors in nude mice, *J. Natl. Cancer Inst.,* 56, 1251, 1976.
6. **Giovanella, B. C., Stehlin, J. S., Williams, L. J., Lee, S. S., and Shephard, R. C.,** Heterotransplantation of human cancers into nude mice. A model system for human cancer chemotherapy, *Cancer,* 42, 2269, 1978.
7. **Fogh, J., Tiso, J., Orfeo, F., Sharkey, F. E., Daniels, W. P., and Fogh, J. M.,** Thirty-four lines of six human tumor categories established in nude mice, *J. Natl. Cancer Inst.,* 64, 745, 1980.
8. **Köpf-Mayer, P.,** Dying and regeneration of human tumor cells after heterotransplantation to athymic mice, *Histol. Histopathol.,* 1, 383, 1986.
9. **Hofer, K. G., Prensky, W., and Hughes, W. L.,** Death and metastatic distribution of tumor cells in mice monitored with ^{125}I-iododeoxyuridine, *J. Natl. Cancer Inst.,* 43, 763, 1969.
10. **Lockshin, A., Giovanella, B. C., Kozielski, T., and Stehlin, J. S.,** Determination of cytotoxicity *in vivo* using ^{111}indium-labelled human tumor cells, *Cancer Lett.,* 24, 179, 1984.

11. **Lockshin, A., Giovanella, B. C., Quian, C., Mendoza, J. T., Vardeman, D. M., and Stehlin, J. S.,** Correlation of cytotoxicity with elimination of iodine-125 from nude mice inoculated with prelabeled human melanoma cells, *J. Natl. Cancer Inst.,* 73, 417, 1984.
12. **Romijn, J. C., Verkoelen, C. F., and Schroeder, F. H.,** Measurement of the survival of human tumor cells after implantation in athymic nude mice, *Int. J. Cancer,* 38, 97, 1986.
13. **Romijn, J. C., Verkoelen, C. F., Trapman, J., and Schroeder, F. H.,** Determination of the survival of human tumor cells implanted in nude mice, in *Immune-Deficient Animals in Biomedical Research,* Rygaard, J., Brünner, N., Graem, N., and Spang-Thomsen, M., Eds., Karger, Basel, 1987, 194.
14. **Duprez, A., Guerret, S., Vignaud, J. M., Plenat, F., Hartmann, D., and Grimaud, J. A.,** The interstitial matrix of human carcinomas and sarcomas transplanted to the nude mouse: immunolocalization of some human and murine components, *Cell. Mol. Biol.,* 33, 647, 1987.
15. **Köpf-Maier, P. and Jackel, M.,** Proliferation behavior of xenografted human tumors: a flow cytometric study, *Anticancer Res.,* 8, 1355, 1988.
16. **Damjanov, I., Damjanov, N., Knowless, B. B., and Engvall, E.,** Origin of laminin in the extracellular matrix of human tumor xenografts in nude mice, *Virch. Arch. B,* 49, 45, 1985.
17. **Wilson, E. L., Gartner, M., Campbell, J. A. H., and Dowdle, E. B.,** Growth and behaviour of human melanomas in nude mice: effect of fibroblasts, in *Immune-Deficient Animals,* Sordat, B., Ed., Karger, Basel, 1984, 357.
18. **Picard, O., Rolland, Y., and Poupon, M. F.,** Fibroblast-dependent tumorigenicity of cells in nude mice: implication for implantation of metastases, *Cancer Res.,* 46, 3290, 1986.
19. **Horgan, K., Jones, D. L., and Mansel, R. E.,** Mitogenicity of human fibroblasts *in vivo* for human breast cancer cells, *Br. J. Surg.,* 74, 227, 1987.
20. **Révész, L.,** Effect of tumour cells killed by X-rays upon the growth of admixed viable cells, *Nature (London),* 178, 1391, 1956.
21. **Ziegler, H. W., Frizzera, G., and Bach, F. H.,** Successful transplantation of a human leukemia cell line into nude mice: conditions optimizing graft acceptance, *J. Natl. Cancer Inst.,* 68, 15, 1982.
22. **Milas, L., Iwakawa, M., and Hunter, N.,** Enhancement of lung colony formation by admixing irradiated with viable tumor cells: dependence on host status, *Clin. Exp. Metastasis,* 5, 213, 1987.
23. **Peters, L. J. and Hewitt, H. B.,** The influence of fibrin formation on the transplantability of murine tumour cells: implications for the mechanism of the Révész effect, *Br. J. Cancer,* 29, 279, 1974.
24. **Levy, M. H. and Wheelock, E. F.,** The role of macrophages in defense against neoplastic disease, *Adv. Cancer Res.,* 20, 131, 1974.
25. **Den Otter, W., Dullens, H. F. J., and De Weger, R. A.,** Macrophages and antitumor reactions, *Cancer Immunol. Immunother.,* 16, 67, 1983.
26. **Adams, D. O. and Somers, S. D.,** The cell biology of tumor cell capture by activated macrophages, *Adv. Exp. Med. Biol.,* 184, 65, 1985.
27. **Reynolds, C. W., Brunda, M. J., Holden, H. T., and Herberman, R. B.,** Role of macrophages in *in vitro* augmentation of rat, mouse, and human natural killer activities, *J. Natl. Cancer Inst.,* 66, 837, 1981.
28. **Olstad, R., Kaplan, G., and Seljelid, R.,** *In vitro* cytotoxicity of mouse macrophages activated by coculture with syngeneic sarcoma cells, *Scand. J. Immunol.,* 16, 421, 1982.
29. **Benomar, A., Gerlier, D., and Doré, J. F.,** *In vivo* activation of nude mouse macrophages by human melanoma cells, *J. Natl. Cancer Inst.,* 79, 131, 1987.
30. **Cheers, C. and Waller, R.,** Activated macrophages in congenitally athymic "nude" mice and in lethally irradiated mice, *J. Immunol.,* 115, 844, 1975.
31. **Meltzer, M. S.,** Tumoricidal responses *in vitro* of peritoneal macrophages from conventionally housed and germ-free nude mice, *Cell. Immunol.,* 22, 176, 1976.
32. **Sharp, A. K. and Colston, M. J.,** The regulation of macrophage activity in congenitally athymic mice, *Eur. J. Immunol.,* 14, 102, 1984.
33. **Hernandez-Delgado, J., Bravo-Cuellar, A., Diaz-Esquivel, P., and Gomez-Estrada, H.,** Phagocytic activity of lung macrophages from nu/nu athymic nude mice, *Arch. Invest. Med.,* 13, 65, 1982.
34. **Vetvicka, V., Fornusek, L., Holub, M., Zidková, J., and Kopecek, J.,** Macrophages of athymic nude mice: Fc receptors, C receptors, phagocytic and pinocytic activities, *Eur. J. Cell Biol.,* 35, 35, 1984.
35. **Jacubovich, R., Cabrillat, H., Gerlier, D., Bailly, M., and Doré, J. F.,** Tumourigenic phenotypes of human melanoma cell lines in nude mice determined by an active antitumour mechanism, *Br. J. Cancer,* 51, 335, 1985.
36. **Benomar, A., Gerlier, D., and Doré, J. F.,** Interactions with host macrophages and ability of human melanoma cell lines to grow in nude mice, *Int. J. Cancer,* 38, 419, 1986.
37. **Herberman, R. B. and Holden, H. T.,** Natural cell-mediated immunity, *Adv. Cancer Res.,* 27, 305, 1978.
38. **Herberman, R. B. and Ortaldo, J. R.,** Natural killer cells: their role in defenses against disease, *Science,* 214, 24, 1981.
39. **Herberman, R. B., Nunn, M. E., and Lavrin, D. H.,** Natural cytotoxic reactivity of mouse lymphoid cells against syngeneic and allogeneic tumors. I. Distribution of reactivity and specificity, *Int. J. Cancer,* 16, 216, 1975.

40. **Kiessling, R., Klein, E., Pross, H., and Wigzell, H.,** "Natural" killer cells in the mouse. II. Cytotoxic cells with specificity for Moloney leukemia cells. Characteristics of the killer cell, *Eur. J. Immunol.,* 112, 1975.
41. **Djeu, J. Y., Heinbaugh, J. A., Vieira, W. D., Holden, H. T., and Herberman, R. B.,** The effect of immunopharmacological agents on mouse natural cell-mediated cytotoxicity and on its augmentation by Poly I:C, *Immunopharmacology,* 1, 231, 1979.
42. **Romijn, J. C.,** Growth of tumor cells with different sensitivities for murine natural killer cells in young and adult athymic nude mice, *Exp. Cell Biol.,* 53, 24, 1985.
43. **Herberman, R. B., Nunn, M. E., Holden, H. T., Staal, S., and Djeu, J. Y.,** Augmentation of natural cytotoxic reactivity of mouse lymphoid cells against syngeneic and allogeneic target cells, *Int. J. Cancer,* 19, 555, 1977.
44. **Kiessling, R., Klein, E., and Wigzell, H.,** "Natural" killer cells in the mouse. I. Cytotoxic cells with specificity for Moloney leukemia cells. Specificity and distribution according to genotype, *Eur J. Immunol.,* 5, 112, 1975.
45. **Nunn, M. E., Herberman, R. B., and Holden, H. T.,** Natural cell-mediated cytotoxicity in mice against non-lymphoid tumor cells and some normal cells, *Int. J. Cancer,* 20, 381, 1977.
46. **Haller, O., Kiessling, R., Örn, A., Kärre, K., Nilsson, K., and Wigzell, H.,** Natural cytotoxicity to human leukemia mediated by mouse non-T cells, *Int. J. Cancer,* 20, 93, 1977.
47. **Warner, N. L., Woodruff, M. F. A., and Burton, R. C.,** Inhibition of the growth of lymphoid tumors in syngeneic athymic (nude) mice, *Int. J. Cancer,* 20, 146, 1977.
48. **Kristensen, E.,** Natural cytotoxicity: relationship to tumorigenicity of human cell lines in nude mice, *Proc. Soc. Exp. Biol. Med.,* 162, 467, 1979.
49. **Klein, A. S., Plata, F., Jackson, M. J., and Shin, S.,** Cellular tumorigenicity in nude mice. Role of susceptibility to natural killer cells, *Exp. Cell Biol.,* 47, 430, 1979.
50. **McCormick, K. J., Giovanella, B. C., Klein, G., Nilsson, K., and Stehlin, J. S.,** Diploid human lymphoblastoid and Burkitt lymphoma cell lines: susceptibility to murine NK cells and heterotransplantation to nude mice, *Int. J. Cancer,* 28, 455, 1981.
51. **Hanna, N.,** Expression of metastatic potential of tumor cells in nude mice is correlated with low levels of natural killer cell-mediated cytotoxicity, *Int. J. Cancer,* 26, 675, 1980.
52. **Hanna, N. and Burton, R. C.,** Definitive evidence that natural killer (NK) cells inhibit experimental tumor metastasis *in vivo, J. Immunol.,* 127, 1754, 1981.
53. **Hanna, N. and Schneider, M.,** Enhancement of tumor metastasis and suppression of natural killer cell activity by beta-estradiol treatment, *J. Immunol.,* 130, 974, 1982.
54. **Roder, J. and Duwe, A.,** The beige mutation in the mouse selectively impairs natural killer cell function, *Nature (London),* 278, 451, 1979.
55. **Talmadge, J. E., Meyers, K. M., Prieur, D. J., and Starkey, J. R.,** Role of NK cells in tumour growth and metastasis in beige mice, *Nature (London),* 284, 622, 1980.
56. **Kärre, K., Klein, G. O., Kiessling, R., Klein, G., and Roder, J. C.,** *In vitro* NK-activity and *in vivo* resistance to leukemia: studies of beige, beige/nude and wild-type hosts on C57BL background, *Int. J. Cancer,* 26, 789, 1980.
57. **Fodstad, Ø., Hansen, C. T., Cannon, G. B., Statham, C. N., Lichtenstein, G. R., and Boyd, M. R.,** Lack of correlation between natural killer activity and tumor growth control in nude mice with different immune defects, *Cancer Res.,* 44, 4403, 1984.
58. **Naito, S., Giavazzi, R., Walker, S. M., Itoh, K., Mayo, J., and Fidler, I. J.,** Growth and metastatic behavior of human tumor cells implanted into nude and beige nude mice, *Clin. Exp. Metastasis,* 5, 135, 1987.
59. **Bader, A., Fortmeyer, H. P., Lapp, C., and Bastert, G.,** Zum Wachstumsverhalten menschlichen Malignome nach Transplantation auf nu/nu Mäuse. I. Abhängigkeit vom Alter der Empfängertiere, *Z. Versuchstierk.,* 23, 161, 1981.
60. **Maruo, K., Ueyama, Y., Hioki, K., Saito, M., Nomura, T., and Tamaoki, N.,** Strain-dependent growth of a human carcinoma in nude mice with different genetic backgrounds. Selection of nude mouse strains useful for anticancer agent screening system, *Exp. Cell Biol.,* 50, 115, 1982.
61. **Auerbach, R., Morrissey, L. W., and Sidky, Y. A.,** Gradients in tumour growth. *Nature (London),* 274, 697, 1978.
62. **Auerbach, R., Morrissey, L. W., and Sidky, Y. A.,** Regional differences in the incidence and growth of mouse tumors following intradermal or subcutaneous inoculation, *Cancer Res.* 38, 1739, 1978.
63. **Kyriazis, A. A. and Kyriazis, A. P.,** Preferential sites of growth of human tumors in nude mice following subcutaneous transplantation, *Cancer Res.,* 40, 4509, 1980.
64. **Dipersio, L. P.,** Regional growth differences of human tumour xenografts in nude mice, *Lab. Animals,* 15, 179, 1981.
65. **Allison, A. C., Harington, J. S., and Birbeck, M.,** An examination of the cytotoxic effects of silica on macrophages, *J. Exp. Med.,* 124, 141, 1966.
66. **Levy, M. H. and Wheelock, E. F.,** Effects of intravenous silica on immune and non-immune functions of the murine host, *J. Immunol.,* 115, 41, 1975.
67. **Keller, R.,** Discrimination between macrophage- and NK-type tumoricidal activities via anti-asialo-GM1 antibody, *Exp. Cell Biol.,* 51, 158, 1983.

68. **Keller, R.,** Promotion of tumor growth *in vivo* by antimacrophage agents, *J. Natl. Cancer Inst.*, 57, 1355, 1976.
69. **Hopper, D. G., Pimm, M. V., and Baldwin, R. W.,** Silica abrogation of mycobacterial adjuvant contact suppression of tumor growth in rats and athymic mice, *Cancer Immunol. Immunother.*, 1, 143, 1976.
70. **Lapis, K., Kopper, L., Timár, J., van Hanh, T., and Hegedüs, C.,** Effect of macrophages on the growth of syn- and xenogeneic tumors, *Arch. Geschwulstf.*, 51, 475, 1981.
71. **Fowler, E. F. and Thomson, A. W.,** Effect of carrageenan on activity of the mononuclear phagocyte system in the mouse, *Br. J. Exp. Path.*, 59, 213, 1978.
72. **Thomson, A. W. and Fowler, E. F.,** Carrageenan: a review of its effect on the immune system, *Agents Actions*, 11, 265, 1981.
73. **Kopper, L., van Hanh, T., Lapis, K., and Timár, J.,** Increased take rate of human tumour xenografts after carragheenan treatment, *Eur. J. Cancer*, 16, 671, 1980.
74. **Lotzova, E. and Richie, E. R.,** Promotion of incidence of adenovirus type 12 transplantable tumors by carrageenan, a specific antimacrophage agent, *J. Natl. Cancer Inst.*, 58, 1171, 1977.
75. **Thomson, A. W. and Fowler, E. F.,** Potentiation of tumour growth by carrageenan, *Transplantation*, 24, 397, 1977.
76. **Kopper, L., van Hanh, T., Hegedüs, C., and Lapis, K.,** Growth of human colorectal tumor xenografts in immunosuppressed mice reconstituted with normal cells, *Exp. Cell Biol.*, 49, 141, 1981.
77. **Quan, P. C., Kolb, J. P., and Lespinats, G.,** NK activity in carrageenan-treated mice, *Immunology*, 40, 495, 1980.
78. **Kasai, M., Iwamori, M., Nagai, Y., Okumura, K., and Tada, T.,** A glycolipid on the surface of mouse natural killer cells, *Eur. J. Immunol.*, 10, 175, 1980.
79. **Habu, S., Fukui, H., Shimamura, K., Kasai, M., Nagai, Y., Okumura, K., and Tamaoki, N.,** *In vivo* effects of anti-asialo GM1. I. Reduction of NK activity and enhancement of transplanted tumor growth in nude mice, *J. Immunol.*, 127, 34, 1981.
80. **Kawase, I., Urdal, D. L., Brooks, C. G., and Henney, C. S.,** Selective depletion of NK cell activity *in vivo* and its effect on the growth of NK-sensitive and NK-resistant tumor cell variants, *Int. J. Cancer*, 29, 567, 1982.
81. **Pollack, S. B. and Hallenbeck, L. A.,** *In vivo* reduction of NK activity with anti-NK serum: direct evaluation of NK cells in tumor clearance, *Int. J. Cancer*, 29, 203, 1982.
82. **Gershwin, M. E., Ikeda, R. M., Erickson, K., and Owens, R.,** Enhancement of heterotransplanted human tumor graft survival in nude mice treated with antilymphocyte serum in congenitally athymic-asplenic (lasat) mice, *J. Natl. Cancer Inst.*, 61, 245, 1978.
83. **Ohsugi, Y., Gershwin, M. E., Owens, R. B., and Nelson-Rees, W. A.,** Tumorigenicity of human malignant lymphoblasts: comparative study with unmanipulated nude mice, anti-lymphocyte serum-treated nude mice, and X-irradiated nude mice, *J. Natl. Cancer Inst.*, 65, 715, 1980.
84. **Cudkowicz, G. and Hochman, P. S.,** Do natural killer cells engage in regulated reactions against self to ensure homeostasis?, *Immunol. Rev.*, 44, 13, 1979.
85. **Gorelik, E. and Herberman, R. B.,** Depression of natural antitumor resistance of C57BL/6 mice by leukogenic doses of radiation and restoration of resistance by transfer of bone marrow or spleen cells from normal, but not beige, syngeneic mice, *J. Natl. Cancer Inst.*, 69, 89, 1982.
86. **Prehn, L. M. and Outzen, H. C.,** Primary tumor immunity in nude mice, *Int. J. Cancer*, 19, 688, 1977.
87. **Watanabe, S., Shimosato, Y., Kuroki, M., Sato, Y., and Nakajima, T.,** Transplantability of human lymphoid cell line, lymphoma, and leukemia in splenectomized and/or irradiated nude mice, *Cancer Res.*, 40, 2588, 1980.
88. **Sudo, K.,** Improvement of transplantability of human neoplastic cells to nude mice, *Jpn. J. Exp. Med.*, 57, 189, 1987.
89. **Shouval, D., Schuger, L., Levij, I. S., Reid, L. M., Neeman, Z., and Shafritz, D. A.,** Comparative morphology and tumourigenicity of human hepatocellular carcinoma cell lines in athymic rats and mice, *Virchows Arch. (A)*, 412, 595, 1988.
90. **Turk, J. L. and Parker, D.,** Effect of cyclophosphamide on immunological control mechanisms, *Immunol. Rev.*, 65, 99, 1892.
91. **Mantovani, A., Luini, W., Peri, G., Vecchi, A., and Spreafico, F.,** Effect of chemotherapeutic agents on natural cell-mediated cytotoxicity in mice, *J. Natl. Cancer Inst.*, 61, 1255, 1978.
92. **Riccardi, C., Barlozzari, T., Santoni, A., Herberman, R. B., and Cesarini, C.,** Transfer to cyclophosphamide-treated mice of natural killer (NK) cells and *in vivo* natural reactivity against tumors, *J. Immunol.*, 126, 1284, 1981.
93. **Braakhuis, B. J. M., Nauta, M. M., Romijn, J. C., Rutgers, D. H., and Smink, T.,** Enhanced success rate of transplantation with human tumors in cyclophosphamide-treated nude mice, *J. Natl. Cancer Inst.*, 76, 241, 1986.
94. **Kolb, J. P. B., Poupon, M. F. M., Lespinats, G. M., Sabolovic, D., and Loisillier, F.,** Splenic modifications induced by cyclophosphamide in C3H/He, nude, and "B"-mice, *J. Immunol.*, 118, 1595, 1977.
95. **Hunninghake, G. W. and Fauci, A. S.,** Divergent effects of cyclophosphamide administration on mononuclear killer cells: quantitative depletion of cell numbers versus qualitative suppression of functional capabilities, *J. Immunol.*, 117, 337, 1976.

96. **Hunninghake, G. W. and Fauci, A. S.,** Immunological reactivity of the lung. IV. Effect of cyclophosphamide on alveolar macrophage cytotoxic effector function, *Clin. Exp. Immunol.*, 27, 555, 1977.
97. **Nauta, M. M., Boven, E., Schlüper, H. M. M., Erkelens, C. A. M., and Pinedo, H. M.,** Enhanced transplantability of human ovarian cancer lines in cyclophosphamide-pretreated nude mice, *Br. J. Cancer*, 54, 331, 1986.
98. **Borel, J. F., Feurer, C., Magnee, C., and Stahelin, H.,** Effects of the new anti-lymphocyte peptide cyclosporin A in animals, *Immunology*, 32, 1017, 1977.
99. **Paavonen, T. and Hayry, P.,** Effect of cyclosporin-A on T-dependent and independent immunoglobulin synthesis *in vitro*, *Nature (London)*, 287, 542, 1980.
100. **Yanagihara, R. H. and Adler, W. H.,** Inhibition of mouse natural killer activity by cyclosporin A, *Immunology*, 45, 325, 1982.
101. **Yanagihara, R. H. and Adler, W. H.,** Cyclosporin A inhibits T cell-mediated augmentation of mouse natural killer cell activity, *Immunopharmacology*, 4, 243, 1982.
102. **Floersheim, G. L.,** Comparative growth of human tumors in pharmacologically immunosuppressed, immune-deprived, Cyclosporin A-treated and nude mice, *Eur. J. Cancer Clin. Oncol.*, 18, 589, 1982.
103. **Hochman, P. S. and Cudkowicz, G.,** Suppression of natural cytotoxicity by spleen cells of hydrocortisone-treated mice, *J. Immunol.*, 123, 968, 1979.
104. **Brunda, M. J., Herberman, R. B., and Holden, H. T.,** Inhibition of murine natural killer cell activity by prostaglandins, *J. Immunol.*, 124, 2682, 1980.
105. **Freund, I. R. and Blair, P. B.,** Depression of natural killer activity and mitogen responsiveness in mice treated with pristane, *J. Immunol.*, 129, 2826, 1982.
106. **Emmanuel, F. X. S., Vaughan, A. T. M., and Catty, D.,** Mice treated with strontium 90: an animal model deficient in NK cells, *Br. J. Cancer*, 44, 160, 1981.
107. **Reid, L. M., Minato, N., Gresser, I., Holland, J., Kadish, A., and Bloom, B. R.,** Influence of anti-mouse interferon serum on the growth and metastasis of tumor cells persistently infected with virus and of human prostatic tumors in athymic nude mice, *Proc. Natl. Acad. Sci. U.S.A.*, 78, 1171, 1981.
108. **Schultz, R. M.,** Factors limiting tumoricidal function of interferon-induced effector systems. Activated macrophages and natural killer cells, *Cancer Immunol. Immunother.*, 10, 61, 1981.
109. **Lozzio, B. B.,** The lasat mouse: a new model for transplantation of human tissues, *Biomedicine*, 24, 144, 1976.
110. **Tropé, C., Wennerberg, J., Alm, P., and von Eyben, F. E.,** Heterotransplantation of two human tumours in athymic mice and asplenic-athymic mice, *Acta Path. Microbiol. Immunol. Scand. (A)*, 90, 75, 1982.
111. **Sawa, H., Takeshita, I., Kuramitsu, M., Egami, H., Fukui, M., and Sato, Y.,** Neuronal and glial proteins in medulloblastomas. II. Heterotransplantation of human medulloblastoma, *Anticancer Res.*, 6, 911, 1986.
112. **Stutman, O., Lattime, E. C., and Figarella, E. F.,** Natural cytotoxic cells against solid tumors in mice. A comparison with natural killer cells, *Fed. Proc.*, 40, 2699, 1981.
113. **Kraemer, P. M., Travis, G. L., Saunders, G. C., Ray, F. A., Stevenson, A. P., Bame, K., and Cram, L. S.,** Tumorigenicity assays in nude mice: analysis of the implanted gelatin sponge method, in *Immune-Deficient Animals*, Sordat, B., Ed., Karger, Basel, 1984, 214.
114. **Wells, R. S., Campbell, E. W., Swartzendruber, D. E., Holland, L. M., and Kraemer, P. M.,** Role of anchorage in the expression of tumorigenicity of untransformed mouse cell lines, *J. Natl. Cancer Inst.*, 69, 415, 1982.

Chapter 6

TRANSPLANTABILITY OF HUMAN TUMORS

Øystein Fodstad

TABLE OF CONTENTS

I. INTRODUCTION

During the 20 years that have passed since Rygaard and Povlsen[1] demonstrated that human colon cancer tissue could be grown s.c. and serially transferred in athymic, nude mice, the usefulness of human tumor xenografts in such animals has become increasingly apparent. The use of nude mice in oncology research includes studies on the characteristics of tumor development and progression, aspects of metastasis formation, and tumor response to treatment. With the major advances made in basic research in the last few years, as well as with the advent of new approaches for the diagnosis and therapy of cancer, the list of research areas wherein the nude mouse-human tumor model is of value has continued to grow.

Not all human tumors grow in nude mice,[2-9] and there is a need for better understanding of the mechanisms underlying this fact, and for identifying the means by which optimal propagation of human malignancies can be achieved. A variety of factors have been identified which may influence the growth and biology of transplanted tumors, and in this chapter some of these factors will be discussed in the light of 15 years of personal experience with such models, and by citing relevant literature.

II. TUMOR TAKE

In a discussion on the transplantability of tumors, it is necessary first to define "tumor take", particularly as the meaning of this term seems to vary considerably between different laboratories. This variation makes it difficult to prepare a meaningful list of tumor take rates, the percentage of tumors of the different histological types that have been found to grow in the immunodeficient animals. Moreover, the results obtained in the various laboratories are not easy to compare because of differences with regard to the mouse strain used, the quality of animal husbandry, and the methodology employed when preparing and implanting the tumor cells and tissues.

In our laboratory, the criterion for tumor take is considered as fulfilled when the tumor tissue can be successfully and reproducibly transferred to recipient mice for at least three to four passages, regardless of the transplantation procedure and site. It is important that the animals can be observed for a sufficiently long period, as several transplanted tumors may have a latency time of up to 6 months from the day of transplantation until progressively growing xenografts can be observed.[10] During the first few passages, the latency period decreases and usually stabilizes after three to five subsequent passages.[4,6,8]

The morphology of the established xenografts must be compared with that of the parent tumor, and wherever some characteristics of the xenograft seem inconsistent with its supposed origin, additional testing should be performed (Chapter 8). The human nature of the tumor should be demonstrated; for example, by studying isoenzyme or antigen profiles, or by checking the presence of human chromosomes or DNA (Alu)-sequences.[8,11,12] We have recently introduced DNA-fingerprinting techniques[12,13] to ascertain the authenticity of the tumor, both upon initial establishment and in cases where the possibility of a mixture with other tumor tissue cannot be otherwise excluded.

III. HOST FACTORS OF INFLUENCE ON TUMOR TAKE RATE

The nude gene may have a different impact on different mouse strains, which should not come as a surprise. The importance of the genetic background of the nu/nu mice used as hosts for human tumor tissue has, however, not received a great deal of attention. Moreover, for various practical reasons, work with xenografts has often been performed with animals of variable health. Genetic dissimilarities, age, and environmental factors may all have an effect on the immune system and, hence, on the ability of the nude mouse to sustain human tumor growth.

A. MOUSE GENETICS

1. Background Strain

The nu gene has been bred into a variety of mouse strains,[14] an issue dealt with in Chapter 3 of this book.

Since most investigators have been working with only one strain of nude mice, direct comparisons of the tumorigenicity of human tumors in mice with different genetic backgrounds is not readily available. However, it is the experience of several laboratories that mice obtained from various commercial sources may sometimes differ in their ability to sustain growth of established tumor lines. When publishing results obtained in nude mice, it is therefore necessary to provide adequate information on the genetic constitution of the animals.

2. Additional Immune Defects

Since many tumors are difficult to grow in regular (nu/nu) nude mice, there has been great interest in obtaining animals that, in addition to their T cell deficiency, have other immune defects. In regular nude mice, the B cell response is intact,[15,16] and it was thought that tumor take could be improved in nude animals with reduced B cell activity. The early reports[17] on increased tumor take in B cell-defective NIH-II (nude-Xid) mice were, unfortunately, not confirmed in a later study.[18] Because the activity of natural killer (NK) cells is much higher in nu/nu compared to the parent mouse strains,[8,16,17] it was then expected that a better take rate could be obtained by transplanting human tumors to NK cell-deficient beige-nude mice.[19-22] This was not confirmed in our studies, however, as we did not find any correlation between NK cell activity and tumor growth control (Tables 1 and 2).[18]

Triple immunodeficient nude mice NIH-III (beige-nude-Xid)[6] were later developed that, in addition to their lack of functional T cells, also showed very low NK and B cell activities (Table 1). During 7 years of experience with these mice, we have observed a slight tendency to increased tumor take, whereas the growth rate of established tumors is similar to that in regular nude animals (unpublished results). The minor advantage of using these mice for the establishment of tumor xenografts is countered by the disadvantage that they are slightly more difficult to breed, and also more vulnerable to minor infections. The increased susceptibility to tumor transplants may, however, be significant for some experimental purposes. Thus, it has recently been demonstrated that growth and differentiation of normal human bone marrow stem cells can be obtained in irradiated beige-nude-Xid mice.[23] In fact, for this particular purpose, these mice seem to be superior to the severe combined immunodeficiency (scid) mice that have come available only recently.[23,24]

B. EFFECT OF ANIMAL AGE

It is a common experience that tumor inoculation into newborn mice results in increased tumor take. For obvious reasons, this approach is of limited usefulness. Evidence has been provided that young animals (2 to 3 weeks old) are also more susceptible than adult mice as hosts for tumor xenografts,[25,26] and higher rates of metastasis formation have also been observed.[25] The inconsistency between these results and our own data[18] represents one of the many examples that point out the inherent risk of errors when drawing general conclusions based on results in one particular laboratory with one or only a few tumors.

TABLE 1
Summary of Immunological Response in Nude Mice

Mouse genotype	Cytotoxicity assay[a] NK(11-19)[c]	Cytotoxicity assay[a] ADCC(5)	Lymphocyte proliferation assay[b] Phytohemagglutinin (6-12)	Lymphocyte proliferation assay[b] Concanavalin A (6-11)	Lymphocyte proliferation assay[b] LPS(6-12)	PFC assay[d](4-9)
NIH (6 wk old)	25 ± 3[e]	22 ± 5	1520 ± 530	4400 ± 179	41,300 ± 6400	2400 ± 312
NIH (2-3 wk old)	15 ± 2	14 ± 5	330 ± 200	180 ± 70	14,600 ± 1960	210 ± 40
Beige (5-7 wk old)	6 ± 1	7 ± 3	650 ± 130	960 ± 90	15,600 ± 3630	1930 ± 700
NIH-II (5-7 wk old)	18 ± 4	5 ± 2	1350 ± 780	1750 ± 1260	570 ± 120	0
NIH-III (5-7 wk old)	9 ± 3	N.T.	0	800 ± 800	4030 ± 800	0

Note: Abbreviations: NK - natural killer; LPS - lipopolysaccharide; ADCC - antibody-dependent cell-mediated cytotoxicity; PFC - plaque-forming cell.

[a] % of cytotoxicity = Test cpm – spontaneous cpm/cpm initially incorporated
[b] Net cpm = cpm in stimulated cultures – cpm in cultures with no mitogen
[c] Numbers in parentheses, number of mice of each genotype tested for each assay
[d] PFC per spleen
[e] Mean ± SE

TABLE 2
Lethality of Nude Mice Given i.v. Injections of Human (LOX) and Murine (B16F10) Melanoma Cells

		Deaths[a]			
Tumor	No. of cells injected	NIH (6 wk old)	NIH (2-3 wk old)	Beige	NIH II
LOX	1×10^6	48/55 (1.0)[b]	11/1 (1.0)	5/17 (1.2)	9/15 (1.0)
B16F10	5×10^4	20/20 (1.0)	15/15 (0.96)	17/17 (0.96)	15/17 (1.2)

[a] The ratio of mice that died of lung colonies per total number of mice given injections.
[b] Numbers in parentheses, mean survival time of mice in each group, normalized to that of 6-week-old NIH nude mice. This was 38 d for mice given injections of LOX cells and 27 d for mice given injections of B16F10 cells.

C. EFFECTOR MECHANISMS

Although the animals lack mature T lymphocytes, it has been shown that T cell-like activity can be induced.[15,27] Moreover, other types of cell-mediated immune responses, as well as the B cell antibody response, are present at normal or even higher levels in nu/nu mice as compared to those in the corresponding conventional animals.[15,16]

The importance of the immune system in tumor rejection is indicated by the evidence of high tumor take rates and metastasis formation in newborn[28] and, possibly, in young nude mice.[25,29] At birth, the immune defense is not yet fully developed and, in the time needed to reach the mature stage, the animals may have less ability to resist tumor transplants.

Several reports suggest an important role for NK cells[19,22,26] and macrophages[26] in preventing growth of human tumors in nude mice. It is possible that these cells may have a different impact on tumor take in different regions. Thus, macrophage activity may be more important in the lungs and in the peritoneal cavity than in the s.c. space. It is also well known that tumor cells of various origin show a dissimilar sensitivity to these effector cells.

In addition to immunological factors, variables such as the hormonal status and enzyme activities may differ between mice and man. Such differences might influence tumor growth, which also depends on the site of tumor implantation.

D. HEALTH STATUS

The fact that immunodeficient nude animals are prone to attract various viral and bacterial infections is well known to be an important determinant in tumor take and growth.[8,14,26] Fulminant infections with mouse hepatitis virus, or other agents in a nude mouse colony, were early recognized to have far-reaching consequences for the life span of the animals.[8] Subsequently, the infections may be transmitted with tumor transplants obtained from infected mice. Also, less dramatic infections may affect tumor take, and all possible measures should be employed to maintain a healthy stock of animals. Appropriate facilities, personnel restrictions, and regular microbiological quality control are key elements.

After transplantation of biopsied tumor tissue from patients, it may, for some tumors, take more than 6 months before a xenograft starts to grow,[10] a fact which emphasizes the importance of keeping the mice alive for a sufficiently long period. The effect of poor animal health on the observed take rate is, however, not only due to the reduced life span of the animals, but also because infections may trigger the immune system to reject the graft.

IV. EFFECTS RELATED TO METHODOLOGY

Upon implantation of human tumor tissue or cells into nude mice, the probability of success is related to the selection and viability of the tumor cells, and the site and route of cell/tissue inoculation.

A. TISSUE CUBES OR CELL SUSPENSIONS

In our experience, higher take rates are achieved with well-selected fragments of tumor tissue than when a single cell suspension prepared from the same tissue is injected s.c. This is most readily observed when xenografts are transferred which have slow growth rates and long latency times.

Several investigators have reported that invasive growth and metastases, particularly to abdominal organs such as the pancreas and the liver, occur frequently after s.c. injection of tumor cells,[8,9,30] in contrast to implantation of tissue cubes.[6,8] It is possible that when injecting cells, tumor spread may be a result of direct transfer of tumor cells into small blood vessels during the injection. In many studies, the cell suspensions used are obtained from *in vitro* cultures,[5,9,30,31] commonly from cell lines that have been in serial passage for several years, and which may not be representative for the parent tumor.[6]

B. SITES AND ROUTES OF TUMOR INOCULATION

Tumor tissue has been transplanted to various s.c. sites in the mouse, including the fat-pads in the axilla or in the mammary region.[8,32,33] The importance of a favorable microenvironment for the transplant is well recognized, but the requirements may differ significantly within and between tumor types. In some reports,[33] higher tumor takes were described for the axilla than for other regions, a finding we have not been able to confirm.

For the establishment of s.c. xenografts, it is recommended to implant small ($3 \times 3 \times 3$ mm) tumor cubes in one or both flanks of the animals through a dorsal incision, which afterwards may be easily closed with a physiological glue[34] or a small metal clamp. Also, from an animal ethics point of view, this area is of advantage since growing tumors in the flanks affect the mice in mobility much less than transplants in the axilla or under the tight skin on the back. With this procedure, possible metastases in the regional lymph nodes are also well detectable.

To achieve a high take rate, most investigators implant at least two pieces of tumor tissue. Experiments with murine tumors have, however, shown that tumors growing at different sites in the host may exert a mutually inhibitory effect on each other's growth.[35] Moreover, Shoemaker et al.[36] have found that transplantation of lung tissue from nude mice carrying s.c. LOX human melanoma xenografts resulted in the development of new LOX tumors in the recipient animals, even though spontaneous lung metastases were not observed in the donor mice (unpublished). These findings may suggest that the growth of human xenografts may also inhibit the development of tumors implanted into other sites in nude mice.

Single tumor cell suspensions can be introduced in the nude host by various routes of injection and in most tissues. Thus, cells have been injected s.c., i.v., i.p., intradermally, intra-arterially, intrabronchially, into the left ventricle of the heart, the spleen, the gut, the brain, the foot-pad, and the kidney.[5,8,26,32,37-46] Data have been presented indicating that tumor take is more likely to occur after "orthotopic" injection of the cells, for example, when lung cancer cells are injected intrabronchially,[37] and cells of renal cell cancer into the kidney.[42] Unfortunately, some of these types of experiments have involved a less than adequate number of control cell lines.

General experience supports the view that interaction between the tumor cells and their microenvironment is a decisive factor in determining tumor growth. For example, the variety in metastatic patterns seen when cells of the same tumor line are injected by different routes can only be explained by such interactions. Similarly, the tissue specificity of metastatic tumor cells observed in nude mice[26,41] and nude rats[43] seems predominantly to be caused by microenvironmental factors. However, the finding that several tumors may grow after injection directly into the brain,[39] but hardly when other routes are used, may simply reflect the fact that the brain represents an immunologically secluded or privileged area. In conclusion, the diversity in tumor-host interactions makes it possible to obtain models where human tumors can be grown as "primary" tumors or as metastases in most tissues relevant for the clinical situation.[23,26,37,40-47]

V. TUMOR-RELATED FACTORS

A. GENERAL CONSIDERATIONS

1. Tumor Growth Fraction

In patients, malignant tumors differ in their content of viable tumor cells, which depends on factors such as the tumor type, size, and blood supply. The tumor take rate in nude mice is closely related to the tumor growth fraction (see Chapter 20), the fraction of viable proliferating cells. In this context, it is noteworthy that the observed growth of a tumor is the result of the net increase in tumor cells, that is, the number of new cells minus the number of cells that die (the cell loss).[48]

Commonly, metastases have higher growth fractions than the primary tumors. At the time of diagnosis, the primary tumors may already have reached a considerable size, resulting in inadequate blood supply and central necrosis. Moreover, the tumor may be growing at a site of limited vascularity and under space constraints. In contrast, metastases often develop in favorable environments in the lungs, liver, lymph nodes, bone marrow, and brain. The tumor growth fraction is also determined by the tumor's content of stroma and intercellular substance, variables that strongly differ between individual tumors and between tumor types. As an example, in breast cancer, the fibrous tissue content is much more abundant than in melanomas.

2. Gompertzian Growth

Tumors are known to display Gompertzian growth patterns, that is, sigmoid-shaped growth curves.[49] This pattern reflects inherent characteristics of the tumor, resulting in a variation between different tumors in the start and the duration of exponential growth.[6,50] If tissue is taken from a tumor that has reached a plateau-phase growth stage, the transplantability is reduced. It should be noted that some tumor lines reach their Gompertzian plateau when the xenografts are still very small. Thus, in several cases, we have observed that lines can be kept alive by continuous transfer, but the tumors never grow larger than a few millimeters in diameter.

3. Implications for the Selection of Tumor Tissue

The practical implications of the above considerations are that the chances to obtain successful growth with clinical material are higher if the tissue is derived from a relatively small lymph node or lung metastasis originating from a melanoma than from a large primary breast cancer. For continuous transfer of tumor xenografts, the most viable tumor tissue at the periphery of the nodules that are still in a rapid growth phase should be used, the latter judged by the growth curves of the previous passages.

B. DIFFERENTIATION AND HORMONE DEPENDENCY

Generally, an inverse correlation exists between the aggressiveness of a malignant tumor and the degree of differentiation of the tumor cells. In agreement with this clinical experience, the take rate of human tumors in nude mice is higher when tissue is used from poorly, rather than from differentiated, cancers.[8] As an example, melanin production in melanomas is a parameter of differentiation. The most rapidly growing melanoma xenografts are amelanotic or show very low degrees of differentiation,[6,8] whereas the most slowly growing lines appear as tiny black-colored tumor nodules that may never reach a size above 1 cm in diameter.

In contrast to the general experience of an inverse relationship between tumor growth and the degree of differentiation, some interesting exceptions have recently been reported. Thus, a correlation between a differentiated phenotype and the ability of tumor cells to form metastasis has been observed.[51] Similarly, during an attempt in our laboratory to select *in vivo* for a more aggressively growing subline of a metastasizing human melanoma,[47] repeated injection of cells from each new generation of s.c. metastases resulted in more differentiated tumor cells (manuscript in preparation).

Human tumors of several types of cancer contain cells with hormone receptors, and tumors such as breast, endometrial, and prostate cancer are at least partially under hormonal control. As

TABLE 3
Relationship Between Progression of Disease in Melanoma Patients and the Growth Rate of the Corresponding Xenografts in Nude Mice

	Patient: Time (months) from appearance of:		Xenograft: Volume-doubling time[a] (days)
Designation	Primary to first relapse	First relapse to death	
LO	3	5	2.0
SES	32	1	3.0
KF	60	10	4.0
GRE	18	13	4.8
EF	18	13	6.3
RV	13	17	7.4
KA	120	16	8.0
NSR	144	14	8.0
MF	3—4	15	8.9
GE	41	32	9.0

[a] Time required to double mean tumor volume (50—100 mm^3); median of values in at least three experiments.

will be discussed below, several tumor lines have been established from these tumor types, the growth and transplantability of which can be clearly hormone dependent.[8,52-56] Recently, sublines of originally hormone-dependent tumors have been established which seem to have escaped hormonal control.[53] Pairs of hormone-dependent and -independent tumors provide excellent tools to study the mechanisms underlying the evolvement of hormone independent growth in the clinic.

C. REPRESENTATIVITY

An overwhelming body of evidence now exists demonstrating that human tumor xenografts generally retain the characteristics of the original tumors in patients. This topic is discussed in several review articles. Here, only one example will be mentioned. We have found a correlation between the progression of the disease in melanoma patients and the growth rate of the corresponding xenografts in nude mice.[6] As demonstrated in Table 3, a relationship seems to be present between the tumor volume-doubling time in the mice and the time from first relapse to death of the patients.

The percentage of the tumors that can be transplanted continuously as xenografts varies between the different tumor types. Breast cancer, testicular cancer, lymphomas, and leukemias are examples of tumors that are difficult to establish in nude mice.[8] However, even for tumors that are regarded as easy to grow as xenografts, the representativity of the tumor panels should be considered before extensive experiments are initiated. Thus, our experience, and that of others, shows that although a high percentage of human melanomas are transplantable into nude mice, l5 to 20% of the samples will either not grow or give tumors that have such a slow growth that they can never be used for experimentation.[4,6-9]

The question about the extent to which xenografts can be considered as truly representative of their parent tumors is partially raised because only tiny pieces of tumor tissue are inoculated into each recipient mouse. The possibility therefore exists that subpopulations of the tumor cells present in the primary tumor may be selected. The chances of such a selection, or induction of new characteristics, might be particularly high in cases where tumors develop in the mice only after a long latency period. The number of reports presenting firm evidence that growth of particular subpopulations has occurred is surprisingly low. For murine tumors, however, ample evidence exists that, during progression, a selection of pre-existing tumor subpopulations may take place.[57] These data cannot be used to validate the conclusion that such selection is the main cause of heterogeneity in human tumors. One might speculate that at least part of the diversity in cell characteristics may not necessarily represent different subpopulations of cells with finite characteristics. It is conceivable that tumor cells may retain the ability of their stem cells to develop various phenotypic characteristics, which will depend on a complex interaction between the tumor and normal cells. Epigenetic alterations representing what has been described as "dynamic heterogeneity"[58] may also take place. An example of changes in tumor characteristics in nude mice may illustrate this point. We attempted to select sublines of the LOX human melanoma by injecting cells into different tissues and by transfer of the developing tumors to recipient mice. In only one case, a selection appeared to take place. Cells injected directly into the brain resulted in a subline with a reduced growth rate, but we were not able to demonstrate that this change was the result of a selection.

VI. TRANSPLANTABILITY OF INDIVIDUAL TUMOR TYPES

Some of the data obtained from the transplantability of various types of human cancers in nude mice will be summarized in this section. The brief outline does not intend to be chronological or comprehensive. Moreover, as discussed earlier, the limitations related to the term "tumor take rate" should be kept in mind.

A. GASTROINTESTINAL TUMORS

The first tumor successfully grown in nude mice was a human colon cancer.[1] Because of this, and because of the clinical importance of gastrointestinal tumors, numerically and biologically, much effort has been put into the establishment of tumor lines of this category. In three different reports, the take rate of colorectal tumors varied between 44 and 72%.[4,8,59] It seems that with the higher stages of colon cancers (Dukes C and D), at least 70% of the transplanted tumors will grow; whereas, the take rate is significantly lower for less advanced cases. Reports on chemosensitivity testing and biological studies in panels of colorectal cancer xenografts are numerous.[59-64]

Gastric cancers show a lower take rate than colorectal tumors.[8] In a study of 49 tumor specimens, Susuki and Sudo[65] found that 16 of 35 primary tumors and 5 of 16 lymph node metastases could be successfully transplanted. Nakatani et al.[66] obtained a 45% take rate with 33 inoculated gastric tumors. Since ascites is a common finding in this group of tumors, it is of interest that ascites models in nude mice may also be established.[67]

B. MELANOMAS

Melanoma tissue was among the tumor types first established as xenografts. In 1976, Povlsen[38] reported the successful take of 14 of 32 melanomas. Of these, 10 of 13 cases of skin metastases, 4 of 6 primaries in the eye, and 0 of 13 skin primaries were found to grow. Notably, only half of the transplantable tumors could be transferred serially. For the purpose of testing the antitumor efficacy of new and established drugs, we and others[34,68-74] established panels of xenograft lines. Melanoma cell lines were grown *in vivo* as s.c. xenografts by several investigators.[7,68,69] The take rate for melanomas is estimated to be approximately 70%, and slightly higher if only tissue from metastases is used.

C. LUNG TUMORS

Most tumors originating from cells in the lungs can be serially transplanted to nude mice. Histological subtypes seem to vary in transplantability, that is, adenocarcinomas and large cell tumors have higher take rates than the squamous cell and, possibly also, the small cell carcinomas.[4,8,37,75-80] However, significant differences exist between laboratories. Mattern et al.[79] reported an overall take rate of 44% in a large study on 255 tumor specimens, with the squamous cell tumors and adenocarcinomas having the highest success rates. In our experience, small cell carcinoma cells obtained from bone marrow are more difficult to grow s.c. than biopsied tissue from solid tumor metastases.

D. BREAST CANCER

Breast cancer represents one of the most common and challenging clinical tumors, but, in spite of all efforts, only a very limited number of tumor lines are available. The take rate is estimated to be about 10%, but xenografts with growth rates and tumor characteristics that permit extensive studies are even more scarce.[8,32,52,81] Most commonly used are tumors developing after the injection of cells from *in vitro* cultures. In 1974, Ozello et al.[82] successfully inoculated BT 20 cells. Sebesteny et al.[83] obtained a take of 3 of 19 tissue specimens obtained from patients and were able to maintain these lines for several passages. Rae-Venter and Reid[84] established 4 permanently transplantable lines out of 32 attempts. Otake[85] described a line that metastasized to regional lymph nodes, frequently observed in female and rarely in male mice. Several studies have addressed the requirements of hormones, such as estrogen, pituitary factors, and antiestrogens for the growth of breast cancers.[52,53]

E. GYNECOLOGICAL CANCER

Whereas ovarian tumors were found to grow relatively well in nude mice,[4,8,86-90] in our experience squamous cell carcinomas of the cervix showed a very low take rate. Kullander et al.[86] obtained a take of 10 of 16 ovarian tumors when small tissue fragments were transplanted, but had a much lower success rate with injection of suspended tumor cells. Hamilton et al.[44] and Massazza et al.[90] established ovarian cancer lines that form ascitic tumors in the mice.

F. HEAD AND NECK TUMORS

In spite of the general tendency for squamous cell carcinomas to be difficult to establish as permanent lines in nude mice,[4,8,91] a significant number of such xenografts from the head and neck region have been reported. In a study of 130 cases, Braakhuis et al.[92] showed that tissue from metastases of undifferentiated tumors had the highest take rates. Wennerberg et al.[93] reported successful establishment of tumor lines in 19% of 85 tissue transplantations.

G. SARCOMAS

The heterogeneous group of sarcomas initially appeared to be of poor transplantability into nude mice. Later, the success rates improved, and many soft tissue sarcoma[94] and osteosarcoma lines[95-98] are now available. We have established several lines, many of which had a very long latency time before progressive growth was observed. With this in mind, the take rate may be estimated as high as 50%. Hajdu et al.[94] observed an even higher percentage of progressive growth in their series of soft tissue sarcomas. Because of the many histological subtypes of this relatively rare malignancy, take rates for each of them are difficult to determine. Whereas rhabdomyosarcomas of children show reasonably good growth in nude mice, liposarcomas may be more difficult to establish, and malignant fibrous histiocytomas have, in our experience, long volume-doubling times.

H. BLADDER, PROSTATE, AND TESTICULAR CANCER

For transitional cell carcinomas, initial growth was achieved in 40% of specimens transplanted by Sufrin et al.[99] and progressive growth in 20 of 53 biopsies (38%) in a study by Kovnat et al.[100] Prostate cancers are difficult to grow as continuous tumor lines, but several individual lines have been described.[101] Testicular tumors, particularly seminomas, show low take rates as xenografts. Tveit et al.[102] obtained growth in only 1 of 19 cases of testicular cancer, whereas Steel et al.[103] reported tumor take in 30% of testicular teratoma specimens transplanted to immunosuppressed mice.

I. RENAL CELL CANCER

Relatively few tumor lines of renal cell cancer have been reported,[42,104,105] and tumors with a nuclear grade 3 or 4 seem to be the easiest to establish. Naito et al.[42] have suggested that orthotopic injection of tumor cells into the kidney improves the growth of renal cell cancer xenografts.

J. BRAIN TUMORS

Bradley et al.[106] obtained a take of 13 of 55 glioma specimens and found a better transplantability for adult than for pediatric cases. Shapiro et al.[107] had very high take rates by injecting brain tumor tissue both directly into the brain and s.c. into nude mice. Gliomas seem easier to grow than medulloblastomas.[108,109] Several neuroblastoma lines have also been described.[110]

K. LYMPHOMAS AND LEUKEMIAS

Epstein et al.[39] obtained a 90% take rate with 11 primary lymphomas after intracranial inoculation of the cells into nude mice. In contrast, only very few lymphomas can be grown s.c. This difference is explained by the blood-brain barrier that renders the brain an immunological sanctuary. The s.c. take rate may be improved in splenectomized, irradiated animals[111,112] and Igarashi et al.[113] reported that 6 of 22 (27%) of transplanted lymphomas could be established as xenografts in such mice. The successful lines were all derived from diffuse lymphomas of advanced stage in patients.

In 1977, Ueyama et al.[67] reported a myeloblastic Philadelphia chromosome-positive tumor that could be serially transferred to nude mice. Nilsson et al.[114] found that some Burkitt lines, but not other EBV-positive cell lines, could be established as s.c. tumors. A B cell line reported by Leonard et al.[115] was also of the Burkitt type. Several T cell leukemia lines in nude mice have been reported.[116-118] Recently, an acute lymphoblastic leukemia cell line was established after systemic inoculation into scid mice.[23]

VII. CONCLUSIONS

A wide variety of host, tumor, and methodological factors determine the extent to which human tumors can be serially transplanted to nude mice. Obviously, the residual immune response of nu/nu animals has a variable impact on the transplantability of different tumors and tumor types. Mice with immune defects in the function of B cells, NK cells, or both, as well as scid mice, may in some cases be better hosts for transplants than regular nude mice. Manipulation of the immune system with irradiation and drugs may also be used to increase the tumor take (Chapter 2).

There is ample evidence for the importance of using healthy animals for transplantation experiments. Tumor take is also dependent on whether pieces of tumor tissue or cell suspensions are used, and on the site and route of tumor implantation. The local microenvironment provided for the tumor cells in the various tissues is important, and it is possible that orthotopic transplantation in some cases may increase tumor growth and metastasis formation.

In general, tumor tissue obtained from a metastatic lesion of a poorly differentiated cancer gives the best chance of a tumor take. The growth fraction of a tumor in that patient and in the tissue as a xenograft is one indicator of the probability for successful transplantation. Viable tissue in the periphery of the xenograft should be used for serial transfer. Some tumor types known to be hormone dependent in man show the same characteristics as xenografts. In such cases, hormone supplementation may be necessary to support tumor growth in the mice.

With or without the addition of estrogens, breast cancer is particularly difficult to establish as xenografts. This is also the case for leukemias and lymphomas, testicular, and prostate cancer. Even in the tumor types where the great majority of transplanted specimens will grow, the xenografts may not represent a clinically important subgroup of tumors.

The knowledge accrued with time concerning the factors limiting tumor transplantability seems to have resulted in improved tumor take rates. Further refinement and innovative manipulations may give additional progress in the years to come.

REFERENCES

1. **Rygaard, J. and Povlsen, C. O.,** Heterotransplantation of a human malignant tumor to "nude" mice, *Acta. Path. Microbiol. Scand.*, 77, 758, 1969.
2. **Povlsen, C. O.,** Heterotransplants of human tumors in nude mice, *Antibiotics Chemother.*, 28, 15, 1980.
3. **Jewett, M. A. S.,** Testis carcinoma: transplantation into nude mice, *Natl. Cancer Inst. Monogr.*, 49, 65, 1970.
4. **Shimosato, Y., Kameya, T., Nagai, K., Hiroshashi, S., Koide, T., Hayashi, H., and Nomura, T.,** Transplantation of human tumors in nude mice, *J. Natl. Cancer Inst.*, 56, 1251, 1976.
5. **Fogh, J., Tiso, J., Orfeo, T., Sharkey, F. E., Daniels, W. P., and Fogh, J. M.,** Thirty-four lines of six human tumor categories established in nude mice, *J. Natl. Cancer Inst.*, 64, 745, 1980.
6. **Fodstad, Ø.,** Representativity of xenografts for clinical cancer. Tumour and host characteristics as variables of tumor take rate, in *Human Tumour Xenografts in Anticancer Drug Development*, Winograd, B., Pinedo, H. M., and Peckham, M. J., Eds., Springer-Verlag, Heidelberg, 1988, 15.
7. **Giovanella, B. C., Stehlin, J. S., Fogh, J., and Sharkey, F. E.,** Serial transplantation of human malignant tumors in nude mice and their use in experimental chemotherapy, in *Proc. Symp. Use of Athymic (Nude) Mice in Cancer Res.*, Houchens, D. P. and Ovejera, A. A., Eds., Gustav Fischer, New York, 1978, 163.
8. **Sharkey, F. E. and Fogh, J.,** Considerations in the use of nude mice for cancer research, *Cancer Metast. Rev.*, 3, 341, 1984.
9. **Fogh, J., Fogh, J. M., and Orfeo, T.,** One hundred and twenty-seven cultured human tumor cell lines producing tumors in nude mice, *J. Natl. Cancer Inst.*, 59, 221, 1977.
10. **Fodstad, Ø., Brøgger, A., Bruland, Ø., Solheim, Ø. P., Nesland, J. M., and Pihl, A.,** Characteristics of a cell line established from a patient with multiple osteosarcoma, appearing 13 years after treatment for bilateral retinoblastoma, *Int. J. Cancer*, 77, 1247, 1986.
11. **Fargion, S., Carney, D., Mulshine, J., Rosen, S., Bunn, P., Jewett, P., Cuttitta, F., Gazdar, A., and Minna, J.,** Heterogeneity of cell surface antigen expression of human small cell lung cancer detected by monoclonal antibodies, *Cancer Res.*, 46, 2633, 1986.
12. **Van Helden, P. D., Wiid, I. J. F., Albrecht, C. F., Theron, E., Thornley, A. L., and Hoal-van Helden, E. G.,** Cross-contamination of human esophageal squamous carcinoma cell lines detected by DNA fingerprint analysis, *Cancer Res.*, 48, 5660, 1988.
13. **Thacker, J., Webb, M. B. T., and Debenham, P. G.,** Fingerprinting cell lines: use of human hypervariable DNA probes to characterize mammalian cell cultures, *Somatic Cell Mol. Genet.*, 14, 519, 1988.
14. **Hansen, C. T.,** The nude gene and its effects, in *The Nude Mouse in Experimental and Clinical Research*, Fogh, J. and Giovanella, B. C., Eds., Academic Press, London, 1978, 1.
15. **Kindred, B.,** Nude mice in immunology, *Prog. Allergy*, 26, 137, 1979.
16. **Fodstad, Ø., Hansen, C. T., Cannon, G. B., and Boyd, M. R.,** Immune characteristics of the beige-nude mouse, a model for studying immune surveillance, *Scand. J. Immunol.*, 20, 267, 1984.
17. **Azar, H. A., Hansen, C. T., and Costa, J.,** N:NIH(S)II-nu/nu mice with combined immunodeficiency. A new model for human tumor heterotransplantation, *J. Natl. Cancer Inst.*, 65, 421, 1980.
18. **Fodstad, Ø., Hansen, C. T., Cannon, G. B., Statham, C. N., Lichtenstein, G., and Boyd, M. R.,** Lack of correlation between natural killer activity and tumor growth control in nude mice with different immune defects, *Cancer Res.*, 44, 4403, 1984.
19. **Kärre, K., Klein, G. O., Kiessling, R., Klein, G., and Roder, J. C.,** *In vitro* NK-activity and *in vivo* resistance to leukemia: studies of beige, beige/nude and wild-type hosts on C57BL background, *Int. J. Cancer*, 26, 789, 1980.
20. **Roder, J. C. and Duwe, A.,** The beige mutation in the mouse selectively impairs natural killer cell function, *Nature (London)*, 278, 451, 1979.
21. **Talmadge, J. E., Meyers, K. M., Prieur, D. J., and Starkey, J. R.,** Role of NK cells in tumour growth and metastasis in beige mice, *Nature (London)*, 284, 622, 1980.
22. **Hanna, N.,** Role of natural killer cells in control of cancer metastasis, *Cancer Metast. Rev.*, 1, 45, 1982.
23. **Kamel-Reid, S., Letarte, M., Sirard, C., Doedens, M., Grunberger, T., Fulop, G., Freedman, M. H., Phillips, R. A., and Dick, J. E.,** A model of human acute lymphoblastic leukemia in immune-deficient SCID mice, *Science*, 246, 1597, 1989.
24. **Phillips, R. A., Jewett, M. A. S., and Gallie, B. L.,** Growth of human tumors in immune-deficient scid mice and nude mice, *Curr. Top. Microbiol. Immunol.*, 152, 259, 1989.
25. **Hanna, N. and Fidler, I. J.,** Expression of metastatic potential of allogeneic and xenogeneic neoplasms in young nude mice, *Cancer Res.*, 41, 438, 1981.
26. **Fidler, I. J.,** Rationale and methods for the use of nude mice to study the biology and therapy of human cancer metastasis, *Cancer Metast. Rev.*, 5, 29, 1986.
27. **MacDonald, H. R.,** Phenotypic and functional characteristics of "T"-like cells in nude mice, *Exp. Cell Biol.*, 50, 305, 1982.

28. **Sordat, B., Merenda, C., and Carrel, S.,** Invasive growth and dissemination of human solid tumors and malignant cell lines grafted subcutaneously to newborn nude mice, in *Proc. 2nd Int. Workshop on Nude Mice,* Nomura, T., Ohsawa, N., Tamaoki, N., and Fujiwara, K., Eds., University of Tokyo Press, 1977, 313.
29. **Sordat, B. C. M., Ueyama, Y., and Fogh, J.,** Metastasis of tumor xenografts in the nude mouse, in *The Nude Mouse in Experimental and Clinical Research,* Fogh, J. and Giovanella, B. C., Eds., Academic Press, New York, 1982, 95.
30. **Fogh, J., Orfeo, T., Tiso, J., Sharkey, F. E., Fogh, J. M., and Daniels, W. P.,** Twenty-three new human tumor lines established in nude mice, *Exp. Cell Biol.,* 48, 229, 1980.
31. **Kovnat, A., Buick, R. N., Connolly, J. G., Jewett, M. A., Keresteci, A. G., and Tannock, I. F.,** Comparison of growth of human bladder cancer in tissue culture or as xenografts with clinical and pathological characteristics, *Cancer Res.,* 44, 2530, 1984.
32. **White, A. C., Levy, J. A., and McGrath, C. M.,** Site-selective growth of a hormone-responsive human breast carcinoma in athymic mice, *Cancer Res.,* 906, 1982.
33. **Kyriazis, A. A. and Kyriazis, A. P.,** Preferential sites of growth of human tumors in nude mice following subcutaneous transplantation, *Cancer Res.,* 40, 4509, 1980.
34. **Fodstad, Ø., Aamdal, S., Pihl, A., and Boyd, M. R.,** Activity of mitozolomide (NSC 353451), a new imidazotetrazine, against xenografts from human melanomas, sarcomas, and lung and colon carcinomas, *Cancer Res.,* 45, 1778, 1985.
35. **Fisher, B., Gundez, N., Coyle, J., Rudock, C., and Saffer, E.,** Presence of a growth-stimulating factor in serum following primary tumor removal in mice, *Cancer Res.,* 49, 1996, 1989.
36. **Shoemaker, R. H., McLemore, T. L., Abbott, B. J., Fine, D. L., Gorelik, E., Mayo, J. G., Fodstad, Ø., and Boyd, M. R.,** Human tumor xenograft models for use with an *in vitro*-based, disease-oriented antitumor drug screening program, in *Human Tumour Xenografts in Anticancer Drug Development,* Winograd, B., Pinedo, H. M., and Peckham, M. J., Eds., Springer-Verlag, Heidelberg, 1988, 115.
37. **McLemore, T. L., Eggleston, J. C., Shoemaker, R. H., Abbott, B. J., Bohman, M. E., Liu, M. C., Fine, D. L., Mayo, J. G., and Boyd, M. R.,** Comparison of intrapulmonary, percutaneous intrathoracic, and subcutaneous models for the propagation of human pulmonary and nonpulmonary cancer cell lines in athymic nude mice, *Cancer Res.,* 48, 2880, 1988.
38. **Povlsen, C. O.,** Heterotransplantation of human malignant melanomas to the mouse mutant nude, *Acta Path. Microbiol. Scand.,* 84, 9, 1976.
39. **Epstein, A. L., Herman, M. M., Kim, H., Dorfmann, R. F., and Kaplan, H. S.,** Biology of the human malignant lymphomas. III. Intracranial heterotransplantation in the nude, athymic, mouse, *Cancer,* 37, 2159, 1976.
40. **Fodstad, Ø., Aamdal, S., McMenamin, M., Nesland, J. M., and Pihl, A.,** A useful new experimental metastasis model in athymic, nude mice. Characteristics of the LOX human malignant melanoma, *Int. J. Cancer,* 41, 442, 1988.
41. **Giavazzi, R., Campbell, D. E., Jessup, J. M., Cleary, K., and Fidler, I. J.,** Metastatic behavior of tumor cells isolated from primary and metastatic human colorectal carcinomas implanted into different sites in nude mice, *Cancer Res.,* 46, 1928, 1986.
42. **Naito, S., von Eschenbach, A. C., Giavazzi, R., and Fidler, I. J.,** Growth and metastasis of tumor cells implanted into different organs of nude mice, *Cancer Res.,* 46, 4109, 1986.
43. **Kjønniksen, I., Nesland, J. M., Pihl, A., and Fodstad, Ø.,** A nude rat model for studying metastasis of human tumor cells to bone and bone marrow, *J. Natl. Cancer Inst.,* 82, 408, 1990.
44. **Hamilton, T. C., Young, R. C., Louie, K. G., Behrens, B. C., McKoy, W. M., Grotzinger, K. R., and Ozols, R. F.,** Characterization of a xenograft model of human ovarian carcinoma which produces ascites and intra-abdominal carcinomatosis in mice, *Cancer Res.,* 44, 5286, 1984.
45. **Kozlowski, J. M., Fidler, I. J., Campbell, D., Xu, X. L., Kaighn, M. E., and Hart, I. R.,** Metastatic behavior of human tumor cell lines grown in the nude mouse, *Cancer Res.,* 44, 3522, 1984.
46. **Aamdal, S., Fodstad, Ø., Nesland, J. M., and Pihl, A.,** Characteristics of human tumour xenografts transplanted under the renal capsule of immunocompetent mice, *Br. J. Cancer,* 51, 347, 1985.
47. **Fodstad, Ø., Kjønniksen, I., Aamdal, S., Nesland, J. M., Boyd, M. R., and Pihl, A.,** Extrapulmonary, tissue-specific metastasis formation in nude mice injected with FEMX-I human melanoma cells, *Cancer Res.,* 48, 4382, 1988.
48. **Fodstad, Ø., Tveit, K. M., Rofstad, E. K., and Pihl, A.,** Spontaneous alteration in growth rate of two human melanoma xenografts. Concurrent change in chemosensitivity, *Eur. J. Cancer Clin. Oncol.,* 19, 1175, 1983.
49. **Rofstad, E. K., Fodstad, Ø., and Lindmo, T.,** Growth characteristics of human melanoma xenografts, *Cell Tissue Kinet.,* 15, 545, 1982.
50. **Steel, G. G.,** Species-dependent growth patterns for mammalian neoplasms, *Cell Tissue Kinet.,* 13, 451, 1980.
51. **Bennett, D. C., Dexter, T. J., Ormerod, E. J., and Hart, I. R.,** Increased experimental metastatic capacity of a murine melanoma following induction of differentiation, *Cancer Res.,* 3239, 1986.

52. **Osborne, C. K., Hobbs, K., and Clark, G. M.,** Effect of estrogens and antiestrogens on growth of human breast cancer cells in athymic nude mice, *Cancer Res.*, 45, 584, 1985.
53. **Clarke, R., Brünner, N., Katzenellenbogen, B. S., Thompson, E. W., Norman, M. J., Koppi, C., Paik, S., Lippman, M. C., and Dickson, R. B.,** Progression of human breast cancer cells from hormone-dependent to hormone-independent growth both *in vitro* and *in vivo, Proc. Natl. Acad. Sci. U.S.A.*, 86, 3649, 1989.
54. **Satyaswaroop, P. G., Zaino, R. J., and Mortel, R.,** Human endometrial adenocarcinoma transplanted into nude mice: growth regulation by estradiol, *Science*, 219, 58, 1983.
55. **Kameya, T., Shimosato, Y., Hayashi, H., and Tsumuraya, M.,** Growth and differentiation of hormone-producing human tumors in nude mice, in *Proc. 2nd Int. Workshop on Nude Mice*, Nomura, T., Ohsawa, N., Tamaoki, N., and Fujiwara, K., Eds., University of Tokyo Press, Tokyo, 1977, 405.
56. **Shin, S., Reid, L. C., Colburn, P. C., Kadish, A. S., and Fleischer, N.,** Development of transplantable functional endocrine tumor lines in nude mice, in *Proc. 2nd Int. Workshop on Nude Mice*, Nomura, T., Ohsawa, N., Tamaoki, N., and Fujiwara, K., Eds., University of Tokyo Press, Tokyo, 1977, 375.
57. **Fidler, I. J.,** The evolution of biological heterogeneity in metastatic neoplasms, in *Cancer Invasion and Metastasis: Biologic and Therapeutic Aspects*, Nicolson, G. L. and Milas, L., Eds., Raven Press, New York, 1984, 5.
58. **Harris, J. F., Chambers, A. F., Hill, R. P., and Ling, V.,** Metastatic variants are generated spontaneously at a high rate in mouse KHT tumor, *Proc. Natl. Acad. Sci. U.S.A.*, 79, 5547, 1982.
59. **Steel, G. G., Courtenay, V. D., and Peckham, M. J.,** The response to chemotherapy of a variety of human tumor xenografts, *Br. J. Cancer*, 47, 1, 1983.
60. **Povlsen, C. O. and Rygaard, J.,** Heterotransplantation of human adenocarcinomas of the colon and rectum to the mouse mutant nude. A study of nine consecutive transplantations, *Acta Path. Microbiol. Scand., (Sect. A)*, 79, 159, 1971.
61. **Fodstad, Ø.,** Human gastrointestinal cancer grown in nude mice used to test new chemotherapeutic agents, in *Front. Gastrointest. Res.*, Vol. V, van der Reis, L., Ed., Karger, Basel, 1979, 71.
62. **Carrel, S., Sordat, B., and Merenda, C.,** Establishment of a cell line (Co-115) from a human colon carcinoma transplanted into nude mice, *Cancer Res.*, 36, 978, 1976.
63. **Jessup, J. M., Giavazzi, R., Campbell, K., Cleary, K., Morikawa, K., and Fidler, I. J.,** Growth potential of human colorectal carcinomas in nude mice: association with the preoperative serum concentration of carcinoembryonic antigen in patients, *Cancer Res.*, 48, 1689, 1988.
64. **Sordat, B. and Wang, W. R.,** Human colorectal tumor xenografts in nude mice: expression of malignancy, *Behring Inst. Mitt.*, 74, 291, 1984.
65. **Suzuki, K. and Sudo, K.,** Heterotransplantation of human stomach carcinoma in nude mice, *Gann*, 70, 109, 1979.
66. **Nakatani, K., Takahashi, S., Shiratori, T., and Konishi, Y.,** Heterotransplantation of human gastric carcinomas into nude mice, *World J. Surg.*, 3, 641, 1979.
67. **Ueyama, Y., Kondo, Y., Tamaoki, N., and Ohsawa, N.,** Ascites form of a human cancer serially transplantable in nude mice, *J. Natl. Cancer Inst.*, 57, 965, 1976.
68. **Kanzaki, T., Hashimoto, K., and Bath, D. W.,** Heterotransplantation of human malignant melanoma cell lines in athymic nude mice, *J. Natl. Cancer Inst.*, 62, 1151, 1979.
69. **Tropé, C., Johnsson, J. E., Alm, P., Landberg, T., Olsson, H., and Wennerberg, J.,** Human malignant melanoma heterotransplanted to nude mice, *Neoplasma*, 28, 585, 1981.
70. **Tveit, K. M., Fodstad, Ø., Brøgger, A., and Olsnes, S.,** A human melanoma cell line established from xenograft in athymic mice, *Br. J. Cancer*, 41, 724, 1980.
71. **Fodstad, Ø., Olsnes, S., and Pihl, A.,** Effect of abrin on human tumors in nude mice, in *Proc. 2nd Int. Workshop on Nude Mice*, Nomura, T., Ohsawa, N., Tamaoki, N., and Fujiwara, K., Eds., University of Tokyo Press, Tokyo, 1977, 543.
72. **Bellet, R. E., Danna, V., Mastrangelo, M. J., and Berd, D.,** Evaluation of a "nude" mouse-human tumor panel as a predictive secondary screen for cancer chemotherapeutic agents, *J. Natl. Cancer Inst.*, 63, 1185, 1979.
73. **Fodstad, Ø., Aass, N., and Pihl, A.,** Response to chemotherapy of human malignant melanoma xenografts in athymic nude mice, *Int. J. Cancer*, 25, 453, 1980.
74. **Osieka, R.,** Preclinical drug evaluation with human melanoma xenografts, in *Human Tumour Xenografts in Anticancer Drug Development*, Winograd, B., Pinedo, H. M., and Peckham, M. J., Eds., Springer-Verlag, Heidelberg, 1988, 41.
75. **Fiebig, H. H., Neumann, H. A., Heinz, H., Koch, H., Kaiserm, D., and Arnold, H.,** Development of three human small cell lung cancer models in nude mice, *Rec. Res. Cancer Res.*, 97, 77, 1985.
76. **Shorthouse, A. J., Peckham, M. J., Smyth, J. F., and Steel, G. G.,** The therapeutic response of bronchial carcinoma xenografts: a direct patient-xenograft comparison, *Br. J. Cancer*, 41, 142, 1980.
77. **Cuttitta, F., Carney, D. N., Mulshine, J., Moody, T. W., Fedorko, J., Fischler, A., and Minna, J. D.,** Bombesin-like peptides can function as autocrine growth factors in human small cell lung cancer, *Nature (London)*, 316, 823, 1985.

78. **Duchesne, G. M., Eady, J. J., Peacock, J. H., and Pera, M. F.,** A panel of human lung carcinoma lines: establishment, properties and common characteristics, *Br. J. Cancer,* 56, 287, 1987.
79. **Mattern, J., Jäger, S., Sonka, J., Wayss, K., and Volm, M.,** Growth of human bronchial carcinomas in nude mice, *Br. J. Cancer,* 51, 195, 1985.
80. **Shorthouse, A. J., Smyth, J. F., Steel, G. G., Ellison, M., Mills, J., and Peckham, M. J.,** The human tumour xenograft — a valid model in experimental chemotherapy?, *Br. J. Surg.,* 76, 715, 1980.
81. **Bailey, M. J., Gazet, J. C., and Peckham, M. J.,** Human breast cancer xenografts in immune-suppressed mice, *Br. J. Cancer,* 42, 524, 1980.
82. **Ozzello, L., Sordat, B., Merenda, C., Carrel, S., Hurlimann, J., and Mach, J. P.,** Transplantation of a human mammary carcinoma cell line (BT 20) into nude mice, *J. Natl. Cancer Inst.,* 52, 1669, 1974.
83. **Sebesteny, A., Taylor-Papadimitriou, J., Ceriani, R., Millis, R., Schmitt, C., and Trevan, D.,** Primary human breast carcinomas transplantable in the nude mouse, *J. Natl. Cancer Inst.,* 63, 1331, 1979.
84. **Rae-Venter, B. and Reid, L. M.,** Growth of human breast carcinomas in nude mice and subsequent establishment in tissue culture, *Cancer Res.,* 40, 95, 1980.
85. **Otake, Y.,** Establishment of a metastasizing human breast cancer cell line in nude mice, *Fukushima J. Med. Sci.,* 27, 53, 1980.
86. **Kullander, S., Rausing A., and Tropé, C.,** Human ovarian tumours heterotransplanted to "nude" mice, *Acta Obstet. Gynecol. Scand.,* 57, 149, 1978.
87. **Kleine, W.,** Growth of malignant gynecological tumors xenotransplanted into thymusaplastic nu/nu mice, *Gynecol. Oncol.,* 24, 286, 1986.
88. **Nauta, M. M., Boven, E., Schlüper, H. M. M., Erkelens, C. A. M., and Pinedo, H. M.,** Enhanced transplantability of human ovarian cancer lines in cyclophosphamide-pretreated nude mice, *Br. J. Cancer,* 54, 331, 1986.
89. **Boven, E., Nauta, M. M., Schlüper, H. M. M., Elferink, F., Van der Vijgh, W. J. F., and Pinedo, H. M.,** Secondary screening of platinum compounds in human ovarian cancer xenografts in nude mice, *Eur. J. Cancer Clin. Oncol.,* 21, 1253, 1985.
90. **Massazza, G., Tomasoni, A., Lucchini, V., Allavena, P., Erba, E., Colombo, N., Mantovani, A., D'Incalci, M., Mangioni, C., and Giavazzi, R.,** Intraperitoneal and subcutaneous xenografts of human ovarian carcinoma in nude mice and their potential in experimental therapy, *Int. J. Cancer,* 44, 494, 1989.
91. **Meck, R. A., Ingram, M., Meck, J. M., McCullough, J. L., Wu, M. C., and Yunis, A. A.,** Establishment and cell cycle kinetics of a human squamous cell carcinoma in nude mice and *in vitro, Cancer Res.,* 41, 1076, 1981.
92. **Braakhuis, B. J. M., Sneeuwloper, G., and Snow, G. B.,** The potential of the nude mouse xenograft model for the study of head and neck cancer, *Arch. Otorhinolaryngol.,* 239, 69, 1984.
93. **Wennerberg, J., Willén, R., Biörklund, and Tropé, C.,** Histopathological characteristics predictive for growth of squamous cell carcinomas of the head and neck heterotransplanted to nude mice, *Anticancer Res.,* 6, 1165, 1986.
94. **Hajdu, S. I., Lemos, L. B., Kozakewich, H., Helson, L., and Beattie, E. J.,** Growth pattern and differentiation of human soft tissue sarcomas in nude mice, *Cancer,* 47, 90, 1981.
95. **Meyer, W. H., Houghton, J. A., Houghton, P. J., Webber, B. L., Douglas, E. C., and Look, A. T.,** Development and characterization of pediatric osteosarcoma xenografts, *Cancer Res.,* 50, 2781, 1990.
96. **Kuroki, M.,** A new human osteosarcoma strain (Os-Beppu) with high rate of pulmonary metastasis in nude mice, *Jpn. J. Clin. Oncol.,* 14, 49, 1984.
97. **Nakamura, T., Sakahara, H., Hosoi, S., Yamamuro, T., Higashi, S., Mikawa, H., Endo, K., and Toyama, S.,** *In vivo* radiolocalization of antiosteogenic sarcoma monoclonal antibodies in osteogenic sarcoma xenografts, *Cancer Res.,* 44, 2078, 1984.
98. **Bruland, Ø., Fodstad, Ø., Skretting, A., and Pihl, A.,** Selective radiolocalization of two radiolabelled antisarcoma monoclonal antibodies in human osteosarcoma xenografts, *Br. J. Cancer,* 56, 21, 1987.
99. **Sufrin, G., McGarry, M. P., Sandberg, A. A., and Murphy, G. P.,** Heterotransplantation of human transitional cell carcinoma in athymic mice, *J. Urol.,* 121, 159, 1979.
100. **Kovnat, A., Armitage, M., and Tannock, I.,** Xenografts of human bladder cancer in immune-deprived mice, *Cancer Res.,* 42, 3696, 1982.
101. **Schroeder, F. H.,** Heterotransplantation of human prostate carcinoma into nude mice, *Natl. Cancer Inst. Monogr.,* 49, 71, 1978.
102. **Tveit, K. M., Fodstad, Ø., Brøgger, A., and Olsnes, S.,** Human embryonal carcinoma grown in athymic mice and *in vitro, Cancer Res.,* 40, 949, 1980.
103. **Steel, G. G., Courtenay, V. D., and Peckham, M. J.,** The response to chemotherapy of a variety of human tumour xenografts, *Br. J. Cancer,* 47, 1, 1983.
104. **Otto, U., Köllermann, M. W., Klöppel, W., Dimigen, J., Linden, W., and Rüdiger, H.,** Transplantation von menschlichem Nierenadenokarzinomgewebe auf die nackte Maus, *Urol. Int.,* 36, 110, 1981.
105. **Knöfel, W. T., Otto, U., Baisch, H., and Klöppel, G.,** Stability of human renal cell carcinomas during long term serial transplantation into nude mice: histopathology, nuclear grade, mitotic rate, and DNA content in thirty tumors, *Cancer Res.,* 47, 221, 1987.

106. **Bradley, N. J., Bloom, H. J. G., Davies, A. J. S., and Swift, S. M.,** Growth of human gliomas in immunodeficient mice: a possible model for pre-clinical therapy studies, *Br. J. Cancer,* 38, 263, 1978.
107. **Shapiro, W. R., Basler, G. A., Chernik, N. L., and Posner, J. B.,** Human brain tumor transplantation into nude mice, *J. Natl. Cancer Inst.,* 62, 447, 1979.
108. **Schold, S. C., Bullard, D. E., Bigner, S. H., Jones, T. R., and Bigner, D. D.,** Growth, morphology, and serial transplantation of anaplastic human gliomas in athymic mice, *J. Neurooncol.,* 1, 5, 1983.
109. **Bullard, D. E., Schold, S. C., Bigner, S. H., and Bigner, D. D.,** Growth and chemotherapeutic response in athymic mice of tumors arising from human glioma-derived cell lines, *J. Neuropathol. Exp. Neurol.,* 40, 410, 1981.
110. **Jones, D. H., Goldman, A., Gordon, I., Pritchard, J., Gregory, B. J., and Kemshead, J. T.,** Therapeutic application of a radiolabeled monoclonal antibody in nude mice xenografted with human neuroblastoma: tumoricidal effects and distribution studies, *Int. J. Cancer,* 35, 715, 1985.
111. **Watanabe, S., Shimosato, Y., Kuroki, M., Sato, Y., and Nakijima, T.,** Transplantability of human lymphoid cell line, lymphoma and leukemia in splenectomized and/or irradiated nude mice, *Cancer Res.,* 40, 2588, 1980.
112. **Arione, R., Jemma, C., Forni, M., Marchese, C., Benetton, G., Giubellino, C., Modesti, A., Martinotti, M. G., Cordero di Montezemolo, L., Musso, T., and Forni, G.,** A new childhood T-cell lymphoma established in nude mice and *in vitro, Cancer Res.,* 48, 1312, 1988.
113. **Igarashi, K., Oka, K., and Miyamoto, T.,** Human non-Hodgkin's malignant lymphomas serially transplanted in nude mice conditioned with whole-body irradiation, *Br. J. Cancer,* 59, 356, 1989.
114. **Nilsson, K., Giovanella, B. C., Stehlin, J. S., and Klein, G.,** Tumorigenicity of human hematopoietic cell lines in athymic nude mice, *Int. J. Cancer,* 19, 337, 1977.
115. **Leonard, J. E., Johnson, D. E., Felsen, R. B., Tanney, L. E., Royston, I., and Dillman, R. O.,** Establishment of a human B-cell tumor in athymic mice, *Cancer Res.,* 47, 2899, 1987.
116. **Ziegler, H. W., Frizzera, G., and Bach, F. H.,** Successful transplantation of a human leukemia cell line into nude mice: conditions optimizing graft acceptance, *J. Natl. Cancer Inst.,* 68, 15, 1982.
117. **Houghton, P. J., Mirro, J., Goorha, R. M., Raimondi, S. C., Fridland, A., and Houghton, J. A.,** Growth and differentiation of a human T-cell leukemia cell line, CCRF-CEM, grafted in mice, *Cancer Res.,* 49, 7124, 1989.
118. **Miyoshi, I., Ota, T., Hiraki, S., Sumida, M., Tanaka, T., and Kimura, I.,** Serial transplantation of a human T-cell acute lymphoblastic leukemia line into nude mice, *Br. J. Cancer,* 44, 124, 1981.

Chapter 7

ALTERNATIVE SITES OF TUMOR DEVELOPMENT

Epie Boven

TABLE OF CONTENTS

I. INTRODUCTION

The major part of this volume is dedicated to the use of the nude mouse as the recipient of human tumor tissue implanted s.c. into the flank. Most solid tumor types will be accepted and eventually grow as non-invasive nodules that are easily visible (Chapter 6). For experimental therapy, the model is amenable to precise measurement of drug effects, has analogy to clinically utilized variables of response, and allows systemic drug administration via either the i.v. or i.p. routes. For large-scale drug screening, the s.c. model is not very practical because of the long duration of each experiment and the multiple individual tumor measurements required. Therefore, short-term xenograft models have been developed, such as the subrenal capsule assay. Although the validity of these models for drug-screening has not yet been determined in all its aspects, this chapter will focus on some of the possiblities explored in the past decade.

As has been pointed out already in Chapter 6, transplantability of tumors is dependent on a number of factors, such as the interaction between the tumor cells and their microenvironment. Data have been presented indicating that the tumor take rate may be increased after "orthotopic" injection of the cells, for example, when lung cancer cells are injected intrabronchially, and cells of renal cell cancer into the kidney. This characteristic has been employed to study metastasis formation, which is more likely to occur after orthotopic transplantation than from s.c. implants (Chapter 10). For particular tumors however, growth is absent after s.c. transplantation, such as leukemias and lymphomas. These tumors may grow after injection into the brain, which may simply reflect that the brain represents an immunologically secluded or privileged area. In addition, tumor systems representing central nervous system malignancy have been designed to study the effects of antitumor agents, because various drugs are known to penetrate poorly into the brain.

In this chapter, alternative routes of tumor development are being discussed, either for a short-term drug screening assay or because the tumor type studied will not be accepted by any route except the brain.

II. THE SUBRENAL CAPSULE ASSAY

In 1978, Bogden, et al.[1] introduced a rapid screening method for testing chemotherapeutic agents against human tumor tissue implanted as small fragments under the kidney capsule of nude mice. The subrenal capsule (SRC) area provides a rich vascular bed for nutrients as well as drug delivery, enabling studies in growth as well as in inhibition of growth. The method was developed because xenografts in nude mice provide an *in vivo* system that more closely resembles the human situation than murine tumor models. In this respect, the s.c. grown xenograft is relatively useless for large-scale drug screening as the take rate is not 100% and the model is time-consuming and expensive. The rationale of the SRC assay was mentioned to be the circumvention of the residual immunocompetence in the nude mouse, the use of fewer animals, a shorter duration of the assay, a realistic screen for quickly ranking compounds in order of their activity for subsequent in-depth testing. Since then, the SRC assay has undergone frequent changes in methodology to optimize drug screening results. Also, the SRC assay has been the subject of careful analysis of its predictive value for clinically active antitumor agents.

A. METHODOLOGY

The approach to the development of a rapid screening assay by Bogden et al.[1] was based on the assumption that measurable changes in tumor size were the net effect of cell division and cell death. In this respect, the initial graft size, determined at the time of implantation, compared to a final size at the termination of the assay, provides an objective measurement of growth. Precise measurements of the xenograft *in situ* made under magnification would permit a more sensitive detection of changes in graft size and quantification of such changes into units of measurements.

Various implantation sites were examined, of which the area under the kidney capsule with its rich vascular supply was most appropriate.

The initial graft size of 1-mm^3 fragments of tumor tissue appeared small enough for implantation and to permit permeation of essential nutrients and distribution of test drugs throughout the tissue. At first, a period of 11 d was chosen because it was long enough to sustain measurable growth or drug-induced regression. Measurements were carried out with a stereoscopic microscope equipped with a micrometer disk inserted into one eyepiece.

In short, the SRC assay protocol was as follows. At day 0, xenografts were implanted, *in situ* graft sizes determined, and the inital body weights taken. At day 1, drug treatment was initiated and continued once daily 10×, but other schedules could also be used. At day 11, animals were sacrificed, their final body weights taken, kidneys excised, and *in situ* graft sizes determined.

In the early experiments with the SRC assay, tissue fragments were obtained from human tumor xenografts grown s.c. in nude mice. The average change in tumor size (product of 2 diameters/2) in the 11-day period was 5.8 mm for colon tumor CX-1, 12.2 mm for lung tumor LX-1, and 17.0 mm for breast tumor MX-1. The tumors were used to study the efficacy of various antitumor agents, and it was shown that these drugs could be ranked in order of activity. In this way, new compounds of greatest clinical potential were to be identified for advanced testing, that is, against well-established s.c. grown xenografts and in phase I clinical trials.

Bogden et al.[2] have extensively analyzed the SRC assay in nude mice as the method became a key test system in the Developmental Therapeutics Program of the Division of Cancer Treatment of the National Cancer Institute (NCI). The three well-known tumor models, CX-1, LX-1, and MX-1, could thus be used in both the conventional s.c. site and in the SRC assay. Growth curves of SRC tissue implants showed a typical Gompertzian pattern and progression inevitably resulted in death of the host. At that time, multiple lymph node metastases and hematogenous metastases to various organs could be found. In the presence of weight loss due to caloric restriction, growth was inhibited. Therefore, for routine drug testing, a body weight loss greater than 30% at day 11 was not to be accepted as a representative test. It was found that the profile of tumor response in the SRC assay duplicated the responses of s.c. implanted tumors from which the SRC implants were excised. As the SRC assay appeared slightly more sensitive to drugs, a compound was only considered active if a test/control value ≤10% was reached. For this ratio, tumor weight in milligrams was measured by the formula (length × width^2)/2 and the weight change between day 11 and day 1 taken to calculate T/C%.

B. EVOLUTION OF THE MODEL

With the basic techniques of the SRC assay proven feasible in the nude mouse, the question arose whether the system could be applied in normal, immunocompetent mice with shortening of the assay time-frame. Therefore Bogden et al.[3] investigated a 6-day period as a safe time frame, since the humoral response to a primary antigenic stimulus in a normal animal generally evolves in 7 to 10 d and the cell-mediated response to an initial allograft becomes evident in 9 to 12 d.

In order to complete the SRC assay within 6 d, the growth potential of human tumor tissue in normal mice had to be sufficient and the sensitivity and precision of the measurements adequate to quantitate the small changes in tumor size.[2] Various human tumor xenografts were studied and examples of growth were 3.4 mm for CX-1, 8.1 mm for LX-1, and 5.9 mm for MX-1. In comparison with the nude mouse SRC assay, markedly similar growth potentials were observed regardless of the immune responsiveness of the host.

A next step in the possibilities of the SRC assay was the implantation of fresh tumor tissue fragments obtained from patients and called a "first transplant generation tumor" by Bogden et al.[4] A total of 37 human breast cancers were studied and in 86% a positive change in tumor size was measured at day 6. Seven conventional chemotherapeutic agents were tested and ranked according to their efficacy. In 32 previously untreated breast cancers, it was found that the patterns of response indicated that no single agent was active against all tumors, nor were tumors responsive to one agent necessarily responsive to another. Because of the individual tumor response, it was suggested that the model would have predictive value for clinical activity.

A last step by the group of Bogden[4] was the use of the SRC assay for the prediction of clinical responses in individual patients. Initially, these studies were retrospective and the SRC assay/clinical response correlation was found highly significant. Prospective, decision-aiding clinical trials were initiated and based on the drugs selected by the SRC assay. A response rate of 38% could be obtained in previously treated cancer patients.

Since the pioneering work of the group of Bogden, many years have elapsed. Other institutes have set up the SRC assay and described several flaws which have helped to define the real value of the assay for testing as it is presently used.

C. HISTOLOGY

Similar to s.c. xenografts, SRC implants have been shown morphologically identical to the primary human tumor biopsies or to the human tumor lines from which they were derived.[5,6] In this respect, glandular structures are preserved, as well as cellular functions such as CEA production and antigen expression.

With the use of normal, immunocompetent mice, the question arose as to the effect of host response cells on the growth of the subrenal tumor implant. In 1983, Edelstein et al.[7] examined the histological appearance of the xenograft at day 6 after implantation. Fresh tumor implants from surgical specimens, as well as fragments from human tumor xenografts, were analyzed upon SRC implantation, either as control tumors or after treatment. Variable degrees of lymphocytic infiltration and fibrosis in untreated controls were observed, whereas the effects were even more striking in doxorubicin and cisplatin-treated animals. In none of the tests with doxorubicin and cisplatin could a response be scored macroscopically, while there was clear evidence of activity in the microscopic sections. The discrepancy between the macroscopic and microscopic findings was more pronounced in the primary patient material.

An improvement of the 6-d SRC assay in normal mice was proposed by Edelstein et al.[8] Prevention of the host immunological response was investigated by means of cyclophosphamide pretreatment, wholebody irradiation, plus cortisone and other immunosuppressive techniques. In implants excised from s.c. grown xenografts, host cell reductions could indeed be obtained with cyclophosphamide or irradiation given before implantation. Also, Aamdal et al.[6] were able to prevent the infiltration by mouse inflammatory cells after pretreatment of the animals with cyclophosphamide. However, when first transplant generation tumors were analyzed, irradiation was only rarely able to completely prevent cellular infiltration.[8] In this respect, the use of nude mice did not improve the quality of the SRC assay. The variability observed could be attributed to the distribution of tumor and normal cells within the primary tissue. Stenbäck, et al.[5] have confirmed the observation that supplementary histological examination is necessary to increase the validity of the SRC assay for primary tissue implants. Bennett et al.[9] have advocated against the use of normal mice for first transplant generation xenografts unless these mice were artificially immunodepressed at an early age. Abrams et al.[10] have come to the conclusion that normal mice cannot be used for the 6-d SRC assay, as they could demonstrate an immunological response at the site of the implant as early as day 3. Further, they confirmed the variability in tumor cell content in fresh human tumor specimens of disadvantage to the reliability of the assay. It was felt that s.c. human tumor xenografts serially transplanted to nude mice provided the best material for the SRC assay in order to reduce the uncertainties and to provide a valid tool to assess the chemosensitivity of human tumors.[11]

D. DRUG SCREENING

A number of investigators have analyzed the SRC assay of first transplant generation tumors in normal mice for its predictive value for drugs to be (in)active in a particular tumor type or even in an individual patient. Promising results have been obtained, exemplified by the prospective and retrospective analysis of the chemosensitivity of advanced pelvic malignancies in female patients where the SRC assay showed a sensitivity of 67 to 85% and a specificity of 57 to 71%.[12] For head and neck cancer primary transplants, the sensitivity was 86% and the specificity 78%.[13] In another study of fresh head and neck cancer implants carried out in mice pretreated with the immunosuppressive agent cyclosporin A, a good correlation was demonstrated between the chemosensitivity of the assay and the clinical results.[14] In none of the studies mentioned was the quality of the tumor tissue or the host inflammatory response taken into account to carefully interpret the experimental results.

Based on the disadvantages mentioned for the use of normal mice and fresh tumor implants in the SRC assay, representative drug screening results may only be obtained in nude mice and implants excised from s.c. human tumor xenografts. In this respect, a short-term assay becomes feasible and, as such, has become incorporated in the disease-oriented antitumor drug screening program of the NCI.[15] Apart from the human tumor xenografts CX-1, LX-1, and MX-1, a series of other SRC assay-xenograft systems are under development. Protocols for individual tumors have required tailoring to suit their individual growth characteristics. The present experience of the NCI is that when sterile, viable tumor fragments are employed and the SRC assay carried out according to the tumor-tailored protocol using nude mice maintained under pathogen-free conditions, the test can give reproducible effects of antitumor drugs in these tumor implants.

III. OTHER RAPID SCREENING SYSTEMS

In the early 1980s, the agar diffusion chamber technique was developed which, in theory, was meant to provide a valid solution to some problems encountered *in vitro:* human tumor cells in diffusion chambers implanted into the peritoneal cavity of mice should grow better under the stimulation of diffusible growth factors present *in vivo* and the cells can be exposed to drugs that require metabolization by the host.[16] The agar diffusion chamber was based on the principle of growth of human tumor cell colonies in semisolid media as a method for drug screening *in vitro,* developed originally by Hamburger and Salmon.[17] Differences in drug exposure and in cellular chemosensitivity were found between human melanoma xenografts grown as colonies in chambers or as s.c. tumors.[18] With the evolution of the human tumor colony-forming assay, several technical problems were elucidated specifically related to the poor reliability of colony formation from fresh tumors.[19] This finding, together with the high costs, have been reasons that the agar diffusion chamber technique has never found a widespread application for drug screening.

In order to complete the SRC assay promptly, an adequate supply of tumor material of an s.c. grown xenograft should be available.[15] The requirement to initially grow the tumor cells s.c. as a source of solid tumor fragments for implantation under the kidney capsule still contributes to the complexity, costs, and time for a single test. Fingert et al.[20] have demonstrated that cells derived from solid tumors or leukemias and harvested from cell culture flasks can be placed into a fibrin clot and thus implanted into the SRC site. This method has not yet been investigated further as a potential solution to this problem.

A recently introduced *in vivo* assay is based on the encapsulation of human tumor cells in small (1 mm in diameter) microcapsules with semipermeable membranes injected i.p. into nude mice.[21] The microcapsules are recovered at various time intervals following treatment. Drug effects are based on the differences in the number of tumor cells recovered from the treated and control animals. Preliminary studies indicated that the test is applicable to all histological types of human tumor cells. Further, the drugs cross the capsular membrane of the microcapsules. The antitumor effects were consistent with the relative therapeutic efficacy of drugs or level of resistance of tumor cells detected by other *in vitro* or *in vivo* tests. The microencapsulated tumor assay is relatively inexpensive, can be completed in less than 3 weeks, and may be particularly suitable for use as an initial short-term *in vivo* assay for evaluation of *in vitro* selected compounds.[15]

IV. INTRACRANIAL TRANSPLANTATION

Human lymphomas and leukemias taken directly from patients will usually be rejected upon s.c. implantation into the nude mouse. Cell cultures of such tumor types were reported to have preferential growth in immunologically privileged sites, such as the brain. Epstein et al.[22] have studied inocula of 11 primary lymphomas and 6 derived cell lines, and found a 90% take rate. Tumorigenicity of hematopoietic malignant cells after intracranial inoculation was confirmed by Schaadt et al.[23] They found that infection of primary cells with exogenous Epstein-Barr virus prior to the transplantation procedure significantly increased the take rate. In human lymphoblastoid cell lines carrying the Epstein-Barr virus, Giovanella et al.[24] also found a high take rate; in 18 of 18 lines, intracerebral growth was obtained.

The inoculation of tumor cells into the brain is carried out by injecting 10^5 to 10^6 cells through a thin needle (27-gauge) into the frontal region of the cerebrum. Giovanella et al.[24] have a preference for the anterior left quadrant of the skull, as the corresponding area allows growth without neurological symptoms until the disease has reached macroscopic size. Upon growth, within a particular time period, the animals will exhibit lethargy, sometimes hyperkinesis and convulsions, and will eventually die. Xenografts grown intracranially show the morphological characteristics of the original tumor cells.[22,24] Although confined to the central nervous system and meninges, the tumors appear to be highly invasive with widespread within the parenchyma, ventricles, and leptomeninges.

The intracranial transplantation site has been studied to compare the take rate of tumor types that may be accepted s.c. with their tumorigenicity in the brain. For small-cell lung cancer cells, it was shown by the group of Gazdar[25] that tumorigenicity is increased upon intracranial injection. S.c. tumors of smallcell lung cancer had a longer latency period, and were noninvasive and nonlethal. The growth pattern observed after intracranial injection resembled that reported for hematopoietic malignant cells. Also, the number of cells inducing a tumor upon injection into the brain was 10 to 1000 times lower than necessary for s.c. inoculation. Thus, intracranial transplantation may be useful to study the biology of human tumor cells, especially when small numbers of tumor cells are available.

In order to characterize the potential of a new drug for treatment of an intracranial tumor or leptomeningeal disease, it is important to determine whether the drug can reach the tumor cells.[26] Tumor systems have been made available in which tumor cells are injected into the brain and treatment is instituted on designated schedules following the inoculation of the tumor cells. Determination can be performed whether a drug crosses the blood-brain barrier, as the central nervous system is considered a pharmacologically secluded area. The injection of tumor cells into the brain may therefore be an appropriate model to study the efficacy of new compounds in the treatment of central nervous system metastases. The intracranial site may also be used for orthotopic transplantation of brain tumors, as reported by Shapiro et al.[27] With a medulloblastoma line and a glioma line, Houchens et al.[28] have designed intracranial tumor systems to assess the effects of various cytotoxic agents against brain tumors.

REFERENCES

1. **Bogden, A. E., Kelton, D. E., Cobb, W. R., and Esber, H. J.,** A rapid screening method for testing chemotherapeutic agents against human tumor xenografts, in *Proc. Symp. on the Use of Athymic (Nude) Mice in Cancer Res.,* Houchens, D. P. and Ovejera, A. A., Eds., Gustav Fischer Verlag, New York, 1978, 231.
2. **Bogden, A. E., Houchens, D. P., Ovejera, A. A., and Cobb, W. R.,** Advances in chemotherapy studies with the nude mouse, in *The Nude Mouse in Experimental and Clinical Research,* Fogh, J. and Giovanella, B. C., Eds., Academic Press, New York, 1982, 367.
3. **Bogden, A. E., Haskell, P. M., LePage, D. J., Kelton, D. E., Cobb, W. R., and Esber, H. J.,** Growth of human tumor xenografts implanted under the renal capsule of normal immunocompetent mice, *Exp. Cell Biol.,* 47, 281, 1979.
4. **Bogden, A. E., Cobb, W. R., LePage, D. J., Haskell, P. M., Gulkin, T. A., Ward, A., Kelton, D. E., and Esber, H. J.,** Chemotherapy responsiveness of human tumors as first transplant generation xenografts in the normal mouse: six-day subrenal capsule assay, *Cancer,* 48, 10, 1981.
5. **Stenbäck, F., Kangas, L., and Wasenius, V. M.,** Cell structure and function and response to chemotherapy in tumors heterotransplanted into the subrenal capsule of mice an rats, *Eur. J. Cancer Clin. Oncol.,* 21, 1523, 1985.
6. **Aamdal, S., Fodstad, Ø., Nesland, J. M., and Pihl, A.,** Characteristics of human tumour xenografts transplanted under the renal capsule of immunocompetent mice, *Br. J. Cancer,* 51, 347, 1985.
7. **Edelstein, M. B., Fiebig, H. H., Smink, T., Van Putten, L. M., and Schuchhardt, C.,** Comparison between macroscopic and microscopic evaluation of tumour responsiveness using the subrenal capsule assay, *Eur. J. Cancer Clin. Oncol.,* 19, 995, 1983.
8. **Edelstein, M. B., Smink, T., Ruiter, D. J., Visser, W., and Van Putten, L. M.,** Improvements and limitations of the subrenal capsule assay for determining tumour sensitivity to cytostatic drugs, *Eur. J. Cancer Clin. Oncol.,* 20, 1549, 1984.
9. **Bennett, J. A., Pilou, V. A., and MacDowell, R. T.,** Evaluation of growth and histology of human tumor xenografts implanted under the renal capsule of immunocompetent and immunodeficient mice, *Cancer Res.,* 45, 4963, 1985.
10. **Abrams, J., Jacobovitz, D., Dumont, P., Semal, P., Mommen, P., Klastersky, J., and Atassi, G.,** Subrenal capsule assay of fresh human tumors: problems and pitfalls, *Eur. J. Cancer Clin. Oncol.,* 22, 1387, 1986.
11. **Atassi, G., Dumont, P., and Fournier, J.,** New methods for determining tumor sensitivity, *Eur. J. Cancer Clin. Oncol.,* 21, 1299, 1985.
12. **Stratton, J. A., Kucera, P. R., Micha, J. P. Rettenmaier, M. A., Braly, P. S., Berman, M. L., and DiSaia, P. J.,** The subrenal capsule tumor implant assay as a predictor of clinical response to chemotherapy: 3 years of experience, *Gynec. Oncol.,* 19, 336, 1984.
13. **Panje, R. and McCormick, K. J.,** Murine subrenal capsule assay: prediction of chemoresponsiveness in head and neck cancer, *Laryngoscope,* 99, 41, 1989.
14. **Parnes, S. M., Bennett, J. A., DeConti, R. C., and Sagerman, P.,** Accuracy of the subrenal capsule xenograft assay in predicting the clinical growth and chemosensitivity of human squamous head and neck carcinomas, *J. Surg. Oncol.,* 42, 21, 1989.
15. **Shoemaker, R. H., McLemore, T. L., Abbott, B. J., Fine, D. L., Gorelik, E., Mayo, J. C., Fodstad, Ø., and Boyd, M. R.,** Human tumour xenograft models for use with an *in vitro*-based, disease-oriented antitumour drug screening program, in *Human Tumour Xenografts in Anticancer Drug Screening,* Winograd, B., Peckham, M., and Pinedo, H. M., Eds., Springer-Verlag, Berlin, 1988, 115.
16. **Sobrero, A. F. and Marsh, J. C.,** Chemosensitivity of human tumor clonogenic cells simultaneously arranged in agar diffusion chambers and in a two-layer agar culture system, *Cancer Treat. Rep.,* 68, 615, 1984.
17. **Hamburger, A. W. and Salmon, S. E.,** Primary bioassay of human tumor stem cells, *Science,* 197, 461, 1977.
18. **Selby, P. J. and Steel, G. G.,** Use of the agar diffusion chamber for the exposure of human tumor cells to drugs, *Cancer Res.,* 42, 4758, 1982.
19. **Selby, P., Buick, R. N., and Tannock, I.,** A critical appraisal of the "human tumor stem-cell assay", *N. Engl. J. Med.,* 308, 129, 1983.
20. **Fingert, H. J., Chen, Z., Mizrahi, N., Gajewski, W. H., Bamberg, M. P., and Kradin, R. L.,** Rapid growth of human cancer cells in a mouse model with fibrin clot subrenal capsule assay, *Cancer Res.,* 47, 3824, 1987.
21. **Gorelik, E., Ovejera, A., Shoemaker, R., Jarvis, A., Alley, M., Duff, R., Mayo, J., Herberman, R., and Boyd, M.,** Microencapsulated tumor assay: new short-term assay for *in vivo* evaluation of the effects of anticancer drugs on human tumor cell lines, *Cancer Res.,* 47, 5739, 1987.
22. **Epstein, A. L., Herman, M. M., Kim, H., Dorfman, R. F., and Kaplan, H. S.,** Biology of the human malignant lympomas. III. Intracranial heterotransplantation in the nude, athymic mouse, *Cancer,* 37, 2158, 1976.
23. **Schaadt, M., Kirchner, H., Fonatsch, C., and Diehl, V.,** Intracranial heterotransplantation of human hematopoietic cells in nude mice, *Int. J. Cancer,* 23, 751, 1979.

24. **Giovanella, B., Nilsson, K., Zech, L., Klein, G., and Stehlin, J. S.,** Growth of diploid, Epstein-Barr virus-carrying human lymphoblastoid cell lines heterotransplanted into nude mice under immunologically privileged conditions, *Int. J. Cancer,* 24, 103, 1979.
25. **Gazdar, A. F., Carney, D. N., Sims, H. L., and Simmons, A.,** Heterotransplantation of small-cell carcinoma of the lung into nude mice: comparison of intracranial and subcutaneous routes, *Int. J. Cancer,* 28, 777, 1981.
26. **Goldin, A., Johnson, R. K., and Venditti, J. M.,** Preclinical characterization of candidate antitumor drugs, *Cancer Chemother. Rep.,* 5 (Part 2), 21, 1975.
27. **Shapiro, W. R., Basler, G. A., Chernik, N. L., and Posner, J. B.,** Human brain tumor transplantation into nude mice, *J. Natl. Cancer Inst.,* 62, 447, 1979.
28. **Houchens, D. P., Ovejera, A. A., Riblet, S. M., and Slagel, D. E.,** Human brain tumor xenografts in nude mice as a chemotherapy model, *Eur. J. Cancer Clin. Oncol.,* 19, 799, 1983.

Chapter 8

CHARACTERIZATION AND MONITORING

Epie Boven

TABLE OF CONTENTS

I. INTRODUCTION

Once an s.c. implanted tissue fragment obtained from a patient's tumor has been accepted by the nude mouse, the proliferating nodule can be excised and transplanted to a next series of recipients. By repeated serial transfer, a tumor line may be established. For continuous use in a long-term research program, the tumor line should be monitored frequently for retention of original characteristics and a consistent volume-doubling time. Characterization of the line is essential to demonstrate its identity with or close similarity to the original tumor tissue. Various characteristics can be determined in both the patient's tumor and the respective xenograft, as described in Table 1. Minimal requirements will be the comparison of the histology and the histochemistry between tissue sections of xenografts and the patient's tumor. In addition, the human origin of the xenograft may be demonstrated by chromosome analysis, or lactate dehydrogenase (LDH) isoenzyme electrophoresis. For monitoring of established tumor lines, the histology and the histochemistry should be examined regularly, whereas the volume-doubling time should maintain a constant rate. For those investigators with research related to monoclonal antibodies, hormone sensitivity, chemotherapy, etc., specific requirements for a human tumor line should also be monitored every four to five passages. Lastly, but importantly, in early passages of xenografts, tumor material should be frozen in liquid nitrogen to provide for a continuous supply of tumor tissue with characteristics best resembling the original patient's tumor.

II. TUMOR LINES AND GROWTH

A. THE ESTABLISHMENT OF A TUMOR LINE

Progressive growth of a tissue fragment obtained from a patient's tumor transplanted s.c. to the flank of the nude mouse is recorded as a positive "take". Larger nodules may be removed for serial transfer. Samples are kept sterile in a suitable medium (e.g., Hanks' balanced salt solution) and rapidly processed. Similar to the initial transplantation fragments of 2 to 3 mm in diameter or a tissue suspension are implanted into new nude recipients. The initial growth does not guarantee the successful establishment of a continuous tumor line. This inconsistency can be partially explained by differences between tumor types. Examples of easily transplantable tumors are colorectal cancer, melanoma, sarcoma, and lung cancer, while breast cancer, ovarian cancer, and cancer of the head and neck region are difficult to grow.[1] Leukemias and lymphomas will usually be rejected after s.c. implantation. Other variables influencing the transplantability, as well as the success of serial transfer, are factors related to the host, of which its general condition is the most important.

B. MEASUREMENT OF GROWTH

S.c. xenografts offer a good visualization of tumor growth and measurements can be carried out periodically to follow volume progression or regression. Diameters are determined with slide calipers and results expressed as a measure of size (length × width),[2] tumor weight ($1/2 \times \text{length} \times \text{width}^2$),[3] or tumor volume ($1/6 \times \pi \times \text{mean tumor diameter}^3$ or $1/2 \times \text{length} \times \text{width} \times \text{height}$ or $1/2 \times \text{length} \times \text{width}^2$).[4-6] Growth curves fitted to data points can be constructed for each individual tumor on semi-log graph paper. Usually, for each tumor line, there is a slight variation in growth rate, as well as in latency time, for the individual appearance of palpable nodules. For the initial proliferation period, a growth pattern approximating an exponential volume increase can be noticed, which is followed by a slower growth rate.[4] Overall, the course of tumor growth can be described mathematically with Gompertz growth equation.[7-9]

Obviously, the growth rate of human tumor lines is much faster than is generally observed in lesions of the corresponding tumor type in patients.[10] The minimum detectable tumor lesion in man is usually of the order of 1×10^9 cells (approximately 1000 mg of tissue). Measurement

TABLE 1
Characteristics of Human Tumor Lines

Volume-doubling time
Chromosomes
DNA content
Lactate dehydrogenase
Histology
Histochemistry
Tumor products
Antigen expression
Receptors for hormones
Receptors for growth factors
Sensitivity to drugs

of large lesions includes the sum of proliferating areas next to necrotic cells. In xenografts, volume doubling time can be calculated from 100 to 200 mm^3 (approximately 100 to 200 mg), a period where tissue contains none or few areas with necrotic cells. Mattern et al.[11] have demonstrated a higher proportion of S-phase cells in xenografts, as compared to the S-phase fraction in lung cancer tissue of the donating patients. Lauk et al.[12] showed that in xenografts vasculature is better distributed than in the original tumors and mitotic cells can be preferably found adjacent to blood vessels.

C. VOLUME-DOUBLING TIME

In general, volume-doubling time is calculated in the exponential growth phase of the xenograft, for example, from 100 to 200 mm^3 and expressed in days. Growth rate as well as latency time tend to decrease during the first few serial passages, as has been demonstrated for a variety of lines derived from head and neck cancer,[13] breast cancer,[14] melanoma,[15] pediatric tumors,[16,17] colorectal cancer,[18] and gastrointestinal cancer.[19] After the third or fourth serial passage in the mouse, a tumor line usually maintains a consistent doubling time.

A spontaneous change in volume-doubling time in an established tumor line has to be handled with care. Additional characteristics should be determined, such as the histological pattern, the LDH isoenzymes, or the karyotype, in order to exclude the rare occurrence of a murine tumor overgrowing the human tissue. In this respect, Beattie et al.[20] have described two cases of induction of sarcomas in the murine stromal tissue associated with the human tumor xenografts, which were detected by analysis of the LDH isoenzyme profile and histology in each passage. The sarcoma cells were found to contain large amounts of murine leukemia virus. In an earlier report, the same authors[21] described the higher incidence of murine lymphomas in antigenically stimulated nude mice, p.e. by human tumor xenografts. Figures 1A to 1D show a human ovarian cancer line, poorly differentiated serous adenocarcinoma overgrown by mouse stromal tissue. Figure 2 visualizes the accelerated growth in the same tumor line at the time of mouse tissue overgrowth. An example of induction of a mouse lymphoma is shown in Figures 3A to 3D. In this tumor line of fibrosarcoma origin, the xenografts were infiltrated by mononuclear cells and, in the subsequent passage, fast growing nodules of lymphoproliferative cells induced early animal death. LDH isoenzyme analysis indicated the sole presence of mouse cells.

III. HUMAN CHARACTERISTICS

A. CHROMOSOMES AND DNA CONTENT

A basic requirement for a human tumor line established in the nude mouse is the retention of a human karyotype. In an early report on human tumor xenografts, it was already demonstrated that chromosomes of the human type were found in between mouse stromal cells containing a normal murine karyotype.[22] Moreover, in subsequent passages, a total chromoso-

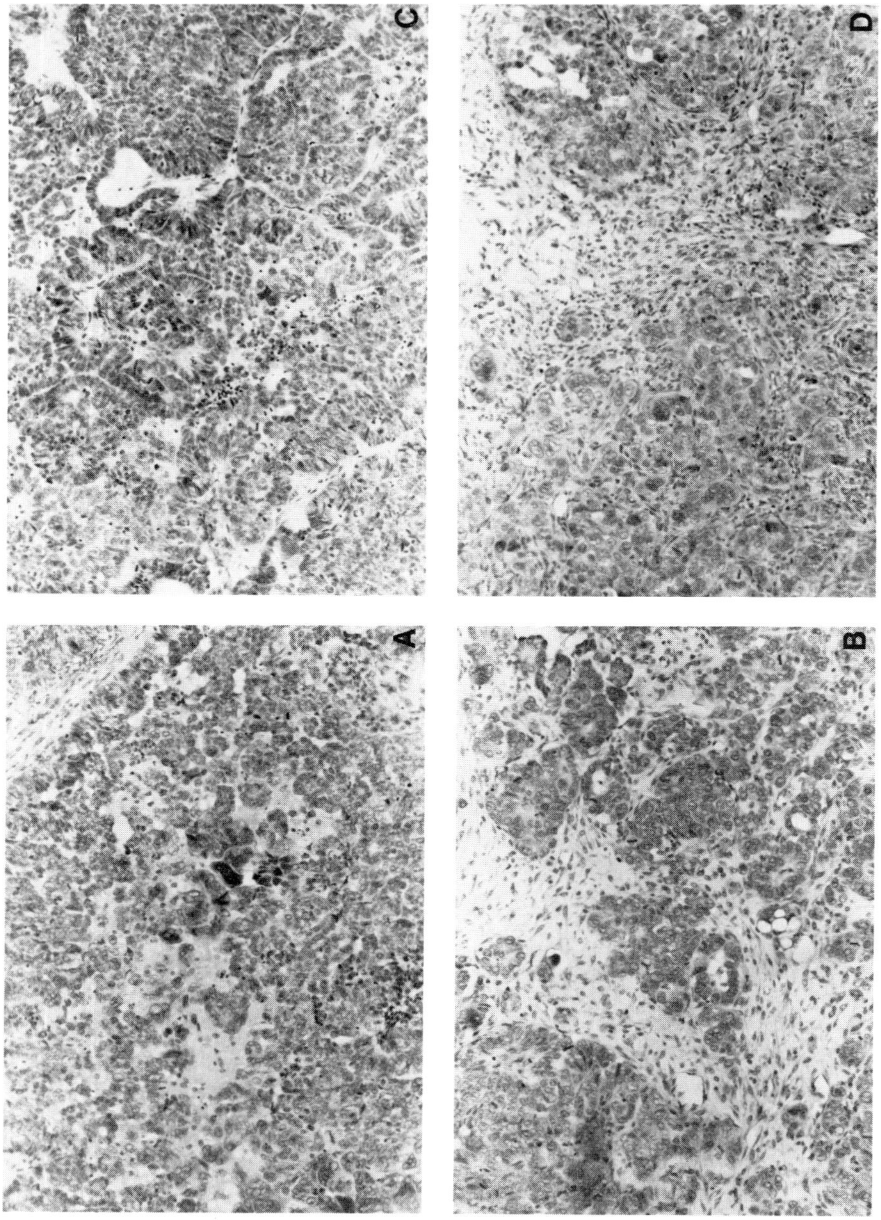

FIGURE 1. A: human ovarian cancer xenograft, poorly differentiated serous type, passage 14; B: xenograft in passage 16 used for serial transfer showing infiltration by mouse stroma; C: xenograft in passage 16 of another mouse than 1B showing a similar pattern to that of Figure 1A; D: serial transfer of xenograft shown in Figure 1B results in further increase of mouse stromal tissue overgrowth (passage 17).

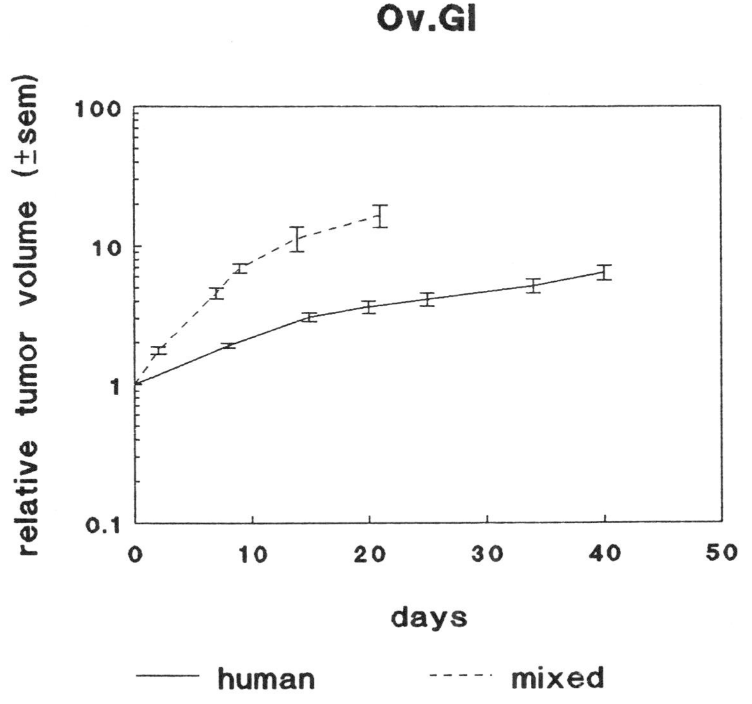

FIGURE 2. Growth rate of human ovarian cancer xenografts before (—human) and at the time of overgrowth by mouse stromal tissue (----mixed tissue).

mal species shift or interspecies hybridization could not be detected.[23] During serial transfer, progressive changes in the human chromosome constitution may occur, possibly based on clonal variation in the tumor cell population.[24] In tumor lines derived from the same tumor type, aneuploid numbers of chromosomes are found without consistent karyotypic abnormalities. Upon analysis of the chromosome constitution in human colorectal cancer lines compared with tumor tissue of the donating patients, it was demonstrated by Lefrançois et al.[25] that karyotypic changes were grossly similar. However, polyploidization appeared to be more pronounced in xenografts.

Cytogenetic analysis of solid tumors is a difficult task because a good single-cell preparation and a sufficient number of viable tumor cells in metaphase are required. An easier approach to obtain insight in individual specific genome variations to be retained in serial transfer is provided by the DNA "fingerprinting" technique.[26]

Abnormal nuclear DNA content is a conclusive marker of malignancy and is found in approximately 75% of patients with solid tumors. The proportion of cells in the S phase of the cell cycle increases with increasing DNA excess in a number of different solid tumors and is suggested to have an adverse impact on prognosis.[27] Unlike mitotic karyotyping, ploidy analysis by DNA flow cytometry can be easily performed after the introduction of various cell preparative and staining procedures and different flow instruments. Although flow cytometric analysis of DNA content of a human tumor xenograft will not give insight into the possible presence of human chromosomes, it may well be used to identify different cell populations causing heterogeneity within human tumors. In this respect, Van Haaften-Day et al.[28] have described that in human ovarian cancer lines aneuploid cell growth was predominant in the mouse regardless of the presence of diploid tumor cell populations in the original tumor tissue. In contrast, Massazza et al.[29] have reported similar DNA histograms for human ovarian cancer

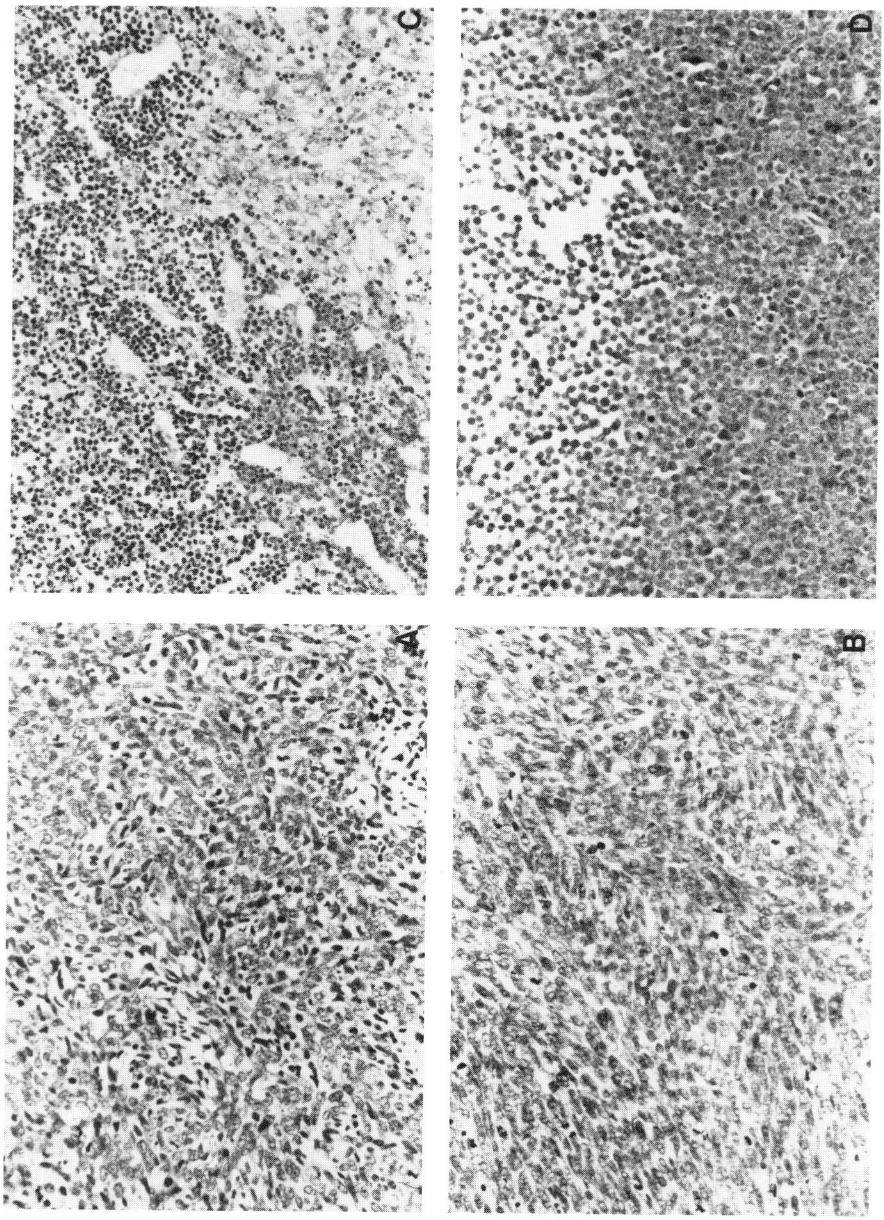

FIGURE 3. A: fibrosarcoma, original tumor tissue as obtained from the patient; B: fibrosarcoma xenograft in passage 2; C: the same xenograft in passage 4, but infiltrated by mononuclear cells; D: xenograft in passage 5 completely replaced by lymphoproliferative cells.

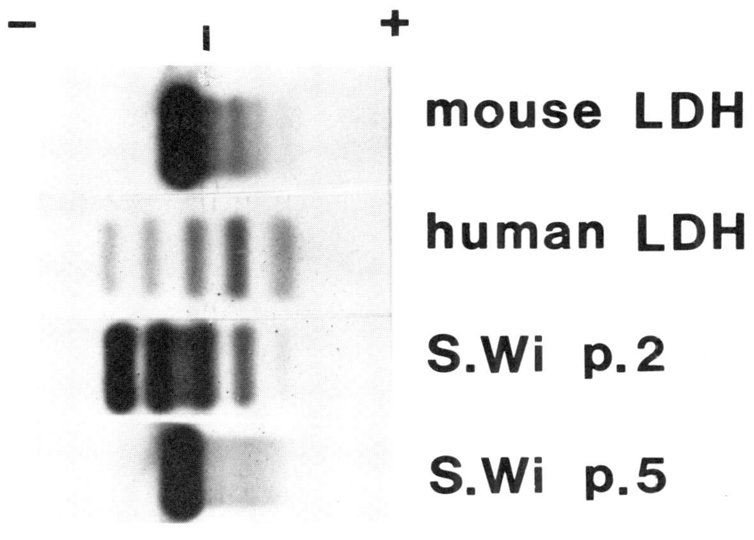

FIGURE 4. Lactate dehydrogenase isoenzyme pattern of mouse and human serum. LDH obtained from tumor tissue extracts of fibrosarcoma xenograft in passage 2, showing a mixture of mouse (faint LDH-5 band) and human isoenzymes. At the time of overgrowth by lymphoproliferative cells, only mouse LDH (clear LDH-5 band) is left.

at surgery and as xenografts in nude mice. The aneuploid value remained unchanged during serial transfer. For small cell lung cancer lines in serial transfer, Engelholm et al.[30] have shown that genetic instability may be detected by flow cytometry. Subsequent DNA content analysis over various passages may well discriminate the outgrowth of new, possibly more aggressive cell populations.

B. LACTATE DEHYDROGENASE

The lactate dehydrogenase (LDH) enzyme has been found to exist in species-specific forms. Human tissues demonstrate five LDH variants which are released into the circulation. Five LDH variants can also be found in mice. Electrophoretic separation of these LDH isoenzymes shows that human LDH-IV and LDH-V move further to the cathode than any of the mouse fractions. In nude mice bearing human tumors, human LDH can be found in the serum as detected by electrophoresis. Mouse LDH-5 is the most prominent band in such a mixture (Figure 4). As the release of human LDH from xenografts is directly related to growth, especially in small tumors, it is recommended to perform LDH analysis of tumor tissue extracts. The detection of human LDH in mouse serum or in a tissue extract is an easy procedure to demonstrate the human nature of a tumor line.[31] Especially in those lines with an undifferentiated or sarcomatous histological pattern, this assay is an indispensable tool for origin control. In the case of the fibrosarcoma line described earlier, where in subsequent passages human tissue was overgrown by mouse cells, the tissue extract only contained mouse LDH isoenzymes (Figure 4).

Mouse serum-human LDH enzyme levels are grossly proportional to the size of the xenograft. Some investigators have used the enzyme as a marker to monitor the efficacy of anticancer drugs in human tumor lines.[32,33] A transient increase may be expected if tumors are sensitive to the given drug followed by a decrease in LDH level after volume reduction has been obtained.[34] However, fluctuations in serum levels of biochemical tumor markers may vary remarkably in xenografts with a short volume-doubling time or with cystic or necrotic changes.

IV. CHARACTERISTICS OF ORIGIN

A. HISTOLOGY AND HISTOCHEMISTRY

Histological and histochemical characterization of the xenograft tissue is essential in order to ensure its relationship with the original tumor tissue. Light microscopy should be preferably repeated in each passage to monitor the retention of the specific morphological pattern. Most investigators have shown that the histology of origin is usually well maintained in xenografts.[1,35] In a review by Steel et al.,[35] a frequent tendency for the xenografts to be more poorly differentiated and pleomorphic than the original tumor tissue was reported. Often, the stromal elements are less pronounced in xenografts and of mouse origin. Specific staining procedures for epithelial mucins,[7,31,36] melanin,[7] etc. may provide additional information on functional capacities of tumor cells. Histological monitoring of human tumor lines is an easy approach to identify the induction of a murine tumor invading the human tumor cell areas at an early stage (Figures 1A to 1D and Figures 3A to 3D). For human tumor lines of solid tumors with an undifferentiated pattern, electron microscopy may identify ultrastructural features related to a particular tumor type, for example, melanoma.[7]

B. TUMOR PRODUCTS

Over the past years, a number of tumor markers have been identified with specificity for particular tumor types in the clinical situation. In some instances, the monitoring of serum marker levels will predict progressive disease or response to treatment, as is the case for α-fetoprotein (AFP) in testicular cancer, human chorionic gonadotropin (HCG) in testicular cancer and choriocarcinoma, carcinoembryonic antigen (CEA) in colon cancer, and CA125 in ovarian cancer, serous subtype. In addition, a number of functioning tumors can be found producing humoral factors resulting in paraneoplastic syndromes in patients.

Upon characterization of tumor lines derived from human tumor tissue known to release cellular products, it has been demonstrated that these biological functions can be retained in the nude mouse. In 1977, Ohsawa et al.[37] described the successful transplantation and serial transfer of malignant functioning tumors. Transplantation of benign tumors was unsuccessful. The production of granulocyte-macrophage colony-stimulating factor, resulting in an increase of peripheral blood granulocytes in tumor-bearing mice, was described in detail by Asano et al.[38] The group of Ohsawa[37] were able to establish an erythropoietin-producing renal cell cancer line, a hypercalcemia-producing lung cancer line, and an HCG-producing choriocarcinoma line.

HCG production and release by human tumor xenografts can be detected in a tissue extract or in mouse serum by immunoprecipitation with antisera against complete HCG, α-HCG or β-HCG with the use of commercially available test systems. Procedures have been described by various investigators in choriocarcinoma lines and other human tumors producing HCG.[39-41] As in patients, the extent of HCG release appears to be related to the size of the xenograft.[40]

The release of AFP into the circulation has mostly been studied in human hepatomas and hepatoblastomas grown in nude mice.[42-44] Serum AFP levels, as measured by a radioimmunoassay, were also shown to correlate with the tumor growth, indicating AFP as a good tumor marker.[43,44] Of interest, Hirohashi et al.[43] have found that the rapidly growing hepatomas were moderately differentiated and produced a large amount of AFP, while a slowly growing hepatoma line was well differentiated and produced less AFP.

CEA production in tumor-bearing nude mice has mainly been studied for human colon cancer.[31,45-47] For measurement of CEA levels in serum or in extracts of homogenized tumor tissue, various immunoassay test kits are presently available. The production of CEA does not always parallel tumor growth. Stragand et al.[46] did not find a correlation for the LoVo colon cancer line, while other investigators confirmed a linear relationship between CEA levels and enlargement of tumors. Martin and Halpern[47] have added the observation that the concentration of CEA per gram of extracted tumor tissue remained constant throughout the weight range studied in the T-380 colon cancer line.

With the advent of the generation of monoclonal antibodies, new serological tumor markers have been identified, such as CA125[48] with OC125. CA125 can be detected in the circulation of nude mice bearing CA125-positive human ovarian cancer lines of the serous subtype.[49] As in patients, Molthoff et al.[49] have not found CA125 elevation in human ovarian cancer lines of the mucinous subtype. In a recent report of Klapdor et al.,[50] human pancreatic cancer lines were described with elevated CA19-9 levels in mouse serum. This tumor marker level correlated not only with xenograft volume, but also with CA19-9 levels in the donating pancreatic cancer patients.

C. ANTIGEN EXPRESSION

To date, numerous reports have come forth dealing with tumor localization of human tumor xenografts with the use of monoclonal antibodies against tumor cell-associated antigens. Such antibodies were selected for their favorable reaction pattern with human malignant tissue as opposed to normal tissue. Xenografts have been widely shown to possess similar antigens as can be found in the same tumor type in the clinical situation. In the previous section on the release of HCG, AFP, and CEA into the circulation of the nude mouse, it was described that these products can be easily detected by immunoassay test systems in extracts of homogenized tumor tissue. The indirect immunoperoxidase staining of tissue sections is another technique to demonstrate the presence and the distribution of these and other specific antigens. In an early report, Bailey et al.[14] have shown the retention of CEA and an epithelial membrane antigen in tumor lines obtained from breast cancer. The same was reported by Friedlander et al.[51] for CA125, AFP, and CEA. Additional antibodies were used to confirm the retention of the antigen expression in human ovarian cancer xenografts and their serial passages (Molthoff et al.[49,52]).

The nude mouse-human tumor model offers an ideal *in vivo* system to further study the fate of the monoclonal antibody at the tumor cell level.[53] In addition, monoclonal antibody-guided therapy in this system using the antibody as a carrier of radionuclides or cytotoxic agents can be investigated to evaluate its possible clinical usefulness.[54,55] Further details on monoclonal antibody-guided tumor targeting in the nude mouse bearing human tumor xenografts will be given in Chapter 17.

D. RECEPTORS FOR HORMONES AND GROWTH FACTORS

Considerable information has been acquired on the presence and the function of receptors for hormones or growth factors on tumor cells. Steroid sex hormone receptors may be found in a number of tumor types, such as cancer of the breast, ovary, uterus, and prostate. Detection can be carried out by biochemical assays in homogenized tumor tissue or by immunohistochemical staining of tumor tissue sections. Growth factor receptors are present in a variety of tissues, and most cells have receptors for more than one growth factor. In malignant cells, the loss of or decreased requirement for specific growth factors is a common occurrence which may lead to a growth advantage. Evidence has been derived for autocrine secretion in which the cell produces its own growth factor. The ability of tumor cells to produce and to respond to their own growth factors has become a central concept linking oncogene and growth factor research. Examples of the autocrine release of polypeptide growth factors by tumor cells include type α-transforming growth factor, peptides related to platelet-derived growth factor, bombesin, and type β-transforming growth factor.[56,57] Radioimmune and radioreceptor assays have been developed to measure growth factor concentrations or to detect the presence of receptors in tumor tissue.

The use of human tumor xenografts to obtain more insight into the influence of hormones or growth factors on regulation of cell growth requires the retention of specific biological properties of the tumor type involved. For breast cancer, it has been found difficult to establish tumor lines in the nude mouse. Bailey et al.[58] were able to develop 8 serially transplantable human breast cancer lines out of 102 specimens from primary and secondary lesions of 80 patients upon implantation into female immunodeprived mice. In three of the eight lines, the original tumor tissue was estrogen-positive, but none of the xenografts proved to be hormone sensitive or were demonstrated to contain receptors. Possibly, clonal selection of aggressive hormone-independent cells had caused this loss of hormone dependence. Michel et al.[58] demonstrated the retention of the receptor status in early passages of human breast cancer xenografts in female nudes, but this characteristic was not stable upon further serial transfer. For endometrial cancer, Horvath et al.[69] obtained a better take rate for progesterone-negative tumors. Also in their study, receptor content for both estrogen and progesterone could be lost upon serial transfer. In an extensive description of the transplantability of malignant gynecological tumors by Kleine,[61] it was again concluded that receptor-negative tumor lines were easier to establish than xenografts from tissue with a positive estrogen and progesterone receptor status. Of interest, Friedl et al.[62] were able to establish 2 human endometrial cancer lines out of 16 primary tumors using estradiol-treated ovariectomized mice.

Despite the difficulty in establishing hormone-sensitive human tumor lines in the nude mouse and retention of the receptor content upon serial transfer, several human tumor xenografts have been described as being useful for testing the effects of endocrine manipulations. Examples are the T61 human breast cancer line,[63] MCF-7 xenografts grown from the MCF-7 estrogen-positive breast cancer cell line,[64] and the PC-82 prostate cancer line,[65] where the latter xenografts are grown in male nude mice.

The retention of growth factor receptors upon transplantation of human malignant tissue to the nude mouse has not yet been studied widely. Most experiments have been carried out with respect to the epidermal growth factor receptor (EGF), known as the high-affinity receptor for both EGF and type α-transforming growth factor. Human tumor xenografts with elevated levels of EGF receptors have been successfully used to study inhibition of growth by anti-EGF receptor monoclonal antibodies,[66] and tumor localization by radiolabeled monoclonal antibodies against the EGF receptor.[67,68]

E. SENSITIVITY TO DRUGS

Because of their extensive use as models for screening of cytotoxic drugs, human tumor xenografts have been repeatedly characterized for their chemosensitivity. It was clearly shown that xenografts maintain the chemotherapy response characteristics of the tumor type from which they are derived.[35] In those instances where a direct comparison could be made between xenograft response and the clinical response of the donating patient, a positive correlation was apparent. Winograd et al.[1] have analyzed a large number of comparisons between laboratory data and patients' treatments. In general, a clinical response or resistance to drug treatment was also found in the respective human tumor xenograft. In addition, Mattern et al.[69] have indicated a good correlation between the chemosensitivity of a series of human tumor lines derived from the same tumor type and the clinical response in that particular disease.

With the purpose to characterize a human tumor line for its chemosensitivity both clinically active and inactive drugs should be included. As part of the monitoring of a line, reference drugs should be chosen for testing every four to five passages to guarantee the stability of the chemosensitivity. Further details on the treatment with conventional antitumor agents and the use of the nude mouse model in preclinical phase II drug screening are given in Chapters 13 and 23.

V. TUMOR BANK

To ensure the stability of a human tumor line established in the nude mouse, usual requirements include the measurement of growth and the determination of the histological pattern in each passage. As mentioned earlier, after the third or fourth serial transfer, a tumor line maintains a consistent doubling time. It is of utmost importance to construct a tumor bank for sufficient supply of tumor material from earliest passages stored in liquid nitrogen. As was previously shown, a human tumor line may be lost from overgrowth by mouse cells.[20,21] Changing characteristics in serial transfer, such as a decreased melanin production[7] or an altered chemosensitivity pattern concurrent with accelerated growth,[15] may also be a reason to re-establish the line from an early passage.

For storage in liquid nitrogen, tumor tissue will be cut in fragments of 2- to 3-diameter in a similar manner to serial transplantation. Six to 8 fragments are placed in sterile freshly prepared 7.5% dimethylsulphoxide, 20% fetal calf serum, tissue culture medium (RPMI-1640 or DMEM) in a freezing vial. Immediately thereafter, vials are cooled at 1°C/min to 196°C and stored in a liquid nitrogen bank. Vials removed from the bank have to be thawed at 37°C as quickly as possible and kept in Hank's balanced salt solution (RPMI-1640 or DMEM) at room temperature for 15 min. After suction of the medium, fresh Hank's is added. After 30 min, implantation can be carried out in the usual way. Latency time after transplantation of tissue fragments from liquid nitrogen may be twice as long than for fresh samples of the tumor line involved.

REFERENCES

1. **Winograd, B., Boven, E., Lobbezoo, M. W., and Pinedo, H. M.,** Human tumor xenografts in the nude mouse and their value as test models in anticancer drug development, *In Vivo,* 1, 1, 1987.
2. **Fiebig, H. H.,** Comparison of tumor response in nude mice and in patients, in *Human Tumour Xenografts in Anticancer Drug Development,* Winograd, B., Peckham, M. J., and Pinedo, H. M., Eds., Springer-Verlag, Berlin, 1988, 25.
3. **Ovejera, A. A. and Houchens, D. P.,** Human tumor xenografts in athymic nude mice as a preclinical screen for anticancer agents, *Semin. Oncol.,* 8, 386, 1981.
4. **Houghton, J. A. and Taylor, D. M.,** Growth characteristics of human colorectal tumours during serial passage in immune-deprived mice, *Br. J. Cancer,* 37, 213, 1978.
5. **Boven, E., Van der Vijgh, W. J. F., Nauta, M. M., Schlüper, H. M. M., and Pinedo, H. M.,** Comparative activity and distribution studies of five platinum analogues in nude mice bearing human ovarian cancer xenografts, *Cancer Res.,* 45, 86, 1985.
6. **Boven, E., Winograd, B., Fodstad, Ø., Lobbezoo, M. W., and Pinedo, H. M.,** Preclinical phase II studies in human tumor lines: a European multicenter study, *Eur. J. Cancer Clin. Oncol.,* 24, 567, 1988.
7. **Selby, P. J., Thomas, J. M., Monaghan, P., Sloane, J., and Peckham, M. J.,** Human tumour xenografts estabished and serially transplanted in mice immunologically deprived by thymectomy, cytosine arabinoside and whole-body irradiation, *Br. J. Cancer,* 41, 52, 1980.
8. **Spang-Thomsen, M., Nielsen, A., and Visfeldt, J.,** Growth curves of three human malignant tumors transplanted to nude mice, *Exp. Cell Biol.,* 48, 138, 1980.
9. **Meck, R. A., Ingram, M., Meck, J. M., McCullough, J. L., Wu, M. C., and Yunis, A. A.,** Establishment and cell cycle kinetics of a human squamous cell carcinoma in nude mice and *in vitro, Cancer Res.,* 41, 1076, 1981.
10. **Shackney, S. E., McCormack, G. W., and Cuchural, G. J.,** Growth rate pattern of solid tumors and their relation to responsiveness to therapy, *Ann. Int. Med.,* 89, 107, 1978.
11. **Mattern, J., Wayss, K., Haag, D., Toomes, H., and Volm, M.,** Different growth rates of lung tumours in man and their xenografts in nude mice, *Eur. J. Cancer Clin. Oncol.,* 16, 289, 1980.
12. **Lauk, S., Zietman, A., Skates, S., Fabian, R., and Suit, H. D.,** Comparative morphometric study of tumor vasculature in human squamous cell carcinomas and their xenotransplants in athymic nude mice, *Cancer Res.,* 49, 4557, 1989.
13. **Wennerberg, J.,** Changes in growth pattern of human squamous-cell carcinomas of the head and neck during serial passages in nude mice, *Int. J. Cancer,* 33, 245, 1984.
14. **Bailey, M. J., Ormerod, M. G., Imrie, S. F., Humphreys, J., Roberts, J. D. B., Gazet, J. C., and Neville, A. M.,** Comparative functional histopathology of human breast carcinoma xenografts, *Br. J. Cancer,* 43, 125, 1981.
15. **Fodstad, Ø., Rofstad, E. K., Tveit, K. M., and Pihl, A.,** Spontaneous alteration in growth rates of two human melanoma xenografts. Concurrent changes in chemosensitivity, *Eur. J. Cancer Clin. Oncol.,* 19, 1175, 1983.
16. **Neeley, J. E., Ballard, E. T., Britt, A. L., and Workman, L.,** Characteristics of 85 pediatric tumors heterotransplanted into nude mice, *Exp. Cell Biol.,* 51, 217, 1983.

17. **Yeger, H., Baumal, R., Pawlin, G., and Phillips, M. J.,** Relationship of histology of Wilms' tumor to growth characteristics of nude mouse heterotransplants, *Cancer Res.,* 45, 2340, 1985.
18. **Nowak, K., Peckham, M. J., and Steel, G. G.,** Variation in response of xenografts of colorectal carcinoma to chemotherapy, *Br. J. Cancer,* 37, 576, 1978.
19. **Fiebig, H. H., Widmer, K. H., Fiedler, L., Wittekind, O., and Löhr, G. W.,** Development and characterization of 51 human tumor models for large bowel, stomach and esophageal cancers, *Dig. Surg.,* 1, 225, 1984.
20. **Beattie, G. M., Knowles, A. F., Jensen, F. C., Baird, S. M., and Kaplan, N. O.,** Induction of sarcomas in athymic mice, *Proc. Natl. Acad. Sci., U S.A.,* 79, 3033, 1982.
21. **Baird, S. M., Beattie, G. M., Lannom, R. A., Lipsick, J. S., Jensen, F. C., and Kaplan, N. O.,** Induction of lymphoma in antigenically stimulated athymic mice, *Cancer Res.,* 42, 198, 1982.
22. **Visfeldt, J., Povlsen, C. O., and Rygaard, J.,** Chromosome analyses of human tumours following heterotransplantation to the mouse mutant nude, *Acta Path. Microbiol. Scand. Sect. A,* 80, 169, 1972.
23. **Povlsen, C. O., Visfeldt, J., Rygaard, J., and Jensen, G.,** Growth patterns and chromosome constitutions of human malignant tumours after long-term serial transplantation in nude mice, *Acta Path. Microbiol. Scand. Sect. A,* 83, 709, 1975.
24. **Reeves, B. R. and Houghton, J. A.,** Serial cytogenetic studies of human colonic tumour xenografts, *Br. J. Cancer,* 37, 612, 1978.
25. **Lefrançois, D., Olschwang, S., Delattre, O., Muleris, M., Dutrillaux, A. M., Thomas, G., and Dutrillaux, B.,** Preservation of chromosome and DNA characteristics of human colorectal adenocarcinomas after passage in nude mice, *Int. J. Cancer,* 44, 871, 1989.
26. **Thacker, J., Webb, M. B. T., and Debenham, P. G.,** Fingerprinting cell lines: use of human hypervariable DNA probes to characterize mammalian cell cultures, *Som. Cell Mol. Gen.,* 14, 519, 1988.
27. **Barlogie, B., Raber, M. N., Schumann, J., Johnson, T. S., Drewinko, B., Swartzendruber, D. E., Göhde, W., Andreeff, M., and Freireich, E. J.,** Flow cytometry in clinical cancer research, *Cancer Res.,* 43, 3982, 1983.
28. **Van Haaften-Day, C., Russell, P., Rugg, C., Wills, E. J., and Tattersall, M. H. N.,** Flow cytometric and morphological studies of ovarian carcinoma cell lines and xenografts, *Cancer Res.,* 43, 3725, 1983.
29. **Massazza, G., Tomasoni, A., Lucchini, V., Allavena, P., Erba, E., Colombo, N., Mantovani, A., D'Incalci, M., Mangioni, C., and Giavazzi R.,** Intraperitoneal and subcutaneous xenografts of human ovarian carcinoma in nude mice and their potential in experimental therapy, *Int. J. Cancer,* 44, 494, 1989.
30. **Engelholm, S. A., Spang-Thomsen, M., Vindeløv, L. L., and Brünner, N. A.,** Expression of heterogeneity of human tumors grown in nude mice, in *Immune-Deficient Animals in Biomedical Research,* Rygaard, J., Brünner, N., Græm, N., and Spang-Thomsen, M., Eds., Karger, Basel, 1987, 183.
31. **Houghton, J. A. and Taylor, D. M.,** Maintenance of biological and biochemical characteristics of human colorectal tumours during serial passage in immune-deprived mice, *Br. J. Cancer,* 37, 199, 1978.
32. **Pesce, A. J., Bubel, H. C., DiPersio, L., and Michael, J. G.,** Human lactic dehydrogenase as a marker for human tumor cells grown in athymic mice, *Cancer Res.,* 37, 1998, 1977.
33. **Dipersio, L., Kyriazis, A. P., Michael, J. G., and Pesce, A. J.,** Monitoring the therapy of human tumor xenografts in nude mice by the use of lactate dehydrogenase, *J. Natl. Cancer Inst.,* 62, 375, 1979.
34. **Nakamura, T. and Kitagawa, T.,** Anticancer drug screening test with LDH in nude mouse bearing bone and soft part sarcoma, *Cancer,* 56, 1112, 1985.
35. **Steel, G. G., Courtenay, V. D., and Peckham, M. J.,** The response to chemotherapy of a variety of human tumour xenografts, *Br. J. Cancer,* 47, 1, 1983.
36. **Boven, E.,** Biology of human ovarian cancer xenografts, in *Human Tumour Xenografts in Anticancer Drug Development,* Winograd B., Peckham, M. J., and Pinedo, H. M., Eds., Springer-Verlag, Berlin, 1988, 11.
37. **Ohsawa, N., Ueyama, Y., Morita, K., and Kondo, Y.,** Heterotransplantation of human functioning tumors to nude mice, in *Proc. 2nd Int. Workshop on Nude Mice,* Nomura, T., Ohsawa, N., Tamaoki, N., and Fujiwara, K., Eds., University of Tokyo Press, Tokyo, 1977; Gustav Fischer Verlag, Stuttgart, 1977, 395.
38. **Asano, S., Sato, N., Mori, M., Ohsawa, N., Kosaka, K., and Ueyama, Y.,** Detection and assessment of human tumours producing granulocyte-macrophage colony-stimulating factor (GM-CSF) by heterotransplantation into nude mice, *Br. J. Cancer,* 41, 689, 1980.
39. **Kameya, T., Shimosato, Y., Hayashi, H., and Tsumuraya, M.,** Growth and differentiation of hormone-producing human tumors in nude mice, in *Proc. 2nd Int. Workshop on Nude Mice,* Nomura, T., Ohsawa, N., Tamaoki, N., and Fujiwara, K., Eds., University of Tokyo Press, Tokyo, 1977; Gustav Fischer Verlag, Stuttgart, 1977, 405.
40. **Kim, W., Takahashi, T., Nisselbaum, J. S., and Lewis, J. L.,** Heterotransplantation of human choriocarcinoma in nude mice. I. Morphologic and biologic characteristics, *Gynecol. Oncol.,* 6, 165, 1978.
41. **Ruddon, R. W., Bryan, A. H., Meade-Cobun, K. S., and Pollack, V. A.,** Production of human chorionic gonadotropin and its subunits by human tumors growing in nude mice, *Cancer Res.,* 40, 4007, 1980.
42. **Hirohashi, S., Shimosato, Y., Kameya, T., Koide, T., Mukojima, T., Taguchi, Y., and Kageyama, K.,** Production of α-fetoprotein and normal serum proteins by xenotransplanted human hepatomas in relation to their growth and morphology, *Cancer Res.,* 39, 1819, 1979.

43. **Hata, Y., Uchino, J., Sato, K., Sasaki, F., Une, Y., Naito, H., Manabe, K., Kuwahara, T., and Kasai Y.,** Establishment of an experimental model of human hepatoblastoma, *Cancer,* 50, 97, 1982.
44. **Bassendine, M. F., Wright, N. A., Thomas, H. C., and Sherlock, S.,** Growth characteristics of α-fetoprotein secreting human hepatocellular carcinoma in athymic (nude) mice, *Clin. Sci.,* 64, 643, 1983.
45. **Miwa, M., Hirohashi, S., Shimosato, Y., Mukojima, T., Ohkura, H., Sakura, H., Kawachi, T., and Sugimura, T.,** Coproduction of carcinoembryonic antigen and α-fetoprotein in transplantable human colon cancer in nude mice, in *Proc. 2nd Int. Workshop on Nude Mice,* Nomura, T., Ohsawa, N., Tamaoki, N., and Fujiwara, K., Eds., University of Tokyo Press, Tokyo, 1977; Gustav Fischer Verlag, Stuttgart, 1977, 435.
46. **Stragand, J. J., Yang, L. Y., and Drewinko, B.,** Serum CEA levels in a human colonic adenocarcinoma (LoVo) xenograft system, *Cancer Lett.,* 10, 45, 1980.
47. **Martin, K. W. and Halpern, S. E.,** Carcinoembryonic antigen production, secretion, and kinetics in BALB/c mice and a nude mouse-human tumor model, *Cancer Res.,* 44, 5475, 1984.
48. **Bast, R. C., Klug, T. L., St. John, E., Jenison, E., Niloff, J. M., Lazarus, H., Berkowitz, R. S., Leavitt, T., Griffiths, C. T., Parker, L., Zurawski, V. R., and Knapp, R. C.,** A radioimmunoassay using a monoclonal antibody to monitor the course of epithelial ovarian cancer, *N. Engl. J. Med.,* 309, 883, 1983.
49. **Molthoff, C. F. M., Calame, J. J., Pinedo, H. M., and Boven, E.,** Human ovarian xenografts in nude mice: characterization and analysis of antigen expression, *Int. J. Cancer,* 47, 72, 1991.
50. **Klapdor, R., Harms, F., Bahlo, M., Arps, H., and Dietel, M.,** Tumor marker secretion (CA19-9, CEA, CA125) by xenografts of eight different human pancreatic carcinomas compared to the human tumors, *Strahlenther. Onkol.,* 165, 496, 1989.
51. **Friedlander, M. L., Russell, P., Taylor, I. W., and Tattersall, M. H. N.,** Ovarian tumour xenografts in the study of the biology of human epithelial ovarian cancer, *Br. J. Cancer,* 51, 319, 1985.
52. **Molthoff, C. F. M., Boven, E., Calame, J. J., Van Beek, P. J., and Pinedo, H. M.,** Characterization of human ovarian cancer lines with a panel of monoclonal antibodies, *Strahlenther. Onkol.,* 165, 494, 1989.
53. **Schlom, J.,** Basic principles and applications of monoclonal antibodies in the management of carcinomas: the Richard and Hinda Rosenthal Foundation Award Lecture, *Cancer Res.,* 46, 3225, 1986.
54. **Wessels, B. W.,** Current status of animal radioimmuno-therapy, *Cancer Res.,* 50 (Suppl.), 970s, 1990.
55. **Marks, A., Ettenson, D., Bjorn, M. J., Lei, M., and Baumal, R.,** Inhibition of human tumor growth by intraperitoneal immunotoxins in nude mice, *Cancer Res.,* 50, 288, 1990.
56. **Sporn, M. B. and Roberts, A. B.,** Autocrine growth factors and cancer. *Nature (London),* 313, 745, 1985.
57. **Goustin, A. S., Leof, E. B., Shipley, G. D., and Moses, H. L.,** Growth factors and cancer, *Cancer Res.,* 46, 1015, 1986.
58. **Bailey, M. J., Gazet, J.-C., and Peckham, M. J.,** Human breast cancer xenografts in immune-suppressed mice, *Br. J. Cancer,* 42, 524, 1980.
59. **Michel, R.-T., Schmidt-Matthiesen, H., Bastert, G., and Fortmeyer, H. P.,** Human breast cancer in nude mice: a model for testing endocrine treatment, in *Endocrine Treatment of Breast Cancer: a New Approach,* Henningsen, B., Linder, F., and Steichele, C., Eds., Springer, Berlin, 1980, 59.
60. **Horvath, G., Tropé, C., Alm, P., and Fernö, M.,** Heterotransplantation of human malignant endometrial tumors into nude mice. *In Vivo,* 2, 361, 1988.
61. **Kleine, W.,** Growth of malignant gynecological tumors xenotransplanted into thymusaplastic nu/nu mice, *Gynecol. Oncol.,* 24, 286, 1986.
62. **Friedl, A., Gottardis, M. M., Pink, J., Buchler, D. A., and Jordan, V. C.,** Enhanced growth of an estrogen receptor-negative endometrial adenocarcinoma by estradiol in athymic mice, *Cancer Res.,* 49, 4758, 1989.
63. **Brünner, N., Spang-Thomsen, M., Skovgaard-Poulsen, H., Engelholm, S. A., Nielsen, A., and Vindeløv, L.,** Endocrine sensitivity of the receptor-positive T61 human breast carcinoma serially grown in nude mice, *Int. J. Cancer,* 35, 59, 1985.
64. **Osborne, C. K., Hobbs, K., and Clark, C. M.,** Effect of estrogens and antiestrogens on growth of human breast cancer cells in athymic nude mice, *Cancer Res.,* 45, 584, 1985.
65. **Van Steenbrugge, G. J., Blankenstein, M. A., Bolt-de Vries, J., Romijn, J. C., Schröder, F. H., and Vihko, P.,** Effect of hormone treatment on prostatic acid phosphatase in a serially transplantable human prostatic adenocarcinoma (PC-82), *J. Urol.,* 129, 630, 1983.
66. **Masui, H., Kawamoto, T., Sato, J. D., Wolf, B., Sato, G., and Mendelsohn, J.,** Growth inhibition of human tumor cells in athymic mice by anti-epidermal growth factor receptor monoclonal antibodies, *Cancer Res.,* 44, 1002, 1984.
67. **Takahashi, H., Herlyn, D., Atkinson, B., Powe, J., Rodeck, U., Alavi, A., Bruce, D. A., and Koprowski, H.,** Radioimmunodetection of human glioma xenografts by monoclonal antibody to epidermal growth factor receptor, *Cancer Res.,* 47, 3847, 1987.
68. **Goldenberg, A., Masui, H., Divgi, C., Kamrath, H., Pentlow, K., and Mendelsohn, J.,** Imaging of human tumor xenografts with an indium-111-labeled anti-epidermal growth factor receptor monoclonal antibody, *J. Natl. Cancer Inst.,* 81, 1616, 1989.
69. **Mattern, J., Bak, M., Hahn, E. W., and Volm, M.,** Human tumor xenografts as model for drug testing, *Cancer Metast. Rev.,* 7, 263, 1988.

Chapter 9

STUDIES IN TUMOR CELL BIOLOGY

Simon P. Langdon and John F. Smyth

TABLE OF CONTENTS

I. INTRODUCTION

Xenografts grown in nude mice have been used extensively to model the behavior of human tumors *in vivo*. While transplantation of a human tumor to an immunosuppressed mouse does not provide the same host environment as the *de novo* tumor, it represents an improved model relative to *in vitro* culture. Thus, the influence of host interactions, for example, effects of stroma and host endocrinology, can be studied and information complementary to data from *in vitro* studies can be obtained.

As the model provides an environment for basic experimental investigation of tumor growth under reproducible and controlled conditions, the effects of growth modulators such as hormones and growth factors have been conveniently studied. Certain features of the original tumor, such as oncogene expression, drug resistance, production of tumor markers, and heterogeneity, are often maintained in the xenograft and these have also been examined in the nude mouse model. Examples of these studies are described in the following sections. The model has also been used to investigate the potential of monoclonal antibodies for tumor imaging and therapy and studies of this nature are described in Chapter 17. The purpose of this chapter is not to attempt a comprehensive review of the numerous biological studies which have used nude mice, but to illustrate, with selected examples, the diverse ways in which nude mice have been used to help extend our knowledge of human tumor biology.

II. HORMONE DEPENDENCY

A. BREAST CANCER

The growth of several tumor types including breast, prostate, and endometrial cancer is, to some degree, under hormonal control. In the case of breast cancer, one in three tumors is thought to be regulated by estrogen.[1] The estrogen-dependent MCF-7 mammary xenograft has provided a useful model to study the estrogen regulation of breast cancer cells *in vivo*. Initial studies in nude mice indicated the importance of estrogen supplementation for continued growth of this xenograft.[2-7] Tumorigenicity and growth rate varied at different implantation sites and were shown to be directly related to the ability of the host tissue to concentrate estrogen.[5,6] Thus, the periuterine region was the only site sufficiently estrophilic to support tumor growth without the need for exogenous estrogen.[4] Other hormones, including prolactin and insulin are also capable of promoting MCF-7 growth *in vivo*.[8] The requirement of estrogen supplementation for optimal growth has been demonstrated for other estrogen receptor (ER)-positive breast cancer xenografts, including the ZR 75-1[9] and T47D lines.[10] In contrast, exogenous estrogen has no effect on the growth of the ER-negative MDA-MB-231 xenograft.[5,9]

The effects of anti-estrogen therapy have been studied against the MCF-7 xenograft. Treatment with the anti-estrogens tamoxifen and LY156758 inhibited estrogen stimulated growth in nude mice, but did not cause tumor regression or loss of cell viability.[9,11-16] These anti-estrogens also inhibit growth of the ZR 75-1, but not MDA-MB-231 xenografts, which is consistent with the level of ER present.[9] Most studies with these tumors followed growth for a period of 1 to 2 months after initial treatment. Since tamoxifen may be given to patients for prolonged periods as adjuvant therapy, studies were conducted to investigate the effect of long-term treatment on xenografts. In 2 studies, over 4 and 6 months respectively,[11,13] MCF-7 tumors which had been treated continuously with tamoxifen and whose growth was inhibited could still be stimulated by estrogen. However, after 3 to 4 months of endocrine therapy, tumors began to regrow, despite continued treatment, implying the development of hormone independence.[11,13] Some reports have even indicated stimulation of growth by tamoxifen,[13,16] and MCF-7 variants with the tamoxifen-stimulated phenotype have been isolated.[16]

The responses of these lines to estrogen and tamoxifen and presence or absence of receptor are summarized in Table 1.

TABLE 1
Hormone dependency of Breast and Endometrial Cancer Xenografts

Xenograft	Tumor type	Estrogen receptor status	Progesterone receptor status	PR Induced by estrogen	Effect on Growth of: Estrogen	Effect on Growth of: Tamoxifen	Ref.
MCF-7	Breast	+ve	+ve	+ve	Stimulatory	Inhibitory	1–7, 9, 11, 12–16, 26
ZR 75-1	Breast	+ve	+ve	NR[a]	Stimulatory	Inhibitory	9
T47D	Breast	+ve	NR	NR	Stimulatory	NR	10
T61	Breast	+ve	+ve	+ve	Inhibitory	Inhibitory	27–29
MDA-MB–231	Breast	–ve	–ve	NR	No effect	No effect	5, 9
EnCa X	Endometrial	+ve	–ve	+ve	Stimulatory	Stimulatory	17–19
EnCa 101	Endometrial	+ve	+ve	+ve	Stimulatory	Stimulatory	14, 19
EnCa V	Endometrial	–ve	–ve	–ve	No effect	No effect	17–19
EnCa K	Endometrial	–ve	–ve	–ve	No effect	NR	19

[a] NR = Not recorded.

B. ENDOMETRIAL CANCER

The first models developed to study the hormonal regulation of endometrial cancer were the nude mouse xenografts of Satyaswaroop et al. (Table 1).[17-20] This series of xenografts includes the ER-positive EnCa X and EnCa 101 tumors, wherein estrogen stimulates growth and induces expression of the progesterone receptor (PR) in the former and increases PR in the latter.[17-19] In contrast, growth of the ER-negative EnCa V and EnCa K xenografts was not stimulated by estrogen and PR was not induced in either tumor. These findings were therefore analogous to those obtained in breast cancer xenografts. However, while tamoxifen had no effect on the growth of the ER-negative EnCa V xenograft,[19] it stimulated growth of the ER-positive EnCa X and EnCa 101 xenografts.[18-20] Tamoxifen is not a pure anti-estrogen and is known to possess estrogenic as well as anti-estrogenic activity,[13,16] and it is feasible that it acts as an estrogen in these models. The contrasting effects of tamoxifen against breast and endometrial cancer xenografts were well illustrated in an experiment wherein the tamoxifen-stimulated EnCa 101 tumor was implanted s.c. on one side of a nude mouse and the estrogen-stimulated MCF-7 tumor was transplanted on the other side of the same animal.[14] Tamoxifen treatment resulted in stimulation of the endometrial tumor growth, but control of the breast tumor growth.

Recently, a study has been conducted with two xenografts newly derived from primary endometrial cancers. These xenografts were used in their second and third passages in mice.[20] One xenograft contained both ER and PR and was stimulated by estrogen or tamoxifen or a mixture of these two. The other xenograft, although ER-negative, was stimulated by estradiol and tamoxifen also. This study concluded that estradiol and tamoxifen may stimulate tumor growth in the nude mouse independently of ER, possibly through a host mechanism.

Clearly, the effects of tamoxifen on endometrial cancer are complex. Clinical studies suggest that tamoxifen may have some benefit in the treatment of endometrial cancer.[21] However, there is also a report of the detection of endometrial cancer in three women undergoing tamoxifen treatment for breast cancer,[22] suggesting tumor promotion. Further studies with the above models may help unravel these paradoxes.

C. PROSTATE CANCER

About 75% of patients with prostate cancer respond to endocrine therapy in the form of orchiectomy or treatment with estrogens, progestins, or anti-androgens.[23] One of the first human prostate cancer xenografts shown to be hormonally dependent was the LNCaP model established from a metastatic lesion of this disease.[24] This xenograft contains both androgen and estrogen receptors. The frequency of tumor development and the mean time of tumor appearance in nude mice was significantly different between the sexes and related to serum androgen levels. Thus, castration in male mice reduced the frequency of tumor development, while testosterone treatment enhanced growth. Frequency of tumor appearance was reduced in females, while testosterone treatment markedly increased the rate of tumor appearance.

Other well-characterized models of this disease include the HONDA xenograft described by Ito et al.[23] and the PC82 tumor of Hoehn et al.[25] These models respond to hormonal manipulation in a manner similar to the LNCaP xenograft.

III. GROWTH FACTORS

Evidence that polypeptide growth factors can regulate the growth of tumor cells has come from *in vitro* studies. With several tumor types, these studies have been extended *in vivo* using nude mice.

Growth factors have been proposed to mediate the estrogen regulation of growth in estrogen-sensitive breast tumors.[1] Evidence from xenograft studies supports this concept.[26] When medium conditioned by MCF-7 cells grown *in vitro* in the presence of estrogen was infused into oophorectomized nude mice (after acid dialysis to remove estrogen), it supported growth of

MCF-7 xenografts without the need for estrogen supplementation. This implied that cultivated human breast cancer cells under estrogen stimulation release tumor growth promoting factors which can act *in vivo*. Levels of transforming growth factor alpha (TGF-α) and insulin-like growth factor I (IGF-I) have been shown to be increased in MCF-7 cells after estrogen stimulation *in vitro*.[1] When epidermal growth factor (which acts through the same receptor as TGF-α) or IGF-I were infused into oophorectomized nude mice, they supported MCF-7 tumor growth *in vivo*, in the absence of exogenous estrogen.[26] The significance and generality of these observations are, however, unclear. Thus, while levels of TGF-α protein and mRNA are elevated *in vivo* after estrogen treatment in MCF-7 xenografts and this occurs with stimulation of tumor growth, levels are also elevated in the T61 breast cancer xenograft after estrogen treatment, but this occurs with inhibition of growth.[27-29]

Gastrin-releasing peptide, the mammalian homolog of amphibian bombesin, has been shown to be an autocrine factor in small cell lung cancer (SCLC) cells *in vitro*.[30] Bombesin, when injected i.p., stimulated growth of NCI-H69 SCLC cells implanted s.c. in nude mice.[31] The monoclonal antibody, 2A11, which had been raised against an analog of bombesin, inhibited s.c. growth of the NCI-N592 SCLC xenograft.[30] Both of these SCLC tumors had previously been shown to possess bombesin-like immunoreactivity. This data provided the rationale for an ongoing clinical trial of the 2A11 antibody for the treatment of SCLC.[32] The NCI-N592 xenograft has also been used to test antagonists of bombesin.[33] One of these antagonists, Psi-13,14, inhibited tumor growth if injected s.c. next to the tumor.

IGF-I is another proposed autocrine growth factor for SCLC.[34] Somatostatin and its analogs inhibit the endogenous production of IGF-I and a somatostatin analog has demonstrated growth inhibition of NCI-H69 SCLC cells *in vivo*.[35]

Growth modulating polypeptides may be inhibitory as well as stimulatory. One factor commonly inhibitory in epithelial systems is transforming growth factor beta (TGF-β). In a recent report, both TGF-β types I and II, when injected peritumorally, inhibited the growth of the A549 lung adenocarcinoma xenograft in a dose-dependent manner.[36]

The presence and expression of growth factor receptors has been studied in xenograft models. In a report of Humphrey et al.,[37] xenografts from eight malignant gliomas were established to study the amplification and expression of the EGF receptor gene. EGF receptor gene amplification was found in six of the eight glioma biopsies studied and also in the corresponding xenografts. Significant expression of the EGF receptor gene was only detectable in xenografts with gene amplification. Furthermore, five of the six xenografts with amplification demonstrated structural alterations of the EGF receptor gene and this was associated with low molecular weight EGF receptor proteins. Thus, the xenograft models represented an excellent tissue source of material and corresponded well with the original tumors.

The presence of a growth factor receptor also provides the possibility of inhibiting tumor growth. Administration of anti-EGF receptor monoclonal antibodies inhibited tumor formation of the A431 and T222 epidermoid carcinoma cell lines when injected into nude mice. Both these tumors contain EGF receptors.[38] Similarly, an anti-transferrin receptor monoclonal antibody has been shown to inhibit tumor formation of a human melanoma cell line in athymic mice.[39]

IV. ONCOGENES

Since 1972, nude mice have been used to assay the tumorigenicity of human cell lines.[40] Soon thereafter, the system was used to help detect the transformed phenotype in cells after infection with oncogenic viruses such as SV40.[41-43] Cell lines such as the murine embryonic 3T3 line have commonly been used as recipients of the virus in these experiments. These cells are non-tumorigenic[41,42] (or, at best, very poorly tumorigenic[44,45]) before transformation; however, once transformed, form undifferentiated fibrosarcomas at the site of injection.[41,42] More recently, the assay has been used to help detect dominant transforming genes in cellular DNA.[46,47] The

TABLE 2
Myc Amplification and Expression in Xenografts

Myc gene	Xenograft type	Amplification (amp), overexpression (overexp), or allelic deletion (del)	Ref.
c-myc	Lung cancer	Amp	55, 56
	Breast cancer	Amp, overexp	57
	Gastric cancer	Amp, overexp	58, 59
	Hepatoblastoma	Overexp	60
	Osteosarcoma	Amp, overexp	61
	Intestinal cancer	Overexp	62
	Renal cell cancer	Overexp	62
	Ewing sarcoma	Overexp	62
N-myc	Neuroblastoma	Amp	63
L-myc	Lung cancer	Amp, del	56

presence of transforming genes in the DNA of chemically transformed human cell lines and cell lines derived from human teratocarcinomas, pancreatic, and bladder cancers have been screened using the nude mouse assay.[46,47] In these experiments, DNA from tumor cells was transfected into NIH/3T3 cells and this population of cells were then injected into mice. Thus, for example, DNA from the EJ bladder cancer cell line produced tumors in 6 of 6 mice, while DNA derived from NIH (A1165)-Tl pancreatic tumor cells produced tumors in 12 of 28 mice.

The malignant nature of oncogenic transformation has primarily been demonstrated by the *in vitro* focus induction assay, but in many cases has also been verified by the use of the nude mouse. One example of where this has been done was the demonstration by Chang et al.[48] that the c-Ha-ras l gene was oncogenic when ligated to a long terminal repeat sequence. Transfection of this gene into NIH/3T3 cells, in addition to producing foci *in vitro,* resulted in tumorigenicity in nude mice. Since then, many studies using nude mice have examined the involvement of ras in tumorigenesis, progression, hormone-dependence, and metastasis. For example, NIH/3T3 cells transfected with activated c-Ha-ras or N-ras are not only highly tumorigenic in nude mice, but also form metastasis very efficiently.[44,49,50] In contrast, parent NIH/3T3 cells were tumorigenic only after a very prolonged period and rarely metastasized.[44] Spontaneous 3T3 transformants were tumorigenic, but non-metastatic.[49] The effects of ras on other tissue types have also been examined. For instance, v-Ha-ras can transform human breast epithelial cells which produce tumors after injection into mice.[51] The estrogen-dependent MCF-7 cancer cell line (described earlier), when transfected with v-Ha-ras, becomes hormone independent.[52] These hormone-independent cells secrete factors which not only support their own growth, but can support the growth of hormone-dependent tumors without the need for exogenous estrogen. Introduction of the activated c-Ha-ras or N-ras gene into non-tumorigenic liver or teratocarcinoma cells also produces tumorigenicity.[53,54]

The presence of over-expression and amplification of members of the myc gene family has been demonstrated in xenografts from a number of tumor types (Table 2). A relationship between c-myc or L-myc gene abnormalities and xenotransplantibility has been proposed and data from lung, breast, and gastric cancer xenograft studies supports this proposal.[56-58] In a study of 32 lung tumors (covering all histologies), 11 of 13 cancers with abnormalities of the myc family genes (including all tumors with myc gene amplification) were transplantable to nude mice.[56] In contrast, only 5 of 19 tumors without myc gene abnormality were tumorigenic. It is feasible, however, that amplification and over-expression of c-myc might occur during serial transfer in nude mice and, thus, may not truly reflect the situation in primary tumors.

V. DRUG RESISTANCE

Xenograft models have been used to investigate mechanisms of drug resistance *in vivo*. Repair of DNA interstrand and DNA-protein cross-linking is thought to be one mechanism via which tumors become resistant to DNA interacting agents. In support of this, the group of Kohn et al.,[64,65] using colon cancer xenograft models sensitive or resistant to 1-(2-chloroethyl)-3-(4-methylcyclohexyl)-1-nitrosourea (MeCCNU) demonstrated that the resistant xenograft exhibited less cross-linking relative to the sensitive xenograft after exposure to a single i.p. injection of MeCCNU. A mitomycin C-resistant colon cancer xenograft derived from a sensitive line by repeated exposure to mitomycin C also had decreased cross-linking relative to the sensitive xenograft after treatment.[66] One of the enzymes involved in DNA repair is O^6-alkylguanine-DNA alkyltransferase (AT), the primary intracellular repair mechanism for lesions at the O^6 position of guanine. Schold et al.[67] have shown, using human brain tumor xenografts, that in tumors where levels of AT were greater than 100 fmol/mg protein, procarbazine had little effect; in contrast, where levels of AT were undetectable, procarbazine produced growth delays greater than 30 d.

Recently, the importance of multidrug or pleitropic drug resistance has been identified. As resistance is induced in a tumor cell to an agent such as doxorubicin, collateral resistance to agents such as vincristine and actinomycin D also develops. This phenomenon is associated with increased expression of an MW 170,000 cell surface glycoprotein (P-glycoprotein), which is believed to function as a drug efflux pump.[68]

One of the best descriptions of the multidrug resistance phenomenon in a xenograft system is that by Mattern et al.[69-71] using human lung squamous cell cancers. These studies, in a panel of eight xenografts, demonstrated that xenograft lines derived from human lung tumors not previously treated with chemotherapy exhibit a similar pattern of cross-resistance to the drugs doxorubicin, actinomycin D, and vincristine, and collateral sensitivity to alkylating agents and metabolites. The level of expression of P-glycoprotein detected by a monoclonal antibody 265/F4 correlated with the degree of drug resistance.[71] Verapamil, the calcium channel blocker capable of modifying vincristine resistance *in vitro*, was able to partially overcome resistance in one of the resistant xenografts.[69] Sublines resistant to vincristine, actinomycin D, and cisplatin were derived from a xenograft that was initially sensitive to these agents.[70] Resistance developed rapidly to the former two agents, while repeated administration was required to derive a gradual decrease in response to cisplatin. The vincristine resistant subline demonstrated cross-resistance to actinomycin D and doxorubicin, and expression of P-glycoprotein was induced.[71]

The above studies have confirmed that many of the observations relating to mechanisms of drug resistance observed *in vitro* also apply for the *in vivo* situation.

VI. PRODUCTION OF TUMOR MARKERS

The nude mouse has provided a useful system in which to study the *in vivo* synthesis, secretion, and excretion of tumor markers. The potential of tumor markers as a target for tumor imaging and radioimmunotherapy has also been investigated.

The most extensively studied tumor-associated markers are the oncodevelopmental antigens. These include carcinoembryonic antigen, α-fetoprotein, and human chorionic gonadotropin. Other proteins, such as CA125[72] (in ovarian cancer) and LDH isoenzymes, have also been studied.[73]

A. CARCINOEMBRYONIC ANTIGEN (CEA)

The occurrence of CEA in human colorectal cancer was first described in 1965[74] and the presence of CEA in a human colon xenograft grown in nude mice reported in 1974.[75] A great deal of attention has focused on a possible correlation between serum CEA levels and colorectal

TABLE 3
Radiolabeled Anti-CEA Antibodies Used in Colorectal Cancer Xenografts

Antibody	Monoclonal or polyclonal	Isotope	Localization or radioimmunotherapy	Xenograft model	Ref.
Unnamed	Polyclonal	^{131}I	Localization	115	75
38S1	Monoclonal	^{131}I	Localization	LS174T and HT-29	83
1H12	Monoclonal	^{125}I	Localization	MAWI	84
T84.66.A3.1H11	Monoclonal	^{111}In	Localization	LS174T	85
CEJ-326	Monoclonal	^{111}In	Localization	T380	86
Mixture	Monoclonals	^{131}I	Radioimmunotherapy	T380	87
NP-2	Monoclonal	^{90}Y	Radioimmunotherapy	GW-39	88

tumor size since this might provide a useful means of monitoring the progress of the disease. Several studies have been conducted in nude mice to try and establish whether such a relationship exists.[76-79] The studies by Martin et al.[76] and Miwa et al.[77] showed a correlation between tumor size and serum CEA levels, though not with tumor CEA content. This data contrasts with that of Stragand et al.[78] and Lewis et al.,[79] who did not find a correlation. These results indicated that the measurement of circulating CEA can provide (in some cases) an indicator of CEA secretion, but gives no indication of the extent of its production.

A relationship between the growth potential of xenograft and preoperative serum concentrations of CEA in patients from whom the material was obtained has been suggested.[80] In a recent study, 26 of 33 cancers from patients where serum levels exceeded 5 ng/ml formed xenografts in nude mice, whereas only 8 of 22 tumors from patients with normal CEA levels (less than 5 ng/ml) were tumorigenic. However, data from other studies does not support this presumption. For instance, in a study with the HCT-l colorectal cancer cell line, a variant was cloned which produced about eight times the level of CEA compared to the parent line; however, both clones were equally tumorigenic.[81] In a study of three esophageal cancer cell lines, the secretion of CEA in a non-tumorigenic line was intermediate between that of the two tumorigenic lines, again indicating the lack of a relationship.[82]

A number of monoclonal and polyclonal antibodies raised against CEA have been shown to localize in colorectal cancer xenografts. These antibodies have been labeled with a variety of radionuclides, including ^{131}I, ^{125}I, and ^{111}In (Table 3). The xenograft model is particularly useful for studying the amounts of specific antibody retained by the tumor and antibody-residence time. This information is necessary for helping define effective half-lives and doses for optimal therapeutic administration.[84] For radioimmunotherapy, both ^{131}I and ^{90}Y-labeled monoclonal anti-CEA antibodies have been shown to inhibit colorectal cancer xenograft growth (Table 3).

B. ALPHA-FETOPROTEIN (AFP)

AFP is a serum protein produced by the fetal liver and yolk sac during prenatal development and is often found in the corresponding tumors.[89-95] Neiderberger et al.[89] have established three human testicular teratocarcinoma xenografts — an embryonal cell carcinoma and two yolk sac tumors — and compared various characteristics of these xenografts when grown either s.c. or in the peritoneal cavity.[89] AFP was detected in the serum and tissue of the yolk sac tumor xenografts, but not in the transplants of the embryonal cell carcinoma, although it had been detected in the serum of all three donating patients. Attachment of tumor cells to a surface appeared to be necessary for production of AFP since it was not secreted by spheroids produced from these tumors. AFP was also detected in the plasma and three different yolk sac tumor xenografts described by Hata et al.[90]

The relationship between tumor size and serum AFP levels in liver tumor xenografts has been explored. In one detailed study, Hirohashi et al.[91] established five liver cell cancers and two hepatoblastoma xenografts in nude mice, and showed the presence of AFP in the tumor tissue of six of these xenografts and in the serum of these mice. A positive correlation was found between AFP levels and tumor size and cytological differentiation towards liver cells. This finding has been supported by data from other studies.[92,93] Interestingly, this marker is often lost when cells are cultured *in vitro,* but re-expressed when cells are transferred back to an in vivo environment.[93,94]

C. HUMAN CHORIONIC GONADOTROPIN (HCG)

HCG is secreted by the normal placenta and various trophoblastic tumors. Small amounts of HCG are also found in the serum of approximately 16% of patients with malignancies that do not contain trophoblastic elements.[96] The synthesis, secretion, and excretion of HCG and its subunits by eutopic and ectopic hormone-producing tumors has been studied in nude mouse models.[97-100] In several of these studies, correlations between serum or plasma levels of HCG and tumor size have been demonstrated.[98-100] Comparison of a variety of tumor cell types *in vitro* and *in vivo* revealed that there was a shift towards complete HCG production and a decrease in free subunit production in nude mice[97,98] as compared to the pattern observed for cultured cells.[97,101]

VII. HETEROGENEITY

The proliferating cells of a malignant neoplasm may be markedly heterogeneous with respect to morphology, karyotype, metastatic ability, the presence and levels of hormone and growth factor receptors, sensitivity to cytotoxic drugs and irradiation, expression of surface antigens, and tumorigenic capacity.[102] Many of these features have been examined in xenograft systems.

Heterogeneity of metastatic ability within a human tumor cell population has been demonstrated by isolating subpopulations of cells with increased metastatic potential.[103-106] This has been done by injecting a population of tumor cells i.v. and then establishing cell lines from the pulmonary metastases formed. These metastatic lines were then re-injected and produced markedly more lung metastases than the parent population, indicating the presence of highly metastatic clones in the initial population. Metastatic sublines have been derived in this way in nude mice from human malignant melanoma (MeWo, A375-M),[103,104] colorectal cancer (HT29-M),[104] and prostate cancer (PC-3-M)[104] cell lines. Metastatic clones from murine (Bl6, UV-2237, K-1735)[105] and hamster (MMl)[106] cell lines have also been selected in nude mice by the same process.

The diversity of tumorigenic potential within breast cancer and sarcoma models has been shown by the use of clones derived from the parent population.[5,107] In a study examining heterogeneity within a sarcoma cell line, 25 clonal strains were isolated from the parent line and their tumorigenicity tested.[107] The clones could be divided into two types based on morphology and *in vitro* growth behavior. The type I clones were tumorigenic in nude mice and possessed modal chromosomal numbers ranging from 5l to 6l. Type II clones were not tumorigenic and possessed modal chromosome numbers from 47 to 54. When clones derived from the MCF-7 breast cancer cell line were implanted into mice, the take rate percentage varied from 0 to 94%.[5] Karyotypic analysis of these breast cancer sublines indicated both common, but also unique chromosome markers, while progesterone receptor levels varied10-fold. These results indicate that cells with very different tumorigenic potential exist within the parental cell population.

VIII. CONCLUSIONS

Nude mice have proven invaluable as a research tool to aid the study of the *in vivo* biology of human tumors. Most types of human tumors have now been transplanted and established as continuous lines in these animals. Many characteristics of the original tumor are maintained after transfer into mice, such as morphology, hormone secretion, oncogene expression, and expression of receptors for growth factors and hormones, thus allowing more detailed study of the original tumor than would otherwise be possible. Since nude mice provide the link between the tumor growing in its original environment and under *in vitro* conditions, the relevance of mechanisms of hormonal and growth factor control and resistance to cytotoxic agents, suggested on the basis of *in vitro* experimentation, can now be tested in an *in vivo* situation.

REFERENCES

1. **Lippman, M. E., Dickson, R. B., Gelmann, E. P., Knabbe, C., Kasid, A., Bates, S., and Swain, S.,** Mechanisms of estrogenic and antiestrogenic regulation of growth of human breast carcinoma, in *Hormonal Manipulation of Cancer: Peptides, Growth Factors and New (Anti)Steroidal Agents,* Klijn, J. G. M., Ed., Raven Press, New York, 1987, 381.
2. **Ozzello, L. and Sordat, M.,** Behaviour of tumors produced by transplantation of human mammary cell lines in athymic nude mice, *Eur. J. Cancer Clin. Oncol.,* 16, 553, 1980.
3. **Soule, H. D. and McGrath, C. M.,** Estrogen responsive proliferation of clonal human breast carcinoma in athymic mice, *Cancer Lett.,* 10, 177, 1980.
4. **White, A. C., Levy, J. A., and McGrath, C. M.,** Site-selective growth of a hormone-responsive human breast carcinoma in athymic mice, *Cancer Res.,* 42, 906, 1982.
5. **Seibert, C., Shafie, S. M., Triche, T. J., Whang-Peng, J. J., O'Brien, S. J., Toney, J. T., Huff, K. K., and Lipmann, M. E.,** Clonal variation of MCF-7 breast cancer cells *in vitro* and in athymic nude mice, *Cancer Res.,* 43, 2223, 1983.
6. **Levy, J. A., White, A. C., and McGrath, C. M.,** Growth and histology of a human mammary carcinoma cell line at different sites in the athymic mouse, *Br. J. Cancer,* 45, 375, 1982.
7. **Russo, J., McGrath, C. M., Russo, I., and Rich, M. A.,** Tumoral growth of a human breast cancer cell line (MCF-7) in athymic mice, in *3rd Int. Symp. on Detection and Prevention of Cancer,* Vol. 2, Nieburgs, N. H., Ed., Marcel Dekker, New York, 1975, 617.
8. **Shafie, S. M. and Grantham, F. H.,** Role of hormones in the growth and regression of human breast cancer cells (MCF-7) transplanted into athymic nude mice, *J. Natl. Cancer Inst.,* 67, 51, 1981.
9. **Osborne, C. K., Hobbs, K., and Clark, G. M.,** Effect of estrogens and antiestrogens on growth of human breast cancer cells in athymic nude mice, *Cancer Res.,* 45, 584, 1985.
10. **Leung, C. K. H. and Shiu, R. P. C.,** Required presence of both estrogen and pituitary factors for the growth of human breast cancer cells in athymic nude mice, *Cancer Res.,* 41, 546, 1981.
11. **Osborne, C. K., Coronado, E. B., and Robinson, J. P.,** Human breast cancer in the athymic nude mouse: cytostatic effects of long term anti-estrogen therapy, *Eur. J. Cancer Clin. Oncol.,* 23, 1189, 1987.
12. **Osborne, C. K., Hobbs, K., and Trent, J. M.,** Biological differences among MCF-7 human breast cancer cell lines from different laboratories, *Breast Cancer Res. Treat.,* 9, 111, 1987.
13. **Gottardis, M. M. and Jordan, V. C.,** Development of tamoxifen-stimulated growth of MCF-7 tumors in athymic nude mice after long term anti-estrogen administration, *Cancer Res.,* 48, 5183, 1988.
14. **Gottardis, M. M., Robinson, S. P., Satyaswaroop, P. G., and Jordan, V. C.,** Contrasting actions of tamoxifen on endometrial and breast tumor growth in the athymic mouse, *Cancer Res.,* 48, 812, 1988.
15. **Brünner, N., Bronzert, D., Vindeløv, L. L., Rygaard, K., Spang-Thomsen, M., and Lippman, M. E.,** Effect on growth and cell cycle kinetics of estradiol and tamoxifen on MCF-7 human breast cancer cells grown *in vitro* and in nude mice, *Cancer Res.,* 49, 1515, 1989.
16. **Gottardis, M. M., Wagner, R. J., Borden, E. C., and Jordon, V. C.,** Differential ability of antiestrogens to stimulate breast cancer cell (MCF-7) growth *in vivo* and *in vitro, Cancer Res.,* 49, 4765, 1989.
17. **Satyaswaroop, P. G., Zaino, R. J., and Mortel, R.,** Human endometrial adenocarcinoma transplanted into nude mice: growth regulation by estradiol, *Science,* 219, 58, 1983.
18. **Satyaswaroop, P. G., Zaino, R. J., and Mortel, R.,** Estrogen-like effects of tamoxifen on human endometrial carcinoma transplanted into nude mice, *Cancer Res.,* 44, 4006, 1984.
19. **Clarke, C. L. and Satyaswaroop, P. G.,** Photoaffinity labeling of the progesterone receptor from human endometrial carcinoma, *Cancer Res.,* 45, 5417, 1985.
20. **Friedl, A., Gottardis, M. M. Pink, J., Buchler, D. A., and Jordan, V. C.,** Enhanced growth of an estrogen receptor-negative endometrial adenocarcinoma by estradiol in athymic mice, *Cancer Res.,* 49, 4758, 1989.
21. **Bonte, J., Ide, P., Billiet, G., and Wynants, P.,** Tamoxifen as a possible chemotherapeutic agent in endometrial adenocarcinoma, *Gynecol. Oncol.,* 11, 140, 1981.

22. **Killackey, M. A., Hakes, T. B., and Pierce, V. K.,** Endometrial adenocarcinoma in breast cancer patients receiving antiestrogens, *Cancer Treat. Rep.*, 69, 237, 1985.
23. **Ito, Y. Z., Mashimo, S., Nakazato, Y., and Takikawa, H.,** Hormone dependency of a serially transplantable human prostatic cancer (HONDA) in nude mice, *Cancer Res.*, 45, 5058, 1985.
24. **Horoszewicz, J. S., Leong, S. S., Kawinski, E., Karr, J. P., Rosenthal, H., Chu, T. M., Mirand, E. A., and Murphy, G. P.,** LNCaP model of human prostatic carcinoma, *Cancer Res.*, 43, 1809, 1983.
25. **Hoen, W., Schroeder, F. H., Riemann, J. F., Joebis, A. C., and Hermanek, P.,** Human prostatic adenocarcinoma: some characteristics of a serially transplantable line in nude mice (PC 82), *Prostate*, 1, 95, 1980.
26. **Dickson, R. B., McManaway, M., and Lippman, M. E.,** Estrogen-induced growth factor activities from MCF-7 human breast cancer cells replace estrogen as *in vivo* stimuli of tumor formation. *Science*, 232, 1540, 1986.
27. **Brünner, N., Ciardiello, F., Spang-Thomsen, M., Salomon, D., Lippman, M. E., Gelmann, E., and Kern, F. G.,** Effect of estradiol (E2) on transforming growth factor alpha (TGF-α) mRNA expression and protein secretion in the E_2-inhibited T61 human breast carcinoma xenograft, *Proc. Am. Ass. Cancer Res.*, 30, 306, 1989.
28. **Brünner, N., Bastert, G. B., Poulsen, H. S., Spang-Thomsen, M., Engelholm, S. A., Vindeløv, L., Nielsen, A., Tommerup, N., and Elling, F.,** Characterisation of the T61 human breast carcinoma established in nude mice, *Eur. J. Cancer Clin. Oncol.*, 21, 833, 1985.
29. **Brünner, N., Spang-Thomsen, M., Vindeløv, L., Wolff, J., and Engelholm, S. A.,** Effect of tamoxifen on the receptor-positive T61 and the receptor-negative T60 human breast carcinomas grown in nude mice, *Eur. J. Cancer Clin. Oncol.*, 21, 1349, 1985.
30. **Cuttitta, F., Carney, D. N., Mulshine, J., Moody, T. W., Fedorka, J., Fischler, A., and Minna, J. D.,** Bombesin-like peptides can function as autocrine growth factors in human small-cell lung cancer, *Nature (London)*, 316, 823, 1985.
31. **Alexander, R. W., Upp, J. R., Poston, G. J., Gupta, V., Townsend, C. M., and Thompson, J. C.,** Effects of bombesin on growth of human small cell lung carcinoma *in vivo*, *Cancer Res.*, 48, 1439, 1988.
32. **Gazdar, A. F., Cuttitta, F., and Mulshine, J. L.,** Bombesin-like peptides and other growth factors in lung cancer, in *Proc. 6th NCI-EORTC Symp. on New Drugs in Cancer Therapy*, Amsterdam, 1989, 153.
33. **Mahmoud, S., Palaszynski, E., Cuttitta, F., Mulshine, J., Taylor, J., Coy, D., Koros, A. M. C., and Moody, T. W.,** Bombesin antagonists (Psi-13,14) inhibit the growth of small cell lung cancer, *Proc. Am. Ass. Cancer Res.*, 30, 1151, 1989.
34. **Macaulay, V. M., Teale, J. D., Everard, M. J., Joshi, G. P., Smith, I. E., and Millar, J. L.,** Somatomedin-C/insulin-like growth factor-I is a mitogen for human small cell lung cancer, *Br. J. Cancer*, 57, 91, 1988.
35. **Taylor, J. E. Bogden, A. E., Moreau, J.-P., and Coy, D. H.,** *In vitro* and *in vivo* inhibition of human small cell lung cancer (NCI-H69) by a somatostatin analogue, *Biochem. Biophys. Res. Commun.*, 153, 81, 1988.
36. **Twardzik, D. R., Ranchalis, J. E., McPherson, J. M., Ogawa, Y., Gentry, L., Purchio, A., Plata, E., and Todaro, G. J.,** Inhibition and promotion of differentiated-like phenotype of a human lung carcinoma in athymic mice by natural and recombinant forms of transforming growth factor-B, *J. Natl. Cancer Inst.*, 81, 1182, 1989.
37. **Humphrey, P. A., Wong, A. J., Vogelstein, B., Friedman, H. S., Werner, M. H., Bigner, D. D., and Bigner, S. H.,** Amplification and expression of the epidermal growth factor receptor gene in human glioma xenografts, *Cancer Res.*, 48, 2231, 1988.
38. **Masui, H., Kawamoto, T., Sato, J. D., Wolf, B., Sato, G., and Mendelsohn, J.,** Growth inhibition of human tumor cells in athymic mice by anti-epidermal growth factor receptor monoclonal antibodies, *Cancer Res.*, 44, 1002, 1984.
39. **Trowbridge, I. S. and Domingo, D. L.,** Anti-transferrin receptor monoclonal antibody and toxin-antibody conjugates affect growth of human tumor cells, *Nature (London)*, 294, 171, 1981.
40. **Giovanella, B. C., Yim, S. O., Stehlin, J. S., and Williams, L. J.,** Development of invasive tumors in the nude mouse after injection of cultured human melanoma cells, *J. Natl. Cancer Inst.*, 48, 1531, 1972.
41. **Shin, S., Freedman, V. H., Risser, R., and Pollack, R.,** Tumorigenicity of virus-transformed cells in nude mice is correlated specifically with anchorage-independent growth *in vitro*, *Proc. Natl. Acad. Sci. U.S.A.*, 72, 4335, 1975.
42. **Freedman, V. H. and Shin, S.,** Use of nude mice for studies on the tumorigenicity of animal cells, in *The Nude Mouse in Experimental and Clinical Research*, Fogh, J. and Giovanella, B. C., Eds., Academic Press, New York, 1978, 353.
43. **Stiles, C. D. and Kawahara, A. A.,** The growth behaviour of virus-transformed cells in nude mice, in *The Nude Mouse in Experimental and Clinical Research*, Fogh, J. and Giovanella, B. C., Eds., Academic Press, New York, 1978, 385.
44. **Greig, R. G., Koestler, T. P., Trainer, D. L., Corwin, S. P., Miles, L., Kline, T., Sweet, R., Yokoyama, S., and Poste, G.,** Tumorigenic and metastatic properties of normal and ras-transfected NIH/3T3 cells, *Proc. Natl. Acad. Sci. U.S.A.*, 82, 3698, 1985.
45. **Boone, C. W.,** Malignant hemangioendotheliomas produced by subcutaneous inoculation of Balb/c 3T3 cells attached to glass beads, *Science*, 188, 68, 1975.

46. **Blair, D. G., Cooper, C. S., Oskarsson, M. K., Eader, L. A., and Vande Woude, G. F. V.,** New method for detecting cellular transforming genes, *Science,* 218, 1122, 1982.
47. **Cooper, C. S., Blair, D. G., Oskarsson, M. K., Tainskey, M. A., Eader, L. A., and Vande Woude, G. F.,** Characterisation of human transforming genes from chemically transformed, teratocarcinoma and pancreatic carcinoma cell lines, *Cancer Res.,* 44, 1, 1984.
48. **Chang, E. H., Furth, M. E., Scolnick, E. M., and Lowy, D. R.,** Tumorigenic transformation of mammalian cells induced by a normal gene homologous to the oncogene of Harvey murine sarcoma virus, *Science,* 297, 479, 1982.
49. **Thorgeirsson, U. P., Turpeenniemi-Hujanen, T., Williams, J. E., Westin, E. H., Heilman, C. A., Talmadge, J. E., and Liotta, L. A.,** NIH/3T3 cells transfected with human tumor DNA containing activated ras oncogenes express the metastatic phenotype in nude mice, *Mol. Cell Biol.,* 5, 259, 1985.
50. **Muschel, R. J., Williams, J. E., Lowy, D. R., and Liotta, L. A.,** Harvey ras induction of metastatic potential depends upon oncogene activation and the type of recipient cell, *Am. J. Pathol.,* 121, 1, 1985.
51. **Clark, R., Stampfer, M. R., Milley, R., O'Rourke, E., Walen, K., Kriegler, M., Kopplin, J., and McCormick, F.,** Transformation of human mammary epithelial cells by oncogenic retroviruses, *Cancer Res.,* 48, 4689, 1988.
52. **Kasid, A., Knabbe, C., and Lippman, M. E.,** Effect of v-ras^H oncogene transformation on estrogen-independent tumorigenicity of estrogen-dependent human breast cancer cells, *Cancer Res.,* 47, 5733, 1987.
53. **Braun, L., Goyette, M., Yaswen, P., Thompson, N. L., and Fausto, N.,** Growth in culture and tumorigenicity after transfection with the ras oncogene of liver epithelial cells from carcinogen-treated rats, *Cancer Res.,* 47, 4116, 1987.
54. **Tainsky, M. A., Shamanski, F., Blair, D., and Giovanella, B. C.,** Causal role for an activated N-ras oncogene in the induction of tumorigenicity acquired by a human cell line, *Cancer Res.,* 47, 3235, 1987.
55. **Little, C. D., Nau, M. M., Carney, D. N., Gazdar, A. F., and Minna, J. D.,** Amplification and expression of the c-myc oncogene in human lung cancer cell lines, *Nature (London),* 306, 194, 1983.
56. **Gemma, A., Nakajima, T., Shiraishi, M., Noguchi, M., Gotoh, M., Sekiya, T., Niitani, H., and Shimosato, Y.,** myc Family gene abnormality in lung cancers and its relationship to xenotransplantability, *Cancer Res.,* 48, 6025, 1988.
57. **Modjtahedi, N., Lavialle, C., Poupon, M., Landin, R., Cassingena, R., Monier, R., and Brison, O.,** Increased level of amplification of the c-myc oncogene in tumors induced in nude mice by a human breast carcinoma cell line, *Cancer Res.,* 45, 4372, 1985.
58. **Shibuya, M., Yokoyata, J., and Ueyama, Y.,** Amplification and expression of a cellular oncogene (c-myc) in human gastric adenocarcinoma cells, *Mol. Cell Biol.,* 5, 414, 1985.
59. **Nakasato, F., Sakamoto, H., Mori, M., Hayashi, K., Shimosato, Y., Nishi, M., Takao, S., Nakatani, K., Terada, M., and Sugimura, T.,** Amplification of the c-myc oncogene in human stomach cancers, *Gann,* 75, 737, 1984.
60. **Huber, B. E., Dearfield, K. L., Williams, J. R., Heilman, C. A., and Thorgeirsson, S. S.,** Tumorigenicity and transcriptional modulation of c-myc and N-ras oncogenes in a human hepatoma cell line, *Cancer Res.,* 45, 4322, 1985.
61. **Bogenmann, E., Moghadam, H., DeClerck, Y. A., and Mock, A.,** c-Myc amplification and expression in newly established human osteosarcoma cell lines, *Cancer Res.,* 47, 3808, 1987.
62. **Tatosyan, A. G., Galetzki, S. A., Kiseljova, N. P., Asanova, A. A., Zborovskaya, I. B., Stipkovsky, D. D., Revasova, E. S., Martin, P., and Kisseljov, F. L.,** Oncogene expression in human tumors, *Int. J. Cancer,* 35, 731, 1985.
63. **Kanda, N., Tsuchida, Y., Hata, J., Kohl, J., Alt, F. W., Latt, S. A., and Utakoji, T.,** Amplification of IMR-32 clones 8, G21 and N-myc in human neuroblastoma xenografts, *Cancer Res.,* 47, 3291, 1987.
64. **Thomas, C. B., Osieka, R., and Kohn, K. W.,** DNA cross linking by *in vivo* treatment with 1-(2-chloroethyl)-3-(4-methylcyclohexyl)-1-nitrosourea of sensitive and resistant human colon carcinoma xenografts in nude mice, *Cancer Res.,* 38, 2448, 1978.
65. **Erickson, L. C., Osieka, R., and Kohn, K. W.,** Differential repair of 1-(2-chloroethyl)-3-(4-methylcyclohexyl)-1-nitrosourea-induced DNA damage in two human colon tumor cell lines, *Cancer Res.,* 38, 802, 1978.
66. **Willson, J. K., Long, B. H., Marks, M. E., Brattain, D. E., Wiley, J. E., and Brattain, M. G.,** Mitomycin C resistance in a human colon carcinoma cell line associated with cell surface protein alterations, *Cancer Res.,* 44, 5880, 1984.
67. **Schold, S. C., Jr., Brent, T. P., von-Hofe, E., Friedman, H. S., Mitra, S., Bigner, D. D., Swenberg, J. A., and Kleihuis, P.,** O^6-Alkylguanine-DNA alkytransferase and sensitivity to procarbazine in human brain tumor xenografts, *J. Neurosurg.,* 70, 573, 1989.
68. **Bradley, G., Juranka, P. F., and Ling, V.,** Mechanism of multidrug resistance, *Biochim. Biophys. Acta,* 948, 87, 1988.
69. **Mattern, J., Bak, M., and Volm, M.,** Occurence of a multidrug-resistant phenotype in human lung xenografts, *Br. J. Cancer,* 56, 407, 1987.

70. **Mattern, J., Bak, M., Hoever, K. H., and Volm, M.,** Development of drug resistance in a human epidermoid lung carcinoma xenograft line, *Br. J. Cancer,* 58, 30, 1988.
71. **Mattern, J., Bak, M., Efferth, T., and Volm, M.,** Inherent and induced multidrug resistance in human lung carcinoma xenografts, in *Proc. 6th NCI-EORTC Symp. on New Drugs in Cancer Therapy,* Amsterdam, 1989, 383.
72. **Hamilton, T. C., Young, R. C., Louie, K. G., Behrens, B. C., McKoy, W. M., Grotzinger, K. R., and Ozols, R. F.,** Characterisation of a xenograft model of human ovarian carcinoma which produces ascites and intraabdominal carcinomatosis in mice, *Cancer Res.,* 44, 5286, 1984.
73. **Pesce, A. J., Bubel, H. C., DiPersio, L., and Michael, J. G.,** Human lactate dehydrogenase as a marker for human tumors grown in athymic mice, *Cancer Res.,* 37, 1998, 1977.
74. **Gold, P. and Freedman, S. O.,** Specific carcinoembryonic antigens of the human digestive system, *J. Exp. Med.,* 122, 467, 1965.
75. **Mach, J.-P., Carrels, S., Merenda, C., Sordat, B., and Cerottini, J. C.,** *In vivo* localisation of radiolabelled antibodies to carcinoembryonic antigen in human colon carcinoma grafted into nude mice, *Nature (London),* 248, 704, 1974.
76. **Martin, K. W. and Halpern, S. E.,** Carcinoembryonic antigen production, secretion, and kinetics in Balb/c mice and a nude mouse human tumor model, *Cancer Res.,* 44, 5475, 1984.
77. **Miwa, M., Sakura, H., and Kawachi, T.,** Serum carcinoembryonic antigen level and transplanted colonic tumour size in nude mice, in *Onco-Developmental Gene Expression,* Fishman and Sell, Eds., Academic Press, New York, 1976, 435.
78. **Stragand, J. J., Yang, L., and Drewinko, B.,** Serum CEA levels in a human colonic adenocarcinoma (LOVO) xenograft system, *Cancer Lett.,* 10, 45, 1980.
79. **Lewis, J. C. M. and Keep, P. A.,** Relationship of serum CEA levels to tumour size and CEA content in nude mice bearing colonic-tumour xenografts, *Br. J. Cancer,* 44, 381, 1981.
80. **Jessup, J. M., Giavazzi, R., Campbell, K., Cleary, K., Morikawa, K., and Fidler, I. J.,** Growth potential of human colorectal carcinomas in nude mice: association with the preoperative serum concentration of carcinoembryonic antigen in patients, *Cancer Res.,* 48, 1689, 1988.
81. **Rosenthal, K. L., Tompkins, W. A. F., Frank, G. L., McCulloch, P., and Rawls, W. E.,** Variants of a human colon adenocarcinoma cell line differ in morphology and carcinoembryonic antigen production, *Cancer Res.,* 37, 4024, 1977.
82. **Hu, C., Hsieh, H., Chien, K., Wang, P., Wang, C., Chen, C., Lo, S. J., Wuu, K., and Chang, C.,** Biologic properties of three newly established human esophageal carcinoma cell lines, *J. Natl. Cancer Inst.,* 72, 577, 1984.
83. **Hedin, A., Wahren, B., and Hammarstrom, S.,** Tumor localisation of CEA-containing human tumors in nude mice by means of monoclonal antibodies, *Int. J. Cancer,* 30, 547, 1982.
84. **Harwood, P. J., Pedley, R. B., Boden, J., Rawlins, G., Pentycross, C. R., Rogers, G. T., and Bagshawe, K. D.,** Prolonged localisation of a monoclonal antibody against CEA in a human colon tumour xenograft, *Br. J. Cancer,* 52, 797, 1985.
85. **Jakowattz, J. G., Beatty, B. G., Vlahos, W. G., Porudominsky, D., Philber, V. J., Williams, L. E., Paxton, R. J., Shively, J. E., and Beatty, J. D.,** High-specific activity ^{111}In-labeled anticarcinoembryonic antigen monoclonal antibody: biodistribution and imaging in nude mice bearing human colon cancer xenografts, *Cancer Res.,* 45, 5700, 1985.
86. **Halpern, S. E., Hagan, P. L., Garver, P. R., Koziol, J. A., Chen, A. W. N., Frincke, J. M., Bartholemew, R. M., David, G. S., and Adams, T. H.,** Stability, characterisation and kinetics of ^{111}In-labeled monoclonal antitumor antibodies in normal and nude mouse-human tumor models, *Cancer Res.,* 43, 5347, 1983.
87. **Buchegger, F., Vacca, A., Carrel, S., Schreyer, M., and Mach, J.,** Radioimmunotherapy of human colon carcinoma by ^{131}I-labelled monoclonal anti-CEA antibodies in a nude mouse model, *Int. J. Cancer,* 41, 127, 1988.
88. **Sharkey, R. M., Kaltovich, F. A., Shih, L. B., Fand, I., Govelitz, G., and Goldenburg, D. M.,** Radioimmunotherapy of human colonic cancer xenografts with ^{90}Y-labeled monoclonal antibodies to carcinoembryonic antigen, *Cancer Res.,* 48, 3270, 1988.
89. **Neiderberger, M., DeLozier-Blanchet, C. D., Hedinger, C. E., and Walt, H.,** Differences between subcutaneous and intraperitoneal forms of three human testicular teratocarcinomas in nude mice, *Cancer,* 61, 1571, 1988.
90. **Hata, J., Ueyama, Y., Tamuoki, N., Akatsuka, A., Yoshimura, S., Shimuzy, K., Morikawa, Y., and Furukawa, T.,** Human yolk sac tumor transplanted in nude mice — its morphologic and functional properties, *Cancer,* 46, 2446, 1980.
91. **Hirohashi, S., Shimosato, Y., Kameya, T., Koide, T., Mukojima, T., Taguchi, Y., and Kageyama, K.,** Production of alpha-fetoprotein and normal serum proteins by xenotransplanted human hepatomas in relation to their growth and morphology, *Cancer Res.,* 39, 1819, 1979.
92. **Bassendine, M. F., Wright, N. A., Thomas, H. C., and Sherlock, S.,** Growth characteristics of alpha-fetoprotein secreting human hepatocellular carcinoma in athymic (nude) mice, *Clin. Sci.,* 64, 643, 1983.

93. **Bassendine, M. F., Arbough, B. A. M., Shipton, U., Monjardino, J., Aranguibel, F., Thomas, H. C., and Sherlock, S.,** Hepatitis B surface antigen and alpha-fetoprotein secreting human primary liver cell cancer in athymic mice, *Gastroenterology*, 79, 528, 1980.
94. **Higgins, P. J. and Borenfreund, E.,** Alpha-fetoprotein and albumin synthesis by heterotransplanted rat liver tumor cells, *Exp. Cell Biol.*, 49, 68, 1981.
95. **Hata, Y., Uchino, J., Sato, K., Sasaki, F., Yoshie, U., Naito, H., Manabe, K., Kuwahara, T., and Kasai, Y.,** Establishment of an experimental model of human hepatoblastoma, *Cancer*, 50, 97, 1982.
96. **Kanabus, J., Braunstein, G. D., Emry, P. K., DiSaia, P. J., and Wade, M. E.,** Kinetics of growth and ectopic production of human chorionic gonadotropin by an ovarian cystadenocarcinoma cell line maintained *in vitro*, *Cancer Res.*, 38, 765, 1978.
97. **Ruddon, R. W., Bryan, A. H., Meade-Cobun, K. S., and Pollack, V. A.,** Production of human chorionic gonadotropin and its subunits by human tumors growing in nude mice, *Cancer Res.*, 40, 4007, 1980.
98. **Lieblich, J. M., Rosen, S. W., Weintraub, B. D., Sindelar, W. F., Tasika, T. S., and Rabson, A. S.,** Maintenance of ectopic chorionic gonadotropin and alpha subunit secretion by a human lung cancer cell line (ChaGo) transplanted into nude mice, *J. Natl. Cancer Inst.*, 59, 1285, 1977.
99. **Kameya, T., Shimosato, Y., Tumuraya, M., Ohsawa, N., and Nomura, T.,** Human gastric choriocarcinoma serially transplanted in nude mice, *J. Natl. Cancer Inst.*, 56, 325, 1976.
100. **Wirth, M., Inglin, B., Romen, W., and Ackermann, R.,** Human embryonal cell carcinoma in nude mice, *Cancer Res.*, 43, 5526, 1983.
101. **Ruddon, R. W., Hanson, C. A., Bryan, A. H., Putterman, G. J., White, E. L., Perini, F., Meade, K. S., and Aldenderfer, P. H.,** Synthesis and secretion of human chorionic gonadotropin subunits by cultured human malignant cells, *J. Biol. Chem.*, 255, 1000, 1980.
102. **Woodruff, M. F. A.,** Cellular heterogeneity in tumours, *Br. J. Cancer*, 47, 589, 1983.
103. **Kerbel, R. S., Man, M. S., and Dexter, D.,** A model of human cancer metastasis: extensive spontaneous and artificial metastasis of a human pigmented melanoma and derived sublines in nude mice, *J. Natl. Cancer Inst.*, 72, 93, 1984.
104. **Kozlowski, J. M., Fidler, I. J., Campbell, Z., Xu, Z., Kaighn, M. E., and Hart, I. R.,** Metastatic behaviour of human tumor cell lines grown in the nude mouse, *Cancer Res.*, 44, 3522, 1984.
105. **Hanna, N. and Fidler, I. J.,** Expression of metastatic potential of allogenic and xenogeneic neoplasms in young nude mice, *Cancer Res.*, 41, 438.
106. **Schleicher, R. L., Green, A. W., and Beattie, C. W.,** Growth and metastasis of hamster melanoma following transplantation into athymic mice, *Cancer Res.*, 4465, 1987.
107. **Okabe, T., Suzuki, A., Hirono, M., Tamaoki, N., Oshimura, M., and Takaku, F.,** Establishment of different clonal strains from a human sarcoma of the stomach: tumorigenic heterogeneity in athymic nude mice, *Cancer Res.*, 43, 5456, 1983.

Chapter 10

METASTATIC MODELS

Raffaella Giavazzi

TABLE OF CONTENTS

I. INTRODUCTION

Metastasis formation in cancer is possibly the most difficult aspect to control. A full understanding of the metastatic process in general and specifically of the characteristics of a metastatic cell is extremely important if effective therapies have to be designed. To form metastases, tumor cells must complete all steps involved in the metastatic process, starting with their release from the primary tumor and invasion, and ending with their dissemination to and proliferation in secondary organs.[1] Throughout the metastatic process, malignant cells are subjected to continuous interaction with the immune and non-immune defense mechanisms of the host. This complexity of the pathogenesis of metastasis makes adequate animal models essential for studying tumor cell-host interactions and therapy.

A large body of information has been obtained on the metastatic behavior of syngeneic tumors transplanted to animal hosts. These studies have shown that neoplasms consist of cells varying widely in their biological phenotypes and their metastatic properties. It has been suggested that metastasis is, in fact, a selective event favoring the survival of a minor subpopulation of tumor cells that have the properties to metastasize.[2-4] Important information on the properties of malignant cells has been gained in the course of these studies.

Unfortunately, adequate models for *in vivo* studies of human tumor metastases have been lacking. There are few reports on metastases from human tumors transplanted to animals other than nude mice and these few concern tumor cells implanted in immunodeficient sites of normal animals or in highly immunosuppressed animals. In general, literature is based on the nude mouse as the host for human tumors. The value of a xenograft model depends on the extent to which the characteristics of the original tumor are maintained. In the nude mouse, the capacity of tumor cells to invade and metastasize is one of the properties that seems to be altered on exposure to the new environment. Nevertheless, strenuous efforts have been made to use this *in vivo* model to study the malignant properties of human tumors and the malignant potential of transformed cells. Results on metastases from human tumor xenografts in experimental animals have recently been reviewed.[4-11] There is now an increasing number of investigations showing that human tumors can metastasize in nude mice. The production of metastases in these mice depends not only on the intrinsic properties of the tumor cells and on the host response, but very much on the experimental procedures followed.

In this chapter, we review the current knowledge on the metastatic behavior of human tumors transplanted to nude mice; emphasis is put on the experimental models recently proposed for studying specific aspects of the metastatic process.

II. MALIGNANT POTENTIAL OF HUMAN TUMORS IN NUDE MICE

Not all human tumors grow in nude mice and very few are able to metastasize spontaneously. The frequency of metastases varies according to the tumor characteristics and nude mouse conditions (Table 1). The knowledge and control of these variables have led to successful metastasis formation in a larger number of human cancers transplanted to nude mice.

The classical site for a xenograft in the nude mouse is subcutaneous (s.c.), but the majority of human tumors that successfully grow s.c. suffer from the drawback of limited invasion and rarely produce distant metastases, even after a long period of growth and with a large volume. In contrast to the growth of murine malignant tumors in syngeneic hosts, human xenografts in nude mice are well circumscribed, even at enormous size. The lack of invasion and consequent lack of metastases has often been associated with the presence of a compact layer of fibrous matrix surrounding the growing tumor.[12-14] In fact, a close association has been shown between the incidence of metastases and the presence of invasion in surrounding tissue.[13-15]

TABLE 1
Factors of Influence on Metastasis of Human Tumors in Nude Mice

Tumor	Nude mouse
Tissue preparation	Immunological status
Type	Age
Origin (primary, metastasis, recurrence) (xenograft, culture)	Genetic background
Invasive properties	Health status
Surgery	Life span
Proliferation/size	Transplantation site
Stroma/vascularization	
Hormone dependence	

A. METASTASIS FROM XENOGRAFTS

Only few studies have addressed the metastatic behavior of a large number of human tumor xenografts of different origin. Sharkey and Fogh[15] reported on 106 malignant tumor lines grown s.c. in more than 1000 nude mice, and showed metastases only in 14 cases (1.3%) from 11 different lines, including cancer of the breast, lung, gastrointestinal, and urogenital tract. Retrospective microscopical examination of the organs of 801 mice raised the incidence to 1.7%. The breast cancer line metastasized more frequently than other tumor types. Lungs and lymph nodes were the secondary sites mostly involved. The same investigators mentioned that the presence of invasion of the body wall by the tumor transplant correlated with metastases, xenografts originating from human metastases were not more prone to metastasize than tumors derived from a primary site, and serial transfer in nude mice was not selective for more malignant tumors.[15] Ueyama et al.[10] reported an overall incidence of 5.6% mice with metastases from 66 different human tumors. Kyriazis et al.[16] described microinvasion and metastases as a constant finding for human epithelial tumors of different histological types. In contrast, in an extensive study of direct transplantation of a large series of human tumors, mainly lung, gastric, and breast cancers, metastases were reported to be uncommon.[17] This fits into the early reports on the absence of metastatic spread in the nude mouse.[18] Of interest, by systematic microscopical analysis, it was found that 26.9% of 63 malignant tumors successfully transplanted s.c. in 831 nude mice formed metastases.[14]

In the last few years, human tumor lines have been reported with consistent metastatic properties upon serial transfer. In a study with four colon cancer xenografts in various s.c. passages in nude mice, Sordat et al.[10] have described one line, the CO115, with a high rate of micro- and macro-metastases. Four mammary cell lines were metastatic over several serial passages with different degrees of malignancy.[19] S.c. injection of cells from a human prostate cancer line and of a human pancreatic cancer line resulted in lymph node and lung metastases.[20-22] A high frequency of spontaneous metastases to distant organs (especially the ovaries) in nude mice was described for a neuroblastoma during serial transplantation.[23] Our experience in a 3-year study with human colorectal cancer xenografts established s.c. in nude mice showed that out of more than 500 mice bearing a xenograft, only 1 had macroscopical metastases to the lungs, but this could not be reproduced upon further passages for the same tumor.[8,24] In a few cases, microscopical examination showed lymph nodes infiltrated by tumor cells. Our study also showed that xenografts derived from patients' metastases tended to grow better than those derived from primary tumors, but were not more metastatic, and serial transfer did not induce higher dissemination properties.

It is thus clear that formation of metastasis from s.c. xenografts in nude mice is definitely an exception.[10,20-23] Metastases found were small and often only detectable after careful microscopical examination. The most frequent sites were the lungs and the lymph nodes, indicating that tumor cells can gain access to both the hematogenous and lymphatic routes. Other sites were rare. In case of metastases, in further passages of the tumor line, dissemination was again a rare phenomenon.

B. METASTASIS FROM CULTURED TUMORS

Invasion and metastasis of human tumors in nude mice was first reported by Giovanella et al.[25] for a melanoma metastatic to the regional lymph nodes and the lungs. This study referred to an *in vitro* established cell line. The same group has since described invasive tumors from different cultured human cells with a frequency of metastasis of about 10%.[26,27] Cultured human cells that produce metastases after s.c. growth in nude mice have now been described for several tumor types. The following are some examples. In a study with a series of human cell lines, Kozlowski et al.[28] have described microscopical metastases for melanomas, colon, and prostate cancers. Highly malignant human melanomas metastasizing to the lung[10] and to the diaphragm and the lungs[29] have been reported. Spontaneous metastases have been described for a cultured lung cancer cell line,[30] and microscopical metastases to the lung were found in the case of another human lung cancer cell line and its metastatic variant.[31] Spontaneous metastases from the s.c. site to the lymph nodes and/or the lungs have also been reported for a human prostate cancer cell line,[21] for malignant bladder cancer cells grown in culture for several years,[32] and for renal cell cancer variants recently established *in vitro*.[33,34] Cultured colon cancer cell lines that produce spontaneous metastases have been reported.[10,13,16,28,35,36] Metastatic models have been described for different cell strains (one melanoma and four lung cancers) in use for the screening program of the National Cancer Institute.

In general, cultured cell lines are reported to metastasize more frequently than lines maintained as xenografts, possibly because of the selection of malignant cells after long-term culture.[37] However, no difference in the ability to produce lung colonies was observed for the LOX melanoma in nude mice injected with either monolayer cells or cells from a xenograft.[38] The same group has described another melanoma xenograft with a lower metastatic capacity after culture.[39] Accordingly, lymphatic metastases arose in mice which received s.c. inoculations of PC-3 prostate cancer cells either cultured *in vitro*[21] or from xenograft tissue.[20,21]

C. SURGERY

Metastases in nude mice are usually not observed because of the short life span of the animals related to their poor health status or the large size of the xenografts. By surgical removal of the tumors, their life can be prolonged, thus providing time for metastases to develop and be detected.[5,10] Indeed, after this procedure, metastases in nude mice were described for human breast cancers,[19] colon cancers,[10,40] melanomas,[41] and pancreatic cancers.[22] Sordat et al.[10] showed that the incidence of metastases from CO115 colon cancer was higher in nude mice with tumor recurrence at the site of surgical removal. Wilson et al.[41] found that the incidence of metastases was directly related to the size of the tumor at surgery and not to the period of s.c. growth. This finding is in agreement with the positive correlation of the size of the s.c. tumor with the probability of lymphatic metastases described for a prostate tumor.[42]

We investigated these possibilities by injection of human colorectal cancer cells into the thigh of nude mice followed by amputation of the tumor-bearing leg at a later stage. Although the tumors were quite large at the time of removal and the majority recurred, metastases were found in only a few mice without a significant increase in incidence up to an observation period of 6 months.[24] In another study with different tumor types, partial surgical excision of the s.c. tumor was also not associated with a higher metastatic rate.[14] It is not clear whether surgical resection of the xenograft simply allows the mice to survive long enough for metastases to develop or if it induces the spread of tumor cells either by the cells becoming metastatic, by vascularization of the primary tumor, by release of proteolytic enzymes, or by modulation of the host immune response.

D. HOST FACTORS AND METASTASIS

The poor metastatic behavior of s.c. transplants in nude mice has led to studies on the influence of the host in the control of other than T cell-dependent metastatic spread. Results in nude mice with additional congenital immunodeficiencies (i.e., beige-nude or lasat) or further immunosuppression (i.e., anti-lymphocyte serum-treated or whole-body irradiated) have been controversial in terms of metastasis.[10,43-47] The complexity of the nude mouse host response in the control of metastatic spread can be realized from observations that metastases from human cancers in nude mice can be induced by the addition of xenogeneic T lymphocytes, thus suggesting the T cell deficit is involved in the lack of metastasis.[48] It has also been reported that i.v. injection of metastatic murine tumor cells to nude mice results in a smaller number of metastases as compared to heterozygous littermates or syngeneic mice.[49-51]

The possible role of natural killer (NK) cell activity in the control of tumor spread has been recognized.[52-54] Studies from different laboratories have shown that nude mice have a high level of NK cell-mediated cytotoxicity, particularly active in the control of hematogenously disseminated tumor cells.[55] For example, the incidence of experimental metastases (i.v. injected tumor cells) was higher in 3-week-old nude mice (low NK), and in nude mice NK cell-depleted by treatment with cyclophosphamide, anti-asialo GMl antiserum, or betaestradiol, as compared to adult nude mice (high NK) or mice treated with NK cell-stimulating biological response modifiers.[55-57] In addition, nude mice maintained under specific pathogen-free (SPF) conditions (weak NK) were more susceptible to experimental metastases than mice in conventional housing areas.[58,59] The role of NK cell activity in the regulation of human tumor spread in nude mice appears not to be as important as for rodent tumor metastases in these animals. For example, Kozolowski et al.[28] have shown that the reduction of NK cell activity in nude mice may facilitate the metastatic capacity of some, though not all, human tumors. The introduction of the beige gene (low NK)[60] in nude mice resulting in mutants (nu/nu, bg/bg) deficient in T and NK cell activity would define a model for analysis of the *in vivo* role of NK cells. Surprisingly, the low NK cell activity in beige-nude mice (though they are not completely devoid) was not associated with increased artificial or spontaneous metastases of human renal and colorectal cancer cell lines.[46] Also, there was no correlation between NK cell activity in beige-nude mice and the malignant behavior of tumor cells.[43] Nude mice with various combinations of additional immunodeficiencies are now under investigation in several laboratories, but their utility in the study of human metastases has not yet been proven.

Among the other host variables of influence on metastasis in nude mice is age, another factor with young or newborn mice being more prone to metastases.[10,56,61,62] Newborn nude mice inoculated s.c. with human tumors developed a locally invasive tumor that disseminated to distant organs like lungs and lymph nodes;[61,62] in contrast, in 3 to 4 weeks or older nude mice, the same tumors grew locally and did not metastasize. These findings were obtained with lymphoid and myeloid cell lines[61,62] as well as solid tumors.[10,61] Interestingly, for a colon cancer line, the formation of regional lymph node metastases did not differ significantly between newborn and adult nude mice, suggesting a post-natal control mechanism on tumors from hematological origin.[10] Again, NK cell activity has been indicated as a source of major responsibility for the lack of metastases in adult mice as compared to the newborn.[62]

The maintainance of the nude mouse colony also appears to be a factor influencing the behavior of human tumor xenografts. The SPF environment and the health of nude mice positively influence the metastatic rate of allogeneic and xenogeneic tumors, and this may be partly related to the immunological response of the host.[7,10,58,59] The presence of pathogens, such as murine hepatitis virus (MHV) in the nude mouse colony, could account for resistance to tumor transplantation and metastasis.[7,9]

In some cases, the background strain, but not the sex, of the nude mouse has been indicated to influence the metastatic potency of tumor cells. Hormone-dependent tumors constitute a special situation, because the hormonal status of the host seems more important for local tumor

TABLE 2
Metastasis Formation of Human Tumors in Nude Mice According to the Route of Injection

Route of injection	Major organs for metastasis
Subcutaneous (s.c.)	Lungs, lymph nodes
Intramuscular (i.m.)	Lungs, lymph nodes
Intraperitoneal (i.p.)	Intra-abdominal carcinomatosis, lungs
Intravenous (i.v.)	Lungs, lymph nodes, intrascapular fat
Intracardiac (i.c.)	Adrenal glands, brain
Intrasplenic (i.s.)	Liver, (lungs)
Intracarotid (i.a.)	Brain

growth than for the intrinsic metastatic ability. For instance, the hormone-responsive MCF-7 breast cancer cell line displayed invasive growth at estrophilic sites (intra-uterine and intracerebral), but not in non-estrogenic tissue (s.c. or lung) where it grew as small localized nodules.[63]

E. SITE OF IMPLANTATION

The site of transplantation in nude mice seems to play a major role in the metastatic behavior of human tumors (Table 2). Local and vascular invasion and the eventual occurrence of distant metastases have been described for s.c. grown tumors, but the overall incidence is low and only a few human tumor lines (xenograft or cultured) are consistently metastatic. The injection of tumor cells in a different site may influence the metastasis trophism or select for a metastatic population. In this regard, Fidler[7] and co-workers have focused on the effect of the anatomical tumor site for tumor growth and metastasis spread (the orthotopic transplant model is discussed in Section VI of this chapter). Kozlowski et al.[18] have shown for seven human tumor cell lines (melanomas and colon, renal and prostate cancers) that tumor dissemination could be increased, depending on the site of tumor implantation of which the cranial area to the flank or the spleen were most favorable. Our experience with a panel of colorectal cancer xenografts showed a characteristic metastatic behavior depending on the route of tumor cell inoculation. Tumor cell suspensions were injected s.c., i.m., i.v., and intrasplenically (i.s.);[24,64] metastases from s.c. and i.m. transplants were rare, but i.v. and i.s. injections resulted in lung and liver colonies of which the number was dependent on the xenograft type. Moreover, with regard to the metastatic behavior of both long-term established (the HT-29) or recently established (the KMl2) colon cancer cell variants, a high rate of liver tumor formation was shown following i.s. injection, while a low frequency of lung colonies was demonstrated after i.v. injection, but with macroscopical metastases in the lymph nodes and intrascapular fat.[65,66] Naito et al.[33,34] have shown that renal cell cancer variants have a distinct metastatic pattern upon injection s.c., i.v., i.s., and under the kidney capsule. The major metastatic target was the lung, but quantitative differences were observed depending on the route of injection. In another study, a human colon cancer cell line injected i.v., intraportally, and i.p. produced experimental metastases respectively in the lungs, liver, and abdomen, the latter resulting in intra-abdominal carcinomatosis with ascites.[67] Human colon cancer lines have been described that were invasive at the s.c. site and metastatic at the site of the cecum, whereas splenic and hepatic implants were more encapsulated.[68] Melanoma cell lines were injected s.c., i.p., and i.s.: metastases were not observed for the s.c. xenograft, but some melanomas disseminated to peritoneal organs or occasionally to the lung and infiltrated lymph nodes after i.p. and i.s. injection.[69] The metastatic behavior of the PC-3 prostate cancer line was also related to its inoculation site in nude mice, such as s.c. (lung and axillary lymph node metastases), peritoneum (intra-abdominal tumor growth), spleen (liver colonies), and tail vein (lung colonies).[20] Metastases to the skeleton were observed in 19% of mice where i.v.-injected tumor cells were pushed through the paravertebral venous plexus after occlusion of the inferior caval vein.[70] In the same study, it was found that the incidence of bone metastases increased to 53% using sublines established from bone lesions. A small-cell prostate cancer line was shown to be locally invasive while growing in the muscles and under the kidney capsule, growth was absent in the liver and spleen, and was non-invasive at the s.c. site and in the peritoneum.[71] After i.v. injection, this particular line did not produce lung metastases. Finally, among the various routes of transplantation, the intracardiac injection has been proposed as a reliable method for widespread hematogenous metastases, including the adrenal glands and the brain.[72]

Other methods have been used to overcome the problems related to the lack of invasion from s.c. human tumor xenografts. For example, metastasis formation in nude mice was improved by injecting human tumor cells within a vascularized artificial polyurethane sponge or co-injected with lung cubes.[73,74]

III. LUNG TUMOR COLONIES

Experimental lung metastases signify tumor colonies produced by tumor cells after i.v. injection. Although these cells do not go through the initial steps of the metastatic process, such as detachment from the primary tumor, local spread and invasion of blood vessels, all the other steps must be completed for metastases to be established. It is not well understood whether the low metastatic potential of s.c. xenografts is the result of an intrinsic inability of the tumor cells to spread and enter the circulation or whether human tumor cells are unable to invade tissue of mouse origin.[12] Therefore, it is possible that the injection directly into the circulation (artificial metastases) may allow metastasis formation from tumors that are not metastatic when implanted s.c. (spontaneous metastases). Despite controversy on this matter, i.v. injection of tumor cells has been widely accepted as a means to study the metastatic potential. Fidler et al.[4] have shown that the patterns of metastasis of various rodent tumors were maintained in both syngeneic hosts and nude mice, thus demonstrating that the nudes could be useful for studying the metastatic behavior of tumor cells. These animals could also be used to select high-metastatic subpopulations. Human tumor cells derived from various xenografts or cultures are described to produce lung tumor colonies after i.v. injection in nude mice.[10,20,24,28,33,36,38,39,65,75-80] In general, the major target of tumor cells injected i.v. are the lungs, but metastases have also been found in the kidney,[77] interscapular fat, lymph nodes, adrenal glands,[39,65,76] brain,[76] and, infrequently, in the ovaries and bone marrow.[76] The consistent pattern of extra-pulmonary metastases reported in these studies suggests that the i.v. injection can distinguish the organ preference of cells for metastasis. The i.v. injection model certainly distinguishes human tumors with different metastatic properties and is useful to study heterogeneity with respect to metastatic potential[28,36,65,75,79] or even select high-metastatic subpopulations of tumor cells.[4,7,23,28,70,79] Thus, many of the observations on the metastatic phenotype made in rodent tumor systems are applicable to human tumors in the nude mouse model.

IV. HEPATIC TUMOR COLONIES

Although liver metastases are a common problem in patients, there are very few human tumors that consistently metastasize to the liver of nude mice. There are few reports on liver invasion from human tumors grown i.p.. To establish a model for liver metastases, tumor cells have been injected directly into the portal vein; but this procedure requires considerable technical expertise and animal mortality is high. Another model of hepatic metastases may be developed by implantation of tumor cells in the spleen of mice. From this site of injection, tumor cells can easily gain access to the blood-stream and reach the liver through the portal vein, proliferating in secondary deposits.[64] A pioneering study by Leduc[81] showed that the intrasplenic growth of two murine hepatomas was followed by the formation of secondary nodules in the liver in one hepatoma, thus suggesting the existence of different metastatic properties. Liver metastases from i.s. injection, followed or not by splenectomy, have also been described for a variety of murine tumor systems.[82,83] Kozolowski et al.[28] investigated the metastatic behavior of a series of human cancer lines in nude mice and indicated the i.s. injection as the best route to allow maximum expression of their malignant potential resulting in metastases to the liver, lungs and lymph nodes.[28]

Hepatic metastases can be considered a major cause of death from colorectal cancer.[84] An experimental model that mimics some aspects of the biological behavior of this disease is quite attractive. We have found that the injection of recently established human colorectal cancer xenografts in the spleen of nude mice was followed by the formation of liver colonies; the production of tumor deposits in the liver distinguished tumor cells with different malignant potential and the ability to form liver colonies was stable after several passages in nude mice.[24,64] The most dramatic expression of malignant potential in the liver was observed for two colorectal cancer xenografts derived from hepatic metastases of patients.[24] While in some cases tumor cells did not form lung colonies after i.v. injection, they were found to produce liver colonies after i.s. injection. Furthermore, Morikawa et al.[66] demonstrated a correlation between the production of experimental hepatic metastases from fresh tumor cells injected i.s. and the clinical condition of the patient.[66] In addition, through cycles of "spleen-liver colonies", selected cell lines with high metastatic efficiency in nude mice were obtained.[85] The ability to produce liver colonies does not merely depend on whether tumor cells will reach the liver. Distribution studies of radiolabeled tumor cells showed that shortly after i.s. injection these were arrested in the liver, but only metastatic cells survived and developed into tumor colonies.[64,85] In addition, for the formation of liver colonies, no relation was found with tumor growth in the spleen, and splenectomy after tumor cell injection did not impair their growth.[64,82,83]

The studies on human colon cancer xenografts were followed by several reports of hepatic secondary deposits obtained from implantation of other human tumors in the spleen.[20,33,68,69] Interestingly, in these experiments, lymph node and lung metastases were often described.

The i.s. injection-hepatic metastases model can thus be used to study some aspects of liver metastasis, such as invasion and growth in the liver and the interaction with the host environment. It is also valuable for investigating therapies for human colorectal cancer growing in a relevant metastatic site, the liver.[86]

V. INTRA-ABDOMINAL CARCINOMATOSIS

Intraperitoneal (i.p.) injection of tumor cells from solid human tumor types into nude mice may result in a transplantable form of malignant ascites.[87-89] Often, these tumors producing metastases i.p. do not metastasize after s.c. injection.[13,69,88,90,91] Widespread tumor deposits occurred consistently with preference for the diaphragm, peritoneal wall, omental fat, and the pancreatic region;[13,20,29,33,67,69,88,91] metastases have also been detected in the liver and the lungs.[13,20,29,33,90]

The nude mouse peritoneum is of particular interest for the study of ovarian cancer metastases, as malignant peritonitis is known to be a clinical problem in this type of disease.[92,93] While direct injection of cells into the ovaries of a nude mouse presents a series of technical difficulties, injection into the abdominal cavity is easily performed. Intra-abdominal carcinomatosis has been induced after i.p. injection of *in vitro* established cell lines,[94-97] of tumor cell suspensions from tumor digestion or from ascites,[87,95,97,98] or after implantation of small pieces of xenograft or fresh tissue.[87,97,98] In two of these studies, an i.p. xenograft model was directly established from the material of the patients' tumors.[87,97] The nude mouse i.p. ovarian cancer model parallels human disease by producing ascites, intra-abdominal carcinomatosis with large tumor masses adhering to the peritoneal wall and specifically to the diaphragm and omental fat

TABLE 3
Expression of Metastatic/Invasive Properties of Human Tumor Xenografts Transplanted Orthotopically to Nude Mice

Site of tumor implantation	Tumor type	Metastasis/ Invasion	Ref.
Intradermal	Melanoma	Lungs, lymph nodes local invasion	26, 105
Gut	Colorectal ca.	Infiltration colonic wall, mesenteric lymph nodes, liver	66, 68, 85, 106, 107
Kidney	Renal cell ca.	Local invasion, lungs, lymph nodes, liver, pancreas, diaphragm, seminal vesicle	33, 34
Mammary fat pad	Breast ca.	Lungs, liver, spleen, lymph nodes	110, 111
Pancreas	Pancreatic ca.	Liver, lungs, lymph nodes	112, 113
Stomach	Gastric ca.	Liver, lungs, lymph nodes	114
Transurethral	Bladder ca.	Local invasion, lung	115
Intrabronchial	Lung ca.	Lung invasion	116, 117

and fatal outcome.[87,94-98] Invasion in the organs has been observed and extra-peritoneal metastases were reported to occur late in the course of the disease.[87,95,97] Tumor characteristics, such as the presence of receptors,[95] ovarian cancer-associated antigens,[95,97] histopathological and biochemical markers, as well as chemosensitivity, closely matched those of patients.[87,97]

The i.p. model is potentially useful for the evaluation of experimental therapy and for further knowledge on the biology of ovarian cancer. In contrast to the routinely used s.c. xenograft,[99-101] i.p. growth cannot be measured continuously, but it represents a more relevant site for ovarian tumors. This is important for experiments that involve the evaluation of tumor burden or the analysis of tumor-specific characteristics, as well as in defining the rationale for i.p. therapy. For example, these experimental models have been proposed for investigation of the *in vivo* therapeutic relevance of labeled monoclonal antibodies and for determination of other new forms of therapy.[96,102-104]

The i.p. growth of human ovarian cancer in nude mice provides a tool for studying the invasive properties of tumor deposits that would not be recognized from the classical s.c. growth. In a recent study of human ovarian cancer transplants obtained from fresh tumor tissue or ascites, we indeed confirmed that the site of tumor implantation affects tumor growth behavior.[87,89] The i.p. growth offered a good opportunity to characterize each tumor for growth, spread, and ascites formation. Invasion in the organs and lung metastases were only observed in nude mice bearing intra-abdominal carcinomatosis (Giavazzi et al., unpublished), while none of the ovarian tumors growing s.c. were metastatic.[87,98]

VI. METASTASIS FROM ORTHOTOPIC TRANSPLANTATION

In the previous sections, several methods were described to overcome problems related to the inability of nude mice to reproduce human tumor metastases. However, the majority of these models do not represent the complete picture of the metastatic process. Recently, it was shown that the tumorigenic and the metastatic properties of a human tumor can be maintained after transplantation to its organ or tissue of origin (orthotopic transplantation) (Table 3).

The first report on metastasis in nude mice originates from 1973 for three melanomas growing intradermally with metastatic spread to lymph nodes and, in one case, to the lung.[26] Similarly, Cornil et al.[105] described that subdermal injection of a human melanoma cell line resulted in a highly melanotic, non-encapsulated tumor, more tumorigenic and metastatic (to lymph nodes) than the s.c.-grown tumor being amelanotic and poorly metastatic. Implantation of human colon tumor along the intestines is perhaps the best studied orthotopic transplant in the nude mouse. In an original study, Sordat et al.[11,106] have implanted two human colon cancers in the descending part of the large bowel of nude mice and subsequently found infiltration of the colon wall and mesenteric lymph nodes. In contrast, the injection of two human melanoma cell lines in the gut resulted in only limited invasiveness. More recently, a colon cancer cell line has been described with local growth after injection into the cecal wall and inducing lymph node metastases, vascular invasion, and hepatic metastases.[68,107] Similarly, Morikawa et al.[66,85] showed that the injection of fresh human colorectal cancer tissue into the cecum of nude mice was followed by hepatic metastases and that orthotopic transplantation could select highly metastatic tumor cells. In all these studies, s.c. injection of the same tumors did not result in visceral organ involvement.

Injection of human tumors into the subrenal capsule was originally proposed as a rapid screening method for testing chemotherapeutic agents against human tumor xenografts.[108] This site provides a rich vascular bed and therefore a good support for tumor growth. Human tumors have been grown consistently under the kidney capsule of nude mice, but little information was given on the invasion and spread of tumor cells from that site.[108,109] Only recently, Naito et al.[33] showed that transplantation of a human renal cell cancer to various sites of nude mice resulted in subpopulations of tumor cells that differed in many aspects, including metastatic behavior. They found that tumor cells injected into the renal capsule grew faster and produced more systemic metastases than did cells injected s.c. Tumors in the kidney were invasive, highly vascularized, and non-encapsulated. It was concluded that the implantation of renal cell cancer into the kidney of nude mice provided the most favorable site to examine their malignant potential.[33,34]

A hormone-responsive human breast cancer line grown in the mammary fat pad of the nude mouse was reported to produce metastases to the lungs, liver, and spleen,[110] and estrogen treatment increased the growth rate and frequency of metastases. Moreover, estrogen receptor-negative human breast cancers grew better in the mammary fat pad than s.c. and produced metastases in the lungs and lymph nodes.[111]

A human pancreatic cancer cell line was injected into the duodenal lobe of the pancreas of nude mice and produced local and distant metastases, while no metastases were found in mice bearing s.c. transplants.[112] After orthotopic transplantation of pieces of human pancreatic cancer xenograft, local tumors developed and metastases were observed in the regional and distal lymph nodes, liver, and lungs, without peritoneal deposits and ascites.[113]

A human gastric cancer implanted into the stomach wall of nude mice showed invasive growth and systemic metastases, while the same tumor implanted s.c. had an expansive and faster growth, was well-encapsulated, and did not form metastases.[114] Ahlering et al.[115] reported on a human bladder transitional cell cancer line originally derived from a low-grade, non-invasive tumor, which produced histologically non-invasive tumors after transurethral inoculation without metastases. In contrast, a high-grade invasive bladder tumor-derived line produced invasive tumors in the nude mouse bladder and metastases to the lungs. Both lines gave a similar incidence of lung metastases when injected i.v. and local well-defined growth without metastases when implanted s.c.[115]

The lungs as an orthotopic transplantation site have been described by McLemore et al.[116,117] Human lung cancers were successfully propagated using the intrabronchial injection technique and a high take rate was found for the tumors metastatic to the lung. This method could offer an alternative means to the classical i.v. injection to study metastatic spread to the lungs.

The injection of human cancer cells into the internal carotid artery of nude mice provides a tool for studying the biology of hematogenous brain metastases. Human-type cancer lesions in the brain and a distinct pattern of growth different for the various tumors can be detected.[118] Also, this experimental model only includes the last few steps: release of tumor cells into the circulation, arrest in capillaries, extravasation into the cerebral tissue through the blood-brain barrier, and proliferation in the brain parenchyma.

The various examples show that orthotopic transplantation results in invasion and metastasis formation of tumors that otherwise do not produce spontaneous metastases in nude mice; the organ provides a suitable microenviroment for the tumor cells to exert their intrinsic properties; metastases for a specific tumor type often occur in organs known to be clinically involved. Finally, these nude mouse tumor metastasis models, despite their limitation related to certain manipulations or surgical techniques, may be useful for the study of specific therapeutic approaches.

VII. CONCLUSIONS

The complexity of the metastatic process calls for model systems that simulate *in vivo* conditions as closely as possible. Human tumor xenografts in nude mice retain several of the biological characteristics of the tumor of origin, but their malignant phenotype is often underestimated. Metastases have been reported in nude mice for the majority of human tumor types, but the lack of consistent results makes the model hard to use for experiments on this phenomenon.

In this chapter, some of the variables related to tumor cell and host characteristics of importance to regulate the spread of human tumors in the nude mouse have been discussed. The anatomical location of the tumor implant seems to have a profound effect on the subsequent formation of metastasis. S.c. growth of human tumors in nude mice usually lacks local invasion and distant metastases. I.p. injection offers a better model to study metastases of human tumors and to select invasive tumor cells. Models of experimental metastases by injecting tumor cells into the portal or systemic circulation have enlightened various aspects related to metastasis formation to the liver and the lungs, respectively. Recently, considerable efforts have been put into the design of models corresponding to clinical aspects of metastasis, that is, sites of transplantation and organs that are common targets for metastases by the orthotopic transplantation approach.

Many experimental metastasis models have already been developed. However, they are still incomplete, and the complexity of metastasis calls for additional information obtained from newer models. Therefore, extensive characterization of the tumor models and their appropriate use are prerequisites for the study metastatic properties.

ACKNOWLEDGMENTS

The author thanks Mrs. J. D. Baggott and Mrs. M. T. Suardo for editing and Mrs. L. Piccoli for typing the manuscript. The critical review of Dr. G. Taraboletti is acknowledged.

REFERENCES

1. **Poste, G. and Fidler, I. J.,** The pathogenesis of cancer metastasis, *Nature (London),* 283, 139, 1980.
2. **Fidler, I. J. and Hart, I. R.,** Biological diversity in metastatic neoplasms: origins and implications, *Science,* 217, 998, 1982.
3. **Heppner, G. H.,** Tumor heterogeneity, *Cancer Res.,* 44, 2259, 1984.
4. **Fidler, I. J., Pollack, V. A., and Hanna, N.,** The use of nude mice for studies of cancer metastasis, in *Immune Deficient Animals,* Sordat, B., Ed., Karger, A. G., Basel, 1984, 328.
5. **Giovanella, B. C. and Fogh, J.,** The nude mouse in cancer research, *Adv. Cancer Res.,* 44, 69, 1985.
6. **Doré, J.-F., Bailly, M., and Bertrand, S.,** Metastases of human tumors in experimental animals, *Anticancer Res.,* 7, 997, 1987.
7. **Fidler, I. J.,** Rationale and methods for the use of nude mice to study the biology and therapy of human cancer metastasis, *Cancer Met. Rev.,* 5, 29, 1986.
8. **Giavazzi, R., Jessup, J. M., Pathak, S., Morikawa, K., and Fidler, I. J.,** Tumorigenic and metastatic properties of human colorectal carcinomas transplanted into nude mouse, in *Colon Cancer Cells,* Moyer, N. P. and Poste, G. H., Eds., Academic Press, New York, 1990, 311.
9. **Sharkey, F. E. and Fogh, J.,** Considerations in the use of nude mice for cancer research, *Cancer Met. Rev.,* 3, 341, 1984.
10. **Sordat, B. C. M., Ueyama, Y., and Fogh, J.,** Metastasis of tumor xenografts in the nude mouse, in *The Nude Mouse in Experimental and Clinical Research,* Fogh, J. and Giovanella, B. C., Eds., Academic Press, New York, 1932, 95.
11. **Sordat, B. and Wang, W. R.,** Human colorectal tumor xenografts in nude mice: expression of malignancy, *Behring Inst. Mitt.,* 74, 291, 1984.
12. **DeVore, D. P., Houchens, D. P., Ovejera, A. A., Dill, G. S., and Hutson, T. B., Jr.,** Collagenase inhibitors retarding invasion of a human tumor in nude mice, *Exp. Cell Biol.,* 48, 367, 1980.
13. **Kyriazis, A. P., DiPersio, L., Michael, G. J., Pesce, A. J., and Stinnett, J. D.,** Growth patterns and metastatic behavior of human tumors growing in athymic mice, *Cancer Res.,* 38, 3186, 1978.
14. **Neulat-Duga, I., Sheppel, A., Marty, C., Lacroux, F., Pourrat, J., Caveriviére, P., and Delsol, G.,** Metastases of human tumor xenografts in nude mice, *Inv. Met.,* 4, 209, 1984.
15. **Sharkey, F. E. and Fogh, J.,** Metastasis of human tumors in athymic nude mice, *Int. J. Cancer,* 24, 733, 1979.
16. **Kyriazis, A. P., Kyriazis, A. A., McCombs, W. B., and Kereiakes, J. A.,** Biological behavior of human malignant tumors grown in the nude mouse, *Cancer Res.,* 41, 3995, 1981.
17. **Shimosato, Y., Kameya, T., Nagai, K., Hirohashi, S., Koide, T., Hayashi, H., and Nomura, T.,** Transplantation of human tumors in nude mice, *J. Natl. Cancer Inst.,* 56, 1251, 1976.
18. **Povlsen, C. O. and Rygaard, J.,** Growth of human tumors in the nude mouse, in *In Vitro Methods of Cell-Mediated and Tumor Immunity,* Bloom, B. B. and David, J. R., Eds., Academic Press, New York, 1976, 701.
19. **Ozzello, L. and Sordat, M.,** Behavior of tumors produced by transplantation of human mammary cell lines in athymic nude mice, *Eur. J. Cancer Clin. Oncol.,* 16, 553, 1980.
20. **Shevrin, D. H., Gorny, K. I., and Kukreja, S. C.,** Patterns of metastasis by the human prostate cancer cell line PC-3 in athymic nude mice, *Prostate,* 15, 187, 1989.
21. **Ware, J. L., Paulson, D. F., Mickey, G. H., and Webb, K. S.,** Spontaneous metastasis of cells of the human prostate carcinoma cell line PC-3 in athymic nude mice, *J. Urol.,* 128, 1064, 1982.
22. **Kajiji, S. M., Meitner, P. A., Bogaars, H. A., Dexter, D. L., Calabresi, P., and Turner, M. D.,** Metastasis of a human pancreatic adenocarcinoma (RWP-1) in nude mice, *Br. J. Cancer,* 46, 970, 1982.
23. **Hata, J.-I., Ueyama, Y., Tamaoki, N., Furukawa, T., and Morita, K.,** Human neuroblastoma serially transplanted in nude mice and metastases, *Cancer,* 42, 468, 1978.
24. **Giavazzi, R., Campbell, D. E., Jessup, J. M., Cleary, K., and Fidler, I. J.,** Metastatic behavior of tumor cells isolated from primary and metastatic human colorectal carcinomas implanted into different sites in nude mice, *Cancer Res.,* 46, 1928, 1986.
25. **Giovanella, B. C., Yim, S. O., Stehlin, J. S., and Williams, L. J., Jr.,** Brief communication: development of invasive tumors in the "nude" mouse after injection of cultured human melanoma cells, *J. Natl. Cancer Inst.,* 48, 1531, 1972.
26. **Giovanella, B. C., Yim, S. O., Morgan, A. C., Stehlin, J. S., and Williams, L. J., Jr.,** Brief communication: metastases of human melanomas transplanted in "nude" mice, *J. Natl. Cancer Inst.,* 50, 1051, 1973.
27. **Giovanella, B. C., Stehlin, J. S., and Williams, L. J., Jr.,** Heterotransplantation of human malignant tumors in "nude" thymusless mice. II. Malignant tumors induced by injection of cell cultures derived from human solid tumors, *J. Natl. Cancer Inst.,* 52, 921, 1974.
28. **Kozlowski, J. M., Fidler, I. J., Campbell, D., Xu, Z., Kaighn, M. E., and Hart, I. R.,** Metastatic behavior of human tumor cell lines grown in the nude mouse, *Cancer Res.,* 44, 3522, 1984.

29. **Lockshin, A., Giovanella, B. C., De Ipolyi, P. D., Williams, L. J., Jr., Mendoza, J. T., Yim, S. O., and Stehlin, J. S., Jr.,** Exceptional lethality for nude mice of cells derived from a primary human melanoma, *Cancer Res.,* 45, 345, 1985.
30. **Cole, S. P. C., Campling, B. G., Dexter, D. F., Holden, J. J. A., and Roder, J. C.,** Establishment of a human large cell lung tumor line (QU-DB) with metastatic properties in athymic mice, *Cancer,* 58, 917, 1986.
31. **Varki, N. M., Roome, L., Sparkes, R. S., and Miller, J. E.,** Microscopic metastasis of a human lung carcinoma cell line in athymic nude mice: isolation of a metastatic variant, *Int. J. Cancer,* 40, 46, 1987.
32. **Tropé, C., Wennerberg, J., Johansson, S., and Unsgard, B.,** Metastasis of a cultured human bladder carcinoma cell line transplanted into nude athymic mice, *Neoplasma,* 29, 421, 1982.
33. **Naito, S., von Eschenbach, A. C., Giavazzi, R., and Fidler, I. J.,** Growth and metastasis of tumor cells isolated from a human renal cell carcinoma implanted into different organs of nude mice, *Cancer Res.,* 46, 4109, 1986.
34. **Naito, S., von Eschenbach, A. C., and Fidler, I. J.,** Different growth pattern and biologic behavior of human renal cell carcinoma implanted into different organs of nude mice, *J. Natl. Cancer Inst.,* 78, 377, 1987.
35. **Kimball, P. M. and Brattain, M. G.,** Isolation of a cellular subpopulation from a human colonic carcinoma cell line, *Cancer Res.,* 40, 1574, 1980.
36. **Spremulli, E. N. and Dexter, D. L.,** Human tumor cell heterogeneity and metastasis, *J. Clin. Oncol.,* 1, 496, 1983.
37. **Fine, D. L., Shoemaker, R., Gazdar, A., Mayo, J. G., Fodstad, O., Boyd, M. R., Abbott, B. J., and Donovan, P. A.,** Metastasis models for human tumors in athymic mice: useful models for drug development, *Cancer Dect. Prev. Suppl.,* 1, 291, 1987.
38. **Fodstad, O., Aamdal, S., McMenamin, M., Nesland, J. M., and Pihl, A.,** A new experimental metastasis model in athymic nude mice, the human malignant melanoma lox, *Int. J. Cancer,* 41, 442, 1988.
39. **Fodstad, O., Kjonniksen, I., Aamdal, S., Nesland, J. M., Boyd, M. R., and Pihl, A.,** Extrapulmonary, tissue-specific metastasis formation in nude mice injected with FEMX-I human melanoma cells, *Cancer Res.,* 48, 4382, 1988.
40. **Lundy, J., Greiner, J. W., and Colcher, D.,** Development of a metastatic human colon cancer xenograft model in the nude mouse, *J. Surg. Oncol.,* 31, 260, 1986.
41. **Wilson, E. L., Gartner, M. F. R. M., Campbell, J. A. H., and Dowdle, E. B.,** Metastasis of a human melanoma cell line in the nude mouse, *Int. J. Cancer,* 41, 83, 1988.
42. **Ware, J. L. and DeLong, E. R.,** Influence of tumor size on human prostate tumour metastasis in athymic nude mice, *Br. J. Cancer,* 51, 419, 1985.
43. **Fodstad, O., Hansen, C. T., Cannon, G. B., Statham, C. N., Lichtenstein, G. R., and Boyd, M. R.,** Lack of correlation between natural killer activity and tumor growth control in nude mice with different immune defects, *Cancer Res.,* 44, 4403, 1984.
44. **Gershwin, M. E., Ikeda, R. M., Erickson, K., and Owens, R.,** Enhancement of heterotransplanted human tumor graft survival in nude mice treated with antilymphocyte serum and in congenitally athymic-asplenic (lasat) mice, *J. Natl. Cancer Inst.,* 61, 245, 1978.
45. **Lozzio, B. B., Giovannella, B. C., and Gallie, B. L.,** Nude mice with additional immunodeficiencies: genetically, anatomically, and experimentally induced, in *The Nude Mouse in Experimental Clinical Research,* Fogh, J. and Giovanella, B. C., Eds., Academic Press, New York, 1982, 160.
46. **Naito, S., Giavazzi, R., Walker, S. M., Itoh, K., Mayo, J., and Fidler, I. J.,** Growth and metastatic behavior of human tumor cells implanted into nude and beige nude mice, *Clin. Exp. Met.,* 5, 135, 1987.
47. **Ohsugi, Y., Gershwin, M. E., Owens, R. B., and Nelson-Rees, W. A.,** Tumorigenicity of human malignant lymphoblasts: comparative study with unmanipulated nude mice, antilymphocyte serum-treated nude mice, and X-irradiated nude mice, *J. Natl. Cancer Inst.,* 65, 715, 1980.
48. **Graham, S. D.,** T lymphocyte induction of metastatic tumors in the nude mouse model, *J. Urol.,* 128, 832, 1982.
49. **Fidler, I. J., Caines, S., and Dolan, Z.,** Survival of hematogenously disseminated allogeneic tumor cells in athymic nude mice, *Transplantation,* 22, 208, 1976.
50. **Richie, J. P., McDonald, J., and Gittes, R. F.,** Resistance to intravenous tumor metastases in the athymic mouse: a paradoxic response, *Surgery,* 90, 214, 1981.
51. **Skov, C. B., Holland, J. M., and Perkins, E. H.,** Brief communication: development of fewer tumor colonies in lungs of athymic nude mice after intravenous injection of tumor cells, *J. Natl. Cancer Inst.,* 56, 193, 1976.
52. **Herberman, R. B. and Holden, H. T.,** Natural cell-mediated immunity, *Adv. Cancer Res.,* 27, 305, 1978.
53. **Kiessling, R., Klein, E., Pross, H., and Wigzell, H.,** "Natural" killer cells in the mouse. II. Cytotoxic cells with specificity for mouse moloney leukemia cells. Characteristics of the killer cells, *Eur. J. Immunol.,* 5, 117, 1975.
54. **Talmadge, J. E., Meyers, K. M., Prieur, D. J., and Starkey, J. R.,** Role of NK cells in tumour growth and metastasis in beige mice, *Nature (London),* 284, 622, 1980.
55. **Hanna, N.,** Role of natural killer cells in control of cancer metastasis, *Cancer Met. Rev.,* 1, 45, 1982.
56. **Hanna, N. and Fidler, I. J.,** Expression of metastatic potential of allogeneic and xenogeneic neoplasms in young nude mice, *Cancer Res.,* 41, 438, 1981.

57. **Hanna, N.,** Metastasis of xenogeneic and allogeneic tumors in nude mice, in *Tumor Invasion and Metastasis,* Liotta, L. A. and Hart, I. R., Eds., Martinus Nijhoff, The Netherlands, 1982, 43.
58. **Fujii, G., Sekiguchi, M., Samaru, Y., Miyamoto, Y., Yamamoto, M., Suzuki, K., and Sudo, K.,** Study of tumor metastasis in athymic nude mice, *J. Exp. Med.,* 49, 217, 1979.
59. **Hanna, N., Davis, T. W., and Fidler, I. J.,** Environmental and genetic factors determine the level of NK activity of nude mice and affect their suitability as models for experimental metastasis, *Int. J. Cancer,* 30, 371, 1982.
60. **Roder, J. and Duwe, A.,** The beige mutation in the mouse selectively impairs natural killer cell function, *Nature (London),* 278, 451, 1979.
61. **Sordat, B., Merenda, C., and Carrel, S.,** Invasive growth and dissemination of human solid tumors and malignant cell lines grafted subcutaneously to newborn nude mice, in *Proc. 2nd Int. Workshop on Nude Mice,* Nomura, T., Ohsawa, N., Tamaoki, N., and Fujiwara, R., Eds., University of Tokyo Press, Tokyo, 1977, 313.
62. **Machado, E. A., Lozzio, B. B., Lozzio, C. B., Lair, S. V., and Maxwell, P. A.,** Study of metastases of human malignant cells in nude mice, in *Proc. 3rd Int. Workshop on Nude Mice,* Gustav Fisher Verlag, New York, 1982, 34.
63. **Levy, J. A., White, A. C., and McGrath, C. M.,** Growth and histology of a human mammary-carcinoma cell line at different sites in the athymic mouse, *Br. J. Cancer,* 45, 375, 1982.
64. **Giavazzi, R., Jessup, J. M., Campbell, D. E., Walker, S. M., and Fidler, I. J.,** Experimental nude mouse model in human colorectal cancer liver metastases, *J. Natl. Cancer Inst.,* 77, 1303, 1986.
65. **Price, J. E., Daniels, L. M., Campbell, D. E., and Giavazzi, R.,** Organ distribution of experimental metastases of a human colorectal carcinoma injected in nude mice, *Clin. Exp. Met.,* 7, 55, 1989.
66. **Morikawa, K., Walker, S. M., Nakajima, M., Pathak, S., Jessup, J. M., and Fidler, I. J.,** Influence of organ environment on the growth selection and metastasis of human colon carcinoma cells in nude mice, *Cancer Res.,* 48, 6863, 1988.
67. **Naomoto, Y., Kondo, H., Tanaka, N., and Orita, K.,** Novel experimental models of human cancer metastasis in nude mice: lung metastasis, intraabdominal carcinomatosis with ascites, and liver metastasis, *J. Cancer Res. Clin. Oncol.,* 113, 544, 1987.
68. **Sekikawa, K., Arends, J. W., Verstijnen, C. P. H. J., van der Linden, E., Dinjens, W., Schutte, B., and Bosman, F. T.,** Influence of implantation site on growth antigen expression and metastatic potential of human colonic cancer HT29 and 5583 xenografts in nude mice, *Inv. Met.,* 8, 238, 1988.
69. **Rodolfo, M., Balsari, A., Clemente, C., Parmiani, G., and Fossati, G.,** Tumorigenicity and dissemination of primary and metastatic human melanomas implanted into different sites in athymic nude mice, *Inv. Met.,* 8, 317, 1988.
70. **Shevrin, D. H., Kukreja, S. C., Ghosh, L., and Lad, T. E.,** Development of skeletal metastasis by human prostate cancer in athymic nude mice, *Clin. Exp. Met.,* 6, 401, 1988.
71. **Jelbart, M. E., Russell, P. J., Russell, P., Wass, J., Fullerton, M., Wills, E. J., and Raghavan, D.,** Site-specific growth of the prostate xenograft line UCRU-PR-2, *Prostate,* 14, 163, 1989.
72. **Verschraegen, C. F., Giovanella, B. C., Mendoza, J. T., and Stehlin, J. S., Jr.,** Metastatic spread of human melanoma clones inoculated in the arterial and venous circulation of nude mice, *Proc. Am. Assoc. Cancer Res.,* 28, 75, 1987.
73. **Kerbel, R. S., Man, M. S., and Dexter, D.,** A model of human cancer metastasis: extensive spontaneous and artificial metastasis of a human pigmented melanoma and derived variant sublines in nude mice, *J. Natl. Cancer. Inst.,* 72, 93, 1984.
74. **Thiede, K., Momburg, F., Zangemeister, U., Schlag, P., and Schirrmacher, V.,** Growth and metastasis of human tumors in nude mice following tumor-cell inoculation into a vascularized polyurethane, *Int. J. Cancer,* 42, 939, 1988.
75. **Kozlowski, J. M., Hart, I. R., Fidler, I. J., and Hanna, N.,** A human melanoma line heterogenous with respect to metastatic capacity in athymic nude mice, *J. Natl. Cancer Inst.,* 72, 913, 1984.
76. **Ishikawa, M., Fernandez, B., and Kerbel, R. S.,** Highly pigmented human melanoma variant which metastasizes widely in nude mice, including to skin and brain, *Cancer Res.,* 48, 4897, 1988.
77. **Nanni, P., De Giovanni, C., Nicoletti, G., Del Re, B., Scotlandi, K., and Lollini, P.-L.,** Human rhabdomyosarcoma cells in nude mice as a model for metastasis and differentiation, *Inv. Met.,* 9, 231, 1989.
78. **Price, J. E., Sauder, D. N., and Fidler, I. J.,** Tumorigenicity and metastatic behavior in nude mice of two human squamous cell carcinoma lines that differ in production of the cytokine ETAF/IL-1, *J. Invest. Dermatol.,* 91, 258, 1988.
79. **Zimmerman, R. J., Gaillard, E. T., and Goldin, A.,** Metastatic potential of four human melanoma xenografts in young athymic mice following tail vein inoculation, *Cancer Res.,* 47, 2305, 1987.
80. **Zimmerman, R. J., Gaillard, E. T., and Goldin, A.,** Pulmonary tumor colony formation following i.v. inoculation of six human colorectal carcinoma xenografts in young gnotobiotic athymic mice, *Clin. Exp. Met.,* 6, 27, 1988.
81. **Leduc, E. H.,** Metastasis of transplantable hepatomas from the spleen to the liver in mice, *Cancer Res.,* 19, 1091, 1959.

82. **Kopper, L., Van Hanh, T., and Lapis, K.,** Experimental model for liver metastasis formation using Lewis lung tumor, *J. Cancer Res. Clin. Oncol.*, 103, 31, 1982.
83. **Lafreniere, R. and Rosenberg, S. A.,** A novel approach to the generation and identification of experimental hepatic metastases in a murine model, *J. Natl. Cancer Inst.*, 76, 309, 1986.
84. **August, D. A., Ottow, R. T., and Sugarbaker, P. H.,** Clinical perspective of human colorectal cancer metastasis, *Cancer Met. Rev.*, 3, 303, 1984.
85. **Morikawa, K., Walker, S. M., Jessup, J. M., and Fidler, I. J.,** *In vivo* selection of highly metastatic cells from surgical specimens of different primary human colon carcinomas implanted into nude mice, *Cancer Res.*, 48, 1943, 1988.
86. **Giavazzi, R., Garofalo, A., Damia, G., Garattini, S., and D'Incalci, M.,** Response to flavone acetic acid (NSC 347512) of primary and metastatic human colorectal carcinoma xenografts, *Br. J. Cancer*, 57, 277, 1988.
87. **Massazza, G., Tomasoni, A., Lucchini, V., Allavena, P., Erba, E., Colombo, N., Mantovani, A., D'Incalci, M., Mangioni, C., and Giavazzi, R.,** Intraperitoneal and subcutaneous xenografts of human ovarian carcinoma in nude mice and their potential in experimental therapy, *Int. J. Cancer*, 44, 494, 1989.
88. **Takahashi, S., Konishi, Y., Nakatani, K., Inui, S., Kojima, K., and Shiratori, T.,** Brief communication: conversion of a poorly differentiated human adenocarcinoma to ascites form with invasion and metastasis in nude mice, *J. Natl. Cancer Inst.*, 60, 925, 1978.
89. **Ueyama, Y., Klondo, Y., Tamaoki, N., and Ohsawa, N.,** Brief communication: ascites form of a human cancer serially transplantable in nude mice, *J. Natl. Cancer Inst.*, 57, 965, 1976.
90. **Jin, G., Kexue, X., Baogui, L., Huayi, D., Lisheng, W., Zhaohe, Z., Shen, T., and Zhengue, X.,** Site-dependence of invasiveness of ECA109 human oesophageal carcinoma cells in nude mice, *Clin. Exp. Met.*, 2, 205, 1984.
91. **Kruitwagen, R. F. P. M., Poels, L. G., Willemsen, W. N. P., Croes, H. J. E., Smeets, D. F. C. M., Beck, J. L. M., Hanselaar, A. G. J. M., Ramaekers, F. C. S., Rolland, R., and Jap, P. H. K.,** ETN-1: a new human endometrial carcinoma cell line producing ascites and distant metastases in nude mice, *Int. J. Cancer*, 43, 1098, 1989.
92. **Feldman, G. B.,** Lymphatic obstruction in carcinomatous ascites, *Cancer Res.*, 35, 325, 1975.
93. **Niedbala, M. J., Crickard, K., and Bernacki, R. J.,** Interactions of human ovarian tumor cells with human mesothelial cells grown on extracellular matrix. An *in vitro* model system for studying tumor cell adhesion and invasion, *Exp. Cell Res.*, 160, 499, 1985.
94. **Bénard, J., Da Silva, J., De Blois, M.-C., Boyer, P., Duvillard, P., Chiric, E., and Riou, G.,** Characterization of human ovarian adenocarcinoma line IGROV1 in tissue culture and in nude mice, *Cancer Res.*, 45, 4970, 1985.
95. **Hamilton, T. C., Young, R. C., Louie, K. G., Behrens, B. C., McKoy, W. M., Grotzinger, K. R., and Ozols, R. F.,** Characterization of a xenograft model of human ovarian carcinoma which produces ascites and intraabdominal carcinomatosis in mice, *Cancer Res.*, 44, 5286, 1984.
96. **Kikuchi, Y., Kizawa, I., Oomori, K., Miyauchi, M., Kita, T., Sugita, M., Tenjin, Y., and Kato, K.,** establishment of a human ovarian cancer cell line capable of forming ascites in nude mice and effects of tranexamic acid on cell proliferation and ascites formation, *Cancer Res.*, 47, 592, 1987.
97. **Ward, B. C., Wallace, K., Shepherd, J. H., and Balkwill, F. R.,** Intraperitoneal xenografts of human epithelial ovarian cancer in nude mice, *Cancer Res.*, 47, 2662, 1987.
98. **Giavazzi, R., Massazza, G., and Tomasoni, A.,** Intraperitoneal xenograft of human ovarian carcinoma in nude mice, in *immune-Deficient Animals in Experimental Medicine*, Wu, B-Q. and Zheng, J., Eds., Karger, Basel, 1989, 162.
99. **Boven, E., van der Vijgh, W. J. F., Nauta, M. M., Schlüper, H. M. M., and Pinedo, H. M.,** Comparative activity and distribution studies of five platinum analogues in nude mice bearing human ovarian carcinoma xenografts, *Cancer Res.*, 45, 86, 1985.
100. **Friedlander, M. L., Russell, P., Taylor, I. W., and Tattersall, M. H. N.,** Ovarian tumor xenografts in the study of the biology of human epithalial ovarian cancer, *Br. J. Cancer*, 51, 319. 1985.
101. **Kullander, S., Rausing, A., Tropé, C.,** Human ovarian tumors heterotransplanted to "nude" mice, *Acta Obstet. Gynecol. Scand.*, 57, 149, 1978.
102. **FitzGerald, D. J., Bjorn, M. J., Ferris, R. J., Winkelhake, J. L., Frankel, A. E., Hamilton, T. C., Ozols, R. F., Willingham, M. C., and Pastan, I.,** Antitumor activity of an immunotoxin in a nude mouse model of human ovarian cancer, *Cancer Res.*, 47, 1407, 1987.
103. **Malik, S. T. A., Griffin, D. B., Fiers, W., and Balkwill, F. R.,** Paradoxical effects of tumor necrosis factor in experimental ovarian cancer, *Int. J. Cancer*, 44, 918, 1989.
104. **Ortaldo, J. R., Porter, H. R., Miller, P., Stevenson, H. C., Ozols, R. F., and Hamilton, T. C.,** Adoptive cellular immunotherapy of human ovarian carcinoma xenografts in nude mice, *Cancer Res.*, 46, 4414, 1986.
105. **Cornil, I., Man, S., Fernandez, B., and Kerbel, R. S.,** Enhanced tumorigenicity melanogenesis, and metastases of a human malignant melanoma after subdermal implantation in nude mice, *J. Natl. Cancer Inst.*, 81, 938, 1989.
106. **Wang, W. R., Sordat, B., Pinguet, D., and Sordat, M.,** Human colon tumors in nude mice: implantation site and expression of the invasive phenotype, in *Immune-Deficient Animals. 4th Int. Workshop on Immune-Deficient Animals in Exp. Res.*, Karger, Basel, 1984, 239.

107. **Bresalier, R. S., Raper, S. E., Hujanen, E. S., and Kim, Y. S.,** A new animal model for human colon cancer metastasis, *Int. J. Cancer,* 39, 625, 1987.
108. **Bogden, A. E., Kelton, D. E., Cobb, W. R., and Esber, H. J.,** A rapid screening method for testing chemotherapeutic agents against human tumor xenografts, in *Proc. Symp. on the Use of Athymic (Nude) Mice in Cancer Res.,* Houchens, D. P. and Ovejera, A. A., Eds., Fisher, New York, 1978, 231.
109. **Bennett, J. A., Pilon, V. A., and MacDowell, R. T.,** Evaluation of growth and histology of human tumor xenografts implanted under the renal capsule of immunocompetent and immunodeficient mice, *Cancer Res.,* 45, 4963, 1985.
110. **Shafie, S. M. and Liotta, L. A.,** Formation of metastasis by human breast carcinoma cells (MCF-7) in nude mice, *Cancer Lett.,* 11, 81, 1980.
111. **Price, J. E., Polyzos, A., Dan Zhang, R., and Daniels, L. M.,** Tumorigenicity and metastasis of human breast carcinoma cells lines in nude mice, *Cancer Res.,* 50, 717, 1990.
112. **Tan, M. H. and Chu, T. M.,** Characterization of the tumorigenic and metastatic properties of a human pancreatic tumor cell line (ASPC-1) implanted orthotopically into nude mice, *Tumour Biol.,* 6, 89, 1985.
113. **Vezeridis, M. P., Doremus, C. M., Tibbetts, L. M., Tzanakakis, G., and Jackson, B. T.,** Invasion and metastasis following orthotopic transplantation of human pancreatic cancer in the nude mouse, *J. Surg. Oncol.,* 40, 261, 1989.
114. **Yamashita, T.,** Manifestation of metastatic potential in human gastric cancer implanted into the stomach wall of nude mice, *J. Cancer Res.,* 79, 945, 1988.
115. **Ahlering, T. E., Dubeau, L., and Jones, P. A.,** A new *in vivo* model to study invasion and metastasis of human bladder carcinoma, *Cancer Res.,* 47, 6660, 1987.
116. **McLemore, T. L., Liu, M. C., Blacker, P. C., Gregg, M., Alley, M. C., Abbott, B. J., Shoemaker, R. H., Bohlman, M. E., Litterst, C. C., Hubbard, W. C., Brennan, R. H., McMahon, J. B., Fine, D. L., Eggleston, J. C., Mayo, J. G., and Boyd, M. R.,** Novel intrapulmonary model for orthopic propagation of human lung cancers in a athymic nude mice, *Cancer Res.,* 47, 5132, 1987.
117. **McLemore, T. L., Eggleston, J. C., Shoemaker, R. H., Abbott, B. J., Bohlman, M. E., Liu, M. C., Fine, D. L., Mayo, J. G., and Boyd, M. R.,** Comparison of intrapulmonary, percutaneous intrathroracic and subcutaneous models for the propagation of human pulmonary and nonpulmonary cancer cell lines in athymic nude mice, *Cancer Res.,* 48, 2880, 1988.
118. **Schackert, G., Price, J. E., Bucana, C. D., and Fidler, I. J.,** Unique patterns of brain metastasis produced by different human carcinomas in athymic nude mice, *Int. J. Cancer,* 44, 892, 1989.

Chapter 11

EXPERIMENTAL THERAPY IN SUBCUTANEOUS TRANSPLANTS

David P. Houchens and Artemio A. Ovejera

I. INTRODUCTION

A variety of tumor models have been developed for tumor studies over the past 30 to 40 years.[1-4] These models have included spontaneous as well as chemical- and viral-induced tumors and transplantable tumors which have been in serial transfer for many generations. Each of these models has its particular use for the biologist, chemotherapist, and immunologist, and each model has its limitations.

The major question which may be asked in studying these tumor models is: "How close is the model to malignant disease in man?" The desired answer would include the ability of the model to mimic the growth and spread of tumors in patients. At the same time, the model should serve as a test system in which various therapeutic modalities can be assessed and selection of optimal treatment regimens for the clinic can be determined. Some researchers believe that syngeneic rodent tumors are not really predictive of what may occur clinically. The transplantation of human tumors to animals which have been artificially immunosuppressed requires manipulation or treatment which could compromise the results of tumor growth.[5,6] With the advent of the nude mouse for the growth of human tumors,[7] the next step in using this model for therapeutic assessment became obvious. Other chapters in this book are dedicated to the history and development of the nude mouse model for growth and maintenance of human tumors. This chapter will address the background and use of the s.c. implanted xenografts as a model which has been employed successfully for the evaluation of various therapeutic modalities by the U.S. National Cancer Institute (NCI) and by a variety of researchers in pharmaceutical companies, universities, and research institutes throughout the world.

II. APPLICATIONS OF THE SUBCUTANEOUS ASSAY

A. PAST AND CURRENT APPLICATIONS

Following the report by Rygaard and Povlsen[7] that human tumors could be successfully transplanted and grown in nude mice, many other researchers began their own studies and reported the growth of a variety of human tumors in these animals.[8-10] These early studies were all conducted using s.c. implantation of tumors because of the ease of the procedures and the ability to grossly visualize the tumor growth. Within a short time, it became apparent that the model could be applied to the evaluation of potential antitumor agents. Tumor inhibition and tumor regression were taken as a measure of the effects of various chemotherapeutic compounds. Numerous laboratories throughout the world published initial papers relating to the use of the model.[11-15] The NCI realized the potential value of this model in the anticancer drug screening effort of the Developmental Therapeutics Program of the Division of Cancer Treatment. One of the most important areas of the NCI program was that of "the development of the means to cure cancer patients and to maintain control of cancer in patients who are not cured".[16] Embodied in this objective were the identification and evaluation of antitumor agents for the cure and palliation of cancer.

Starting in the late 1950s in the NCI program, various murine tumor models were incorporated for the screening of compounds. Such a program ideally should select agents that give a good correlation between preclinical and clinical data. The tumor systems chosen should select "true positive" compounds with as few "false positive" and "false negative" compounds as possible. With the evolution of the program, a tumor panel was developed and, in 1975, human tumor xenografts were added to the panel (Table 1).[17] These xenografts included one line each of breast (MX-1), colon (CX-1), and lung (LX-1)[18] as these types account for 40 to 45% of all cancers in the U.S. The obvious question that was raised is whether more than one tumor of a given histological type should be used in a panel. Studies by Osieka et al.[19] and Bellet et al.[20] had demonstrated that xenografts of the same histological type responded differently to the same drugs. It was determined that a well-designed panel of murine and human tumors can select drugs and families of drugs. Assessment of a number of clinically used antitumor agents in the

TABLE 1
National Cancer Institute Preclinical Tumor Panel

Year	Prescreen	Panel: Murine tumors	Panel: Human tumor Xenografts
1975—1982	P388 leukemia	Leukemia L1210 Melanoma B16 Lewis lung carcinoma C38 colon $CD8F_1$ breast	LX-1 lung CS-1 colon MX-1 breast
After 1982	P388 leukemia	Leukemia L1210 Melanoma B16 M5076 sarcoma	MX-1 breast

TABLE 2
Response of Subcutaneously Implanted Tumors in Nude Mice to Anticancer Drugs

	Human tumor xenograft[a]		
Drug	MX–1	LX–1	CX–1
Actinomycin	—	—	—
BCNU	—	—	—
Bleomycin	+	—	—
CCNU	—	—	—
Chlorambucil	++	—	—
Cyclophosphamide	++	—	—
Cytosine arabinoside	—	—	—
Daunomycin	—	—	—
Doxorubicin	—	—	—
DTIC	—	—	—
5-Fluorouracil	—	—	—
5-FUdR	—	—	—
Melphalan	++	—	—
6-Mercaptopurine	—	—	—
Methotrexate	—	—	—
Mitomycin C	++	—	—
cis-Platinum	++	—	—
Procarbazine	—	++	—
Thioguanine	—	—	—
Vincristine	++	—	—
Total responses / Total tested	7/20	1/20	0/20

[a] ++, tumor regression; +, growth inhibition; —, no response.

s.c. model showed that the MX-1 tumor was the most sensitive to therapy, while the LX-1 and CX-1 tumors were refractory to chemotherapy (Table 2).[21] When experimental compounds were evaluated, several agents were selected by each tumor model (Table 3).[22] Examples of compounds that were selected from their activity profile in xenografts and introduced into clinical trials were D-O-norleucine[23] and acivicin.[24] This panel was employed from 1975 until 1982 when an extensive evaluation of the program was conducted, resulting in a change in the panel to include only one xenograft model (MX-1).[18] It was determined that this murine and xenograft panel could select 94% of the drugs indicated to have activity by the much larger panel

TABLE 3
Number of Drugs Exhibiting Therapeutic Activity in the Subcutaneous Human Tumor Xenograft Systems

Tumor System	Total drugs tested	Number inducing regressions (%)	Number retarding growth (%)
MX-1	142	6 (4)	9 (6)
LX-1	143	8 (6)	7 (5)
CX-1	126	2 (2)	3 (2)

in the previous 6 years. This panel was then used until 1987. In those years, many of the xenograft studies were conducted using the subrenal capsule as the site of implantation (Chapter 7).[25] However, the need to assess drugs against an established tumor still required that the s.c. assay be used in some studies.

In 1984, the NCI began an extensive effort to evaluate an *in vitro* "disease-oriented" approach for the screening of antitumor drugs.[26] In this approach, a panel of approximately 10 human cell lines each of the major tumor types are used to assess the response to potential antitumor drugs by means of various *in vitro* assays. Since there would still be a need to analyze active compounds in animals, the model includes the evaluation of compounds against an *in vivo* tumor panel of s.c. xenografts which include the same tumor lines that the compound showed activity against in the *in vitro* screen (Figure 1).[27]

Since the initial reports on the evaluation of chemotherapeutic agents in human tumor xenografts, hundreds of reports have appeared in the literature. Many of these studies are presented and reviewed in the various *Proceedings of the Workshops on the Use of Nude Mice and Immune-Deficient Animals*[28-33] which have been held every 3 years and in other review volumes.[34-36] The studies include data on the effects of all major groups of drugs and all major tumor types. The s.c. xenograft model has been and continues to be useful for the evaluation of various compounds. Additionally, other types of therapy have been used and are being used in the model with great success.

The s.c. location of the tumors has been especially useful in radiotherapy studies[37-40] since the tumors are easily accessible to the radiation equipment. The tumor site has also served a useful purpose in hyperthermia studies[41,42] and in combination radiation-hyperthermia studies.[43] Rofstad[43] reported on five separate clones from a single melanoma xenograft having a heterogeneous response to both radiation and hyperthermia. Senapati et al.[44] demonstrated that synergistic effects were obtained by the use of ultrasonic hyperthermia in combination with melphalan or procarbazine. Thus, the model may prove valuable in developing methods that predict responsiveness of tumors to these forms of therapy.

Numerous monoclonal antibodies directed against solid human tumor types have been developed and a number of these have been conjugated to radionuclides for both diagnostic tumor imaging and for therapeutic trials.[45,46] Several of these antibodies have been labeled with radionuclides, such as ^{131}I and ^{90}Y, to investigate in preclinical efficacy trials in the s.c. nude mouse-human tumor model.[47,48] Some studies have demonstrated that by administering the labeled monoclonal antibody using a dose fractionation schedule, the bone marrow toxicity which usually results from high doses of irradiation can be overcome.[49]

Other forms of therapy which have been evaluated in the s.c. xenograft model include immunotoxins,[50,51] interferons,[52] other cytokines,[53] and cell differentiation agents.[54] Ultimately, it may be shown that certain types of therapy are more appropriate for a tumor with specific histological or biological characteristics than for an individual patient's tumor.

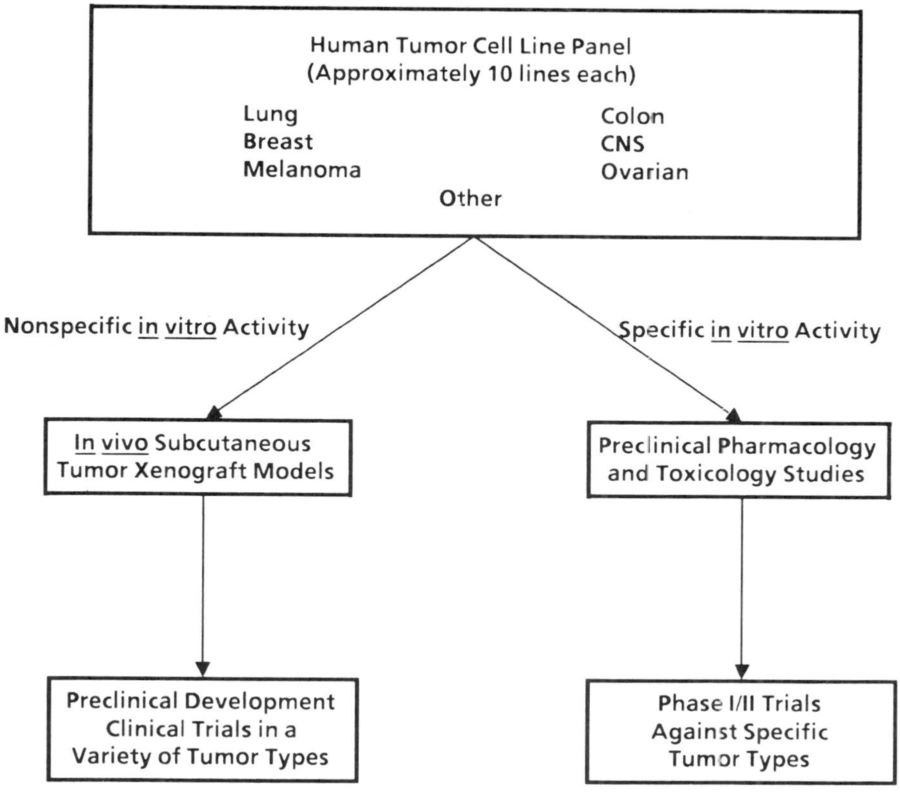

FIGURE 1. Disease-oriented approach for antitumor evaluation of drugs.

B. ADVANTAGES OF THE ASSAY

The most obvious advantage of the s.c. nude mouse-human tumor model is the ability to establish and grow human tumor tissue in a living host which has all of the physiological and metabolic conditions present for cell nutrition, host-tumor interaction, and drug metabolism and detoxification. The procedures required for implantation (as described in Chapter 4 and in Section III of this chapter) are relatively easy and allow for good visualization and measurement of tumor growth. Also, the tumor is readily accessible for direct injection of drugs, radiotherapy, or surgical procedures to obtain tumor tissue or remove the primary implant. Additionally, the ability to assess a certain "stage" with respect to the tumor size at the start of treatment allows the investigator to mimic the clinical situation more readily. It has also been reported that the s.c. implanted tumors maintain their histological and karyotypic characteristics for many transplant generations.[55]

C. LIMITATIONS OF THE ASSAY

The primary limitation of the s.c. xenograft assay is the time required for the determination of a treatment effect. After tumor implantation, there is usually a 2 to 6-week tumor growth period before the therapy can be initiated. After treatment, a 20 to 30-d evaluation period will follow. Therefore, the entire experiment could take from 6 to 12 weeks. If a drug screening program is to be used for relatively rapid decisions, this assay may not be appropriate. There is also a need to implant as many as 30 to 40% more animals than are necessary for the study, since tumors are "staged" by size. The high costs of nude mice coupled with the high costs of maintaining the animals in the proper environment to prevent infections[56] makes the total expenses for evaluating potential therapy very high. Lastly, since most human tumors do not readily metastasize in nude mice,[57] the model is not particularly good for assessing the effects of therapy on metastatic lesions. A metastasis model may be obtained by further manipulation of the animals or the implantation of a metastatic tumor line (Chapter 10).

III. METHODOLOGY

A. TUMOR IMPLANTATION, GROWTH, AND ACCEPTABILITY FOR TESTING

The s.c. nude mouse-human tumor model has been utilized extensively by the NCI in preclinical evaluation of hundreds of drugs.[2,58,59] The procedure employed may slightly differ among institutes and involves implanting 2 to 3 mm^3 fragments of a given tumor type into the lateral thoracic region of nude mice using a trocar/cannula or a skin incision. The s.c. growing tumors are measured periodically with vernier calipers. The length (L), width (W), and height (H) (in mm) of the tumors are taken and used to calculate the volume (V), using the formula $V = (W^2 \times L)/2$ or $V = (W \times L \times H)/2$. This volume (in mm^3) can be considered a valid estimate of weight (in mg), assuming unit density. When the tumors reach 100 to 300 mm^3 in volume or 100 to 300 mg in weight, mice are randomized into test groups of 5 to 10 mice each. For established, rapidly growing tumor lines, these weights are attained in 2 to 3 weeks after implantation. At this point, the tumors are growing progressively and do not regress spontaneously. It is not uncommon that some mice would possess very small or very large tumors at the time of staging. Only about 70 to 80% of the implanted mice usually have a tumor size that falls within the acceptable size range.[58,59] This should be taken into consideration when planning an experiment and sufficient mice in excess of the actual number needed should be inoculated.

While fragments of tumor tissue are ordinarily used for s.c. implantation, especially for primary human tumors from surgical resections, other forms of inocula have been utilized. S.c. tumors have also been grown in nude mice by implanting single cell suspensions of cultured cells.[8,60,61] In each mouse, 10^6 to 10^8 cells are injected s.c., the number depending on the tumor line's rate of growth. It is our experience that tumor configuration shows a more variable pattern when a single cell suspension is utilized, as compared to that obtained with fragments. Because of the liquid medium, cells tend to spread upon s.c. injection which may result in elongated, flattened, or multilobular tumor growth. Such tumors are not only difficult to measure but the estimation of volume or weight poses some problems. Minced tumor tissues[62,63] suspended in physiological salt solution or tissue culture medium (1:5 or 1:10 v/v) have also been used as s.c. inoculation. These preparations usually offer a homogeneous deposit upon implantation.

The lateral thoracic region is the most common site used for s.c. implantation for reasons of convenience in transplantation and measurement of the tumor. Other areas have been used successfully for growing tumors. The dorsoscapular region[61] offers loose skinfolds that would promote optimal tumor growth, but tumors in such a location are slightly inconvenient to measure. Likewise, the inguinal or pelvic region, although easily inoculated, will present difficulties in tumor measurements because of the curvature of the trunk and upper legs. It has been shown that the site of implantation affects the rate of tumor growth.[64,65] A marked cranial-caudal growth gradient exists, that is, s.c. tumors in the anterior region grow at a more rapid rate than those in the posterior area of the trunk. A three- to four-fold difference in tumor volume has been observed.[64]

B. THERAPY

1. Chemotherapy

It has become abundantly clear that the s.c. human tumor xenograft model is a useful tool for assessing the efficacy of antitumor agents. After all, it is not unreasonable to expect that human tumor systems would predict for clinical activity more efficiently than animal tumors. The convenience and reproducibility of the s.c. xenograft make this model very popular among investigators in the field of experimental chemotherapy.

For established human tumor lines, tumor "take" in mice implanted s.c. is virtually 100% and, once tumors reach palpable size, they do not regress spontaneously. As indicated earlier, some variation exists in tumor size among mice at staging and only those mice with the desired tumor size should be selected for the study. It is prudent to pool the selected mice in a large cage and then randomize them into smaller cages of five mice each. Further random assignment of the cages to the different test groups is a good practice. A test group normally consists of 5 to 10 mice each. Identification of each mouse in a cage or group by ear punch or other means is necessary in order to keep track of each individual's tumor size at all measurement points. This is particularly important when account needs to be made for accidental deaths. Tumor measurements and body weights are recorded at the initiation of treatment and at subsequent predetermined time points.

Customarily, the dose of drugs used in chemotherapy studies centers around the LD_{10} dose or maximum tolerated dose (MTD). Obviously, the LD_{10} value or MTD depends on the schedule of treatment. A diversity of treatment schedules has been employed, such as an intermittent schedule (e.g., q4d × 3, q7d × 4),[12,23,58,63] daily (qd × 5),[66,67] or single injections.[66,67] It is desirable to determine the LD_{10} dose or MTD for each treatment schedule to be used. An extensive compilation of these doses for different drugs in various strains of conventional mice has been published by the Southern Research Institute,[68] which can serve as a guide in choosing doses to use, keeping in mind that, for some drugs, nude mice are able to tolerate higher doses than are conventional mice.[69]

The use of a "rational dose" in mice has been proposed, which is supposed to be more appropriate and will improve clinical predictability as the dose studied is pharmacokinetically equivalent to the clinical dose.[67] In a study to compare the MTD and the rational dose (RD) in mice, Inaba et al.[67] reported an RD/MTD ratio of 0.20 to greater than 1.0 for a spectrum of drugs. They concluded that the use of RD in the nude mouse-human tumor model reflects well the clinically observed relative effectiveness of selected antitumor agents.

Ideally, a series of incremental doses of a drug should be used, ranging from an ineffective dose to a frankly lethal dose. This scheme will permit the establishment of dose-response curves and the specification of the minimum effective dose (MED) and the maximum tolerated dose (MTD; usually taken as LD_{10}). The ratio MTD/MED offers a measure of the margin of safety with which a drug can be administered; the higher this ratio, termed "therapeutic index",[70] the greater the latitude in drug usage. Different drugs can be compared using the therapeutic index.

There are general routes by which drugs can be administered into experimental animals. The most widely used are i.p. and i.v. injections. The route of drug administration may play an important role in the expression of the effectiveness of a drug. It has been shown, for example, that doxorubicin is more effective when administered i.v. than i.p.[71]

Depending on the treatment schedule employed, the effects of therapeutic agents are usually manifested within 1 to 4 weeks after the initiation of treatment, at which time evaluation for activity should be made.

2. Other Therapy

Assignment to treatment groups is the same for other forms of therapy as described in the previous section. Just as it is important to determine the optimal dose range for chemotherapy, it is also necessary in the evaluation of immunotoxins,[50] cytokines,[53] and radiotherapy. As with chemotherapy, most other forms of therapy are administered by various routes of injection. External radiation[37] and hyperthermia[44] are the exception, and calibrations of doses must be determined for each experimental model to be used.

C. MEASUREMENT OF TUMOR SIZE

Quantitative changes in tumor size are used to assess the effects of therapy on s.c. xenografts. Undoubtedly, the optimal measure of tumor size is its actual tumor weight, which can only be obtained by killing the tumor-bearing host; this process might not be appropriate for studies where periodic determination of tumor size *in situ* is required to generate a growth curve. The size of an s.c. growing tumor has been variously represented, such as one-dimensional measurement,[72-74] the area calculated from two diameters,[75,76] or volume estimated from two or more linear measurements[58,59,77] or from plaster cast of intact tumor.[78] Various methods have been used to determine tumor volume,[79,80] including application of different formulae for estimating volumes of an ellipsoid and a spheroid.[19,58,59,77,81-83] Studying several methods of tumor size determination, Enhus et al.[79] reported that estimates of area and volume correlated well with actual tumor volume or mass, and that length alone was not a good indicator of tumor size. In a more comprehensive study, Tomayko and Reynolds[80] compared diameters, areas, and estimated volumes with actual weight of s.c. xenografts and concluded that the ellipsoid volume calculated from two or three linear measurements consistently yielded accurate estimates of the actual tumor mass.

It is clear that there is no standard procedure for quantitating the size of an s.c. growing tumor. The most common practice is to measure the length and the width (and the height) of the tumor using a vernier caliper. These measurements, taken perpendicular to each other, are then used to estimate the tumor volume or mass using the formulae shown earlier (Section III. A). It is customary to make periodic measurements, usually once or twice a week, in order to be able to plot a growth curve which could illustrate prominently the pattern of growth and/or response to therapy of a given tumor. The growth curve is useful for intelligent assessment of the therapeutic effects of potential antitumor agents, as compared to the growth curve of untreated xenografts.

While the caliper measurement technique itself is easy and simple to perform, one should be aware of the fact that variation exists between different individual measurements of the same tumor. Tumor measurements made by any two individuals may differ by as much as 15%.[79] It is desirable, therefore, that only one person will perform tumor measurements in a given experiment.

IV. CRITERIA FOR EVALUATION

A. TUMOR SIZE

1. Inhibition and Regression

Tumor growth inhibition, which reflects the retardation in tumor growth of the treated group relative to the control group, is widely used to assess the efficacy of therapeutic agents against s.c. xenografts. Using tumor weight as a measure of size for illustration, tumor growth inhibition (TGI) expressed as a percentage is calculated as follows:

$$\%TGI = 100\,(W_c - W_t)/W_c = 100\,(1 - W_t/W_c)$$

where W_c is the mean tumor weight of the control group and W_t is the mean tumor weight of the treated group. Often, $\%T/C = 100(W_t/W_c)$, which is complementary to %TGI, is simply used to measure treatment effects.[12,58,84] Usually, some variations exist among the mean tumor weights for different groups at the initiation of treatment. The differences in initial weights may be great enough to affect the validity of the computed %TGI or %T/C if they are not taken into account. To standardize these variations, relative tumor weights (RW) are obtained by the formula $RW = W_i/W_o$, where W_i is the mean tumor weight for a group at any given time and W_o is the mean initial tumor weight when treatment was started.[58,63] Another way to correct for the variations is to calculate the change in tumor weight for each group; that is, $\Delta W = W_i - W_o$.[85,86] RWs or ΔWs for treated and control groups are then used to compute %TGI or %T/C.

Some therapeutic agents are sufficiently effective to elicit regression of the s.c. xenografts. In these cases, a more appropriate measure of the treatment effect is the percentage of regression calculated as follows:

$$\%REG = 100\ (W_o - W_i)/W_o$$

where W_o is the mean tumor weight for a treated group at the initiation of treatment and W_i is the mean tumor weight for that group at evaluation time or any given time. Needless to say, regression exists when W_o is greater than W_i. RW, as defined previously, can be used in the above formula and, since the initial RW_o is 1, the equation is reduced to $\%REG = 100(1 - RW_i)$. The NCI employs $\%T/C = (\Delta W/W_o)100$ for evaluation, which gives negative %T/C values.[85,86] Despite this T/C designation, no measurements pertaining to the control group are actually used in the equation.

There is no standard determinant for assessing the activity of therapeutic agents. The NCI proposes a %T/C value less than or equal to 10% to be indicative of significant activity and a %T/C value less than or equal to 20% is considered to demonstrate moderate activity.[85,86] Other groups declare activity with T/C less than or equal to 50%.[84,87] Apart from these arbitrary chosen assignations, some parametric and non-parametric statistical tests have been employed in determining therapeutic efficacy.[63,87,88]

2. Growth Delay

Tumor growth delay is another frequently used measure of therapeutic effects on the s.c. xenograft.[89-91] Defined simply, growth delay is the difference in time for treated and control tumors to reach a given size. The procedure for determining growth delay (GD) is straightforward. A tumor size is chosen that is a multiple of the size of the tumor at the initiation of treatment. If a two-fold increase in volume is selected, then the time it takes for each of the control tumor (t_c) and treated tumor (t_t) to double in volume is determined. Thus,

$$GD = t_t - t_c$$

When therapeutic effects on tumors with widely different growth rates are assessed, the comparison of absolute GDs may not be satisfactory. For valid comparisons, standardized GDs, termed specific growth delay, should be used. Specific growth delay (SGD) is calculated as follows:

$$SGD = (t_t - t_c)t_c$$

This value could be regarded as the number of volume doubling times by which growth is delayed.

3. Surviving Cells

The effectiveness of therapeutic agents against s.c. xenografts can also be assessed by measuring the surviving fraction of clonogenic cells in the tumor nodule. The procedure involves excising the tumor after treatment and preparing single-cell suspensions for each test group. Appropriate dilutions are then made and the cells plated in semi-solid medium,[92,93] incubated, and allowed to form colonies. Colonies are counted when they attain suitable size, and cloning efficiency (CE) is determined by

$$CE = \frac{\text{no. of tumor colonies counted}}{\text{no. of tumor cells plated}}$$

The surviving fraction (SF) is calculated as

$$SF = \frac{\text{CE treated}}{\text{CE control}}$$

SF is used in assessing and comparing the efficacy of different treatments.

The time between drug treatment and tumor excision is an important factor that may affect the outcome of an experiment. As a rule, this time must be sufficiently long that the cytotoxic activity of the drug is complete, usually 24 to 48 h after drug administration.[94,95] Another factor to consider is the disaggregation of the solid tumor mass into a single-cell suspension. It is essential that a satisfactory dispersal procedure is employed so that adequate cell yield is obtained and the morphology and clonogenic capacity of the cells are retained. Additionally, a high quality, truly "single-cell" suspension must be prepared for plating. Clonogenic efficiencies vary widely, so prior knowledge of the characteristics of a tumor cell population is required to determine the appropriate dilution for seeding. The length of incubation time depends upon the proliferation rate of the cells. Treated cells tend to divide slower than untreated cells, and may require a longer incubation time to develop into colonies of at least 50 cells.[98] It is therefore necessary that the human tumor line used for this assay is well characterized for clonogenic properties.

B. TOXICITY — WEIGHT CHANGE AND DEATH

In addition to the effects that experimental drugs may have on the tumor, it is important to determine what, if any, effects the compound may have on the host. It was assumed in early studies with nude mice that they would be more susceptible to the toxic effects of cancer chemotherapeutic agents. However, Houchens et al.[69] conducted a study in which the LD_{50} doses of a number of antitumor drugs were determined in nude mice and in conventional mice of the same genetic background. It was found that nude mice could tolerate similar or higher doses than conventional mice to some of these agents. Freudenthal et al.[97] have observed that the hepatic microsomes from nude mice were able to metabolize some compounds more efficiently than those from conventional mice. Thus, it is very important to determine that the dose range for experimental therapy has been properly selected for work in nude mice. In most cases, the same doses used in conventional mice cover a wide enough range that is appropriate for nude mice. If a limited number of doses will be used or a new class of compounds is to be tested, a preliminary toxicity assessment should be conducted in non-tumor-bearing nude mice.

Since the toxicity of a test compound may affect normal tissue as well as tumor tissue, a regular assessment of animal weight change should be made during the course of a study. The minimum for this should be the weekly determination of a mean group weight for each treatment and control group. Although there are no formal guidelines for weight loss limits in xenograft models, a loss of greater than 20% of body weight (as used in murine tumor models by the NCI[98]) should be considered as a significant sign of toxicity. Additionally, death of greater than 15% of the animals in a group should be considered as a sign of toxicity.

V. CONCLUSIONS

The advent of the nude mouse has stimulated great interest and facilitated the *in vivo* study of human tumors, especially in the area of experimental therapy. Indeed, the nude mouse-human tumor model is perhaps as close as any to approximating the *in vivo* behavior of human tumors under laboratory setting and should offer distinct advantages as compared to the use of murine tumors. Prior to the advent of nude mice, other animal models were used for human tumor transplants with moderate success, but the required artificial immunosuppression is laborious and compromises the pattern of tumor growth. Only genetically immunodeficient animals, such as the nude mouse, provide a model that is convenient to use and does not require intervening treatment or manipulation to render the animals immunologically incompetent.

The extensive use of s.c. grown human tumor xenografts in Europe, the U.S., and Japan has demonstrated their ability to aid in the selection of antitumor drugs for further development. The current use of the model for secondary evaluation in the *in vitro* "disease-oriented" tumor panel[26,27] and for preclinical phase II evaluation of various drugs and the selection of active analogs[99] has proven that the model is valid in treatment regimen selection. The use of clinically equivalent doses[66,67] in some laboratories has made the ultimate selection of drugs even more reliable and evaluation of therapeutic modalities, other than chemotherapy, has shown the value of the model in a variety of protocols. After more than 20 years of use, the s.c. human tumor xenograft model has proven itself to be of real benefit to both researchers and clinicians in an effort to develop more effective treatments in malignant diseases.

REFERENCES

1. **Rothenberg, L. and Terselic, R. A.,** Management of the National Cancer Institute's drug research program through application of the linear array concept, *Cancer Chemother. Rep.*, 54, 303, 1970.
2. **Venditti, J. M.,** Preclinical drug development: rationale and methods, *Semin. Oncol.*, 8, 349, 1981.
3. **Zubrod, C. G.,** Drug development at the National Cancer Institute: a historical perspective, in *Fundamentals of Cancer Chemotherapy*, Hellmann, K. and Carter, S. K., Eds., McGraw-Hill, New York, 1987, 101.
4. **Goldin, A.,** Screening at the National Cancer Institute: basic concepts, in *Fundamentals of Cancer Chemotherapy*, Hellman, K. and Carter, S. K., Eds., McGraw-Hill, New York, 1987, 141.
5. **Berenbaum, M. C., Sheard, C. E., Reittier, J. R., and Bundick, R. V.,** The growth of human tumors in immunosuppressed mice and their response to chemotherapy, *Br. J. Cancer*, 30, 13, 1974.
6. **Steel, C. G.,** Therapeutic response of human tumor xenografts in immune-suppressed mice, in *Immune-Deficient Animals*, Sordat, B., Ed., Karger, Basel, 1984, 395.
7. **Rygaard, J. and Povlsen, C. O.,** Heterotransplantation of a human malignant tumor to "nude" mice, *Acta Pathol. Microbiol. Scand.*, 77, 758, 1969.
8. **Giovanella, B. C., Stehlin, J. S., and Williams, L. J.,** Heterotransplantation of human malignant tumors in "nude" thymusless mice. II. Malignant tumors induced by injection of cell cultures derived from human solid tumors, *J. Natl. Cancer Inst.*, 52, 921, 1974.
9. **Shimosato, Y., Kameya, T., Nagai, K., Hirobashi, S., Koide, T., Hayashi, H., and Nomura, T.,** Transplantation of human tumors in nude mice, *J. Natl. Cancer Inst.*, 56, 1251, 1976.
10. **Fogh, J., Fogh, J. M., and Orfeo, T.,** One hundred and twenty-seven cultured human tumor cell lines producing tumors in nude mice, *J. Natl. Cancer Inst.*, 59, 221, 1977.
11. **Povlsen, C. O. and Jacobsen, G. K.,** Chemotherapy of a human malignant melanoma transplanted in the nude mouse, *Cancer Res.*, 35, 2790, 1975.
12. **Ovejera, A. A., Houchens, D. P., and Barker, A. D.,** Chemotherapy of human tumor xenografts in genetically athymic mice, *Ann. Clin. Lab. Sci.*, 8, 50, 1978.
13. **Lee, S. S., Giovanella, B. C., Stehlin, J. S., and Brunn, J. C.,** Regression of human tumors established in nude mice after continuous infusion of thymidine, *Cancer Res.*, 39, 2928, 1979.
14. **Shimosato, Y., Nagai, K., Kubota, T., Hirohashi, S., Hayashi, H., Ikeuchi, S., Kameya, T., Koide, T., and Kitahara, T.,** Trials of various therapeutic experiments on human tumors transplantable in nude mice, *Jpn. J. Cancer Clin.*, 22, 699, 1976.
15. **Inaba, M., Tashiro, T., Kabayashi, T., Fujimoto, S., Sakurai, Y., Maruo, K., Saito, M., Ueyama, Y., and Nomura, T.,** Responsiveness of human tumor xenografts to chemotherapy with special reference to the clinical dose, in *Immune-Deficient Animals*, Sordat, B., Ed., Karger, Basel, 1984, 421.
16. National Cancer Program, Object 6, Publication No. NCPOTI-R6-75, *National Cancer Institute*, 1971.
17. **Venditti, J. M.,** Foreward, in *Proc. Symp. Use of Thymic (Nude) Mice in Cancer Res.*, Houchens, D. P. and Ovejera, A. A., Eds., Gustav Fischer, New York, 1978, ix.
18. **Houchens, D. P. and Ovejera, A. A.,** Heterotransplantation models for screening: current status and future prospects, in *Fundamentals of Cancer Chemotherapy*, Hellmann, K. and Carter, S. K., Eds., McGraw-Hill, New York, 1987, 167.

19. **Osieka, R., Houchens, D. P., Goldin, A., and Johnson, R. K.,** Chemotherapy of human colon cancer xenografts in athymic mude mice, *Cancer,* 40, 2640, 1977.
20. **Bellet, R. E., Danna, V., Mastrangelo, M. J., and Berd, D.,** Evaluation of a "nude" mouse-human tumor panel as a predictive secondary screen for cancer chemotherapeutic agents, *J. Natl. Cancer Inst.,* 63, 1185, 1979.
21. **Ovejera, A. A. and Houchens, D. P.,** Human tumor xenografts in athymic nude mice as a preclinical screen for anticancer agents, *Sem. Oncol.,* 8, 386, 1981.
22. **Houchens, D. P. and Ovejera, A. A.,** Models for human tumor therapy in nude mice, in *Immune-Deficient Animals,* Sordat, B., Ed., Karger, Basel, 1984, 364.
23. **Ovejera, A. A., Houchens, D. P., Catane, R., Sheridan, M. A., and Muggia, F. M.,** Efficacy of 6-diazo-5-oxo-1-norleucine and N-[γ-glutamyl-6-diazo-5-oxo-norleucinyl]-6-diazo-5-oxo-norleucine against experimental tumors in conventional and nude mice, *Cancer Res.,* 39, 3220, 1979.
24. **Houchens, D. P., Ovejera, A. A., Sheridan, M. A., Johnson, R. K., Bogden, A. E., and Neil, G. L.,** Therapy for mouse tumors and human tumor xenografts with the antitumor antibiotic AT-125, *Cancer Treat. Rep.,* 63, 473, 1979.
25. **Bogden, A. E., Kelton, D. E., Cobb, W. R., and Esber, H. J.,** A rapid screening method for testing chemotherapeutic agents against human tumor xenografts, in *Proc. Symp. Use of Athymic (Nude) Mice in Cancer Res.,* Houchens, D. P. and Ovejera, A. A., Eds., Gustav Fischer, New York, 1978, 231.
26. **Alley, M. C., Scudiero, D. A., Monks, A., Hursey, M. L., Czerwinski, M. J., Fine, D. L., Abbott, B. J., Mayo, J. G., Shoemaker, R. H., and Boyd, M. R.,** Feasibility of drug screening with panels of human tumor cells lines using a microculture tetrazolium assay, *Cancer Res.,* 48, 589, 1988.
27. **Boyd, M. R.,** Workshop on "Disease-oriented antitumor drug discovery and development", *National Cancer Institute,* Bethesda, MD, 1985.
28. **Rygaard, J. and Povlsen, C. O.,** Eds., *Proc. 1st Int. Workshop on Nude Mice,* Gustav Fischer Verlag, Stuttgart, 1974.
29. **Nomura, T., Oshawa, N., Tamaoki, N., and Fujiwara, K.,** Eds., *Proc. 2nd Int. Workshop on Nude Mice,* University of Tokyo Press, Tokyo, 1977.
30. **Reed, N. D.,** Ed., *Proc. 3rd Int. Workshop on Nude Mice,* Gustav Fischer, New York, 1982.
31. **Sordat, B.,** Ed., *4th Int. Workshop on Immune-Deficient Animals in Exp. Res.,* Karger, Basel, 1984.
32. **Rygaard, J., Brünner, N., Græm, N., and Spang-Thomsen, M.,** *5th Int. Workshop on Immune-Deficient Animals,* Karger, Basel, 1987.
33. **Wu, B.-Q. and Zheng, J.,** Eds., *6th Int. Workshop on Immune-Deficient Animals,* Karger, Basel, 1989.
34. **Houchens, D. P. and Ovejera, A. A.,** Eds., *Proc. Symp. Use of Athymic (Nude) Mice in Cancer Res.,* Gustav Fischer, New York, 1978.
35. **Fogh, J. and Giovanella, B. C.,** Eds., *The Nude Mouse in Experimental and Clinical Research,* Vol. 1, Academic Press, New York, 1978.
36. **Fogh, J. and Giovanella, B. C.,** Eds., *The Nude Mouse in Experimental and Clinical Research,* Vol. 2, Academic Press, New York, 1982.
37. **Rofstad, E. K.,** Radiobiological studies using the nude mouse, in *The Nude Mouse in Experimental and Clinical Research,* Vol. 2, Fogh, J. and Giovanella, B. C., Eds., Academic Press, New York, 1987, 401.
38. **van Oostrum, E. A. and Rutgers, D. H.,** Cell kinetics in xenotransplanted cervical tumors after low dose-rate irradiation, in *Immune-Deficient Animals in Biomedical Research,* Rygaard, J., Brünner, N., Græm, N., and Spang-Thomsen, M., Eds., Karger, Basel, 1987, 335.
39. **Slagel, D. E., Feola, J., Houchens, D. P., and Ovejera, A. A.,** Combined modality treatment using radiation and/or chemotherapy in an athymic mouse-human medulloblastoma and glioblastoma xenograft model, *Cancer Res.,* 42, 812, 1982.
40. **Rofstad, E.,** Local tumor following single dose irradiation of human melanoma xenografts: relationship to cellular radiosensitivity and influence of an immune response by the athymic mouse, *Cancer Res.,* 49, 3163, 1989.
41. **Giovanella, B. C., Stehlin, J. S., Shepard, R. C., and Williams, L. J.,** Hyperthermia treatment of human tumors heterotransplanted in nude mice, *Cancer Res.,* 39, 2236, 1979.
42. **Rofstad, E. K.,** Influence of cellular, microenvironmental, and growth parameters on thermotolerance kinetics *in vivo* in human melanoma xenografts, *Cancer Res.,* 49, 5027, 1989.
43. **Rofstad, E. K.,** Heterogeneous responses to radiation and hyperthermia of cell subpopulations of a human melanoma xenograft, in *Immune-Deficient Animals in Biomedical Research,* Rygaard, J., Brünner, N., Græm, N., and Spang-Thomsen, M., Eds., Karger, Basel, 1987, 329.
44. **Senapati, N., Houchens, D., Ovejera, A., Beard, R., and Nines, R.,** Ultrasonic hyperthermia and drugs as therapy for human tumor xenografts, *Cancer Treat. Rep.,* 66, 1635, 1982.
45. **Mach, J.-P., Buchegger, F., Forni, M., Ritschard, J., Berche, C., Lumbroso, J.-D., Schreyer, M., Girardet, C., Accola, R. S., and Carrel, S.,** Use of radiolabelled monoclonal anti-CEA antibodies for the detection of human carcinomas by external photoscanning and tomoscintigraphy, *Immunol. Today,* 2, 239, 1981.

46. **Larson, S. M., Carrasquillo, J. A., Krohn, K. A., Brown, J. P., McGuffin, R. W., Ferens, J. M., Graham, M. M., Hill, L. D., Beaumier, P. L., Hellström, K. E., and Hellström, I.,** Localization of ^{131}I-labeled p^{97}-specific Fab fragments in human melanoma as a basis for radiotherapy, *J. Clin. Invest.*, 72, 2101, 1983.
47. **Lee, Y. S., Bullard, D. E., Zalutsky, M. R., Coleman, R. E., Wikstrand, C. J., Friedman, H. S., Colapinto, E. V., and Bigner, D. D.,** Therapeutic efficacy of antiglioma mesenchymas extracellular matrix ^{131}I-radiolabeled murine monoclonal antibody in a human glioma xenograft model, *Cancer Res.*, 48, 559, 1988.
48. **Anderson-Berg, W. T., Squire, R. A., and Strand, M.,** Specific radioimmunotherapy using ^{90}Y-labeled monoclonal antibody in erythroleukemic mice, *Cancer Res.*, 47, 1905, 1987.
49. **Schlom, J., Molinolo, A., Simpson, J. F., Siler, K., Roselli, M., Hinkle, G., Houchens, D. P., and Colcher, D.,** Advantage of dose fractionation in monoclonal antibody-targeted radioimmunotherapy, *J. Natl. Cancer Inst.*, 82, 763, 1990.
50. **Brusa, P., Pietribiasi, F., Bussolati, G., Dosio, F., Arione, R., Comoglio, P. M., Prat, M., and Cattel, L.,** Blocked and not blocked whole-ricin-antibody immunotoxines: intraperitoneal therapy of human tumor xenografted in nude mice, *Cancer Immunol. Immunother.*, 29, 185, 1989.
51. **Pearson, J. W., Fitzgerald, D. J., Willingham, M. C., Wiltrout, R. H., Pastan, I., and Longo, D. L.,** Chemoimmunotoxin therapy against a human colon tumor (HT-29) xenografted into nude mice, *Cancer Res.*, 49, 3562, 1989.
52. **Balkwill, F. R., Goldstein, L., and Stebbing, N.,** Differential action of six interferons against two human carcinomas growing in nude mice, *Int. J. Cancer*, 35, 613, 1985.
53. **Das, A. K., Walther, P. J., Buckley, N. J., and Poulton, S. H.,** Recombinant human tumor necrosis factor alone and with chemotherapeutic agents. Effect on nude mouse-supported human bladder cancer heterografts, *Arch. Surg.*, 124, 107, 1989.
54. **Hooper, W. C., Barth, R. F., Houchens, D. P., and Nines, R.,** Effect of dimethylsulfoxide and hexamethylene bisacetamide on the JOK-1 hairy cell leukemia derived cell line propagated in the nude mouse, *Exp. Cell Biol.*, 55, 93, 1987.
55. **Povlsen, C. O. and Rygaard, J.,** Heterotransplantation of human adenocarcinomas of the colon and rectum to the mouse mutant nude. A study of nine consecutive transplantations, *Acta Path. Microbiol. Scand.*, 79, 159, 1971.
56. **Ovejera, A. A., Timm, F. M., Houchens, D. P., and Barker, A. D.,** Nude mouse facility designed for large-scale drug evaluation and other cancer research activities, in *Proc. Symp. Use of Athymic (Nude) Mice in Cancer Res.*, Houchens, D. P. and Ovejera, A. A., Eds., Gustav Fischer, New York, 1978, 39.
57. **Rygaard, J.,** Thymus and Self, *Immunobiology of the Mouse Mutant Nude*, F.A.D.L. Publishers, Copenhagen; John Wiley & Sons, New York, 1973.
58. **Houchens, D. P., Ovejera, A. A., and Barker, A. D.,** The therapy of human tumors in athymic (nude) mice, in *Use of Athymic (Nude) Mice in Cancer Research*, Houchens, D. P. and Ovejera, A. A., Eds., Gustav Fischer, New York, 1978, 267.
59. **Ovejera, A. A. and Houchens, D. P.,** Human tumor xenografts in athymic nude mice as a preclinical screen for anticancer agents, *Sem. Oncol.*, 8, 386, 1981.
60. **Fogh, J., Fogh, J. M., and Orfeo, T.,** One hundred and seventy-seven cultured human well lines producing tumors in nude mice, *J. Natl. Cancer Inst.*, 59, 221, 1977.
61. **Tomayko, M. M. and Reynolds, C. P.,** Determination of subcutaneous tumor size in athymic (nude) mice, *Cancer Chemother. Pharmacol.*, 24, 148, 1989.
62. **Giovanella, B. C., Stehlin, J. G., Shephard, R. C., and Williams, L. J.,** Correlation between response to chemotherapy of human tumors in patients and in nude mice, *Cancer*, 52, 1146, 1983.
63. **Taetle, R., Rosen, F., Abramson, I., Venditti, J., and Howell, S.,** Use of nude mouse xenografts as preclinical drug screens: *in vivo* activity of established chemotherapeutic agents against melanoma and ovarian carcinoma xenografts, *Cancer Treat. Rep.*, 71, 297, 1987.
64. **Auerbach, R., Morrissey, C. W., and Sidky, Y. A.,** Regional differences in the incidence and growth of mouse tumors following intradermal or subcutaneous inoculation, *Cancer Res.*, 38, 1739, 1978.
65. **Kyriazis, A. A. and Kyriazis, A. P.,** Preferential sites of growth of human tumors in nude mice following subcutaneous transplantation, *Cancer Res.*, 40, 4509, 1980.
66. **Inaba, M., Tashiro, T., Kobayashi, T., Sakurai, Y., Maruo, K., Onishi, Y., Ueyama, Y., and Nomura, T.,** Responsiveness of human gastric tumors implanted in nude mice to clinically equivalent doses of various antitumor agents, *Jpn. J. Cancer Res.*, 79, 517, 1988.
67. **Inaba, M., Kobayashi, T., Tashiro, T., and Sakurai, Y.,** Pharmacokinetic approach to rational therapeutic doses for human tumor-bearing nude mice, *Jpn. J. Cancer Res.*, 79, 509, 1988.
68. *LD_{10} Summary, Toxicity Testing of Anticancer Drugs in Small Experimental Animals,* Southern Research Institute, 1975.
69. **Houchens, D. P., Johnson, R. K., Gaston, M. R., Goldin, A., and Marks, T.,** Toxicity of cancer chemotherapeutic agents in athymic nude mice, *Cancer Treat. Rep.*, 61, 103, 1977.

70. **Goldin, A., Johnson, R. K., and Venditti, J. M.,** Preclinical characterization of candidate antitumor drugs, *Cancer Chemother. Rep., Part 2,* 5, 21, 1975.
71. **Ovejera, A. A., Johnson, R. K., and Goldin, A.,** Growth characteristics and chemotherapeutic response of intravenously implanted Lewis lung carcinoma, *Cancer Chemother. Rep., Part 2,* 5, 111, 1975.
72. **Gurland, J. and Johnson, R. D.,** Case for using only maximum diameter in measuring tumors, *Cancer Chemother. Rep.,* 50, 119, 1966.
73. **Rubin, H., Arnstein, P., and Chu, B. M.,** Tumor progression in nude mice and its representation in cell culture, *J. Natl. Cancer Inst.,* 77, 1125, 1986.
74. **Rubin, H., Chu, B. M., and Arnstein, P.,** Selection and adaptation for rapid growth in culture of cells from delayed sarcomas in nude mice, *Cancer Res.,* 47, 486, 1987.
75. **Sumiyoshi, H., Yasui, W., Ochiai, A., and Tahara, E.,** Effects of gastrin on tumor growth and cyclic nucleotide metabolism in xenotransplantable human gastric and colonic carcinomas in nude mice, *Cancer Res.,* 44, 4276, 1984.
76. **Balkwill, F. R. and Proietti, E.,** Effects of mouse interferon on human tumor xenografts in the nude mouse host, *Int. J. Cancer,* 38, 375, 1986.
77. **Steel, G. G.,** The growth and therapeutic response of human tumours in immune deficient mice, *Bull. Cancer,* 65, 465, 1978.
78. **Fiennes, A. E. T. W.,** Growth rate of human tumor xenografts measured in nude mice by *in vivo* cast modelling, *Br. J. Surg.,* 75, 23, 1988.
79. **Euhus, D. M., Hudd, C., Lareging, M. C., and Johnson, F. E.,** Tumor measurement in the nude mouse, *J. Surg. Oncol.,* 31, 229, 1986.
80. **Tomayko, M. M. and Reynolds, C. P.,** Determination of subcutaneous tumor size in athymic nude mice, *Cancer Chemother. Pharmacol.,* 24, 148, 1989.
81. **Ware, J. C. and DeLong, E. R.,** Influence of tumor size in human prostate tumor metastasis in athymic nude mice, *Br. J. Cancer,* 51, 419, 1985.
82. **Schmid, F. A., Sirotnek, F. M., Otter, G. N., and DeGraw, J. I.,** New folate analogs of the 10-deaza-aminopterin series: markedly increased antitumor activity of the 10-ethyl-analog compared to the parent compound and methotrexate against some human tumor xenografts in nude mice, *Cancer Treat. Rep.,* 65, 551, 1985.
83. **Santon, J. B., Cronin, M. T., MacLeod, C. L., Mendelsohn, J., Masui, H., and Gill, G. N.,** Effects of epidermal growth factor receptor concentrations on tumorigenicity of A431 cells in nude mice, *Cancer Res.,* 46, 4701, 1986.
84. **Inaba, M., Tashiro, T., Kobayashi, T., Fujimoto, S., Sakurai, Y., Maruo, K., Onishi, Y., Ueyama, Y., and Nomura, T.,** Evaluation of response rates to various antitumor agents of human gastric tumors implanted in nude mouse, *Jpn. J. Cancer Res.,* 77, 190, 1986.
85. **Bogden, A. E., Houchens, D. P., Ovejera, A. A., and Cobb, W. R.,** Advances in chemotherapy studies with the nude mouse, in *The Nude Mouse in Experimental and Clinical Research,* Vol. 2, Fogh, J. and Giovanella, B. C., Eds., Academic Press, New York, 1982, 367.
86. *In Vivo Cancer Models 1976-1982,* NIH Publication, No. 84-2635, 1984.
87. **Kubota, T., Fukutomi, T., Isobe, Y., Kikuyama, S., Ishibiki, K., and Abe, O.,** Experimental endocrinotherapy of hormone-dependent human breast carcinomas transplanted into nude mice, in *Immune-Deficient Animals in Biomedical Research,* Rygaard, J., Brünner, N., Græm, N., and Spang-Thomsen, M., Eds., Karger, Basel, 1987, 353.
88. **Fujita, F., Fujita, M., Shimozuma, K., Sakamoto, Y., Kimoto, Y., Hirai, T., and Taguchi, T.,** Analyses of factors influencing chemosensitivity of human tumor serially transplanted into nude mice, in *Immune-Deficient Animals in Experimental Medicine,* Wu, B.-Q. and Zhang, J., Eds., Karger, Basel, 1989, 284.
89. **Guichard, M., Chavaudra, N., and Malaise, E. P.,** Tumor growth delay and cell survival as endpoints for studying the fraction of hypoxic cells in a human melanoma transplanted into nude mice, *Int. J. Radiat. Oncol. Biol. Phys.,* 38, 349, 1980.
90. **Fodstad, Ø., Aass, N., and Pihl, A.,** Response to chemotherapy of human malignant melanoma xenografts in athymic nude mice, *Int. J. Cancer,* 25, 453, 1980.
91. **Steel, G. G., Courtenay, V. D., and Peckham, M. J.,** The response to chemotherapy of a variety of human tumor xenografts, *Br. J. Cancer,* 47, 1, 1983.
92. **Courtenay, V. D. and Mills, J.,** An *in vitro* colony assay for human tumors grown in immune-suppressed mice and treated *in vivo* with cytotoxic agents, *Br. J. Cancer,* 37, 261, 1978.
93. **Selby, P. J., Courtenay, V. D., McElwain, T. J., Peckham, M. J., and Steel, G. G.,** Colony growth and clonogenic cell survival in human melanoma xenografts treated with chemotherapy, *Br. J. Cancer,* 42, 438, 1980.
94. **Twentyman, P. R.,** Timing of assays: an important consideration in the determination of clonogenic cell survival both *in vitro* and *in vivo, Int. J. Radiat. Oncol. Biol. Phys.,* 5, 1213, 1979.

95. **Twentyman, P. R.,** Experimental chemotherapy studies: intercomparison of assays, *Br. J. Cancer,* 41 (Suppl. IV), 279, 1980.
96. **Wheeler, Jr., K. T. and Wallen, C. A.,** Timing: an important variable in colony-formation assays, in *Rodent Tumor Models in Experimental Cancer Therapy,* Kallman, R. F., Ed., Pergamon Press, New York, 1987, 84.
97. **Freudenthal, R. I., Leber, A. P., Emmerling, D. C., Kerchner, G. A., and Ovejera, A. A.,** Comparison of the drug metabolizing enzymes in the liver and kidneys from homozygous nude Swiss mice, heterozygous normal Swiss, homozygous normal Swiss and DBA/2 mice, *Res. Comm. Chem. Path. and Pharm.,* 15, 267, 1976.
98. Instruction 14, Screening data summary interpretation and outline of current screen, National Cancer Institute, 1984.
99. **Boven, E., Nauta, M. M., Schluper, H. M., Elferink, F., Van der Vijgh, W. J., and Pinedo, H. M.,** Secondary screening of platinum compounds in human-ovarian cancer xenografts in nude mice, *Eur. J. Cancer Clin. Oncol.,* 21, 1253, 1985.

Chapter 12

RADIOTHERAPY

Einar K. Rofstad

TABLE OF CONTENTS

I. INTRODUCTION

Transplantable murine and rat tumors grown in syngeneic hosts have been used frequently as models for studies aimed at improving the radiotherapy of human cancer. Valuable information about the radiation biology of solid tumors has been obtained. However, it has not been possible to explain the underlying mechanisms for inter-tumor differences in radioresponsiveness and to develop clinically useful predictive assays from such studies.

Experimental models for *in vivo* studies of human tumors may in some ways be preferable to rodent tumor systems. Recently, new and improved procedures for immunosuppression of conventional mice and subsequent heterotransplantation of human tumors have been developed. However, congenitally athymic nude mice and rats have been shown to be even more useful as hosts for human tumors. Moreover, athymic mice with additional immune deficiencies are also being used in human tumor heterotransplantation studies.

The main purpose of this chapter is to review the radiation biology of human tumor xenografts and to discuss the clinical relevance and potential usefulness of data obtained from heterotransplants. Since clinical investigations have indicated that hyperthermia used in combination with radiotherapy has a considerable potential in the treatment of cancer, the thermal biology of human tumor xenografts is also reviewed. The discussion is based partly on general biological properties and partly on radio- and thermobiological characteristics of xenografts.

II. GENERAL BIOLOGY

A. TUMOR GROWTH RATE

Human tumor xenografts frequently show an increase in their rate of volume growth during the first three to five passages *in vivo,* possibly as an adaptation to the new environment. On subsequent transplantations, many tumor lines retain relatively constant growth rates, even for several years. Growth characteristics of human melanoma xenografts have been studied in our laboratory.[1] After the growth rate had stabilized, the various lines showed individual and characteristic volume-doubling times, ranging from 3.8 to 64.6 d at a volume of 200 mm^3 (Table 1). Similar observations have also been reported for xenografts of other histological types, indicating that intrinsic properties of the tumor cells partly determine the growth rate of heterotransplanted tumors. However, the volume-doubling times of the majority of established xenografts are considerably shorter than those reported for tumors in patients. Detailed studies of the cell proliferation kinetics have shown that the discrepancy between the volume-doubling times is mainly a consequence of a reduced cell loss factor in the xenografts. Some tumor lines may also show an increased growth fraction and a reduced cell cycle time compared to the parent tumor.

The response to fractionated irradiation of human tumor xenografts may differ from that of tumors in man because of differences in the volume-doubling time. The repopulation, reoxygenation, and redistribution in the cell cycle between each fraction are probably influenced by the tumor growth rate. On the other hand, a panel of xenografts of a given histological category would generally cover a wide range of volume-doubling times compared to a panel of murine tumors, and hence could be useful models in studies of relationships between tumor growth rate and radiobiological parameters.

B. TUMOR PATHOPHYSIOLOGY

Heterotransplanted human tumor cells have the ability to promote neovascularization in the new host, probably via endogenous tumor angiogenesis factors. The vascular network and the supporting stroma in xenografts thus originate from the host, whereas only the tumor parenchyma is of human origin. In addition, the transport of oxygen to xenografts is dependent on the hematological system of the host, which in many respects differs from that of patients, for

TABLE 1
Volume-Doubling Times (T_d) for Human Melanoma Xenografts

Melanoma	T_d(days) ($V = 50\ mm^3$)[a]	($V = 200\ mm^3$)[a]
B.E.	12.7	22.5
E.E.	3.0	4.4
E.F.	8.6	21.6
E.K.	15.3	64.6
G.E.1	10.0	36.5
G.E.2	3.1	4.2
G.R.E.	6.1	12.7
K.A.	4.4	6.0
K.F.	3.4	7.7
K.J.A.	6.8	10.1
M.F.	7.6	20.0
R.A.	16.5	25.3
T.H.	2.8	3.8
V.N.	3.7	6.2

[a] Determined by Gompertzian analysis of tumor growth curves.

example, the oxyhemoglobin dissociation curve for mouse blood is shifted to the right relative to that for human blood. Consequently, the pathophysiological conditions in xenografts may be different from those of tumors in patients.

Although the building blocks of the vascular system in xenografts originate from the host, there is significant evidence that the architecture of the capillary network and the blood perfusion are determined, at least partly, by biological characteristics of the human parenchymal tumor cells. Solesvik et al.[2] studied the vascular architecture in five melanoma xenograft lines in nude mice and found that the melanomas exhibited individual and characteristic microvascular structures, that is, total vessel length, surface, and volume per unit histologically intact tumor volume differed significantly among the lines. Similarly, Vaupel et al.[3] studied blood flow in tissue-isolated breast cancer xenografts in athymic nude rats and found a significantly higher average rate of blood flow in medullary than in squamous cell carcinomas at comparable tumor sizes. Moreover, the pathophysiological conditions in xenografts are not just a consequence of the efficiency of the vascular supply, but depend also directly on genetic and biochemical properties of the human parenchymal tumor cells, for example, metabolic activity, rate of oxygen consumption and their dependence on oxygen and glucose availability, and ability to survive under hypoxic stress, acid pH, and low glucose concentrations. Thus, xenografts of a specific histological category derived from different patients have been found to differ significantly with respect to ^{31}P nuclear magnetic resonance spectra (Figure 1a,b),[4] histograms for intracapillary oxyhemoglobin saturation (Figure 1c,d),[5] pO_2 and pH histograms measured with microelectrodes,[3,6] rate of oxygen and glucose uptake,[3,6] and glucose, lactate, and ATP bioluminescence images.[6]

The radioresponsiveness of tumors depends on the pathophysiological conditions in the tumor microenvironment, particularly the oxygen tension, as well as the radiation sensitivity of the vascular network; that is, the endothelial cells and the supporting connective tissue. Thus, the radioresponsiveness of xenografts is not necessarily identical to that of the corresponding tumors in patients. However, since the micromilieu in xenografts is governed to a significant extent by biological porperties of the human parenchymal tumor cells, pathophysiological and related radiobiological studies using xenograft may give clinically relevant information not achievable from rodent tumor systems.

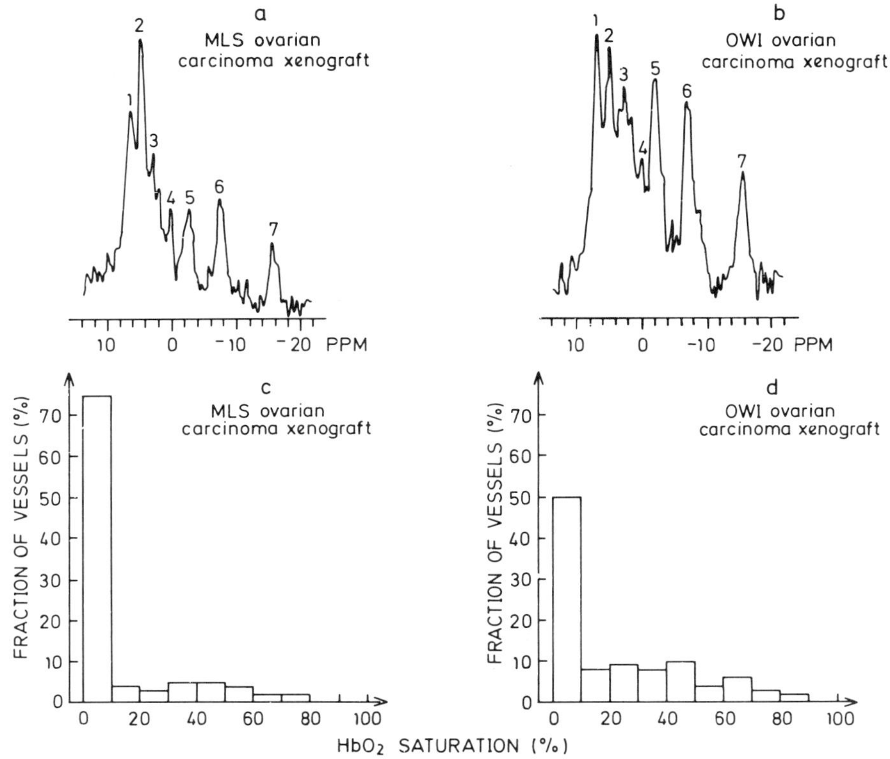

FIGURE 1. ^{31}P-NMR spectra (a,b) and frequency distributions for intracapillary HbO_2 saturation (c,d) for human ovarian cancer xenografts. Data from 2 tumors of approximately the same volume (2000 mm^3), one from the MLS line and the other from the OWI line, are shown. The OWI tumor showed significantly higher energy and oxygenation status than the MLS tumor. The resonance assignment in the upper panels is: Peak 1: phosphomonoesters (PME); Peak 2: inorganic phosphate (P); Peak 3: phosphodiesters (PDE); Peak 4: phosphocreatine (PCr); Peak 5, 6, and 7: γ, α, and β nucleoside triphosphates (NTPγ, NTPα and NTPβ).

III. ENDPOINTS FOR TREATMENT RESPONSE

Single cell survival determined *in vitro* using conventional cell culture techniques is a useful endpoint for many rodent tumor lines treated *in vivo* with radiation or hyperthermia. Much effort has been put into the development of similar colony assays applicable to human tumors. Recently, Courtenay and Mills[7] have published a new soft agar colony assay based on observations of the growth requirements of cells isolated directly from tumors. The most important feature of this assay is the use of tissue-mimicking oxygen concentrations in the gas phase, together with the admixture of rat erythrocytes and the replenishment of the nutrients in the agar by added liquid medium. Comparative studies have indicated that the Courtenay assay is superior to other colony assays commonly used in studies of human tumors. Radiobiological studies of cells from human tumor xenografts have provided survival curves of similar shape to those obtained with established cell lines. Following treatment *in vivo,* this assay has been used to determine the chemosensitivity, the radiation sensitivity, the fraction of hypoxic cells, and the capacity for PLD-repair (repair of potentially lethal damage) of xenografts of many histological categories.[8,9] However, although the Courtenay assay is useful for many xenograft lines, it cannot be used for all types of xenografts, either because the plating efficiency is too low or because of difficulties in obtaining a single cell suspension of sufficient quality, that is, with a high yield of viable cells and without clumps and aggregates.

Tumor growth delay is the only available conventional endpoint for some human tumor xenografts. This endpoint is easy to use in practice and discriminates adequately between the effect of different treatment doses for a given tumor line. However, there is no generally accepted way of analyzing growth delay data in relation to volume-doubling times in order to compare the treatment sensitivity of various xenografts. Specific growth delay (i.e., actual growth delay divided by the volume-doubling time of untreated controls) has been suggested to be a useful parameter for ranking tumor lines according to treatment sensitivity. Good correlations have been found between specific growth delay and cell surviving fraction measured *in vitro* for human melanoma xenografts.[10,11] However, in order to base comparisons of treatment sensitivity on differences in specific growth delay, it is a strict prerequisite that the average doubling time of the surviving clonogenic cells equals that of the clonogenic cells in untreated tumors. There is good evidence from studies of melanoma xenografts that the doubling time of the clonogenic cells is shortened during the repopulation period. Some data have suggested that actual growth delay divided by potential doubling time may also be a useful parameter for comparing treatment sensitivity or xenografts.[12] Further studies involving large series of xenografts need to be carried out before the most appropriate method of analyzing growth delay data can be determined.

Local tumor control is used as an endpoint to measure the radiocurability of experimental rodent tumors.[13] Tumor control assessed in laboratory animals is directly analogous to curative intent in clinical radiotherapy. Thus, TCD_{50}-values (radiation doses required to achieve local control of 50% of the tumors irradiated) determined from tumor control curves have greater relevance to clinical radiotherapy than parameters determined from cell survival curves and tumor growth delay curves. However, tumor control has not been used regularly as an endpoint in studies of human tumor xenografts, mainly because there has been some concern that immune reactions by the host might be active against xenografts and hence could enhance the frequency of tumor control artificially (see below).[9]

IV. RADIATION BIOLOGY

The main question in the evaluation of the usefulness of human tumor xenografts as models for human cancer in radiotherapeutic research is: does the radiation response of xenografts reflect the clinical radioresponsiveness of the parent tumors in the donating patients? Well-documented comparative studies of the radiation response of xenografts and the source tumors have not yet been reported. However, several *in vitro* studies on the radiosensitivity of cell suspensions isolated from xenografts have been carried out. Xenografts of several histological categories have also been irradiated *in vivo* and the radiation response has been measured using cell survival *in vitro*, tumor growth delay, and local tumor control as endpoints. In the following sections, the results from some relevant studies using these endpoints will be discussed in relation to experiences from clinical radiotherapy. Then, xenograft studies of radiobiological parameters and phenomena that may have significant influence on tumor response to fractionated irradiation, for example, α/β ratio, surviving fraction at 2.0 Gy, PLD-repair, contact repair, hypoxia, and reoxygenation, will be reviewed.

A. CELL SURVIVAL *IN VITRO*

Duchesne et al.[14] studied the radiation biology *in vitro* and *in vivo* of human lung cancer xenografts of different histology using the Courtenay assay. Significant differences in radiation

TABLE 2
Radiation Sensitivity of Hypoxic Cells (D_0) and Hypoxic Cell Fraction (HF) of Human Melanoma Xenograft Lines

Melanoma	D_0 (Gy)	HF (%)	Ref.
BE 11	5.85	40	34
Ma 11	4.22	62	34
Na 11	3.36	85	35
E.E.	2.46	6	23, 36
E.F.	3.64	22	23, 36
G.E.	3.17	31	23, 36
M.F.	3.15	45	23, 36
V.N.	3.03	15	23, 36

sensitivity among the various histologies were found; the largecell phenotype exhibited radioresistance, whereas the smallcell carcinomas and the adenocarcinomas were radiosensitive. The authors concluded that the xenograft data were in agreement with the clinical radioresponsiveness of these histologies.

Malignant melanoma is the tumor type that has been subjected to the most extensive radiation studies as xenografts, and clonogenic cell survival has been the most commonly used endpoint after irradiation *in vitro,* as well as *in vivo.* Clinically, melanomas are known to constitute an extremely heterogeneous tumor category with highly variable radioresponsiveness. This heterogeneity is reflected in the radiation biology of melanoma xenografts. Cellular radiosensitivity measured *in vitro* after irradiation *in vivo* is presented in Table 2 for 8 melanoma xenograft lines. The D_0 values range from 2.46 to 5.85 Gy, that is, they vary within a factor of about 2.4. When cells from the same xenograft lines were irradiated under aerobic conditions *in vitro,* the D_0 values varied within a factor of about 2.8. Moreover, studies in our laboratory have shown that melanoma xenograft lines established from the primary tumor and metastases obtained from the same patient may differ significantly in cellular radiosensitivity.[15] Also, cell subpopulations isolated from a single tumor of a melanoma xenograft line were found to be heterogeneous in radiosensitivity.[16]

Cellular radiosensitivity has also been measured *in vitro* after irradiation *in vivo* for xenografts of other histological categories. Deacon et al.[17] studied the radiobiology of neuroblastoma xenografts and found high radiosensitivity with limited repair capacity, as is in agreement with the high clinical radioresponsiveness of neuroblastoma. Xenografts from pancreatic cancer,[18] colon cancer,[19,20] bladder cancer,[21] and ovarian cancer[5] irradiated *in vivo* have been found to show *in vitro* survival curve parameters similar to those characteristic for the majority of mammalian tumor cells.

Comparative studies of the cellular radiosensitivity of human tumor xenografts and the corresponding donor tumors have only been performed *in vitro.*[15] In our laboratory, single cell suspensions were isolated directly from human tumor surgical specimens, irradiated *in vitro,* and assayed for cell survival using the Courtenay assay. Cells from the same suspensions were injected subcutaneously into nude mice. When tumors developed, they were subjected to radiobiological studies in later passages. Figure 2 shows survival curves for three patients' tumors differing significantly in radiation sensitivity, one melanoma, one ovarian cancer, and one cervix (squamous cell) cancer. The xenografts were studied in the second passage. The survival curves were almost identical for the xenografts and the surgical specimens, suggesting that the cellular radiosensitivity of human tumors remains unchanged during growth in immune-deficient animals.

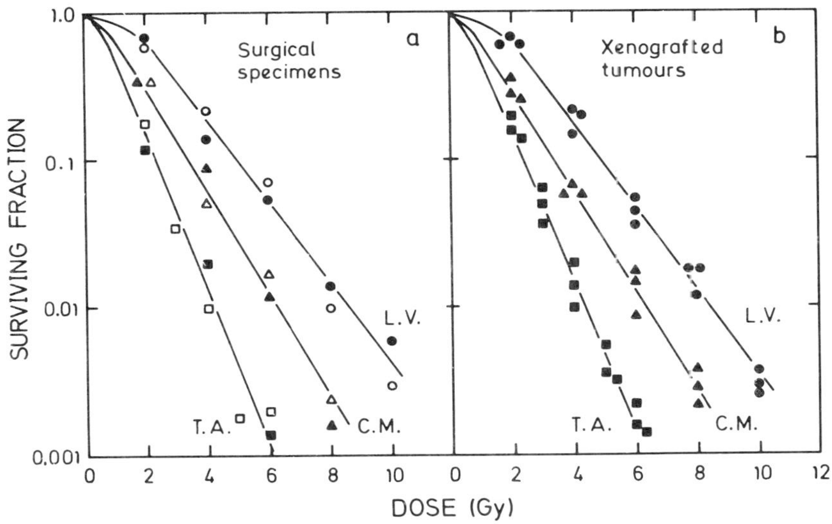

FIGURE 2. Survival curves for cells from the L.V. melanoma, C.M. ovarian cancer, and T.A. cervix cancer irradiated *in vitro*. The cells were isolated directly from surgical specimens (a) or from xenografts in the second passage (b). Circles, triangles, and squares refer to different tumors. Closed and open symbols refer to cells from newly prepared suspensions and cells stored in liquid nitrogen, respectively.

B. TUMOR GROWTH DELAY

Growth delay studies of human tumor xenografts following single-dose irradiation have been carried out at several institutes and include different histological types of tumors. Steel[22] has reviewed these, from which two interesting conclusions can be drawn. First, the radiation response varied considerably among tumor lines. Second, there was some evidence that the radiation response correlated with the clinical radioresponsiveness of tumors of corresponding histology. Thus, smallcell lung cancer and testicular teratoma xenografts appeared to be relatively radiosensitive, whereas bronchial adenocarcinoma and bladder cancer xenografts were found to be radioresistant. A heterogeneous radiation response in melanoma xenografts was also observed when specific growth delay was used as the endpoint, which is in agreement with clinical observations.[10]

C. TUMOR RADIOCURABILITY

Five melanoma xenograft lines were subjected to tumor control studies in our laboratory.[23] Tumor radiocurability was found to differ considerably among the lines; the TCD values ranged from 29.6 ± 2.1 Gy to 67.9 ± 3.5 Gy after single-dose irradiation in air-breathing nude mice. A possible immune reaction against the melanoma xenografts was investigated by comparing the radiocurability data with cell survival data measured *in vitro* after irradiation *in vivo*, and by performing quantitative tumor transplantability studies. The radiocurability and the cell survival data were found to agree well for three of the melanoma lines. Moreover, the TD_{50} values (number of tumor cells required to achieve tumors in 50% of the inoculation sites) in untreated and in whole-body irradiated mice were similar for these lines. These results suggest that the TCD_{50} values were not significantly influenced by a host immune response. On the other hand, the remaining two lines showed significantly lower TCD_{50} values than predicted from the *in vitro* cell survival data and significantly lower TD_{50} values in whole-body irradiated than in untreated mice, suggesting that the radiocurability of these lines was enhanced due to an immune response.

Nude mice may thus express a significant immune reaction against some xenograft lines but not against others. Other studies have also indicated an immune reaction against human tumors in mice. Zietman et al.[24] studied a pharyngeal squamous cell cancer line and found lower TD_{50} values in whole-body irradiated than in untreated control mice. The pharyngeal squamous cell cancer line was also reported to show higher TCD_{50} values in further immunosuppressed than in control nude mice.

In conclusion, TCD_{50} values for human tumor xenografts can only be considered of clinical relevance if it has been verified that the radioresponsiveness of the tumor has not been influenced by an immune reaction of the host.

D. α/β RATIOS

Late responding normal tissues usually show α/β ratios in the range from 6 to 14 Gy.[25] If the α/β ratio is higher for a tumor than for the surrounding late responding normal tissue, hyperfractionation to doses smaller that the conventional 2.0-Gy fractions should result in sparing of that normal tissue relative to the tumor tissue. By correct adjustment of the total dose, hyperfractionation should then provide a true therapeutic gain with an increased probability of tumor control, although the acute normal tissue reactions probably would be more severe than in the case of conventional fractionation.

Lindenberger et al.[26] have studied two xenograft lines derived from squamous cell cancers of the head and neck using tumor growth delay as the endpoint. They found α/β ratios in the range of 6.4 to 16.0 Gy, depending on the experimental conditions. Zietman et al.[27] used tumor growth delay, as well as tumor control as endpoints, to determine α/β ratios for two xenograft lines. The α/β ratio was dependent on the endpoint and was found to be 29.5 to 37.5 Gy for a malignant glioma and 20.1 to 41.8 Gy for a pharyngeal squamous cell cancer line. Both groups concluded that the use of hyperfractionation could result in a therapeutic gain for the tumors studied.

E. SURVIVING FRACTION AT 2.0 GY

Analysis of radiation survival curves for human tumor cell lines treated *in vitro* has led to the suggestion that the clinical radioresponsiveness of tumors may be correlated to the initial slope of the cell survival curve, for example, the surviving fraction at 2.0 Gy.[28] This suggestion has been tested under well-defined experimental conditions in our laboratory using human tumor xenografts.[11] Five different melanoma lines were studied and specific growth delay, as well as cell survival *in vitro*, were used as endpoints after treatment *in vivo*. Superfractionation (3 fractions of 2.0 Gy with 4-h intervals each day) as well as conventional fractionation (1 fraction of 2.0 Gy each day) were used. The respective total doses varied within the range of 12 to 30 Gy and 10 to 30 Gy. The rankings of the melanomas in radioresponsiveness were found to be almost identical, irrespective of the endpoint and the fractionation regimen applied. The cellular radiosensitivity was studied under aerobic conditions *in vitro* at conventional and low-dose rates. The radioresponsiveness *in vivo* was found to be positively correlated to the surviving fraction at 2.0 Gy (conventional dose rate) and the D_0 (low-dose rate) *in vitro* (Figure 3). It was concluded that the initial slope of the *in vitro* cell survival curve may be a useful parameter for predicting the clinical radioresponsiveness of a tumor.

F. POTENTIAL LETHAL DAMAGE-REPAIR

PLD-repair following single dose and fractionated irradiation has been studied *in vivo* in our laboratory using five human melanoma xenografts.[29] Tumors given single graded radiation doses were excised immediately after or 24 h after the radiation exposure for determination of cell survival *in vitro*. All melanoma lines showed PLD-repair after single-dose irradiation; the PLD-repair factors, (i.e., the ratio of the D_0 values for tumors excised 24 h after and immediately after irradiation) ranged from 1.2 ± 0.1 to 1.4 ± 0.1. These PLD-repair factors are similar to those reported for rodent tumors and human colon cancer xenografts given single-dose irradiation.[30]

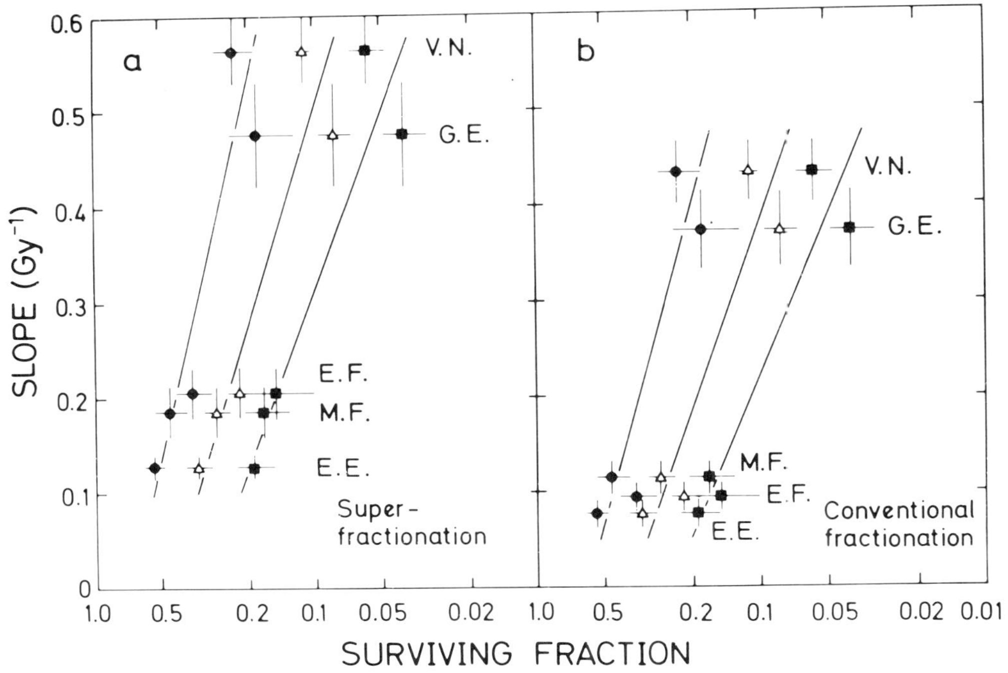

FIGURE 3. Radioresponsiveness of human melanoma xenografts *in vivo*, measured as the slope of specific growth delay curves, as a function of cell surviving fraction at 2.0 Gy (single-dose irradiation) (●), 2 × 2.0 Gy (split-dose irradiation) (■), and 4.0 Gy (low dose rate irradiation) (Δ). All three parameters were measured after irradiation of single cells *in vitro*. The dose per fraction *in vivo* was 2.0 Gy: (a) superfractionation and (b) conventional fractionation (see text). The bars represent standard errors.

PLD-repair following fractionated irradiation was studied by giving the melanomas 7 fractions of 2.0 Gy over 7 d and then, after an interval of 24 h, single graded radiation doses in the range 6 to 21 Gy. Cell survival was assayed *in vitro* immediately after or 24 h after the last radiation exposure. The D values, as well as the surviving fractions, were approximately equal after immediate and delayed cell seeding; that is, none of the melanoma lines showed significant PLD-repair after fractionated irradiation. The lack of PLD-repair after fractionated irradiation was possibly a consequence of radiation-induced recruitment of quiescent tumor cells into the cell cycle.

G. CONTACT REPAIR

Contact repair has been studied in four human melanomas grown as xenografts as well as multicellular spheroids. There was agreement between the xenograft and the spheroid studies; the Na 11,[31] the E.E.,[32] and the V.N.[32] melanomas did not show contact repair in any of the systems, whereas the Be 11 melanoma,[19] studied only as xenograft, showed significant contact repair. Recently, Dertinger et al.[31] and Dertinger and Hülser[33] demonstrated that contact repair occurred only in cells that were electrically coupled via gap junctions. Such electrical couplings were demonstrated to be present in the Be 11 melanoma and absent in the Na 11 melanoma.[31]

H. HYPOXIA AND REOXYGENATION

Most experimental rodent tumors contain a significant fraction of hypoxic cells and these cells may limit the radiocurability of the tumors. The hypoxic fraction has been measured to be within the range of 5 to 50% in the majority of the murine tumor lines. Similar hypoxic fractions have been reported for several human tumor xenografts, for example, pancreatic cancer,[18] colorectal cancer,[19,20] bladder cancer,[21] squamous cell cancer,[26] lung cancer of various histologies,[14] and ovarian cancer.[5] Melanoma xenografts show considerable differences in hypoxic fractions;[34-36] mean values from 6 to 85% have been reported (Table 2). Data consistent with the existence of oxygen tensions causing partial radioresistance in a significant fraction of the tumor cells have been reported for a neuroblastoma[17] and a colorectal cancer[37] xenograft line.

Human tumor xenografts have also been subjected to reoxygenation studies. In our laboratory, the reoxygenation pattern in 5 melanoma lines following single-dose (10 Gy) and fractionated (7 fractions of 2.0 Gy, 1 fraction per day) irradiation was investigated using the paired survival curve method.[36] The hypoxic fractions in untreated tumors differed considerably among the melanomas (Table 2). The reoxygenation was rapid and extensive in all lines; 12 to 24 h after the single-dose irradiation or the last fraction of the fractionated irradiation, the hypoxic fractions were similar to those in untreated tumors and stayed at that level up to at least 10 d after irradiation. There was a positive correlation between the hypoxic fractions after irradiation and reoxygenation. Because of the extensive reoxygenation, the radioresponsiveness of the melanomas was not significantly influenced by hypoxia when treated with daily radiation fractions of 2.0 Gy.[11]

V. THERMAL BIOLOGY

Hyperthermia in the temperature range of 42 to 45°C has a considerable potential as a treatment modality for cancer. Heat is cytotoxic to tumor cells, potentiates the effects of ionizing radiation and enhances the uptake and cytotoxicity of some chemotherapeutic agents. Human tumor xenografts can be subjected to studies of heat responsiveness using the same endpoints as are being used in studies of radioresponsiveness. However, the heat responsiveness of xenografts has not been studied as extensively as the radioresponsiveness. Little is known about the extent to which the heat response of xenografts reflects the clinical heat responsiveness of the parent tumors in the donating patients. Few xenograft studies have dealt with interactions between hyperthermia and radiotherapy or chemotherapy. However, it has been demonstrated that the response to radiation and some chemotherapeutic agents can be enhanced significantly by heat treatment.[38,39] Most hyperthermia studies involving xenografts have focused on heterogeneity in heat responsiveness among tumor lines and the mechanisms by which tumor cells are inactivated *in vivo.*

A. HEAT SENSITIVITY

Five human melanoma xenograft lines have been subjected to hyperthermic treatment *in vitro* and *in vivo* in our laboratory.[40,41] Tumors grown subcutaneously in the hind leg of the mice were heated *in vivo* in a water bath and specific growth delay was used as the endpoint. Tumor cells isolated from the same xenograft lines were heated *in vitro* and the colony-forming ability of the cells was measured using the Courtenay assay. The response to heat treatment differed considerably among the xenograft lines, both *in vivo* and *in vitro.* At a treatment temperature of 42.5°C, for example, the slopes of the *in vivo* dose response curves (specific growth delay vs. heating time) varied within a factor of 3.0.[40] Similarly, the D values of the heat survival curves *in vitro* at 42.5°C ranged from 44 to 123 min; that is, the heat sensitivity *in vitro* varied within a factor of 2.8.[40] However, there was no clear correlation between the response to heat treatment *in vivo* and *in vitro,* suggesting that the mechanisms of heat inactivation *in vivo* was different from those *in vitro* (see below). In contrast to our observations, Giovanella et al.,[38] who treated four human tumor xenograft lines, one squamous cell cancer, one melanoma, one colon cancer, and one breast cancer, concluded that the response to heat did not seem to vary much between the lines.

Complete dose-response curves have been determined for our five melanoma xenograft lines in the temperature ranges 40.5 to 44.0°C *(in vivo)* and 40.5 to 45.5°C *(in vitro)* and subjected to

Arrhenius analysis.[41] The Arrhenius curves were found to have an inflection point at 42.0 to 43.0°C *in vivo* and 41.5 to 42.5°C *in vitro*. The activation energies were in the ranges 426 to 771 kJ/mol *in vivo* and 676 to 739 kJ/mol *in vitro* above the inflection point and 774 to 1661 kJ/mol *in vivo* and 1118 to 2190 kJ/mol *in vitro* below the inflection point. Above the inflection point, the activation energies *in vivo* and *in vitro* were not significantly different for any of the melanomas and were found similar to those reported for rodent tumors, normal tissues, and cells in culture in the same temperature range. Below the inflection point, the activation energies were lower *in vivo* and *in vitro*. This difference was probably caused by differences in the physiological conditions *in vivo* and *in vitro*. The activation energies *in vitro* in this temperature range were comparable to those reported for normal tissues and cells in culture.

B. MECHANISMS OF HEAT CYTOTOXICITY

Our five melanoma xenograft lines have also been used in studies designed to elucidate the mechanisms of heat cytotoxicity in tumors *in vivo*.[42,43] Tumors were excised at predetermined time points after hyperthermic treatment (42.5°C for 60 min) *in vivo* and colony formation was measured *in vitro*. The experiments showed that some tumor cells were inactivated during treatment, while others lost clonogenicity only after completion of treatment. Examinations of histological sections from tumors fixed 1 h after completion of treatment revealed considerable vascular occlusion in all five melanomas, suggesting that the observed delay in cell death might be due to a number of factors, for example, insufficient supply of oxygen and nutrients, increased tumor acidity, and accumulation of toxic metabolic products. It was concluded that at least two different mechanisms governed the overall heat response of the xenografts: the primary cell death induced after treatment was due to direct cytotoxic effects of the heat; and the secondary cell death induced after completion of treatment was due to heat-induced vascular damage. The differences among the melanomas in overall heat responsiveness appeared mainly to be a consequence of differences in secondary cell death. The secondary cell death was shown to be least pronounced for those melanomas in which most of the larger vessels were embedded in broad bands of connective tissue.

C. THERMOTOLERANCE

Hyperthermic treatments are known to induce a transient, nonheritable resistance to subsequent heat exposure, a phenomenon called thermotolerance. Thermotolerance has been demonstrated for cells cultured *in vitro* as well as for normal tissues and experimental tumors *in vivo*, and it may limit the success of clinical hyperthermia. The kinetics of thermotolerance in both normal and malignant tissues has been shown to depend on the conditioning heat dose and the type of tissue.

The kinetics of thermotolerance in our five melanoma xenograft lines has been studied *in vivo*.[44] Specific growth delay was used as the endpoint for tumor response. The thermotolerance ratio (TTR) [i.e., the ratio of the slopes of dose response curves (specific growth delay vs. heating time) for single-heated and preheated tumors] was used as a quantitative measure of thermotolerance. All melanoma lines developed thermotolerance (Figure 4); TTR reached a maximum (TTR_{max}) 16 to 24 h after the conditioning heat treatment and then decayed slowly. TTR_{max} and the time necessary to reach TTR_{max} tended to increase with increasing conditioning heat dose, whereas the half-time of thermotolerance decay did not change with the conditioning heat dose. The kinetics of thermotolerance differed considerably among the melanoma lines. After a conditioning heat treatment of 43.5°C for 30 min, TTR_{max} ranged from 2.3 ± 0.5 to 7.0 ± 1.2, the half-time of thermotolerance decay from 53 ± 13 h to 142 ± 30 h and the time necessary to reach complete decay of the thermotolerance from 5 d to more than 14 d. TTR_{max} showed no correlation with heat sensitivity or any known growth or microenvironmental parameter of the melanoma lines. On the other hand, TTR_{max} was positively correlated with TTR_{max} measured *in vitro* when cells from the melanomas were studied in soft agar.[45] However, TTR_{max} *in vivo* was always

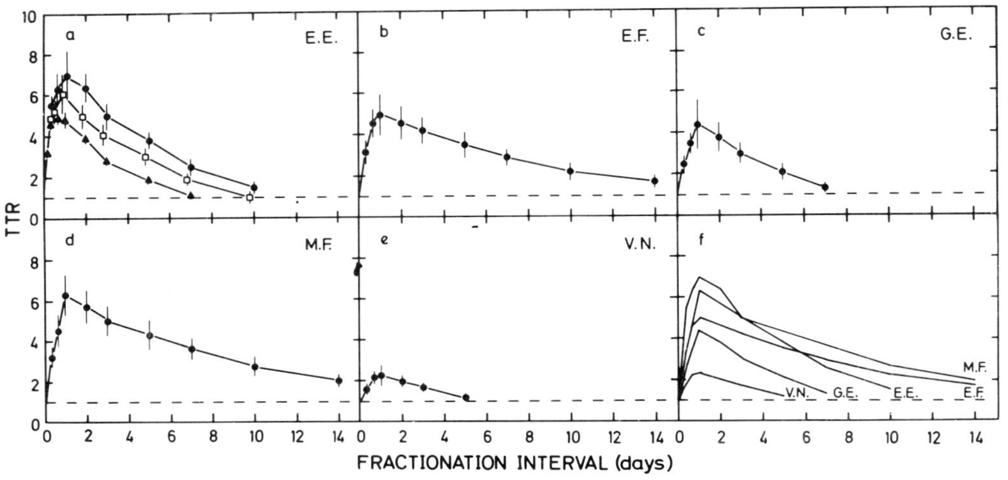

FIGURE 4. Thermotolerance ratio as a function of fractionation interval for the E.E. (a), E.F. (b), G.E. (c), M.F. (d), and V.N. (e) melanoma xenografts. Thermotolerance was induced by conditioning heat treatments of 42.5°C for 30 min (▲), 43.0°C for 30 min (❑) and 43.5°C for 30 min (●). The bars represent standard errors —, TTR = 1.0. The curves pertaining to a conditioning heat treatment of 43.5°C for 30 min are redrawn without experimental points in (f) for comparison.

somewhat lower than TTR_{max} *in vitro*. Consequently, the development of thermotolerance in the melanomas *in vivo* was governed mainly by the intrinsic ability of the tumor cells to develop thermotolerance and was just slightly modified by the tumor microenvironment. The rate of thermotolerance decay was independent of TTR_{max}. The half-times of thermotolerance decay *in vivo* were longer than, and not correlated with, those measured *in vitro*. The decay half-time *in vivo* tended to increase with increasing tumor volume-doubling time and to decrease with increasing growth fraction and vascular density. There was no relationship between decay half-time and fraction of radiobiologically hypoxic cells. Consequently, the rate of thermotolerance decay in the melanomas *in vivo* was governed mainly by tumor growth parameters and not by intrinsic characteristics of the tumor cells. The considerable difference in the kinetics of thermotolerance observed among the melanoma lines indicated that fractionated hyperthermia cannot be expected to give optimum clinical results until individualized treatment regimens are being used. Moreover, clinical treatment protocols should probably not prescribe more than one hyperthermic treatment per week.[44]

VI. CONCLUSIONS

This review has revealed at least three main disadvantages with xenografts as models for human tumors in radiotherapeutic research.

1. Human tumor xenografts usually show shorter volume-doubling times than tumors in patients
2. The hematological system and the vascular network of xenografts originate from the host
3. Immune reactions by the host are active against some xenograft lines, complicating the use of local tumor control as an endpoint

In spite of these disadvantages, several observations indicate that xenografts may be useful in studies aimed at (1) elucidating the underlying mechanisms for inter-tumor differences in clinical radioresponsiveness, (2) developing short-term *in vitro* or *in vivo* assays for clinical radiosensitivity testing, and (3) investigating the potential usefulness of hyperthermia as an adjunct to radiotherapy:

1. Human tumor xenografts show an individual and characteristic growth, and specific pathophysiological, radiobiological and thermobiological parameters, suggesting that these parameters are determined, at least partly, by biological characteristics of the human parenchymal tumor cells.
2. The radiation and heat response of xenografts can be measured with sufficient accuracy by using cell survival *in vitro* and tumor growth delay as endpoints.
3. The radiation response of xenografts seems to mirror the clinical radioresponsiveness of the corresponding tumors in patients.

ACKNOWLEDGMENTS

Financial support was received from The Norwegian Cancer Society.

REFERENCES

1. **Rofstad, E. K., Fodstad, Ø., and Lindmo, T.,** Growth characteristics of human melanoma xenografts, *Cell Tissue Kinet.*, 15, 545, 1982.
2. **Solesvik, O. V., Rofstad, E. K., and Brustad, T.,** Vascular structure of five human malignant melanomas grown in athymic nude mice, *Br. J. Cancer*, 46, 557, 1982.
3. **Vaupel, P., Fortmeyer, H. P., Runkel, S., and Kallinowski, F.,** Blood flow, oxygen consumption, and tissue oxygenation of human breast cancer xenografts in nude rats, *Cancer Res.*, 47, 3496, 1987.
4. **Rofstad, E. K., DeMuth, P., Fenton, B. M., and Sutherland, R. M.,** ^{31}P nuclear magnetic resonance spectroscopy studies of tumor energy metabolism and its relationship to intracapillary oxyhemoglobin saturation status and tumor hypoxia, *Cancer Res.*, 48, 5440, 1988.
5. **Rofstad, E. K., Fenton, B. M., and Sutherland, R. M.,** Intracapillary HbO_2 saturations in murine tumours and human tumour xenografts measured by cryospectrophotometry: relationship to tumour volume, tumour pH and fraction of radiobiologically hypoxic cells, *Br. J. Cancer*, 57, 494, 1988.
6. **Kallinowski, F., Vaupel, P., Runkel, S., Berg, G., Fortmeyer, H. P., Baessler, K. H., Wagner, K., Mueller-Klieser, W., and Walenta, S.,** Glucose uptake, lactate release, ketone body turnover, metabolic micromilieu, and pH distributions in human breast cancer xenografts in nude rats, *Cancer Res.*, 48, 7264, 1988.
7. **Courtenay, V. D. and Mills, J.,** An *in vitro* colony assay for human tumours grown in immune-suppressed mice and treated *in vivo* with cytotoxic agents, *Br. J. Cancer*, 37, 261, 1978.
8. **Steel, G. G., Courtenay, V. D., and Peckham, M. J.,** The response to chemotherapy of a variety of human tumour xenografts, *Br. J. Cancer*, 47, 1, 1983.
9. **Rofstad, E. K.,** Human tumour xenografts in radiotherapeutic research, *Radiother. Oncol.*, 3, 35, 1985.
10. **Rofstad, E. K. and Brustad, T.,** Tumour growth delay following single dose irradiation of human melanoma xenografts. Correlations with tumour growth parameters, vascular structure and cellular radiosensitivity, *Br. J. Cancer*, 51, 201, 1985.
11. **Rofstad, E. K. and Brustad, T.,** Radioresponsiveness of human melanoma xenografts given fractionated irradiation *in vivo* — relationship to the initial slope of the cell survival curves *in vivo*, *Radiother. Oncol.*, 9, 45, 1987.
12. **Rofstad, E. K.,** Radioresponsiveness of human tumor xenografts: growth delay and vascular damage, in *Rodent Tumor Models in Experimental Cancer Therapy*, Kallman, R. F., Ed., Pergamon Press, Oxford, 1987, 209.
13. **Suit, H. D., Sedlacek, R., and Thames, H. D.,** Radiation dose-response assays of tumor control, in *Rodent Tumor Models in Experimental Cancer Therapy*, Kallman, R. F., Ed., Pergamon Press, Oxford, 1987, 138.
14. **Duchesne, G. M., Peacock, J. H., and Steel, G. G.,** The acute *in vitro* and *in vivo* radiosensitivity of human lung tumour lines, *Radiother. Oncol.*, 7, 353, 1986.

15. **Rofstad, E. K., Zaffaroni, N., and Hystad, M.,** Heterogeneous radiation and heat sensitivity *in vitro* of human melanoma xenograft lines established from different lesions in the same patient: comparisons with the radiation and heat sensitivity of cells isolated from the donor patient's surgical specimens, *Int. J. Radiat. Oncol. Biol. Phys.,* submitted.
16. **Rofstad, E. K. and Brustad, T.,** Differential responses to radiation and hyperthermia of cloned cell lines derived from a single human melanoma xenograft, *Int. J. Radiat. Oncol. Biol. Phys.,* 10, 857, 1984.
17. **Deacon, J. M., Wilson, P. A., and Peckham, M. J.,** The radiobiology of human neuroblastoma, *Radiother. Oncol.,* 3, 201, 1985.
18. **Courtenay, V. D., Smith, I. E., Peckham, M. J., and Steel, G. G.,** *In vitro* and *in vivo* radiosensitivity of human tumour cells obtained from a pancreatic carcinoma xenograft, *Nature (London),* 263, 771, 1976.
19. **Guichard, M., Dertinger, H., and Malaise, E.-P.,** Radiosensitivity of four human tumor xenografts. Influence of hypoxia and cell-cell contact, *Radiat. Res.,* 95, 602, 1983.
20. **West, C. M. L. and Sutherland, R. M.,** The radiation response of a human colon adenocarcinoma grown in monolayer, as spheroids, and in nude mice, *Radiat. Res.,* 112, 105, 1987.
21. **Tannock, I., Choo, B., and Buick, R.,** The radiation response of human bladder cancer assessed *in vitro* or as xenografts in immune-deprived mice, *Int. J. Radiat. Oncol. Biol. Phys.,* 10, 1897, 1984.
22. **Steel, G. G.,** Therapeutic response of human tumour xenografts in immune-suppressed mice, in *Immune-Deficient Animals,* Sordat, B., Ed., Karger, Basel, 1984, 395.
23. **Rofstad, E. K.,** Local tumor control following single dose irradiation of human melanoma xenografts: relationship to cellular radiosensitivity and influence of an immune response by the athymic mouse, *Cancer Res.,* 49, 3163, 1989.
24. **Zietman, A. L., Suit, H. D., Sedlacek, R. S., and Ramsay, J.,** A quantitative comparison of brain and subcutaneous tissue as sites for experimental tumor transplantation (Abstr.), in *Radiation Research Society 36th Annual Meeting,* 1988, 56.
25. **Williams, M. V., Denekamp, J., and Fowler, J. F.,** A review of α/β ratios for experimental tumors: implications for clinical studies of altered fractionation, *Int. J. Radiat. Oncol. Biol. Phys.,* 11, 87, 1985.
26. **Lindenberger, J., Hermeking, H., Kummermehr, J., and Denekamp, J.,** Response of human tumour xenografts to fractionated X-irradiation, *Radiother. Oncol.,* 6, 15, 1986.
27. **Zietman, A. L., Suit, H. D., Tomkinson, K. N., and Sedlacek, R. S.,** The response of two human tumor xenografts to fractionated irradiation. The derivation of α/β ratios from growth delay and tumor control assays, *Int. J. Radiat. Oncol. Biol. Phys.,* in press, 1989.
28. **Fertil, B. and Malaise, E.-P.,** Intrinsic radiosensitivity of human cell lines is correlated with radioresponsiveness of human tumors: analysis of 101 published survival curves, *Int. J. Radiat. Oncol. Biol. Phys.,* 11, 1699, 1985.
29. **Rofstad, E. K.,** PLD-repair in human melanoma xenografts following single dose and fractionated irradiation, *Br. J. Cancer,* 61, 256, 1990.
30. **Guichard, M., Weichselbaum, R. R., Little, J. B., and Malaise, E.-P.,** Potentially lethal damage repair as a possible determinant of human tumour radiosensitivity, *Radiother. Oncol.,* 1, 263, 1984.
31. **Dertinger, H., Guichard, M., and Malaise, E.-P.,** Relationship between intercellular communication and radiosensitivity of human tumor xenografts, *Eur. J. Cancer Clin. Oncol.,* 20, 561, 1984.
32. **Rofstad, E. K. and Brustad, T.,** Intercellular contact and the radiation response of cells of human melanoma xenografts, *Int. J. Radiat. Oncol. Biol. Phys.,* 9, 191, 1983.
33. **Dertinger, H. and Hülser, D.,** Increased radioresistance of cells in cultured multicell spheroids. I. Dependence on cellular interaction, *Radiat. Environ. Biophys.,* 19, 101, 1981.
34. **Chavaudra, N., Guichard, M., and Malaise, E.-P.,** Hypoxic fraction and repair of potentially lethal radiation damage in two human melanomas transplanted into nude mice, *Radiat. Res.,* 88, 56, 1981.
35. **Guichard, M., Gosse, C., and Malaise, E.-P.,** Survival curve of a human melanoma in nude mice, *J. Natl. Cancer Inst.,* 58, 1665, 1977.
36. **Rofstad, E. K.,** Hypoxia and reoxygenation in human melanoma xenografts, *Int. J. Radiat. Oncol. Biol. Phys.,* 17, 81, 1989.
37. **Reynaud-Bougnoux, A., Lespinasse, F., Malaise, E.-P., and Guichard, M.,** Partial hypoxia as a cause of radioresistance in a human tumor xenograft: its influence illustrated by the sensitizing effect of misonidazole and hyperbaric oxygen, *Int. J. Radiat. Oncol. Biol. Phys.,* 12, 1283, 1986.
38. **Giovanella, B. C., Stehlin, J. S., Shepard, R. C., and Williams, L. J.,** Hyperthermia treatment of human tumors heterotransplanted in nude mice, *Cancer Res.,* 39, 2231, 1979.
39. **Senapati, N., Houchens, D. P., Ovejera, A. A., Beard, R., and Nines, R.,** Ultrasonic hyperthermia and drugs as therapy for human tumor xenografts, *Cancer Treat. Rep.,* 66, 1635, 1982.
40. **Rofstad, E. K. and Brustad, T.,** Response to heat treatment (42.5°C) *in vivo* and *in vitro* of five human melanoma xenografts, *Br. J. Radiol.,* 57, 1023, 1984.
41. **Rofstad, E. K. and Brustad, T.,** Arrhenius analysis of the heat response *in vivo* and *in vitro* of human melanoma xenografts, *Int. J. Hypertherm.,* 4, 359, 1986.

42. **Rofstad, E. K., Solesvik, O. V., and Brustad, T.,** Tumour growth delay, cell inactivation and vascular damage following hyperthermic treatment of a human melanoma xenograft, *Eur. J. Cancer Clin. Oncol.*, 20, 1295, 1984.
43. **Rofstad, E. K. and Brustad, T.,** Primary and secondary cell death in human melanoma xenografts following hyperthermic treatment, *Cancer Res.*, 46, 355, 1986.
44. **Rofstad, E. K.,** Influence of cellular, microenvironmental, and growth parameters on thermotolerance kinetics *in vivo* in human melanoma xenografts, *Cancer Res.*, 49, 5027, 1989.
45. **Rofstad, E. K., Midthjell, H., and Brustad, T.,** Heat sensitivity and thermotolerance in cells from five human melanoma xenografts, *Cancer Res.*, 44, 4347, 1984.

Chapter 13

CONVENTIONAL CHEMOTHERAPY

D. P. Berger, B. R. Winterhalter, and H. H. Fiebig

TABLE OF CONTENTS

I. INTRODUCTION

Since the first successful transplantation of a human colorectal cancer to nude mice by Rygaard and Povlsen[1] in 1969, the utilization of these animals in oncology research has continuously increased. Apart from studies in tumor biology, genetics, and immunology, the use of human tumor xenografts for anticancer drug screening has become invaluable.

The validity of a preclinical model for cancer chemotherapy is ultimately established by its sensitivity or resistance to clinically used cytostatic agents. Determination of the efficacy of these conventional drugs may clarify whether the nude mouse-human tumor model is able to identify active and inactive compounds in a variety of tumor types. The predictive value and the clinical relevance of this test system may be assessed by comparison of the activity of established cytostatic agents between tumor-bearing nude mice and cancer patients.

II. METHODOLOGICAL CONSIDERATIONS

A. NUDE MICE AND TUMORS

Human tumor xenograft models based on heterotransplantation into host animals have been developed before athymic nude mice became available. Malignant tissues transplanted to immunologically privileged sites (e.g., the hamster cheek pouch[2,3]) and immunologically incompetent animals (e.g., thymectomized, irradiated, or cortisone-treated mice[4]) have been used as early models for the determination of the effects of anticancer drugs on human tumors *in vivo*. However, these animals show an unstable immune response, recover the ability to reject foreign tissues within a few months,[5] and require time-consuming and technically difficult preparation procedures, because of which they have never been studied for drug testing on a large-scale basis.

The advantages of the nude mouse model for cancer chemotherapy can be summarized as follows:

1. Human solid tumors of a variety of histologies are readily accepted and can be established as a tumor line by serial transfer.
2. Human tumor xenografts retain the characteristics of the original tumor to a large extent.
3. Tumors can be frozen and stored in liquid nitrogen; reestablishment of frozen material in nude mice is possible.
4. Different types of tests are feasible, including studies in tumor biology and cancer chemotherapy; all experiments are reproducible, as tumor material can be provided repeatedly.

Nude mice of various genetic backgrounds are available and in use for anticancer drug screening in a number of institutes. However, the choice of the nude mouse strain, sex, and age are important variables affecting the toxicity and efficacy of the drugs to be tested.

Human tumor lines established in nude mice should be monitored closely in order to demonstrate the likeness of the mouse-grown material to the original patient's tumor and to determine the stability of the biological tumor characteristics. In our laboratory, more than 1300 surgical specimens of human tumors have been implanted into nude mice within the last 10 years. Some 220 of these tumors have been established in serial transfer[6-9] and show a regular growth. The human origin of the xenografts was confirmed by isoenzymatic and immunohistochemical methods. Further characterization of tumor lines included histology, growth behavior, chemosensitivity *in vivo* and *in vitro* (clonogenic assay), isoenzyme phenotype analysis, hormone receptor analysis, and DNA histogram as generated by flow cytometry. Other groups have employed karyotype studies, DNA fingerprinting, cellular enzyme profiles, and determination of tumor-associated antigens or hormones in order to ensure the close similarity of the xenograft tumor to the original patient's tissue.[10-13]

B. DRUG TREATMENT

Drug doses, schedules, and routes applied for treatment of human tumor xenografts show considerable variation between institutes. Tumor characteristics (volume-doubling time, para-neoplastic activity, etc.) and the experimental setting have to be considered for determination of the optimal treatment methodology.

1. Route

Drug administration is possible by the i.v., i.p., s.c., i.m., or the oral route (by gavage), or by continuous i.p., s.c., or i.v. infusion. The solubility, stability, bioavailability, and side effects of a compound, as well as the safety and technical experience, should be considered in the route of choice:

1. I.p. administration is safe, technically easy, and widely used for drug treatment. Bioavailability of most cytostatic agents is similar for i.p. and i.v. injections, LD_{50} values are generally not significantly different.
2. Bolus injections are easily performed by the i.v., i.p., and s.c. route. Multiple applications are possible i.p. and s.c. Repeated treatment per os (p.o.) is technically difficult and should be performed only by experienced technicians. For locally irritating compounds, multiple injections should be avoided.
3. Compounds which are poorly soluble in water or saline can be given either as a solution in dimethylsulfoxide (i.p. injection only) or ethanol, or as a suspension in distilled water or saline (i.p. or p.o.).
4. For conventional cytostatic agents, the clinical situation should be considered. If human and murine pharmacokinetics are similar for a given drug, the route in patients should be mimicked in nude mice.

2. Drug Dose and Schedule

The drug dose and schedule are of crucial importance in xenograft chemotherapy studies, especially if the efficacy of conventional cytostatic agents in nude mice is to be compared with the clinical activity known in patients. The procedure most commonly applied has been the administration of a drug at its maximum tolerated dose (MTD) to mice. The rationale behind this lies in the analysis by Freireich et al.[13] indicating the presence of a linear relationship between the MTD levels in mouse and man based on body surface. For definition of the MTD, two different effects are usually studied:

1. Drug-related mortality, expressed in terms of a lethal dose for 10% (LD_{10}), 20% (LD_{20}), and 50% (LD_{50}) of the animals
2. The median drug-induced body weight loss

Both values vary considerably in different strains of mice, and also differ according to sex, age, nutritional status, and environmental conditions. In general, a median drug-induced body weight loss of 10 to 20% or a drug-related mortality equivalent to the LD_{10} or LD_{20} are accepted at the MTD.

Another methodological approach was recently proposed by Inaba et al.[15,16] suggesting the use of a "rational dose" (RD), which is pharmacokinetically equivalent to the clinical dose. Based on comparative pharmacokinetic studies, the dose of a given drug that can reproduce a plasma level in nude mice similar to that observed in patients was determined for each of 8 compounds. For 5 of 8 agents, the RD was estimated to be about 20 to 25% of the MTD level. In the case of doxorubicin, the MTD was approximately equivalent to the clinical dose. For 2 of 8 agents, 5-fluorouracil and methotrexate, the RD was significantly higher than the MTD.[16] Further studies are necessary to clarify whether RD levels predict more accurately for clinical activity of cytostatic agents than MTDs and whether the use of RD levels is feasible in the drug development of novel compounds.

The drug dose and schedule are related to each other. For example, a treatment schedule may be chosen in which a single LD_{10} dose is employed,[5] or lower doses may be administered repeatedly.[17] Especially in slowly growing tumor lines, frequent injections are preferred in an attempt to mimic the clinical situation.[18] The following considerations might be helpful in the determination of the dose and schedule of a given drug:

1. Dose-finding studies in conventional mice are valuable. However, toxicity and pharmacokinetics of various anticancer agents may differ between nude mice and normal mice.[19-21] Determination of the MTD levels in tumor-bearing nude mice should be carried out more frequently.
2. *In vitro* data are helpful for the determination of MTD levels, especially for experimental agents. A variable outcome for the *in vitro* efficacy of a compound related to different drug exposure times (1 h, 3 h, 48 h, continous exposure) suggests a schedule dependency *in vivo*. Steep dose-response effects in cell culture tests should lead to careful determination of the *in vivo* MTD.
3. Conversion factors (rat-mouse, etc.) might give an estimate of the drug dose and schedule to be used where preclinical data of other species are available.[14]

Treatment regimens used in our laboratory are outlined in Table 1. Doxorubicin was given i.v., as toxicity was disproportionately high by the i.p. route.[22] Similarly, mitomycin C and vindesine had to be injected i.v. due to local induction of peritonitis after i.p. administration. Compounds clinically applied as short-time infusions were given s.c., for example, cisplatin, bleomycin, and etoposide. Nitrosoureas, DTIC, cyclophosphamide, and 5-fluorouracil were injected i.p.. Most compounds were given in 2 treatment cycles with an interval of 7 to 14 d. If a drug was effective, dose-response effects were evaluated using dose levels equivalent to 50 and 25% of the MTD. In general, the MTD levels were reached for all compounds tested, resulting in a drug-induced lethality between the LD_{10} and the LD_{20}. A higher toxicity at day 28 after the start of treatment was partially caused by the tumor-associated death of the animals.

III. EFFECTS OF CONVENTIONAL AGENTS

A. TOXICITY

Difficulties experienced in evaluation of the side effects in nude mice are due to various reasons. In particular, accidental deaths (mainly caused by infections) among nude mice add to the variability of toxic dose levels. Additionally, other variables including the mouse strain, age and sex, nutritional status, hormonal balance, and tumor characteristics, as well as the environment of the animals, contribute to the variation in drug toxicity reported by different laboratories.

Only a few investigators have simultaneously studied anticancer drug toxicity in nude mice and in conventional mice. Houchens et al.[20] compared 10 cytostatic agents in nude and conventional Swiss mice. The LD_{50} doses were significantly higher in nude mice for BCNU, methotrexate, and 5-fluorouracil. For bleomycin, vincristine, doxorubicin, cyclophosphamide, actinomycin D, and vinblastine, no significant differences were observed. Only DTIC was somewhat more toxic in nude than in conventional mice. These discrepancies might be due to

TABLE 1
Schedules for Standard Drug Administration in Nude Mice

Drug[a]	Dose, mg/kg/d	Schedule Days	Route[a]	Lethality/total Day 14 n/total	%	Day 28[c] n/total	%
BLEO	15	1–4,15–18	s.c.	2/ 30	7	10/ 30	33
CCNU	20	1	i.p.	41/275	15	62/275	23
CDDP	6.4	1,15	s.c.	18/183	10	48/183	26
CY	200	1,15	i.p.	21/246	9	57/246	23
DOX	8	1,15	i.v.	35/390	9	75/383	20
DTIC	300	1,15	i.p.	4/ 69	6	7/ 69	10
ETO	24	1–3,15–17	s.c.	23/182	13	31/182	17
5FU	40	1–4,15–18	i.p.	32/257	12	51/257	20
HECNU	10–12	1	i.v.	37/240	15	74/240	31
IFO	130	1–3,15–17	i.p.	5/ 93	5	19/ 93	20
MMC	2	1,15	i.v.	26/250	10	67/250	27
MTX	4	1,4,8,11,15,18	i.p.	0/ 60	0	7/ 60	12
MITOX	4	1,15	i.v.	14/ 95	15	32/ 90	36
VBL	3	1,15	i.v.	9/ 68	13	12/ 62	19
VCR	0.5	1,8,15	i.v.	1/ 29	3	2/ 29	7
VIND	1.5	1,8,15	i.v.	21/361	6	43/356	12

[a] BLEO, Bleomycin; CCNU, Chloroethyl-cyclohexyl-nitrosourea; CDDP, *Cis*-d.amminodichloroplatinum (II); CY, Cyclophosphamide; DOX, Doxorubicin; DTIC, Dacarbazine; ETO, Etoposide; 5FU, 5-Fluorouracil; HECNU, Hydroxyethyl-chloroethyl-nitrosourea; IFO, Ifosfamide; MMC, Mitomycine C; MTX, Methotrexate; MITOX, Mitoxantrone; VBL, Vinblastine; VCR, Vincristine; VIND, Vindesine.

[b] i.v., intravenously; i.p., intraperitoneally; s.c., subcutaneously.

[c] Lethality is due to drug toxicity and tumor progression.

differences in the immune system, drug pharmacokinetics, and metabolism between nudes and conventional mice.[21] In contrast, Povlsen and Jacobsen[19] reported toxicity of several anticancer agents to be higher in nude mice than in normal mice. Of importance, in nude mice bearing tumors toxicity may show further variability as compared to that in non-tumor-bearing nude mice. Drug-induced and tumor-associated effects may superimpose, resulting in a disproportionately high toxicity of anticancer drugs.

B. ANALYSIS OF EFFECTS VS. CLINICAL OBSERVATIONS

Since the first application of bleomycin to a human squamous cell carcinoma grown s.c. in nude mice,[23] a variety of human tumor xenografts have been treated with various established anticancer drugs. A large amount of data has accumulated, allowing the following general conclusions:

1. The chemosensitivity of a tumor line to a spectrum of agents always shows an individual pattern.[24]
2. Individual tumors of the same histological type respond differently to the same drug.[25]
3. Different passages of the same tumor generally respond in the same way to a particular drug.[26]

Drug-oriented validation of the nude mouse-human tumor model is made possible by the determination of the chemosensitivity spectrum of a panel of human tumor xenografts and a comparison of the overall sensitivity to the clinical response obtained in patients with the corresponding tumor types. In view of the variability in antitumor response to conventional

TABLE 2
Efficacy of Bleomycin in Human Tumor Xenografts in Nude Mice

	Drug application			Efficacy		
Tumor type carcinoma	**Dose, mg/kg/d**	**Schedule days**	**Route[a]**	**Activity criteria[b]**	**Active/ total**	**Ref.**
Breast	15	1–4,15—18	s.c.	Tumor regression	0/2	34
Gastrointestinal	15	1–4,15—18	s.c.	Tumor regression	0/5	34
Head and neck	15	0–2	i.p.	SGD >2	4/13	33
Lung, non-small cell	15	1–4,15–18	s.c.	Tumor regression	0/14	34
Melanoma	23.2	1,8,15,22	i.p.	TVOL: T/C ≤20%	0/7	53
Melanoma	15	1–4,15–18	s.c.	Tumor regression	0/5	34
Ovarian	23.2	1,8,15,22	i.p.	TVOL: T/C ≤20%	0/3	53
Testicular	15	1–4,15–18	s.c.	Tumor regression	1/4	34
Uterine cervix	50	1–5	i.v.	Tumor regression	0/2	87

[a] i.v., intravenously; i.p., intraperitoneally; s.c., subcutaneously.

[b] TVOL, tumor volume; SGD, growth delay, mean doubling time of treated tumors/mean doubling time of control tumors – 1; T/C, T/C-value, mean of treated tumors/mean of control tumors * 100 (%).

cytostatic agents between patients with the same tumor type, Bellet et al.[27] suggested the use of a representative panel of tumor lines to adequately evaluate the activity of an anticancer drug. As in clinical phase II trials, a total of 14 xenografts of 1 histological type should be tested in order to allow the rejection of the hypothesis of a true response rate ≥20% with 95% certainty.[28] As this number of xenografts of one tumor type is hard to find in one institute, data have to be accumulated in order to evaluate the reponse rate in specific tumor types.

For the following analysis, data published by a number of investigators have been reviewed with regard to the antitumor efficacy of established cytostatic agents. Only studies performed with at least two different solid human tumors transplanted s.c. to nude mice were evaluated. If possible, a definite tumor regression was required for preclinical activity of a given drug. Alternatively, activity criteria given by the author, for example, the T/C-ratio (test vs. control group value) or the specific growth delay as described by Steel et al.,[29] were evaluated. With regard to clinical studies, drugs giving a response rate (CR+PR) of ≥20% of the patients treated with single agent chemotherapy, were considered to be effective.

1. Bleomycin

Bleomycin is one of a family of antibiotic polypeptides that possess antitumor activity. Major clinical indications for bleomycin in solid tumors include squamous cell carcinoma of the head and neck[30] and uterine cervix, as well as non-seminomatous testicular cancer.[31,32]

Various studies in nude mice confirmed the antitumor activity of bleomycin (Table 2). Van Dongen et al.[33] demonstrated a significant tumor growth inhibition in 4 of 13 tumors of the head and neck; and Fiebig and Berger[34] observed a tumor regression in 1 of 4 different testicular cancer lines. Other tumor types, for example, cancer of the gastrointestinal tract, breast, and ovary, as well as non-small cell lung cancer and melanoma, were not responsive to bleomycin treatment.

2. Cisplatin

Cisplatin [*cis*-diamminodichloroplatinum (II)] is the only heavy metal compound used in cancer chemotherapy. The drug entered clinical trials in 1971 and proved to be highly effective in the treatment of testicular cancer,[32,35] as well as cancer of the head and neck,[36] bladder,[37] and ovary.[38,39] Somewhat lower response rates were achieved in gastric,[40] breast,[41] cervical,[42] and lung cancer.[43-46]

TABLE 3
Efficacy of Cisplatin in Human Tumor Xenografts in Nude Mice

	Drug Application			Efficacy		
Tumor type carcinoma	**Dose, mg/kg/d**	**Schedule days**	**Route[a]**	**Activity criteria[b]**	**Active/ total**	**Ref.**
Bladder	5	1,8,15,22	i.p.	Tumor regression	2/3	88
Bladder	2.5	1,8,15	i.p.	TVOL: T/C ≤20%	1/2	89
Breast	6.4	1,15	s.c.	Tumor regression	2/7	34
Colorectal	10	1	i.p.	SGD ≥2	0/2	26
Colorectal	6.4	1,15	s.c.	Tumor regression	1/5	34
Gastric	10	1	i.v.	TVOL: T/C ≤50%	5/10	15
Gastric	6.4	1,15	s.c.	Tumor regression	3/13	34
Head and neck	2	0-2	i.p.	SGD >2	0/8	90
Head and neck	5	0,7,14	i.p.	SGD >2	2/5	33
Head and neck	6.4	1,15	s.c.	Tumor regression	0/2	34
Lung, non-small cell	10	1	i.p.	SGD ≥2	0/11	26
Lung, non-small cell	6.4	1,15	s.c.	Tumor regression	4/23	34
Lung, small cell	6.4	1,15	s.c.	Tumor regression	2/8	34
Melanoma	9	1	i.p.	SGD ≥2	4/14	59
Melanoma	3.2	1,8,15,22	i.p.	TVOL: T/C ≤20%	0/7	53
Melanoma	6.4	1,15	s.c.	Tumor regression	0/10	34
Mesothelioma	4	1,8,15	i.p.	TVOL: ≤42%	1/2	81
Mesothelioma	6.4	1,15	s.c.	Tumor regression	0/3	34
Ovarian	5	1,8,15,22,29,36	i.p.	Tumor regression	4/10	91
Ovarian	10	1	i.p.	SGD ≥2	1/3	26
Ovarian	3.2	1,8,15,22	i.p.	TVOL: T/C ≤20%	1/3	53
Ovarian	4-5	0,7,14	i.v.	TVOL: T/C ≤25%	4/10	50
Ovarian	6.4	1,15	s.c.	Tumor regression	2/4	34
Renal	6.4	1,15	s.c.	Tumor regression	0/2	34
Sarcoma, bone	15	1,2	i.v.	Tumor regression	0/4	92
Sarcoma, soft tissue	6.4	1,15	s.c.	Tumor regression	1/5	34
Testicular	3	1-5	i.p.	Tumor regression	3/4	93
Testicular	6.4	1,15	s.c.	Tumor regression	4/5	34
Uterine cervix	10	1	i.p.	SGD ≥2	0/1	26
Uterine cervix	10	0,7,14	i.p.	Tumor regression	2/2	94

[a] i.v., intravenously; i.p., intraperitoneally; s.c., subcutaneously.
[b] TVOL, tumor volume; SGD, growth delay, mean doubling time of treated tumors/mean doubling time of control tumors – 1; T/C, T/C-value, mean of treated tumors/mean of control tumors * 100 (%).

Cisplatin has been extensively studied in nude mice. Various dose levels, schedules, and routes have been used for administration to tumor-bearing animals (Table 3). If a tumor regression was required for a particular tumor line to be considered sensitive, cisplatin demonstrated high activity in cancer of the bladder, lung, stomach, ovary, testes, and uterine cervix. For cancer of the breast, response rates in nude mice exceeded the clinical activity of the drug.

3. Cyclophosphamide

Cyclophosphamide is an oxazaphosphorine compound with alkylating properties. After activation by liver P-450 microsomes, the drug undergoes intracellular conversion to phosphoramide mustard and acrolein in peripheral target tissues. Cyclophosphamide is active in patients with cancer of the breast,[41] lung,[43-46] testes and ovary,[39] as well as soft tissue sarcoma and mesothelioma.[47]

Preclinical chemotherapy studies in tumor-bearing nude mice confirmed the broad activity

TABLE 4
Efficacy of Cyclophosphamide in Human Tumor Xenografts in Nude Mice

	Drug Application			Efficacy		
Tumor type carcinoma	**Dose, mg/kg/d**	**Schedule days**	**Route[a]**	**Activity criteria[b]**	**Active/ total**	**Ref.**
Breast	20	1–8	i.p.	TW: T/C ≤20%	3/14	25
Breast	75	1,4,8,11,15	i.p.	Tumor regression	0/3	95
Breast	65	1	i.v.	TVOL: T/C ≤50%	2/8	66
Breast	200	1,15	i.p.	Tumor regression	4/7	34
Colorectal	20	1–8	i.p.	TW: T/C ≤20%	1/14	25
Colorectal	240	1	i.p.	SGD >2	1/2	26
Colorectal	75	1,4,8,11,15	i.p.	Tumor regression	0/3	95
Colorectal	200	1,15	i.p.	Tumor regression	1/8	34
Gastric	75	1,4,8,11,15	i.p.	Tumor regression	1/7	95
Gastric	65	1	i.v.	TVOL: T/C ≤50%	0/11	72
Gastric	200	1,15	i.p.	Tumor regression	1/5	34
Head and neck	100	0,7	i.p.	SGD ≥2	0/7	33
Lung, non-small cell	240	1	i.p.	SGD ≥2	4/13	26
Lung, non-small cell	65	1	i.v.	TVOL: T/C ≤50%	1/10	71
Lung, non-small cell	200	1,15	i.p.	Tumor regression	2/17	34
Lung, small cell	65	1	i.v.	TVOL: T/C ≤50%	1/4	71
Lung, small cell	200	1,15	i.p.	Tumor regression	5/7	34
Melanoma	20	1–8	i.p.	TW: T/C ≤20%	3/14	25
Melanoma	200	1,15	i.p.	Tumor regression	0/2	34
Mesothelioma	200	1,15	i.p.	Tumor regression	0/3	34
Ovarian	200	1,8,15,22,29,36	i.p.	Tumor regression	4/10	91
Ovarian	240	1	i.p.	SGD ≥2	1/6	26
Ovarian	150	0,14	i.p.	TVOL: T/C ≤25%	3/10	50
Ovarian	200	1,15	i.p.	Tumor regression	1/5	34
Pancreatic	75	1,4,8,11,15	i.p.	Tumor regression	0/2	95
Renal	200	1,15	i.p.	Tumor regression	0/2	34
Sarcoma, bone	240	1	i.p.	SGD ≥2	0/1	26
Sarcoma, bone	780	1,2	i.v.	Tumor regression	0/4	92
Sarcoma, soft tissue	150	0,14	i.p.	TVOL: T/C ≤25%	3/6	60
Sarcoma, soft tissue	200	1,15	i.p.	Tumor regression	2/3	34
Testicular	200	1,15	i.p.	Tumor regression	2/3	34
Uterine cervix	100	1–5	i.p.	Tumor regression	0/2	87
Uterine cervix	240	1	i.p.	SGD ≥2	2/2	26

[a] i.v., intravenously; i.p., intraperitoneally.
[b] TVOL, tumor volume; TW, tumor weight; SGD, growth delay, mean doubling time of treated tumors/mean doubling time of control tumors – 1; T/C, T/C-value, mean of treated tumors/mean of control tumors * 100 (%).

spectrum of cyclophosphamide (Table 4). Response rates greater than 20% were observed in small cell lung cancer, breast, ovarian and testicular cancer, as well as in soft tissue sarcoma.

4. Doxorubicin

Doxorubicin has proven to be one of the most effective compounds in the treatment of solid human tumors, such as cancer of the breast,[41,48] lung (small cell),[43] thyroid and ovary,[39] as well as soft tissue sarcoma.[49] Furthermore, some antitumor effects were observed in gastric and testicular cancer[40] and in mesothelioma.[47] As a result of this clinical activity, a variety of analogs have been synthesized in order to increase the activity and to reduce the side effects of doxorubicin (e.g., cardiac toxicity, myelosuppression, and mucositis).

Doxorubicin was studied in detail in a large number of human tumor xenografts in nude mice (Table 5). Several investigators documented activity of the drug in breast and lung cancers.

TABLE 5
Efficacy of Doxorubicin in Human Tumor Xenografts in Nude Mice

	Drug application			Efficacy		
Tumor type carcinoma	**Dose, mg/kg/d**	**Schedule days**	**Route[a]**	**Activity criteria[b]**	**Active/ total**	**Ref.**
Breast	10	1,8,15	i.v.	TW: T/C ≤20%	3/3	96
Breast	1	1–8	i.p.	TW: T/C ≤20%	2/14	25
Breast	5	1,6,11,16	i.v.	Tumor regression	0/3	95
Breast	12	1	i.v.	TVOL: T/C ≤50%	3/8	66
Breast	8	1,15	i.v.	Tumor regression	1/8	34
Colorectal	10	1,8,15	i.v.	TVOL: T/C ≤20%	0/5	97
Colorectal	1	1–8	i.p.	TW: T/C ≤20%	1/14	25
Colorectal	5	1	i.p.	SGD ≥2	0/2	26
Colorectal	5	1,6,11,16	i.v.	Tumor regression	0/3	95
Colorectal	8	1,15	i.v.	Tumor regression	0/5	34
Gastric	5	1,6,11,16	i.v.	Tumor regression	0/7	95
Gastric	12	1	i.v.	TVOL: T/C ≤50%	2/11	72
Gastric	8	1,15	i.v.	Tumor regression	1/13	34
Head and neck	8	1,8	i.p.	SGD >2	0/2	98
Lung, non-small cell	5	1	i.p.	SGD ≥2	0/6	26
Lung, non-small cell	12	1	i.v.	TVOL: T/C ≤50%	4/10	71
Lung, non-small cell	8	1,15	i.v.	Tumor regression	2/19	34
Lung, small cell	12	1	i.v.	TVOL: T/C ≤50%	3/4	71
Lung, small cell	8	1,15	i.v.	Tumor regression	4/9	34
Melanoma	4	1	i.p.	STI (p ≤0.05)	0/9	27
Melanoma	10	1,8,15	i.v.	TW: T/C ≤20%	0/3	96
Melanoma	1	1–8	i.p.	TW: T/C ≤20%	2/14	25
Melanoma	2.6	1,8,15,22	i.p.	TVOL: T/C ≤20%	0/7	53
Melanoma	8	1,15	i.v.	Tumor regression	0/7	34
Mesothelioma	8	1,15	i.v.	Tumor regression	0/3	34
Ovarian	10	1,8,15	i.v.	TW: T/C ≤20%	0/2	96
Ovarian	5	1	i.p.	SGD ≥2	0/3	26
Ovarian	2.6	1,8,15,22	i.p.	TVOL: T/C ≤20%	1/3	53
Ovarian	8–10	0,7	i.v.	TVOL: T/C ≤25%	4/10	50
Ovarian	8	1,15	i.v.	Tumor regression	0/4	34
Pancreatic	5	1,6,11,16	i.v.	Tumor regression	0/2	95
Renal	8	1,15	i.v.	Tumor regression	0/2	34
Sarcoma, bone	20	1,2	i.v.	Tumor regression	0/4	92
Sarcoma, soft tissue	8	0,7	i.v.	TVOL: T/C ≤25%	1/7	60
Sarcoma, soft tissue	8	1,15	i.v.	Tumor regression	0/6	34
Testicular	8	1,15	i.v.	Tumor regression	3/4	34
Uterine cervix	4	1–4	i.v.	Tumor regression	1/2	87

[a] i.v., intravenously; i.p., intraperitoneally.
[b] TVOL, tumor volume; TW, tumor weight; SGD, growth delay, mean doubling time of treated tumors/mean doubling time of control tumors – 1; T/C, T/C-value, mean of treated tumors/mean of control tumors * 100 (%); STI, significant tumor inhibition of treated group as compared to controls.

Boven[50] and Taetle et al.[53] reported growth inhibition in ovarian cancer xenografts, whereas others did not observe responsiveness in this tumor type. Only marginal activity was shown in gastric cancer and soft tissue sarcoma grown in nude mice. Tumors of the colon and rectum, squamous cell cancer of the head and neck, as well as melanoma consistently proved to be resistant to doxorubicin.

5. DTIC

DTIC (diacarbazine) remains the most active compound in the treatment of malignant

TABLE 6
Efficacy of DTIC in Human Tumor Xenografts in Nude Mice

	Drug application			Efficacy		
Tumor type carcinoma	**Dose, mg/kg/d**	**Schedule days**	**Route[a]**	**Activity criteria[b]**	**Active/ total**	**Ref.**
Melanoma	300	1	i.p.	STI ($p \leq 0.05$)	4/9	27
Melanoma	100	1,8,15,22	i.p.	SGD ≥2	5/16	59
Melanoma	250	1,8,15	i.p.	SGD ≥2	1/5	99
Melanoma	300	1,15	i.p.	Tumor regression	1/7	34
Sarcoma, soft tissue	300	1,15	i.p.	Tumor regression	1/3	34

[a] i.p., intraperitoneally.
[b] SGD, growth delay, mean doubling time of treated tumors/mean doubling time of control tumors – 1; STI, significant tumor inhibition of treated group as compared to controls.

melanoma, yielding reponse rates of 15 to 25%.[51] In soft tissue sarcoma, DTIC is frequently used in combination with cyclophosphamide, vincristine, and doxorubicin.[52]

Several groups have described definite preclinical activity for DTIC in human melanoma xenografts (Table 6). For soft tissue sarcoma, only preliminary data are available, indicating some activity of DTIC in this tumor type.

6. Etoposide

The epipodophyllotoxin etoposide is derived by synthetic modification of a product of the mandrake plant. With regard to human solid tumors, it has important clinical activity in small cell cancer of the lung[43] and testicular cancer.[32]

Chemotherapy studies in human tumor xenografts are still limited (Table 7). Taetle et al.[53] described resistance in 4 melanoma xenografts and 1 ovarian tumor line treated with etoposide. Burgers et al.[54] observed progressive growth of a renal cell cancer line in nude mice. In a preclinical phase II study in our laboratory, etoposide was given s.c. at a dose level of 24 mg/kg/d on days 1 to 3 and 15 to 17. Four out of five small cell lung cancer xenografts, as well as 2 of 4 testicular cancer lines showed tumor regression. In other tumor types, etoposide exerted little or no growth inhibitory effect, which agrees well with the clinical activity of the drug.

7. 5-Fluorouracil

The antimetabolite 5-fluorouracil (5-FU) is commonly employed in various adenocarcinoma types including tumors of the breast,[41,48] colon,[55] stomach,[56] and ovary.[39] Response rates vary between 10 and 40%, but complete remissions are rare. Combination of 5-FU with folinic acid allows for slightly higher response rates.

In nude mice, 5-FU did not induce tumor regressions as compared to the initial tumor size at the start of treatment (Table 8). However, various studies demonstrated activity for the drug, if a T/C-value ≤20% or ≤50% was required for a tumor line to be considered sensitive. Inhibition of tumor growth was observed in 1 of 55 colorectal, 2 of 24 gastric, and 6 of 27 breast cancer xenografts. Overall, the activity of 5-FU in nude mice was lower than the clinical efficacy of the compound in single agent phase II studies.

8. Ifosfamide

Ifosfamide is a cyclophosphamide analog with one chloroethyl group shifted to the endocyclic nitrogen atom in the molecule. In clinical studies, ifosfamide has been shown to be highly active in lung cancer, in cancer of the testes, breast, ovary and uterine cervix, as well as in soft tissue sarcoma.[57]

TABLE 7
Efficacy of Etoposide in Human Tumor Xenografts in Nude Mice

	Drug application			Efficacy		
Tumor type carcinoma	**Dose, mg/kg/d**	**Schedule days**	**Route[a]**	**Activity criteria[b]**	**Active/ total**	**Ref.**
Breast	24	1–3,15–17	s.c.	Tumor regression	0/4	34
Colorectal	24	1–3,15–17	s.c.	Tumor regression	0/4	34
Gastric	24	1–3,15–17	s.c.	Tumor regression	0/13	34
Lung, non-small cell	24	1–3,15–17	s.c.	Tumor regression	1/17	34
Lung, small cell	24	1–3,15–17	s.c.	Tumor regression	4/5	34
Melanoma	3.9	1,8,15,22	i.p.	TVOL: T/C ≤20%	0/4	53
Melanoma	24	1–3,15–17	s.c.	Tumor regression	0/4	34
Mesothelioma	24	1–3,15–17	s.c.	Tumor regression	0/2	34
Ovarian	3.9	1,8,15,22	i.p.	TVOL: T/C ≤20%	0/1	53
Ovarian	24	1–3,15–17	s.c.	Tumor regression	1/4	34
Renal	50	1,5,9	i.p.	Tumor regression	0/1	54
Renal	24	1–3,15–17	s.c.	Tumor regression	0/2	34
Testicular	24	1–3,15–17	s.c.	Tumor regression	2/4	34

[a] i.p., intraperitoneally; s.c., subcutaneously.
[b] TVOL, tumor volume; T/C, T/C-value, mean of treated tumors/mean of control tumors * 100 (%).

TABLE 8
Efficacy of 5-Fluorouracil in Human Tumor Xenografts in Nude Mice

	Drug application			Efficacy		
Tumor type carcinoma	**Dose, mg/kg/d**	**Schedule days**	**Route[a]**	**Activity criteria[b]**	**Active/ total**	**Ref.**
Breast	20	1–8	i.p.	TW: T/C ≤20%	3/14	25
Breast	19	1–5	i.v.	TVOL: T/C ≤50%	3/8	66
Breast	40	1–4,15–18	i.p.	Tumor regression	0/5	34
Colorectal	90–100	1,8,15	i.v.	TW: T/C ≤20%	0/9	100
Colorectal	20	1–8	i.p.	TW: T/C ≤20%	1/14	25
Colorectal	40	1–4,15–18	i.p.	Tumor regression	0/32	34
Gastric	19	1–5	i.v.	TVOL: T/C ≤50%	2/11	72
Gastric	40	1–4,15–18	i.p.	Tumor regression	0/13	34
Head and neck	25	0–3	i.p.	SGD >2	0/8	33
Lung, non-small cell	19	1–5	i.v.	TVOL: T/C ≤50%	1/10	71
Lung, non-small cell	40	1–4,15–18	i.p.	Tumor regression	0/4	34
Lung, small cell	19	1–5	i.v.	TVOL: T/C ≤50%	0/4	71
Melanoma	20	1–8	i.p.	TW: T/C ≤20%	2/14	25
Ovarian	60	0,7,14,21	i.p.	TVOL: T/C ≤25%	0/10	50

[a] i.v., intravenously; i.p., intraperitoneally.
[b] TVOL, tumor volume; TW, tumor weight; SGD, growth delay, mean doubling time of treated tumors/mean doubling time of control tumors – 1; T/C, T/C-value, mean of treated tumors/mean of control tumors * 100 (%).

In a preclinical phase II study in human tumor xenografts in nude mice, the observed efficacy of ifosfamide paralleled the clinical situation.[58] At a dose level of 130 mg/kg/d, given days 1 to 3 and 15 to 17 i.p., tumor regressions were observed in 4 of 5 breast cancer, 1 of 3 colon, 1 of 1 gastric, 2 of 7 nonsmall cell lung, and 3 of 4 small cell lung cancer xenografts, in 1 of 2 sarcomas, and 3 of 3 testicular tumors. Two ovarian, two uterine, and six renal cell cancer

TABLE 9
Efficacy of Methotrexate in Human Tumor Xenografts in Nude Mice

	Drug application			Efficacy		
Tumor type carcinoma	**Dose, mg/kg/d**	**Schedule days**	**Route[a]**	**Activity criteria[b]**	**Active/ total**	**Ref.**
Breast	8	1–8	i.p.	TW: T/C ≤20%	2/14	25
Breast	15	1–5	i.v.	TVOL: T/C ≤50%	2/6	66
Breast	4	1,4,8,11,15,18	i.p.	Tumor regression	0/6	34
Colorectal	8	1–8	i.p.	TW: T/C ≤20%	1/14	25
Gastric	15	1–5	i.v.	TVOL: T/C ≤50%	1/11	72
Gastric	4	1,4,8,11,15,18	i.p.	Tumor regression	0/2	34
Head and neck	250	0,7	i.p.	SGD >2	0/13	33
Lung, non-small cell	15	1–5	i.v.	TVOL: T/C ≤50%	1/10	71
Lung, non-small cell	4	1,4,8,11,15,18	i.p.	Tumor regression	0/3	34
Lung, small cell	15	1–5	i.v.	TVOL: T/C ≤50%	0/4	71
Lung, small cell	4	1,4,8,11,15,18	i.p.	Tumor regression	0/3	34
Melanoma	8	1–8	i.p.	TW: T/C ≤20%	1/14	25
Ovarian	150	0,7,14	i.p.	TVOL: T/C ≤25%	0/5	50

[a] i.v., intravenously; i.p., intraperitoneally.
[b] TVOL, tumor volume; TW, tumor weight; SGD, growth delay, mean doubling time of treated tumors/mean doubling time of control tumors – 1; T/C, T/C-value, mean of treated tumors/mean of control tumors * 100 (%).

xenografts, as well as three melanomas were resistant. In contrast, Osieka[59] observed tumor growth inhibition (SGD ≥ 2) in 5 of 15 melanoma xenografts. Boven[60] reported a significant reduction in tumor growth (T/C $\leq 25\%$) in 2 of 5 soft tissue sarcoma lines.

9. Methotrexate

The antimetabolite methotrexate exerts its effects through inhibition of the enzyme dihydrofolate reductase. In single-agent chemotherapy, methotrexate is effective in uterine choriocarcinoma,[61] osteosarcoma,[62,63] squamous cell carcinoma of the head and neck,[64] as well as in breast cancer.[41,48] For gastric cancer[65] and nonsmall cell lung cancer,[46] response rates were observed between 10 and 15%.

In nude mice, only breast cancer xenografts displayed some sensitivity to methotrexate,[25,66] but tumor regressions were not obtained (Table 9). Marginal responses were observed in tumors of the gastrointestinal tract as well as in melanoma and nonsmall cell lung cancer. Some 13 different squamous cell carcinomas of the head and neck were resistant to the drug. Overall, the clinical efficacy of methotrexate could not be reproduced in the nude mouse-human tumor model.

10. Mitomycin C

The first clinical use of the antitumor antibiotic mitomycin C has been in gastrointestinal carcinoma.[40,67] Later studies revealed activity of the drug in breast,[41,48] ovarian,[68] and non-small cell lung cancer.[44-46]

The broad spectrum of activity of mitomycin C was confirmed in nude mice (Table 10). Response rates equal to or greater than 20% were observed in cancer of the stomach, lung, breast, and ovary, as well as in melanoma. Colorectal cancer xenografts demonstrated tumor regressions in 2 of 21 cases. The high activity in small cell lung cancer is of special interest, as this tumor type is not considered clinically responsive to mitomycin C. This may be due to the fact that, in phase II studies of mitomycin C, only heavily pretreated small cell lung cancer patients were evaluated.[69] Further clinical studies with the drug in this tumor type might be indicated.

TABLE 10
Efficacy of Mitomycin-C in Human Tumor Xenografts in Nude Mice

	Drug application			Efficacy		
Tumor type carcinoma	**Dose, mg/kg/d**	**Schedule days**	**Route[a]**	**Activity criteria[b]**	**Active/ total**	**Ref.**
Breast	2	1,8,15,22,29	i.p.	Tumor regression	1/3	95
Breast	1.7	1	i.v.	TVOL: T/C ≤50%	3/8	66
Breast	2	1,15	i.v.	Tumor regression	4/6	34
Colorectal	2	1,8,15,22,29	i.p.	Tumor regression	1/3	95
Colorectal	2	1,15	i.v.	Tumor regression	1/18	34
Gastric	2	1,8,15,22,29	i.p.	Tumor regression	2/7	95
Gastric	1.7	1	i.v.	TVOL: T/C ≤50%	2/11	72
Gastric	2	1,15	i.v.	Tumor regression	4/13	34
Lung, non-small cell	1.7	1	i.v.	TVOL: T/C ≤50%	0/10	71
Lung, non-small cell	2	1,15	i.v.	Tumor regression	6/18	34
Lung, small cell	1.7	1	i.v.	TVOL: T/C ≤50%	3/4	71
Lung, small cell	2	1,15	i.v.	Tumor regression	3/5	34
Melanoma	3	1,8,15,22	i.p.	SGD ≥2	4/9	59
Melanoma	2.4	1,8,15,22	i.p.	TVOL: T/C ≤20%	2/7	53
Melanoma	2	1,15	i.v.	Tumor regression	0/3	34
Mesothelioma	1.5	1,8,15	i.p.	TVOL: T/C ≤42%	1/2	81
Mesothelioma	2	1,15	i.v.	Tumor regression	0/3	34
Ovarian	2.4	1,8,15,22	i.p.	TVOL: T/C ≤20%	1/3	53
Ovarian	2	1,15	i.v.	Tumor regression	1/3	34
Pancreatic	2	1,8,15,22,29	i.p.	Tumor regression	0/2	95
Pancreatic	2.4	1,7,14	i.p.	Tumor regression	1/4	86
Renal	2	1,15	i.v.	Tumor regression	0/2	34
Sarcoma, bone	20	1,2	i.v.	Tumor regression	0/4	92
Testicular	2	1,15	i.v.	Tumor regression	0/2	34
Uterine cervix	4	1–3	i.v.	Tumor regression	1/2	87

[a] i.v., intravenously; i.p., intraperitoneally; s.c., subcutaneously.
[b] TVOL, tumor volume; SGD, growth delay, mean doubling time of treated tumors/mean doubling time of control tumors – 1; T/C, T/C-value, mean of treated tumors/mean of control tumors * 100 (%).

11. Vinca Alkaloids

The vinca alkaloids are complex multiringed molecules derived from the ornamental shrub *Vinca rosea*. Modifications of the original structure yielded various analogs, including the clinically used cytostatic agents vinblastine, vincristine, and vindesine. In clinical studies, vinblastine demonstrated efficacy in breast,[41,48] lung,[46] and testicular cancer.[32] Vincristine is mainly used in combination chemotherapy for soft tissue sarcoma,[52] neuroblastoma, and small cell lung cancer.[70] Vindesine has activity against non-small cell lung cancer[44-46] and hematological malignancies.

Both vinblastine (Table 11) and vincristine (Table 12) have been studied in nude mice by various investigators, whereas reports on vindesine in the treatment of xenografts are not available. In two studies, vinblastine demonstrated activity in breast cancer.[25,66] Using "rational doses", Tashiro et al.[71] and Inaba et al.[72] required a T/C equal to or less than 50% for sensitivity of a particular tumor line. They reported activity of vinblastine in 3 of 11 gastric cancer xenografts and 3 of 14 lung cancer xenografts. Colorectal, ovarian, and renal tumors, as well as melanomas, proved to be resistant to vinblastine. In contrast, vincristine exerted growth inhibition in 6 of 14 melanomas studied by Giovanella et al.[25] Other tumor types showing slight sensitivity to vincristine included breast cancer, small cell lung cancer, and neuroblastoma.

TABLE 11
Efficacy of Vinblastine in Human Tumor Xenografts in Nude Mice

	Drug application			Efficacy		
Tumor type carcinoma	Dose, mg/kg/d	Schedule days	Route[a]	Activity criteria[b]	Active/ total	Ref.
Breast	0.3	1–8	i.p.	TW: T/C ≤20%	4/14	25
Breast	2.6	1	i.v.	TVOL: T/C ≤50%	3/8	66
Colorectal	0.3	1–8	i.p.	TW: T/C ≤20%	0/14	25
Gastric	2.6	1	i.v.	TVOL: T/C ≤50%	3/11	72
Lung, non-small cell	2.6	1	i.v.	TVOL: T/C ≤50%	2/10	71
Lung, small cell	2.6	1	i.v.	TVOL: T/C ≤50%	1/4	71
Melanoma	0.3	1–8	i.p.	TW: T/C ≤20%	2/14	25
Melanoma	0.76	1,8,15,22	i.p.	TVOL: T/C ≤20%	0/7	53
Melanoma	2.5	1,8,15	i.v.	SGD ≥2	0/5	99
Ovarian	0.76	1,8,15,22	i.p.	TVOL: T/C ≤20%	0/3	53
Renal	1.2	1,8,15,22,29,36	i.p.	Tumor regression	0/4	101

[a] i.v., intravenously; i.p., intraperitoneally.
[b] TVOL, tumor volume; TW, tumor weight; T/C, T/C-value, mean of treated tumors/mean of control tumors * 100(%).

TABLE 12
Efficacy of Vincristine in Human Tumor Xenografts in Nude Mice

	Drug application			Efficacy		
Tumor type carcinoma	Dose, mg/kg/d	Schedule days	Route[a]	Activity criteria[b]	Active/ total	Ref.
Breast	0.2	1–8	i.p.	TW: T/C ≤20%	4/14	25
Breast	0.4	1	i.v.	TVOL: T/C ≤50%	1/8	66
Breast	0.5	1,8,15	i.v.	Tumor regression	0/2	34
Colorectal	0.2	1–8	i.p.	TW: T/C ≤20%	0/14	25
Colorectal	0.5	1,8,15	i.v.	Tumor regression	0/2	34
Gastric	0.4	1	i.v.	TVOL: T/C ≤50%	0/11	72
Head and neck	1	1,2	i.p.	SGD ≥2	0/2	98
Lung, non-small cell	2	1	i.p.	SGD ≥2	0/7	26
Lung, non-small cell	0.4	1	i.v.	TVOL: T/C ≤50%	0/10	71
Lung, small cell	0.4	1	i.v.	TVOL: T/C ≤50%	0/4	71
Lung, small cell	0.5	1,8,15	i.v.	Tumor regression	1/4	34
Melanoma	0.2	1–8	i.p.	TW: T/C ≤20%	6/14	25
Neuroblastoma	2.0	1	i.p.	Tumor regression	1/2	102
Ovarian	2	1	i.p.	SGD ≥2	0/2	26
Renal	0.5	1,8,15	i.v.	Tumor regression	0/5	34
Sarcoma, bone	5	1,2	i.v.	Tumor regression	0/4	92

[a] i.v., intravenously; i.p., intraperitoneally.
[b] TVOL, tumor volume; TW, tumor weight; SGD, growth delay, mean doubling time of treated tumors/mean doubling time of control tumors – 1; T/C, T/C-value, mean of treated tumors/mean of control tumors * 100 (%).

12. Absence of a Correlation

Comparison of response rates obtained in clinical trials and *in vivo* studies in nude mice generally reveals a good correlation for the chemosensitivity pattern of a particular drug. However, for some tumor types (e.g., melanoma), the nude mouse model seems to be overpredictive, as a considerable number of drugs gave false positive data (compounds active in the preclinical screening model but without clinical efficacy). The poor preclinical activity of the antimetabolites 5-FU and methotrexate did not reflect the clinical usefulness of these compounds, as tumor regressions in xenografts were hardly observed.

Several factors may contribute to these discrepancies:

1. Differences between mouse and man with respect to drug pharmacokinetics and pharmacodynamics may lead to either overprediction or underprediction for clinical utility. This problem is not unique for xenograft models.[73] Drug doses, methods of administration, bioavailability, metabolism, and elimination may have a different profile in mice as compared to that in man.[18]
2. Freireich et al.[14] have reported a good agreement of the MTD in the mouse with that in man for various cytostatic drugs, if the dose levels were calculated in relation to the body surface. Factors for conversion from mg/kg to mg/m^2 are 3.0 or 3.4 for mice with a body weight of 20 or 28 g, respectively. For a male patient with a body weight of 70 kg, conversion from mg/kg to mg/m^2 is possible by a factor of 39.0. Fiebig[6] compared the MTD levels in mice for several established drugs to the clinically used doses on a mg/m^2 basis. The total dose for nude mice ranged between 44 and 170% of the dose administered to patients. Considerable differences between both species were observed for vindesine (nude mice tolerating 170% of the human dose), DTIC (131%), doxorubicin (51%), and 5-FU (44%). This might add insight as to the reasons for the over- and underprediction of the nude mouse model for some drugs and tumor types. A comparison of plasma levels of standard anticancer agents in nude mice and man has led to the concept of the "rational dose" being applied in mice. For methotrexate and 5-FU, the RD was significantly higher than the MTD.[15,16,72]
3. The nude mouse-human tumor model represents a biological system which certainly cannot reflect all characteristics of the patient's tumor. Treatment studies in nude mice are started at the time of tumor implantation or the s.c. xenografts have reached a median diameter of 5 to 6 mm. In contrast, patients receiving chemotherapy are mainly those with a large tumor burden and metastatic disease. Furthermore, the immune response shows considerable differences between mouse and man.
4. If tests are performed in a panel of xenografts of one histological type, the lines selected do not necessarily represent the clinical heterogeneity of human tumors. Especially for malignancies, where only a minority of patients' samples can be established as permanent xenograft lines (e.g., breast cancer), a selection bias for fast-growing, poorly differentiated tumors may result.[25]
5. In the large majority of nude mouse studies, only a small number of tumors was used. Few authors reported results for a representative panel with equal to or greater than 14 tumor lines of one particular tumor type.
6. The evaluation of tumor response should be similar for nude mice and patients. As illustrated for melanomas, definite tumor regressions should be required for activity of a compound. Use of the growth delay or T/C-values may lead to a false positive assessment of cytostatic drug efficacy.

C. COMBINATION CHEMOTHERAPY

Treatment of cancer patients requires a multidisciplinary approach. In most cases, surgery, irradiation, and chemotherapy are combined in order to achieve an optimal tumor response. In sensitive tumor types, combination chemotherapy regimens of up to seven anticancer drugs may be used with a curative intent. The clinical design of these regimens is primarily based on empirical data; that is, on the results of single agent phase II trials indicating efficacy of the various compounds in a particular tumor type. A number of investigators have used combination chemotherapy in nude mice in order to provide a more rational approach to the use of new drugs and established agents for clinical trials.[74-79]

In general, combined treatment in human tumor xenografts produced superior therapeutic results as compared to single drugs. Drug doses, schedules, and routes for combination chemotherapy were usually determined in a way similar to single agent treatment. In order to avoid drug interactions at the site of application, compounds may be given with an interval of 10 to 20 min using different routes.[80]

Studies on the synergistic activity of established anticancer drugs were performed in malignant mesothelioma[81] and germ cell tumors.[82] Fiebig[80] reported on the efficacy of CCNU, cisplatin, vindesine, etoposide, and the experimental agent dianhydrogalactitol (DAG) alone and in combination in six human gastric cancer xenografts. In 2-drug regimens, the dose for each compound had to be reduced to 70 to 80%, in 3-drug combinations to between 50 and 60%. All possible combinations of 2 drugs were tested systematically ($n = 10$) and resulted in an improved tumor response as compared to single agent chemotherapy. The most effective regimens consisted of CCNU+DAG, vindesine+DAG, CCNU+cisplatin, or CCNU+vindesine. Three-drug combinations resulted in even higher response rates, the most active combinations being CCNU+cisplatin+vindesine, CCNU+vindesine+DAG, and cisplatin+etoposide+DAG. Assessment of the relative contribution of each single agent to the efficacy of the drug combinations gave some evidence for individuality of the therapeutic response of human tumors.

The relationship between tumor response, drug dose, and sequence of application was investigated by Wayss et al.[83] for methotrexate and 5-FU. Pretreatment with methotrexate followed by 5-FU with intervals of 3 h or 24 h showed the highest antitumor effects, as compared to single administration of methotrexate or 5-FU, simultaneous treatment, or the reverse sequence.

A promising application of the nude mouse-human tumor model in drug testing is the evaluation of the synergistic efficacy of standard anticancer drugs and cytokines. In a study performed by Kondo et al.,[84] the inhibitory effect of doxorubicin in a human colon cancer cell line in nude mice was augmented synergistically by daily injections of recombinant human interferon-gamma. Similar effects were described for the combination of cisplatin or mitomycin C with interferon-alpha in xenografts of mesothelioma.[85] The activity of recombinant human tumor necrosis factor-alpha (rTNF-α) and mitomycin C was studied in human pancreatic cancer xenografts by Klapdor et al.[86] Combined therapy resulted in a significant inhibition of tumor growth in 3 of 5 xenografts and in a decrease of tumor volume to less than 50% of the initial values in 2 of 5 tumors. Other investigators found synergistic activity for rTNF-α and etoposide in a metastatic renal cell cancer model.[54] These results support the concept that combinations of cytokines with cytostatic drugs might be of value for treatment of human solid tumors in the clinical situation.

ACKNOWLEDGMENTS

We wish to thank our co-workers E. Wallbrecher, U. Winterhalter, C. Berg, and V. Haberstroh for their excellent assistance in the establishment and characterization of human tumor xenografts in nude mice. K.H. Widmer is thanked for software development and data evaluation.

REFERENCES

1. **Rygaard, J. and Povlsen, C. O.,** Heterotransplantation of a human malignant tumor to "nude" mice, *Acta Pathol. Microbiol. Scand.*, 77, 758, 1969.
2. **Pierce, G. B., Midgley, A. R., and Verney, E. L.,** Therapy of heterotransplanted choriocarcinoma, *Cancer Res.*, 22, 563, 1962.
3. **Burt, F. B., Pavone-Macaluso, M., Horns, J. W., and Kaufman, J. J.,** Heterotransplantation of bladder cancer in the hamster cheek pouch: *in vivo* testing of cancer chemotherapeutic agents, *J. Urol.*, 95, 51, 1966.
4. **Kopper, L. and Steel, G. G.,** The therapeutic response of three human tumor lines maintained in immune-suppressed mice, *Cancer Res.*, 35, 2704, 1975.
5. **Steel, G. G., Courtenay, V. D., and Peckham, M. J.,** The response to chemotherapy of a variety of human tumor xenografts, *Br. J. Cancer*, 47, 1, 1983.
6. **Fiebig, H. H.,** Wachstum und Chemotherapie menschlicher Tumoren - vorwiegend Dickdarm-, Magen- und Bronchialkarzinome - in der thymusaplastischen Nacktmaus, *Habilitationsschrift, Universität Freiburg*, 1983.
7. **Fiebig, H. H., Widmer, K. H., Fiedler, L., Wittekind, C., and Löhr, G. W.,** Development and characterization of 51 human tumor models for large bowel, stomach and esophageal cancers, *Dig. Surg.*, 1, 225, 1984.
8. **Fiebig, H. H. and Löhr, G. W.,** Wachstum menschlicher Malignome in der thymusaplastischen Nacktmaus, *Med. Welt*, 35, 52, 1984.
9. **Fiebig, H. H.,** Comparison of tumor response in nude mice and in patients, in *Human Tumour Xenografts in Anticancer Drug Development*, Winograd, B., Peckham, M. J., and Pinedo, H. M., Eds., Springer-Verlag, Berlin, 1988, 25.
10. **Kameya, T., Shimosato, Y., Hayashi, H., and Tsumuraya, M.,** Growth and differentiation of hormone-producing human tumors in nude mice, in *Proc. 2nd Int. Workshop on Nude Mice*, Nomura, T., Ohsawa, N., Tamaoki, N., and Fujiwara, K., Eds., G. Fischer, Stuttgart, 1977, 375.
11. **Grant, A. G., Duke, D., and Hernon-Taylor, J.,** Establishment and characterization of primary human pancreatic carcinoma in continuous cell culture and in nude mice, *Br. J. Cancer*, 39, 143, 1979.
12. **Hata, J. I., Tamaoki, N., and Ueyama, Y.,** Usefulness of nude mice in detecting function of human tumors, in *Proc. 3rd Int. Workshop on Nude Mice*, Reed, N. D., Ed., G. Fischer, New York, 1982, 557.
13. **Wennerberg, J.,** Changes in growth pattern of human squamous-cell carcinomas of the head and neck during serial passages in nude mice, *Int. J. Cancer*, 33, 245, 1984.
14. **Freireich, E. J., Gehan, E. A., Rall, D. P., Schmidt, L. H., and Skipper, H. E.,** Quantitative comparison of toxicity of anticancer agents in mouse, rat, hamster, dog, monkey and man, *Cancer Chemother. Rep.*, 50, 219, 1966.
15. **Inaba, M., Tashiro, T., Kobayashi, T., Fujimoto, S., Sakurai, Y., Maruo, K., Ohnishi, Y., Ueyama, Y., and Nomura, T.,** Evaluation of response rates to various antitumor agents of human gastric tumors implanted in nude mouse, *Jpn. J. Cancer Res. (Gann)*, 77, 190, 1986.
16. **Inaba, M., Kobayashi, T., Tashiro, T., and Sakurai, Y.,** Pharmacokinetic approach to rational therapeutic doses for human tumor-bearing nude mice, *Jpn. J. Cancer Res. (Gann)*, 79, 509, 1988.
17. **Shorthouse, A. J., Jones, J. M., Steel, G. G., and Peckham, M. J.,** Experimental combination and single agent chemotherapy in human lung tumour xenografts, *Br. J. Cancer*, 46, 35, 1982.
18. **Winograd, B., Boven, E., Lobbezoo, M. W., and Pinedo, H. M.,** Human tumor xenografts in the nude mouse and their value as test models in anticancer drug development, *In Vivo*, 1, 1, 1987.
19. **Povlsen, C. O. and Jacobsen, G. K.,** Chemotherapy of a human malignant melanoma transplanted in the nude mouse, *Cancer Res.*, 35, 2790, 1975.
20. **Houchens, D. P., Johnson, R. K., Gaston, M. R., and Goldin, A.,** Toxicity of cancer chemotherapeutic agents in athymic (nude) mice, *Cancer Treat. Rep.*, 61, 103, 1977.
21. **Litterst, C. L., Sikic, B. I., Mimnaugh, E. G., Guarino, A. M., and Gram, T. E.,** *In vitro* drug metabolism in male and female athymic, nude mice, *Life Sci.*, 22, 1723, 1978.
22. **Bertazolli, C., Chieli, T., Grandi, M., and Ricevuti, G.,** Adriamycin: toxicity data, *Experientia*, 26, 389, 1970.
23. **Povlsen, C. O., Jacobsen, G. K., and Rygaard, J.,** The mouse mutant nude as a model for testing of anticancer agents, in *The Laboratory Animal in Drug Testing, Fifth ICLA Symposium*, Spiegel , A., Ed., Fischer, Stuttgart, 1973, 63.
24. **Houghton, P. J. and Houghton, J. A.,** Evaluation of single agent therapy in human colorectal tumour xenografts, *Br. J. Cancer*, 37, 833, 1978.
25. **Giovanella, B. C., Stehlin, J. S., Shepard, R. C., and Williams, L. J.,** Correlation between response to chemotherapy of human tumors in patients and in nude mice, *Cancer*, 52, 1146, 1983.
26. **Mattern, J., Wayss, K., and Volm, M.,** Effect of five antineoplastic agents on human tumor xenografts with different growth rates, *J. Natl. Cancer Inst.*, 72, 1335, 1984.
27. **Bellet, R. E., Danna, V., Mastrangelo, M. J., and Berd, D.,** Evaluation of a "nude" mouse - human tumor panel as a predictive secondary screen for cancer chemotherapeutic agents, *J. Natl. Cancer Inst.*, 63, 1185, 1979.

28. **Gehan, E. A.,** The determination of the number of patients required in a preliminary and a follow-up trial of a new chemotherapeutic agent, *J. Chronic Dis.*, 13, 346, 1961.
29. **Steel, G. G., Courtenay, V. D., and Peckham, M. J.,** The immune-suppressed mouse as an alternative host for heterotransplantation, in *The Nude Mouse in Experimental and Clinical Research, Vol. 2,* Fogh, J. and Giovanella, B. C., Eds., Academic Press, New York, 1982, 207.
30. **Bonadonna, G., Tancini, G., and Bajetta, E.,** Controlled studies with bleomycin in solid tumors and lymphomas, *Prog. Biochem. Pharmacol.*, 11, 172, 1976.
31. **Blum, R. H., Carter, S., and Agre, K.,** A clinical review of bleomycin: a new antineoplastic agent, *Cancer,* 31, 903, 1973.
32. **Williams, S. D., Birch, R., Einhorn, L. H., Irwin, L., Greco, F. A., and Loehrer, P. J.,** Treatment of disseminated germ-cell tumors with cisplatin, bleomycin and either vinblastine or etoposide, *N. Engl. J. Med.*, 316, 1435, 1987.
33. **Van Dongen, G., Braakhuis, B. J. M., Bagnay, M., Leyva, A., and Snow, G. B.,** Activity of differentiation-inducing agents and conventional drugs in head and neck cancer xenografts, *Acta Otolaryngol. (Stockholm),* 105, 488, 1988.
34. **Fiebig, H. H. and Berger, D. P.,** manuscript in preparation.
35. **Higby, D. J., Wallace, H. J., and Albert, D. J.,** Diaminodichloroplatinum. A Phase I study showing responses in testicular and other tumors, *Cancer,* 33, 1219, 1974.
36. **Wittes, R. E., Cvitkovich, E., and Shah, J.,** *Cis*-dichlorodiammineplatinum (II) in the treatment of epidermoid carcinoma of the head and neck, *Cancer Treat. Rep.*, 61, 359, 1977.
37. **Yagoda, A.,** Chemotherapy of urothelial tract tumors, *Cancer,* 60, 574, 1987.
38. **Ozols, R. F. and Young, R. C.,** Chemotherapy of ovarian cancer, *Semin. Oncol.*, 11, 251, 1984.
39. **Young, R. C., Fuks, Z., and Hoskins, W. J.,** Cancer of the ovary, in *Cancer — Principles and Practice of Oncology,* DeVita, V. T., Hellman, S., and Rosenberg, S. A., Eds., Lippincott, Philadelphia, 1989, 1162.
40. **Wils, J. and Bleiberg, H.,** Current status of chemotherapy for gastric cancer, *Eur. J. Cancer Clin. Oncol.*, 25, 3, 1989.
41. **Henderson, I. C.,** Adjuvant systemic therapy for early breast cancer, *Curr. Probl. Cancer,* 11, 125, 1987.
42. **Thigpen, T., Shingleton, H., Homesley, H., Lagasse, L., and Blessing, J.,** *Cis*-platinum in treatment of advanced or recurrent squamous cell carcinoma of the cervix: a phase II study of the Gynecologic Oncology Group, *Cancer,* 48, 899, 1981.
43. **Greco, F. A., Richardson, R. L., Snell, J. D., Stroup, S. L., and Oldham, R. K.,** Small cell lung cancer — complete remission and improved survival, *Am. J. Med.*, 66, 625, 1979.
44. **Bakowski, M. T. and Crouch, J. C.,** Chemotherapy of non-small cell lung cancer: a reappraisal and a look to the future, *Cancer Treat. Rev.*, 10, 159, 1983.
45. **Joss, R. A., Cavalli, F., Goldhirsch, A., Mermillod, B., and Brunner, K. W.,** New agents in non-small cell lung cancer, *Cancer Treat. Rev.*, 11, 205, 1984.
46. **Kris, M., Cohen, E., and Gralla, R.,** An analysis of 134 Phase II trials in non-small cell lung cancer (NSCLC), *Proc. Fourth World Conf. Lung Cancer,* Toronto, 1985.
47. **Gerner, R. E. and Moore, G. E.,** Chemotherapy of malignant mesothelioma, *Oncology,* 30, 152, 1974.
48. **Carter, S. K.,** The chemical therapy of breast cancer, *Semin. Oncol.*, 1, 131, 1974.
49. **Borden, E. C., Amato, D. A., Rosenbaum, C., Enterline, H. T., Shiraki, M. J., Creech, R. H., Lerner, H. J., and Carbone, P. P.,** Randomized comparison of three adriamycin regimens for metastatic soft tissue sarcomas, *J. Clin. Oncol.*, 5, 840, 1987.
50. **Boven, E.,** Conventional agents in human ovarian cancer xenografts, in *Human Tumour Xenografts in Anticancer Drug Development,* Winograd, B., Peckham, M. J., and Pinedo, H. M., Eds., Springer-Verlag, Berlin, 1988, 33.
51. **Comis, R. L.,** DTIC (NSC-45388) in malignant melanoma: a perspective, *Cancer Treat. Rep.*, 60, 165, 1976.
52. **Pinedo, H. M., Bramwell, V. H. C., Mouridsen, H. T., Somers, R., Vendrik, C. P. J., Santoro, A., Buesa, J., Wagener, T., van Oosterom, A. T., van Unnik, J. A. M., Sylvester, R., de Pauw, M., Thomas, D., and Bonadonna, G.,** CYVADIC in advanced soft tissue sarcoma: a randomized trial comparing two schedules. A study of the EORTC Soft Tissue and Bone Sarcoma Group, *Cancer,* 53, 1825, 1984.
53. **Taetle, R., Rosen, F., Abramson, I., Venditti, J., and Howell, S.,** Use of nude mouse xenografts as preclinical drug screens: *in vivo* activity of established chemotherapeutic agents against melanoma and ovarian carcinoma xenografts, *Cancer Treat. Rep.*, 71, 297, 1987.
54. **Burgers, J. K., Marshall, F. F., and Isaacs, J. T.,** Enhanced anti-tumor effects of recombinant human tumor necrosis factor plus VP-16 on metastatic renal cell carcinoma in a xenograft model, *J. Urol.*, 142, 160, 1989.
55. **Arbuck, S. G.,** Overview of clinical trials using 5-fluorouracil and leucovorin for the treatment of colorectal cancer, *Cancer,* 63(Suppl.), 1036, 1989.

56. **Machover, D., Goldschmidt, E., Chollet, P., Metzger, G., Zittoun, J., Marquet, J., Vandenbulcke, J. M., Misset, J. L., Schwarzenberg, J. L., Fourtillan, J. B., Gaget, H., and Mathé, G.,** Treatment of advanced colorectal and gastric adenocarcinomas with 5-fluorouracil and high dose folinic acid, *J. Clin. Oncol.*, 4, 685, 1986.
57. **Varini, M.,** Ifosfamide in tumor therapy — an overview, in *Contributions to Oncology*, Vol. 26, Brade, W., Nagel, G. A., and Seeber, S., Eds., Karger, Basel, 1987, 12.
58. **Berger, D. P., Fiebig, H. H., Winterhalter, B. R., Wallbrecher, E., and Henss, H.,** Preclinical phase II study of ifosfamide in human tumour xenografts *in vivo*, *Cancer Chemother. Pharmacol.*, 26(Suppl.), 7, 1990.
59. **Osieka, R.,** Studies on drug resistance in a human melanoma xenograft system, *Cancer Treat. Rev.*, 11(Suppl. A), 85, 1984.
60. **Boven, E., Calame, J. J., Molthoff, C. F. M., and Pinedo, H. M.,** Characterization and chemotherapy of human soft tissue sarcoma (STS) lines grown in nude mice, *Strahlenther. Onkol.*, 165, 538, 1989.
61. **Berkowitz, R. S., Goldstein, D. P., and Bernstein, M. R.,** Ten year's experience with methotrexate and folinic acid as primary therapy for gestational trophoblastic disease, *Gynecol. Oncol.*, 23, 111, 1986.
62. **Von Hoff, D. D., Rozencweig, M., Louie, A. C., Bender, R. A., and Muggia, F. M.,** "Single" agent activity of high-dose methotrexate with citrovorin factor rescue, *Cancer Treat. Rep.*, 62, 233, 1978.
63. **Edmonson, J. H., Creagan, E. T., and Gilchrist, G. S.,** Phase II study of high dose methotrexate in patients with unresectable metastatic osteosarcoma, *Cancer Treat. Rep.*, 65, 538, 1981.
64. **Taylor, S. G., McGuire, W. P., Hauck, W. W., Showel, J. L., and Lad, T. E.,** A randomized comparison of high-dose infusion methotrexate versus standard-dose weekly therapy in head and neck squamous cancer, *J. Clin. Oncol.*, 2, 1006, 1984.
65. **Bruckner, H. W., Lockich, J. J., and Stablein, D. M.,** Studies of Baker's antifol, methotrexate and razoxane in advanced gastric cancer: a Gastrointestinal Tumor Study Group report, *Cancer Treat. Rep.*, 66, 1713, 1982.
66. **Inaba, M., Kobayashi, T., Tashiro, T., Sakurai, Y., Maruo, K., Ohnishi, Y., Ueyama, Y., and Nomura, T.,** Evaluation of antitumor activity in a human breast tumor/nude mouse model with a special emphasis on treatment dose, *Cancer*, 64, 1577, 1989.
67. **Crooke, S. T. and Bradner, W. T.,** Mitomycin C: a review, *Cancer Treat. Rev.*, 3, 121, 1976.
68. **Creech, R. H., Shah, M. K., Catalano, R. B., Dierks, K., Dayal, H., and Goldberg-Alberts, R.,** Phase II study of low-dose mitomycin in patients with ovarian cancer previously treated with chemotherapy, *Cancer Treat. Rep.*, 69, 1271, 1985.
69. **Joss, R. A., Obrecht, J. P., Senn, H. J., Alberto, P., Sauter, C., and Cavalli, F.,** Phase II trial of mitomycin-C in patients with small cell carcinoma of the lung after failure on combination chemotherapy, *Eur. J. Cancer Clin. Oncol.*, 20, 1477, 1984.
70. **Dombernowski, P., Hansen, H. H., and Sorenson, P. G.,** Vincristine (NSC-67574) in the treatment of small cell anaplastic carcinoma of the lung, *Cancer Treat. Rep.*, 60, 239, 1976.
71. **Tashiro, T., Inaba, M., Kobayashi, T., Sakurai, Y., Maruo, K., Ohnishi, Y., Ueyama, Y., and Nomura, T.,** Responsiveness of human lung cancer/nude mouse to antitumor agents in a model using clinically equivalent doses, *Cancer Chemother. Pharmacol.*, 24, 187, 1989.
72. **Inaba, M., Tashiro, T., Kobayashi, T., Sakurai, Y., Maruo, K., Ohnishi, Y., Ueyama, Y., and Nomura, T.,** Responsiveness of human gastric tumors implanted in nude mice to clinically equivalent doses of various antitumor agents, *Jpn. J. Cancer Res. (Gann)*, 79, 517, 1988.
73. **Houghton, J. A. and Houghton, P. J.,** The suitability and use of human tumor xenografts, in *Rodent Tumor Models in Experimental Cancer Therapy*, Kallman, R. F., Ed., Pergamon Press, New York, 1987, 199.
74. **Bailey, M. J., Gazet, J. C., Smith, I. E., and Steel, G. G.,** Chemotherapy of human breast carcinoma xenografts, *Br. J. Cancer*, 42, 530, 1980.
75. **Fujita, M., Hayata, S., and Taguchi, T.,** Relationship of chemotherapy on human cancer xenografts in nude mice to clinical response in donor patient, *J. Surg. Oncol.*, 15, 211, 1980.
76. **Fiebig, H. H., Schuchhardt, C., Henss, H., Fiedler, L., and Löhr, G. W.,** Comparison of tumor response in nude mice and in the patients, *Behr. Inst. Mitt.*, 74, 343, 1984.
77. **Fujimoto, S., Igarashi, K., Shrestha, R. D., Miyazaki, M., and Okui, K.,** Antitumor effects of two polyamine antimetabolites combined with mitomycin C on human stomach cancer cells xenotransplanted into nude mice, *Int. J. Cancer*, 35, 821, 1985.
78. **Kikuchi, Y., Oomori, K., Kizawa, I., Hirata, J., Kita, T., Miyauchi, M., and Kato, K.,** Enhancement of antineoplastic effects of cisplatin by calmodulin antagonists in nude mice bearing human ovarian carcinoma, *Cancer Res.*, 47, 6459, 1987.
79. **Ohkawa, K., Tsukada, Y., Dohzono, H., Koike, K., and Terashima, Y.,** The effects of co-administration of selenium and *cis*-platin (CDDP) on CDDP-induced toxicity and antitumour activity, *Br. J. Cancer*, 58, 38, 1988.

80. **Fiebig, H. H.,** Entwicklung neuer synergistischer Zytostatikakombinationen an Magenxenografts, in *Gastrointestinale Tumoren, Aktuelle Onkologie,* Vol. 33, Queiβer, W. and Flechtner, H., Eds., W. Zuckschwerdt, München, 1986, 26.
81. **Chahinian, A. P., Norton, L., Holland, J. F., Szrajer, L., and Hart, R. D.,** Experimental and clinical activity of mitomycin C and *cis*-diamminedichloroplatinum in malignant mesothelioma, *Cancer Res.,* 44, 1688, 1984.
82. **Schlappack, O. K., Bush, C., Delic, J. I., and Steel, G. G.,** Growth and chemotherapy of a human germ-cell tumour line (GCT-27), *Eur. J. Cancer Clin. Oncol.,* 24, 777, 1988.
83. **Wayss, K., Herrmann, R., Mattern, J., and Volm, M.,** Sequential methotrexate (MTX) and 5-fluorouracil (FU) in human tumor xenografts, *Med. Oncol. Tumor Pharmacother.,* 2, 27, 1985.
84. **Kondo, H., Tanaka, N., and Orita, K.,** Antitumor effect of human recombinant interferon gamma alone and in combination with adriamycin on a human colon cancer cell line (RPMI 4788) *in vitro* and in nude mice, *Jpn. J. Cancer Res. (Gann),* 78, 1400, 1987.
85. **Sklarin, N. T., Chahinian, A. P., Feuer, E. J., Lahman, L. A., Szrajer, L., and Holland, J. F.,** Augmentation of activity of *cis*-diamminedichloroplatinum(II) and mitomycin C by interferon in human malignant mesothelioma xenografts in nude mice, *Cancer Res.,* 48, 64, 1988.
86. **Klapdor, R., Franke, N., and Bahlo, M.,** Combined therapy of xenografts of human pancreatic carcinomas with rTNF-α and mitomycin C, *Onkologie,* 12, 143, 1989.
87. **Arai, T., Okamoto, K., Terao, K., Tokita, H., Sekimoto, K., Fukushima, T., and Tanaka, N.,** Effectiveness of antitumor agents against human tumors in nude mice, in *Proc. 2nd Int. Workshop on Nude Mice,* Nomura, T., Ohsawa, N., Tamaoki, N., and Fujiwara, K., Eds., G. Fischer, Stuttgart, 1977, 461.
88. **Kyriazis, A. P., Yagoda, A., Kyriazis, A. A., and Fogh, J.,** Response of nude mouse-grown human urothelial cancer to *cis*-diammine-dichloroplatinum(II), diammine[1,1-cyclobutanedi-carboxylato(2-)-0,0′-platinum] and mitoguazone dihydrochloride, *Cancer Res.,* 45, 2012, 1985.
89. **Russell, P. J., Raghavan, D., Gregory, P., Philips, J., Wills, E. J., Jelbart, M., Wass, J., Zbroja, R. A., and Vincent, P. C.,** Bladder cancer xenografts: a model of tumor cell heterogeneity, *Cancer Res.,* 46, 2035, 1986.
90. **Braakhuis, B. J. M. and Snow, G. B.,** Activity of conventional drugs in head and neck cancer xenografts, in *Human Tumour Xenografts in Anticancer Drug Development,* Winograd, B., Peckham, M. J., and Pinedo, H. M., Eds., Springer-Verlag, Berlin, 1988, 37.
91. **Kleine, W., Teufel, G., Schwörer, D., and Pfleiderer, A.,** Experimental chemotherapy of human malignant ovarian carcinomas xenotransplanted in nude mice, in *Thymus Aplastic Nude Mice and Rats in Clinical Oncology,* Baster, G., Fortmeyer, H. P., and Schmidt-Matthiesen, H., Eds., G. Fischer, Stuttgart, New York, 1981, 117.
92. **Nakamura, T. and Kitagawa, T.,** Anticancer drug screening test with LDH in nude mouse bearing bone and soft tissue sarcoma, *Cancer,* 56, 1112, 1985.
93. **Harstrick, A., Casper, J., Wilke, H., Poliwoda, H., and Schmoll, H. J.,** Comparison of the antitumor activity of cisplatin, carboplatin and iproplatin against established human testicular cancer cell lines *in vivo* and *in vitro, Cancer,* 63, 1079, 1989.
94. **Pillay, C. V., Green-Thompson, R., and Brock-Utne, J. G.,** Efficacy of the anticancer agent cisplatin in the treatment of human cervical squamous carcinoma xenografted in nude mice, *Chemotherapy,* 32, 356, 1986.
95. **Fujita, M., Fujita, F., Kimoto, Y., Shimozuma, K., and Taguchi, T.,** Experimental chemotherapy of human gastrointestinal and breast cancer xenografts in nude mice with drugs in clinical use and under development, and its predictability of clinical effects, in *Recent Advances in Chemotherapy,* Ishigami, J., Ed., University of Tokyo Press, Tokyo, 1985, 349.
96. **Guiliani, F. C., Zirvi, K. A., and Kaplan, N. O.,** Therapeutic response of human tumor xenografts in athymic nude mice to doxorubicin, *Cancer Res.,* 41, 325, 1981.
97. **Guiliani, F. C. and Kaplan, N. O.,** New doxorubicin analogs active against doxorubicin-resistant colon tumor xenografts in the nude mouse, *Cancer Res.,* 40, 4682, 1980.
98. **Boerrigter, G. H., Heinerman, E. C. M., Braakhuis, B. J. M., and Snow, G. B.,** Chemosensitivity of human head and neck cancer xenografts in the clonogenic assay and in nude mice, *Br. J. Cancer,* 54, 53, 1986.
99. **Fodstad, Ø., Aamdal, S., and Pihl, A.,** The usefulness of the nude mouse model for predicting the chemosensitivity of human tumors. Studies in malignant melanoma, in *Human Tumour Xenografts in Anticancer Drug Development,* Winograd, B., Peckham, M. J., and Pinedo, H. M., Eds., Springer-Verlag, Berlin, 1988, 63.
100. **Guiliani, F. C., Zirvi, K. A., Kaplan, N. O., and Goldin, A.,** Chemotherapy of human colorectal tumor xenografts in athymic mice with clinically active drugs: 5-fluorouracil and 1-3-*bis*-(2-chloroethyl)-1-nitrosourea (BCNU). Comparison with doxorubicin derivatives: 4′-deoxydoxorubicin and 4′-0-methyldoxorubicin, *Int. J. Cancer,* 27, 5, 1981.
101. **Otto, U., Klöppel, G., and Baisch, H.,** Transplantation of human renal cell carcinoma into NMRI nu/nu mice, *J. Urol.,* 131, 130, 1984.
102. **Siegel, M. M., Chung, H. S., Rucker, N., Siegel, S. E., Seeger, R. C., Isaacs, H., and Benedict, W. F.,** *In vitro* and *in vivo* preclinical chemotherapy studies of human neuroblastoma, *Cancer Treat. Rep.,* 64, 975, 1980.

Chapter 14

ANALOGS OF CONVENTIONAL AGENTS

Epie Boven

TABLE OF CONTENTS

I. INTRODUCTION

A difficult area for decision making involves the selection of new analogs of conventional cytostatic agents for clinical trials. A number of preclinical criteria may be used to develop these analogs for quantitative or qualitative superiority over the parent coumpound in properties based on pharmacology, therapy, or toxicology. In this determination, specific attention can be focused on those characteristics which are limiting for the parent drug currently in clinical use. Goldin et al.[1] have listed such criteria for the selection of new analogs to proceed to clinical trials:

1. Greater activity in appropriate test systems relative to the parent drug
2. Diminished acute, chronic or delayed host toxicity
3. Broader spectrum of efficacy
4. Lack of cross-resistance to the parent drug
5. Specific organ localization
6. Favorable pharmacological characteristics
7. Superiority over the parent drug in combined modalities

Activity screening of analogs of conventional agents is usually carried out in the well-known transplantable rodent tumor systems, where the percentage increase in life span of treated animals vs. controls is the endpoint of the study.[2] S.c. implanted tumor models, such as human tumor xenografts, may offer better insight into the extent of growth inhibition obtained by the analog as compared to that of the parent compound. As for any promising antitumor agent, and also for analogs, the incorporation of a series of human tumor lines of the same tumor type in the screening panel may indicate specific or additional groups of cancer patients who will potentially benefit from the drug under investigation.[3] In this respect, Harrap et al.[4] have shown that individual tumor response patterns for analogs in a group of human tumor lines may reflect a structural uniqueness in the molecules, that is, the lack of cross-resistance to the parent drug.

II. PLATINUM ANALOGS

Cisplatin [*cis*-diamminodichloroplatinum(II)] has proven to be effective in a variety of human tumor types, such as head and neck cancer, small cell lung cancer, and cancer of the testis, ovary, cervix, and bladder.[5] Major toxicities induced by cisplatin are nausea and vomiting, nephrotoxicity, ototoxicity, and neurotoxicity. To date, numerous modifications have been made to both the chloride leaving groups and amine ligands of platinum (II) and platinum (IV) compounds in a search for alternative active drugs with reduced toxicity. An additional rationale for the development of platinum analogs is the lack of cross-resistance to the parent compound.[6]

In the selection of platinum analogs with a potentially improved therapeutic index, preclinical antitumor activity screening has been generally carried out in the well-known transplantable rodent tumor systems such as L1210 ascites leukemia, Yoshida ascites sarcoma, and the s.c.-grown B16 melanoma, Lewis lung cancer, Madison 109 lung cancer, and colon 26 tumor.[7-9] To provide insight into analogs without cross-resistance to cisplatin, sublines resistant to alkylating agents or to the parent drug were incidentally incorporated into the screening system.[8,9] The endpoint of these studies was either the percentage increase in life span of treated animals vs. controls or merely the determination of surviving tumor cells in the peritoneal cavity. Toxicity studies have included myelosuppression and renal damage. These approaches resulted in the introduction of several analogs in the clinic.

To date, the most important and best of cisplatin analogs is carboplatin [*cis*-diammine-1,1-cyclobutanedicarboxylatoplatinum (II)]. Clinical trials started in 1981 have shown activity in ovarian cancer, small cell lung cancer, head and neck cancer, and testicular cancer.[10] In comparative clinical trials thus far completed, carboplatin shows equal antitumor activity to cisplatin in ovarian cancer.[11,12] Due to its reduced non-hematological toxicity profile, carboplatin is now considered the platinum compound of choice in the first-line treatment of this tumor type. The analog iproplatin [*cis*-dichloro-*bis*-(isopropylamine)-*trans*-dihydroxyplatinum (IV)] has also shown promising results in ovarian cancer patients.[13] Unfortunately, a number of other analogs introduced into clinical trials had to be discarded from general application in cancer patients because of the absence of a better therapeutic index as compared to cisplatin. Examples of such analogs are spiroplatin [*cis*-1,1-di(aminomethyl) cyclohexaneplatinum (II) sulphate] and JM-40 [*cis*-ethylenediaminoplatinum (II) malonate] with preclinical activity in rodent tumors comparable to cisplatin, but both showed nausea and vomiting as well as nephrotoxicity at maximum tolerated doses in patients.[14,15]

The use of the nude mouse-human tumor model may be of help in the secondary screening of platinum analogs in order to better predict a higher response rate in specific tumor types in the clinical situation. In our laboratory, we have compared the activity of promising platinum analogs in human ovarian cancer xenografts for many years.[16-18] In this approach, drugs are administered at doses equitoxic to cisplatin and in the same schedule. Initially, two tumor lines were used, of which one is highly sensitive to cisplatin and in another cisplatin is hardly effective. We found carboplatin at least as effective as cisplatin in both tumor lines, but lower efficacy for another three analogs, that is, JM-40, spiroplatin, and iproplatin. Biodistribution of platinum in tumors and organs of the mice measured 24 h after administration did not predict for antitumor activity or organ-related side effects.[16] Additional studies with TNO-compounds synthesized in the Netherlands[17] indicated TNO-38 [(1,1-*bis*(aminomethyl)-cyclohexane)oxalatoplatinum(II)] as the best active drug, but with an overall activity less than cisplatin. This result was obtained in a comparative study in seven tumor lines of ovarian cancer and a major reason not to consider TNO-38 for clinical evaluation.[19] Also, inferior efficacy was observed with tetraplatin [tetrachloro(*d,l-trans*)1,2-diaminocyclohexane platinum(IV)], an analog with presumed activity against cisplatin-resistant disease.[18,20]

In human tumor xenografts obtained from testicular cancer cell lines, cisplatin proved to be more effective than carboplatin and iproplatin at equitoxic doses, as studied by Harstrick et al.[21] However, the drug schedule was five times daily i.p., while in our laboratory, drugs were given two times weekly i.v. Total doses of drugs given five times daily were 15, 62.5, and 21.5 mg/kg, for cisplatin, carboplatin, and iproplatin, respectively, while our equitoxic doses for weekly injections were, respectively, 5, 60, and 40 mg/kg. The proportionally higher dose of cisplatin in testicular cancer xenografts may have been the reason for the relatively inferior efficacy of carboplatin. Clinical studies of the latter drug in testicular cancer patients are suggestive of high efficacy.[10]

In a large study by the group of Harrap et al.,[4] 23 serially transplantable human ovarian cancer lines were characterized, including their chemosensitivity for cisplatin, carboplatin, iproplatin, and tetraplatin at equitoxic doses i.p. twice weekly to four times weekly in 16 of these lines. Respective doses were 8, 100, 60, and 8 mg/kg, proportionally well related to our doses. Again, for carboplatin, high efficacy was observed which was comparable to that of cisplatin (Table 1). Also, iproplatin was demonstrated to have antitumor activity in the model. Unfortunately, tetraplatin induced growth inhibition only in a limited number of lines (all cisplatin sensitive). Future clinical trials have to elucidate whether tetraplatin has a place in the treatment of cisplatin-resistant (ovarian) cancer.

Recently, Foster et al.[22] have proposed a strategy for the further clinical development of carboplatin and iproplatin based on the vast amount of data obtained from *in vitro* and *in vivo* tumor models, including human tumor xenografts, and from phase I clinical trials. The scientific questions with regard to the relative activity of each analog, as compared to others and to cisplatin, center around the relative therapeutic index (antitumor effects vs. acute and chronic side effects). The relative therapeutic index may only be determined upon combination of the results for relative disease-specific antitumor activity and relative tolerance obtained from phase

TABLE 1
Growth Inhibition (%) by Platinum Compounds in Human Ovarian Cancer Lines

Tumor line	Cisplatin	Carboplatin	Iproplatin	Tetraplatin
PXN/65	96	100	100	100
PXN/100	100	100	100	100
HX/62	55	56	16	60
HX/109	21	65	62	41
PXN/94/C	79	77	81	71
SKOV-3	51	71	65	42
OVCAR-3(O)	49	0	88	67
HX/61	98	74	89	32
PXN/95	90	86	87	0
PA1	83	94	46	32
OVCAR-3(A)	79	86	97	44
PXN/109	44	87	93	68
HX/110	100	99	77	43
PXN/94	85	95	99	44
PXN/87	94	90	94	44
PXN/109/TC	100	91	90	63
Growth inhib. >90%	7/16	7/16	7/16	2/16

From Harrap, K. R., Jones, M., Siracky, J., Pollard, L. A., and Kelland, L. R., *Ann. Oncol.*, 1, 65, 1990. With permission.

II and phase III clinical trials. The proposed methodology should ultimately determine the best platinum analog for each particular disease.

III. ANTHRACYCLINES AND RELATED AGENTS

Doxorubicin is a glycoside antibiotic originally isolated by aerobic fermentation of *Streptomyces peucetius* var. *caesius* and is an analog of daunomycin.[23,24] While daunomycin has shown clinical activity against acute leukemia, doxorubicin is also active against Hodgkin's disease, lymphomas, and a number of solid tumor types such as breast cancer, ovarian cancer, soft tissue sarcomas, and small cell lung cancer. Acute toxicities of the drug include nausea and vomiting, myelosuppression, mucositis, and alopecia. The major long-term, dose-limiting toxicity of doxorubicin is the development of cardiomyopathy to be expected at cumulative doses greater than 550 mg/m^2.

For many years, the search for analogs or related compounds of doxorubicin has aimed at the development of drugs with similar antitumor activity, but diminished cardiotoxicity.[25] As for platinum analogs, the efficacy of these newly synthesized drugs is usually being compared with that of the parent compound in mouse tumor systems, such as P388 and L1210 leukemia, B16 melanoma, and CD8F1 mammary tumor.[23,24] Therapeutic effects, similar to or better than doxorubicin, are usually a reason to proceed with the toxicology studies of the drug, such as the measurement of cardiac lesions induced by chronic dosing in rabbits or beagle dogs.

Presently, a series of doxorubicin and related compounds have been evaluated for toxicity and antitumor activity in clinical trials. Some of these drugs [i.e., epirubicin (4′epidoxorubicin) and mitoxantrone (dihydroxy-anthracenedione dihydrochloride)] have acquired a place in the therapeutic armentarium of antitumor agents.[26-28] Both drugs have shown activity in leukemias, lymphomas, and breast cancer, while epirubicin is also active in ovarian cancer and soft tissue sarcomas. As was predicted from preclinical toxicology analysis of cardiac injury in animals, for epirubicin the incidence of endomyocardial changes in patients occurred at higher cumulative doses and was less pronounced than for doxorubicin.[29] For mitoxantrone, cardiotoxicity is considerably lower when compared to doxorubicin, whereas additional, favorable aspects include a lower incidence of gastrointestinal toxicity and alopecia.[30] Unfortunately, a number of doxorubicin analogs or related agents have been discarded from further clinical evaluation because of minimal or absent responses in patients. An example is given by menogaril (7-con-*O*-methylnogarol), a fermentation-derived antitumor agent of the anthracycline type with low cardiotoxicity in rabbits as compared to doxorubicin-treated controls.[31] Nevertheless, menogaril presents a favorable activity pattern in the well-known mouse tumor systems,[32] and clinical evaluation of menogaril has shown only marginal activity in breast cancer[33] and no responses in platinum-resistant ovarian cancer.[34]

While the nude mouse-human tumor model will not give insight into the potential cardiotoxicity of anthracyclines and related agents, the model may add to information on the differences in antitumor activity for the various drugs to be expected in the clinical situation. Mattern et al.[35] have summarized a large amount of data generated by various investigators from studies with chemotherapy in human tumor xenografts. For doxorubicin, they have shown sensitivity in tumor lines derived from breast cancer (11 of 19 lines), small cell lung cancer (1 of 3), ovarian cancer (1 of 6), stomach cancer (3 of 26), non-small cell lung cancer (1 of 13), melanoma (2 of 4), and inactivity in lines derived from colon cancer (0 of 14). Except for human melanoma lines, the overall pattern of responsiveness reflects that obtained with the drug in phase II trials. Evaluation of doxorubicin in these melanoma xenografts had been carried out by Guiliani et al.,[36] who indicated the lines as probably useless because of the generation of false positive drugs.

The group of Guiliani[37-41] has acquired extensive experience in the comparative analysis of the activity of doxorubicin and analogs or related agents in human tumor xenografts. In an early report in 1980,[37] they compared doxorubicin with epirubicin, esorubicin (4′-deoxydoxorubicin) and 4′-*O*-methyldoxorubicin in five lines of colon cancer and two lines of rectal cancer. While doxorubicin and epirubicin were inactive in all xenografts, both esorubicin and 4′-*O*-methyldoxorubicin appeared effective in 4 of 5 colon cancer lines and inactive in the 2 rectal cancer lines. Overall, the activity of 4′-*O*-methyldoxorubicin was less than that of esorubicin. Additional studies with esorubicin have shown a narrower spectrum of activity in the various tumor types, with xenografts of colon cancer and breast cancer being most sensitive.[38,41] In phase II clinical trials, esorubicin appeared only slightly active in breast cancer.[41,42] With epirubicin as the first doxorubicin analog of clinical importance, a comparative analysis of its activity was carried out with that of the parent drug in xenografts of seven different tumor types.[40] From Table 2 it is clear that epirubicin is effective in the same tumor types as doxorubicin.

For an analog of daunomycin, idarubicin (4-demethoxydaunorubicin), the antitumor activity in human breast cancer xenografts was found less than that of doxorubicin.[41] In contrast, idarubicin was found to have sufficient antileukemic activity in both acute lymphocytic leukemia and acute non-lymphocytic leukemia to warrant a prospective comparison with the conventional anthracyclines, daunomycin and/or doxorubicin, in phase III combination chemotherapy trials in this type of disease.[43]

Based on their experience in anthracycline-analog screening in human tumor lines, the group of Guiliani[41] has found their human breast cancer xenografts as best predictive for tumor responsiveness in patients.

In our laboratory, we have compared doxorubicin, epirubicin, mitoxantrone, and menogaril in eight human tumor lines derived from ovarian cancer.[44] At equitoxic doses, doxorubicin appeared best effective, followed by epirubicin and mitoxantrone. Menogaril induced the lowest growth inhibition. From mitoxantrone and menogaril, proportionally higher doses could be administered to mice as compared to patients. Thus far, epirubicin has also shown responses in ovarian cancer first-line treatment.[27] Minor activity was observed for mitoxantrone in this type of disease,[45-47] whereas menogaril was inactive.[34]

TABLE 2
Antitumor Activity of Doxorubicin and Epirubicin in Human Tumor Lines

Tumor line	Doxorubicin[1]	Epirubicin[1]
Breast	5/5	4/4
Lung		
Non-small cell	3/4	3/4
Small cell	1/1	1/1
Melanoma	3/4	4/4
Ovary	0/2	2/2
Soft tissue sarcoma	1/1	1/1
Rectum	0/2	0/2
Colon	3/10	1/5

[1] Number of sensitive lines (growth inhibition >50%)/lines tested.

From Guiliani, F. C. and Kaplan, N. O., Immune-Deficient Animals, in *Proc. 4th Int. Workshop on Immune-Deficient Animals in Exp. Res.*, Sordat, B., Ed., Karger, Basel, 1984, 374. With permission.

IV. CYCLOPHOSPHAMIDE AND IFOSFAMIDE

Cyclophosphamide and ifosfamide belong to the group of oxazaphosphorine cytostatic agents which have been synthesized to convert the reactive nitrogen mustard group in an inactive transport form.[48] Both drugs must be activated via hydroxylation by liver P-450 microsomes to demonstrate alkylating activity. For cyclophosphamide, the decrease in reactivity of the beta-choroethyl side chain of nitrogen mustard was obtained by substitution of a strongly electrophilic group for a methyl group.[49] While cyclophosphamide is regarded as a true derivative of nitrogen mustard because both chloroethyl groups are attached to the same exocyclic nitrogen, for ifosfamide one of the two chloroethyl groups is shifted to the endocyclic nitrogen in the molecule (Figure 1).

Cyclophosphamide has been recognized widely for its cytostatic activity in a number of tumor types, such as leukemias, lymphomas, breast cancer, soft tissue sarcomas, ovarian cancer, and small cell lung cancer. For ifosfamide, its toxic effects to the urothelial system have initially hampered clinical use of the drug.[49,50] The bladder lesions induced by the oxazaphosphorine compounds are now known to be primarily due to the metabolite acrolein. A number of methods have been investigated to reduce urothelial toxicity. Mesna was found to be the best systemic uroprotector selected from a large number of thiols and thio compounds as agents binding acrolein and the activated drug 4-hydroxy-ifosfamide in the urinary system. Clinical studies have now shown that ifosfamide has advantages over cyclofosfamide, such as a better water-solubility and reduced myelotoxicity. However, toxicity to renal tubules and central nervous system side effects have been incidentally reported. Ifosfamide in combination with mesna can be given in doses at least five times higher than cyclophosphamide doses.[51] Only recently, the drug has undergone wider clinical analysis and has been found more active than cyclophosphamide in soft tissue sarcomas, highly active in testicular cancer, lymphomas, small cell lung cancer, breast cancer, and ovarian cancer.[50,51] Comparative randomized combination chemotherapy trials are presently ongoing in various types of alkylating-agent sensitive malignant diseases.

Cyclophosphamide and ifosfamide have been compared for their antitumor activity in a large number of rodent tumor systems such as in the mouse: L1210 leukemia, Lewis lung cancer, C3H mammary tumor, Ridgway osteosarcoma, colon tumors 26, and 38, B16 melanoma and in the rat (Yoshida ascites sarcoma).[52] The general conclusion for both drugs was their possession of a broad spectrum of activity in these animal tumor systems, while ifosfamide appeared at least

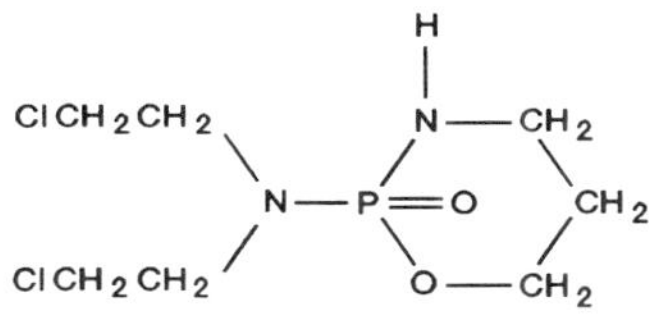

cyclophosphamide

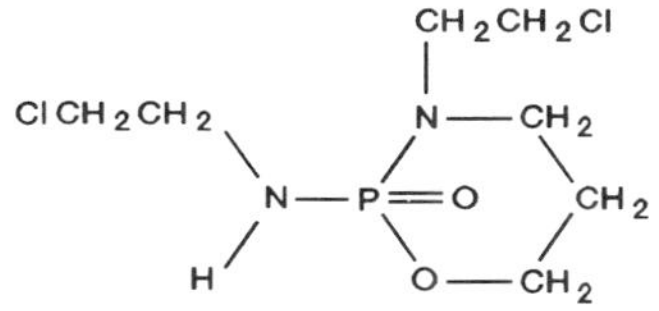

ifosfamide

FIGURE1. Structures of cyclophosphamide and ifosfamide.

similarly active and sometimes more effective than cyclophosphamide. Also, in the animals, it was demonstrated that mesna uroprotection did not decrease antitumor activity and could limit urological lethal toxicity from ifosfamide.

Comparative activity studies of cyclophosphamide and ifosfamide have been carried out in a number of human tumor xenografts. In a large analysis of the chemosensitivity of the human breast cancer xenograft model MX-1 by Inoue et al.,[53] it was found that a single i.v. dose of cyclophosphamide of 220 mg/kg could induce complete tumor responses in all mice. Ifosfamide at a dose of 50 mg/kg i.v. resulted in a maximum growth inhibition of 87%, while 75 mg/kg i.v. could not be given for reasons of excessive toxicity. Activity for ifosfamide was shown in a human medulloblastoma line by Friedman et al.[54] Single i.p. doses for cyclophosphamide and ifosfamide were 1391 and 2014.9 mg/m^2, respectively, resulting in a growth inhibition of 82 and 73%, respectively. Berger et al.[55] have compared the activity of cyclophosphamide 200 mg/kg i.p. given every 2 weeks ¥2 and ifosfamide given i.p. or s.c. in various doses and schedules in 30 different human tumor xenografts, and found tumor regressions or remissions in 33 and 43% of the lines, respectively. The schedule for ifosfamide most often used was 130 mg/kg i.p. or s.c. given on days 1 to 3 and 15 to 17. In our laboratory, we have compared the efficacy of both drugs in human soft tissue sarcoma lines where equitoxic doses of cyclophosphamide and ifosfamide were, respectively, 150 mg/kg and 250 mg/kg given i.p. every 2 weeks ¥2.[56] An update of this study is depicted in Table 3, and activity is shown for both drugs.

In general, the nude mouse-human tumor model has not indicated ifosfamide as a drug more active than cyclophosphamide, which may be contrary to clinical data, for example, in soft tissue sarcoma.[57] However, cyclophosphamide and ifosfamide are examples of drugs where maneuvers to prevent toxicity may explain the discrepancy between preclinical results and data from the clinic. In the studies mentioned, nude mice tolerated ifosfamide at a dose only 1.5 to 2 times higher than that of cyclophosphamide, while with mesna uroprotection, doses in patients can be increased at least 5 times.

TABLE 3
Growth Inhibition (%) at Equitoxic Doses of Cyclophosphamide and Ifosfamide in Human Soft Tissue Sarcoma Xenografts

Tumor line	Histology	Cyclo-phosphamide[1]	Ifosfamide[1]
S.To	Synovial cell sarcoma	35	55 (+)
S.Lt	Synovial cell sarcoma	51 (+)	30
S.Zu	Mal. fibrous histiocytoma	60 (+)	59 (+)
S.Ho	Mal. fibrous histiocytoma	80 (++)	87 (++)
S.La(C)	Mal. fibrous histiocytoma	29	29
S.Ba	Liposarcoma	86 (++)	46
WLS-160	Liposarcoma	81 (++)	76 (++)
S.Hu	Leiomyosarcoma	0	12
S.Hh	Leiomyosarcoma	32	19

[1] Growth inhibition >50% +; >75% ++

V. MITOMYCIN C AND ANALOGS

Mitomycin C is an antitumor antibiotic isolated from *Streptomyces caespitosus* and is clinically active in cancer of the breast, stomach, and uterine cervix.[58,59] Mitomycin C is known to have a delayed and cumulative myelosuppression. Incidentally, a dose-related interstitial pneumonitis and the hemolytic-uremic syndrome may occur.[60,61] Upon analysis of the molecular structure by Bradner et al.,[62] it was found that the intact mitosane was necessary to maintain potency and antitumor activity. Superior antitumor effects for mitomycin analogs in mouse tumor systems, such as L1210 and P388 leukemia, B16 melanoma, and M109 lung cancer, were reported for a number of compounds such as RR-150 (7-cysteaminomitosane) and M-83 [7-N-(*p*-hydroxyphenyl)mitomycin].[62,63] It has been somewhat more difficult to identify an active mitomycin analog with a clearly reduced myelosuppressive action.[64] In a recent summary on the structure activity of more than 600 analogs by Bradner et al.,[65] it was concluded that the best retention of potency and antitumor activity was obtained by modification of the 7-position of the quinone moiety, while leaving the aziridinyl end of the molecule intact. In the past, only three analogs were analyzed in early clinical trials and have not been pursued further.

Mitomycin analogs have not been routinely studied in human tumor xenografts. In a report by Asanuma,[66] the M-83 derivative was found effective in 6 of 8 xenografts of various origins. The activity pattern was slightly inferior to that of mitomycin C. As mentioned before, antitumor activity of mitomycin analogs may be retained, but improvement of the therapeutic index should include reduction of hematological toxicity.

VI. ANTIMETABOLITES

A. ANTIFOLATES

Methotrexate (4-amino-4-deoxy-N^{10}-methyl folic acid) is a folate antagonist with antitumor activity in leukemias, lymphomas, breast cancer, head and neck squamous cell cancer, osteosarcoma, and choriocarcinoma.[67] The drug binds tightly but reversibly to dihydrofolate reductase, thereby inhibiting this enzyme from converting folic acid to tetrahydrofolate. The enzyme block results in depletion of reduced folates necessary for the *de novo* synthesis of thymidylate and purine nucleotides, essential precursors of DNA and RNA. It is the inhibition of thymidylate synthesis necessary for DNA which is thought to be the most important mechanism of methotrexate cytotoxicity.

Improvement of the selectivity of methotrexate by synthesis of analogs has usually been sought in the inhibition of other folate-dependent enzymes, such as thymidylate synthase, to circumvent intrinsic or acquired resistance to methotrexate.[68] Preclinical antitumor efficacy has been assessed in *in vitro* cell lines sensitive to methotrexate or resistant via defective transport mechanisms, as well as murine tumor systems. Trimetrexate, a 2,4-diaminoquinazoline derivative, was shown to have a broader spectrum of activity than methotrexate using this screening approach.[69] As with methotrexate, the drug appeared to be inactive in the human tumor lines MX-1, LX-1, and CX-1. Presently, trimetrexate has entered clinical trials.[70] Of new folate analogs, 10-EdAM (10-ethyl-10-deaza-aminopterin) was found to have antitumor activity in MX-1, LX-1, and CX-1.[71]

In general, it appears that human tumor xenografts may be responsive to folate antagonists and therefore used to extend further information on antitumor properties possibly superior to methotrexate. Braakhuis et al.[72] have studied the efficacy of methotrexate in 16 human tumor lines derived from head and neck cancer. Unexpectedly, they found a lack of effect, while the drug is known to induce a response rate of 50% in this tumor type in the clinic. This discrepancy may be based on differences between the hosts, as described in Chapter 21. The same group[73] have recently reported on 10-EdAM showing activity in 2 of 5 head and neck xenografts, whereas methotrexate was completely inactive. The results of phase III randomized trials comparing both drugs in patients with this tumor type are awaited with interest.

B. CYTIDINE ANALOGS

The cytidine antimetabolites, ara-C (arabinosyl-5-azacytosine) and 5-azacytidine, are effective antitumor agents in the treatment of acute leukemias. Cytidine analogs have been developed to generate compounds with a broader spectrum of activity, that is, against solid tumor types. In this respect, arabinofuranosyl-5-azacytosine showed regression in a variety of human tumor xenografts, including leukemias and solid tumors resistant to ara-C.[74] Recently, 5-aza-2′-deoxycytidine was entered into clinical trials.[75,76] Braakhuis et al.[77] have shown activity of 5-aza-2′-deoxycytidine in 3 of 5 human tumor xenografts derived from head and neck cancer. Except for activity in leukemias, the drug has thus far proven inactive in phase II clinical trials, including head and neck cancer. Another analog, difluorodeoxycytidine (2′,2′-difluorodeoxycytidine), was also introduced into the clinic because of potential activity in solid tumor types grown in rodents. In addition, we have found growth inhibition in 1 of 3 ovarian cancer and 2 of 4 soft tissue sarcoma xenografts.[78] Phase II clinical trials with the drug have yet to be completed.

C. FLUOROPYRIMIDINES

The fluorinated pyrimidine 5-fluorouracil has been found active against many types of solid tumors, including breast, colorectal, and head and neck cancer. For colorectal cancer, 5-fluorouracil is still the drug of choice, but the response rate is only 20 to 25%. The drug has multiple biochemical actions that may be responsible for its cytotoxicity. After the metabolic conversion to fluorouridine monophosphate, either fluorouridine triphosphate will be formed which is incorporated into RNA, or 5-FdUMP is formed which binds tightly to thymidylate synthase and inhibits the eventual formation of deoxythymidine triphosphate, a necessary precursor of DNA.[79]

Several fluoropyrimidines have been developed, such as floxuridine (5-fluoro-2-deoxyuridine) and ftorafur [1-(2-tetrahydrofuranyl)-5-fluorouracil], but extensive clinical trials have not indicated these drugs to be superior to 5-fluorouracil.[80-83] Also, 5′-deoxy-5-fluorouridine, which functions as a prodrug of 5-fluorouracil, has no obvious advantage over the parent compound in clinical practice.

Fluoropyrimidines have usually been selected for clinical trials from their activity profile in several rat and murine tumors. Only a few efficacy studies have been carried out in human tumor xenografts. Fujimoto et al.[84] have evaluated the activity of 5′-deoxy-5-fluorouridine in the breast cancer xenograft MX-1 grown s.c., and found the drug more active than 5-fluorouracil. Also, in 2 of 4 head and neck cancer xenografts, 5′-deoxy-5-fluorouridine was slightly more effective.[85] Using the subrenal capsule assay, the chemosensitivity of 5-fluorouracil and ftorafur was investgated in 20 human malignancies;[86] of 13 gastric cancers, sensitive tumors were respectively 8 and 7, while for 7 colorectal cancers, sensitivity was found in 3 and 4 tumors, respectively. These limited studies do not warrant clear insight into the real value of human tumor xenografts for selecting fluoropyrimidines with clinical activity.

VII. HEXAMETHYLMELAMINE

Hexamethylmelamine is a substituted melamine derived from the reaction of cyanuric chloride with amines.[87] Its mechanism of action appears to be partially related to classical alkylating drugs. Because of aqueous insolubility, treatment in patients has to be carried out with the oral formulation. The drug was incorporated widely in combination chemotherapy regimens for ovarian cancer. However, the contribution of hexamethylmelamine to the antitumor activity of the regimens was ill defined.[88] Considerable side effects included gastrointestinal, hematological, and neurological toxicity. Upon combination with the neurotoxic drug cisplatin, early peripheral motor and sensory deficits hampered adequate treatment of ovarian cancer patients. As a consequence, hexamethylmelamine has been discarded from further clinical use and, in fact, has never found a place in the conventional treatment of other tumor types.

In a search for water-soluble hexamethylmelamine analogs with antitumor activity, the mono-demethylated derivative pentamethylmelamine had to be discarded because of frequent central nervous system disturbances and excessive gastrointestinal toxicity in phase I clinical trials.[89,90] Another water-soluble analog, trimelamol (N^2,N^4,N^6-trimethyl-N^2,N^4,N^6-trimethylol-melamine), was shown to have antitumor activity approaching that of hexamethylmelamine in a human lung cancer xenograft.[91] A further advantage of trimelamol was that, unlike hexamethylmelamine and pentamethylmelamine, the drug did not require metabolic activation to exert its cytotoxic action.[92] We have compared the antitumor activity of hexamethylmelamine and trimelamol at equitoxic doses in eight human ovarian cancer xenografts; growth inhibition greater than 75% for the drugs was obtained in 5 of 8 and 6 of 8 lines, respectively, and complete disappearance of tumors in 1 of 8 and 3 of 8 lines.[93] Activity for hexamethylmelamine in the nude mouse-human ovarian cancer model appears to exceed results from phase II ovarian cancer trials and may be attributed to differences between species with respect to the routes of administration and the metabolism of the drug. A possible improvement of the therapeutic efficacy for trimelamol, as compared to that of hexamethylmelamine in patients, has yet to be determined.

VIII. CONCLUSIONS

The nude mouse-human tumor model is gaining interest for its application in the secondary screening of promising analogs of conventional cytostatic agents. In a general approach, comparative studies involve the extent of growth inhibition induced by the analog and the parent compound administered at equitoxic doses by the same route and in the same schedule to nude mice carrying s.c. well-established tumors. For analogs of cisplatin and anthracyclines, the model has been of use in the decision-making to proceed with these drugs to clinical trials. For other groups of drugs and their analogs, the value of human tumor xenografts has not yet been fully explored.

Several critical remarks have to be added to enlighten some of the difficulties encountered with analogs rather than with unrelated investigational agents. At best, the nude mouse-human tumor model will indicate the extent of activity of an analog relative to the parent drug, may suggest additional, sensitive tumor types, and predict the lack of cross-resistance. The model cannot be applied to select analogs with diminished acute, chronic, or delayed host toxicity.

As for investigational agents (Chapter 15), the variables which may influence the validity of data obtained with analogs should reflect the treatment conditions in the clinic as much as possible. Differences may exist between the parent compound and the analog with respect to the route of administration, schedule, pharmacology, metabolism, and side effects in the mouse as well as in the patient, i.e. hexamethylmelamine administered i.p. to mice and orally to patients. The better bioavailability combined with an advantageous exposure to active metabolites may have caused the high preclinical efficacy of hexamethylmelamine.[93] The higher dose of ifosfamide, if combined with mesna in patients, may have been the reason for actvity superior to cyclophosphamide in soft tissue sarcoma,[57] which was not predicted by comparative studies in human soft tissue sarcoma xenografts.[56]

Future prospective and retrospective analysis of the activity of analogs in human tumor lines and in patients is necessary to further outline the specific contribution of the nude mouse-human tumor model to this area of drug development.

REFERENCES

1. **Goldin, A., Johnson, R. K., and Venditti, J. M.,** Preclinical characterization of candidate antitumor drugs, *Cancer Chemother. Rep.*, (Part 2), 5, 21, 1975.
2. U.S. Department of Health and Human Services, *In vivo* cancer models, *NIH Publication No. 84-2635*, U.S. Government Printing Office, Washington DC, U.S.A., 1984.
3. **Winograd, B., Boven, E., Lobbezoo, M. W., and Pinedo, H. M.,** Human tumor xenografts in the nude mouse and their value as test models in anticancer drug development, *In Vivo*, 1, 1, 1987.
4. **Harrap, K. R., Jones, M., Siracky, J., Pollard, L. A., and Kelland, L. R.,** The establishment, characterization and calibration of human ovarian carcinoma xenografts for the evaluation of novel platinum anticancer drugs, *Ann. Oncol.*, 1, 65, 1990.
5. **Prestayko, A. W., D'Aoust, J. C. D., Issell, B. F., and Crooke, S. K.,** Cisplatin (*cis*-diamminedichloroplatinum II), *Cancer Treat. Rev.*, 63, 17, 1979.
6. **Burchenal, J. H., Kalaher, K., Dew, K., and Lokys, L.,** Rationale for development of platinum analogs, *Cancer Treat. Rep.*, 63, 1493, 1979.
7. **Hill, J. M., Loeb, E., Pardue, A., Khan, A., Kung, J. J., Aleman, C., and Hill, N. O.,** Platinum analogs of clinical interest, *Cancer Treat. Rep.*, 63, 1509, 1979.
8. **Shepherd, R., Kusnierczyk, H., Jones, M., and Harrap, K. H.,** Criteria for the selection of second-generation platinum compounds, *Br. J. Cancer*, 42, 668, 1980.
9. **Rose, W. C., Schurig, J. E., Huftalen, J. B., and Bradner, W. T.,** Antitumor activity and toxicity of cisplatin analogs, *Cancer Treat. Rep.*, 66, 135, 1982.
10. **Muggia, F. M.,** Overview of carboplatin: replacing, complementing, and extending the therapeutic horizons of cisplatin, *Semin. Oncol.*, 16(Suppl. 5), 7, 1989.
11. **Alberts, D. S. and Mason-Liddil, N.,** Carboplatin in the treatment of ovarian cancer, *Semin. Oncol.*, 16(Suppl. 5), 19, 1989.
12. **Mangioni, C., Bolis, G., Pecorelli, S., Bragman, K., Epis, A., Favalli, G., Gambino, A., Landoni, F., Presti, M., Torri, W., Vassena, L., Zanaboni, F., and Marsoni, S.,** Randomized trial in advanced ovarian cancer comparing cisplatin and carboplatin, *J. Natl. Cancer Inst.*, 81, 1464, 1989.
13. **Sessa, C., Vermorken, J., Renard, J., Kaye, S., Smith, D., Ten Bokkel Huinink, W., Cavalli, F., and Pinedo, H.,** Phase II study of iproplatin in advanced ovarian carcinoma, *J. Clin. Oncol.*, 6, 98, 1988.
14. **Van der Vijgh, W. J. F., Elferink, F., Vermorken, J. B., Ten Bokkel Huinink, W. W., Klein, I., Gall, H. E., Simonetti, G., McVie, J. G., and Pinedo, H. M.,** Pharmacokinetics of free and total platinum species after rapid and prolonged infusions of aqua (1,1-*bis*(aminomethyl)cyclohexane)sulfatoplatinum(*IT*) (spiroplatin) during a phase I trial, *Eur. J. Cancer Clin. Oncol.*, 24, 621, 1988.
15. **Winograd, B., Vermorken, J. B., Ten Bokkel Huinink, W. W., Simonetti, G., Gall, H. E., Knobf, M. K. T., Van der Vijgh, W. J. F., McVie, J. G., and Pinedo, H. M.,** Phase I study of ethylenediamine platinum (II) malonate (NSC 146068Q), a second generation platinum analogue, *Cancer Res.*, 46, 2148, 1986.

16. **Boven, E., Van der Vijgh, W. J. F., Nauta, M. M., Schlüper, H. M. M., and Pinedo, H. M.,** Comparative activity and distribution studies of five platinum analogues in nude mice bearing human ovarian carcinoma xenografts, *Cancer Res.,* 45, 86, 1985.
17. **Boven, E., Nauta, M. M., Schlüper, H. M. M., Elferink, F., Van der Vijgh, W. J. F., and Pinedo, H. M.,** Secondary screening of platinum compounds in human ovarian cancer xenografts in nude mice, *Eur. J. Cancer Clin Oncol.,* 21, 1253, 1985.
18. **Boven, E.,** Screening of analogs in human ovarian cancer xenografts, in *Human Tumour Xenografts in Anticancer Drug Development,* Winogra 1, B.. Peckham, M. J., and Pinedo, H. M., Eds., Springer-Verlag, Berlin, 1988, 71.
19. **Lelieveld, P., Van der Vijgh, W. J. F., and Van Velzen, D.,** Toxicology and pharmacokinetics of (1,1-*bis*(aminomethyl)cyclohexane)oxalatoplatinum(II) (TNO-38), *Eur. J. Cancer Clin. Oncol.,* 22, 1467, 1986.
20. **Rahman, A., Roh, J. K., Wolpert-DeFilippes, M. K., Goldin, A., Venditti, J. M., and Woolley, P. V.,** Therapeutic and pharmacological studies of tetrachloro(*d,l-trans*)1,2-diaminocyclohexane platinum (IV) (tetraplatin), a new platinum analogue, *Cancer Res.,* 48, 1745, 1988.
21. **Harstrick, A., Casper, J., and Schmoll, H. J.,** Comparative antitumour activity of cisplatin and two new cisplatin analogues JM8 and JM9 in human testicular carcinoma xenografts, *Int. J. Androl.,* 10, 139, 1987.
22. **Foster, B. J., Harding, B. J., Wolpert-DeFilippes, M. K., Rubinstein, L. Y., Clagett-Carr, K., and Leyland-Jones, B.,** A strategy for the development of two clinically active cisplatin analogs: CBDCA and CHIP, *Cancer Chemother. Pharmacol.,* 25, 395, 1990.
23. **Blum, R. H. and Carter, S. K.,** Adriamycin, a new anticancer agent with significant clinical activity, *Ann. Int. Med.,* 80, 249, 1974.
24. **Young, R. C., Ozols, R. F., and Myers, C. E.,** The anthracycline antineoplastic drugs, *N. Engl. J. Med.,* 305, 139, 1981.
25. **Carter, S. K.,** Anthracycline analogs: re-assessments and prospects, *Cancer Treat. Rep.,* 63, 935, 1979.
26. **Ganzina, F.,** 4′-Epi-doxorubicin, a new analogue of doxorubicin: a preliminary overview of preclinical and clinical data, *Cancer Treat. Rev.,* 10, 1, 1983.
27. **Cersosimo, R. J. and Hong, W. K.,** Epirubicin: a review of the pharmacology, clinical activity, and adverse effects of an adriamycin analogue, *J. Clin. Oncol.,* 4, 425, 1986.
28. **Shenkenberg, T. D. and Von Hoff, D. D.,** Mitoxantrone: a new anticancer drug with significant clinical activity, *Ann. Int. Med.,* 105, 67, 1986.
29. **Torti, F. M., Bristow, M. M., Lim, B. L., Carter, S. K., Howes, A. E., Aston, D. A., Brown, B. W., Hannigan, J. F., Meyers, F. J., Mitchell, E. P., and Billingham, M. E.,** Cardiotoxicity of epirubicin and doxorubicin: assessment by endomyocardial biopsy, *Cancer Res.,* 46, 3722, 1986.
30. **Henderson, I. C., Allegra, J. C., Woodcock, T., Wolff, S., Bryan, S., Cartwright, K., Dukart, G., and Henry, D.,** Randomized clinical trial comparing mitoxantrone with doxorubicin in previously treated patients with metastatic breast cancer, *J. Clin. Oncol.,* 7, 560, 1989.
31. **McGovren, J. P., Neil, G. L., Denlinger, R. H., Hall, T. L., Crampton, S. L., and Swenberg, J. A.,** Chronic cardiotoxicity studies in rabbits with 7-con-*O*-methylnogarol, a new anthracycline antitumor agent, *Cancer Res.,* 39, 4849, 1979.
32. **Neil, G. L., Kuentzel, S. L., and McGovren, J. P.,** Treatment of mouse tumors with 7-con-*O*-methylnogarol and other analogs of the anthracycline antibiotic, nogalomycin, *Cancer Treat. Rep.,* 63, 1971, 1979.
33. **Sessa, C., Gundersen, S., Ten Bokkel Huinink, W., Renard, J., and Cavalli, F.,** Phase II study of intravenous menogaril in patients with advanced breast cancer, *J. Natl. Cancer Inst.,* 80, 1066, 1988.
34. **Sessa, C., Kaye, S. B., Renard, J., Ten Bokkel Huinink, W., and Cavalli, F.,** Phase II study of intravenous menogaril in advanced ovarian carcinoma, *Eur. J. Cancer Clin. Oncol.,* 25, 751, 1989.
35. **Mattern, J., Bak, M., Hahn, E. W., and Volm, M.,** Human tumor xenografts as model for drug testing, *Cancer Metast. Rev.,* 7, 263, 1988.
36. **Giuliani, F. C., Zirvi, K. A., and Kaplan, N. O.,** Therapeutic response of human tumor xenografts in athymic mice to doxorubicin, *Cancer Res.,* 441, 325, 1981.
37. **Guiliani, F. C. and Kaplan, N. O.,** New doxorubicin analogs active against doxorubicin-resistant colon tumor xenografts in the nude mouse, *Cancer Res.,* 40, 4682, 1980.
38. **Guiliani, F. C., Zirvi, K. A., Kaplan, N. O., and Goldin, A.,** Chemotherapy of human colorectal tumor xenografts in athymic mice with clinically active drugs: 5-fluorouracil and 1-3-*bis*-(2-chloroethyl)-1-nitrosourea (BCNU). Comparison with doxorubicin derivatives: 4′-deoxydoxorubicin and 4′-*O*-methyldoxorubicin, *Int. J. Cancer,* 27, 5, 1981.
39. **Taetle, R., Howell, S. B., Giuliani, F. C., Koziol, J., and Koessler, B. S.,** Comparison of the activity of doxorubicin analogues using colony-forming assays and human xenografts, *Cancer,* 50, 1455, 1982.
40. **Guiliani, F. C. and Kaplan, N. O.,** The use of athymic mice in the screening of new anthracycline derivatives, in *Immune-Deficient Animals, 4th Int . Workshop on Immune-Deficient Animals in Exp. Res.,* Sordat, B., Ed., Karger, Basel, 1984, 374.

41. **Pezzoni, G. and Giuliani, F. C.,** Predictability of clinical response of several human and murine tumor models to four anthracycline derivatives, in *Human Tumour Xenografts in Anticancer Drug Development,* Winograd, B., Peckham, M. J., and Pinedo, H. M., Eds., Springer-Verlag, Berlin, 1988, 75.
42. **Hurteloup, P. and Ganzina, F.,** Clinical studies with anthracyclines: epirubicin, idarubicin, esorubicin, *Drug Exp. Clin. Res.,* 12, 233, 1986.
43. **Arlin, Z. A.,** Idarubicin in acute leukemia: an effective new therapy for the future, *Semin. Oncol.,* 16(Suppl. 2), 35, 1989.
44. **Boven, E., Schlüper, H. M. M., Erkelens, C. A. M., and Pinedo, H. M.,** Doxorubicin compared with related compounds in a nude mouse model for human ovarian cancer, *Eur. J. Cancer Clin. Oncol.,* 26, 903, 1990.
45. **Hilgers, R. D., Rivkin, S. E., Von Hoff, D. D., and Alberts, D. S.,** Mitoxantrone in epithelial carcinoma of the ovary, *Am. J. Clin. Oncol.,* 7, 499, 1984.
46. **Muss, H. B., Asbury, R., Bundy, B., Ehrlich, C. F., and Graham, J.,** Mitoxantrone (NSC-301739) in patients with advanced ovarian carcinoma, *Am. J. Clin. Oncol.,* 7, 737, 1984.
47. **Lawton, F., Blackledge, G., Mould, J., Latief, T., Watson, R., and Chetiyawardana, A. D.,** Phase II study of mitoxantrone in epithelial ovarian cancer, *Cancer Treat. Rep.,* 71, 627, 1987.
48. **Brock, N.,** Ideas and reality in the development of cancer chemotherapeutic agents, with particular reference to oxazaphosphorine cytostatics, *J. Cancer Res. Clin. Oncol.,* 111, 1, 1986.
49. **Sarosy, G.,** Ifosfamide-pharmacologic overview, *Semin. Oncol.,* 16(Suppl. 3), 2, 1989.
50. **Zalupski, M. and Baker, L. H.,** Ifosfamide, *J. Natl. Cancer Inst.,* 80, 556, 1988.
51. **Antman, K. H., Elias, A., and Ryan, L.,** Ifosfamide and mesna: response and toxicity at standard- and high-dose schedules, *Semin. Oncol.,* 17(Suppl. 4), 68, 1990.
52. **Goldin, A.,** Ifosfamide in experimental tumor systems, *Semin. Oncol.,* 9(Suppl. 1), 14, 1982.
53. **Inoue, K., Fujimoto, S., and Ogawa, M.,** Antitumor activities of seventeen alkylating agents against human mammary carcinoma (MX-1) in nude mice, *Nagoya J. Med. Sci.,* 43, 89, 1981.
54. **Friedman, H. S., Colvin, O. M., Ludeman, S. M., Schold, S. C., Boyd, V. L., Muhlbaier, L. H., and Bigner, D. D.,** Experimental chemotherapy of human medulloblastoma with classical alkylators, *Cancer Res.,* 46, 2827, 1986.
55. **Berger, D. P., Fiebig, H. H., Winterhalter, B. R., Wallbrecher, E., and Henss, H.,** Preclinical phase II study of ifosfamide in human tumour xenografts *in vivo, Cancer Chemother. Pharmacol.,* 26(Suppl.), s7, 1990.
56. **Boven, E., Calame, J. J., Molthoff, C. F. M., and Pinedo, H. M.,** Characterization and chemotherapy of human soft tissue sarcoma lines grown in nude mice, *Strahlenther. Onkol.,* 165, 538, 1989.
57. **Bramwell, V., Mouridsen, H. T., Santoro, A., Blackledge, G., Somers, R., Verwey, J., Dombernowsky, P., Onsrud, M., Thomas, D., Sylvester, D., and Van Oosterom, A.,** Cyclophosphamide versus ifosfamide: final report of a randomized phase II trial in adult soft tissue sarcomas, *Eur. J. Cancer Clin. Oncol.,* 23, 311, 1987.
58. **Crooke, S. T. and Bradner, W. T.,** Mitomycin C: a review, *Cancer Treat. Rev.,* 3, 121, 1976.
59. **Crooke, S. T.,** Mitomycin C: an overview, in *Mitomycin C: Current Status and New Developments,* Carter, S. K. and Crooke, S. T., Eds., Academic Press, New York, 1979, 1.
60. **Boven, E. and Pinedo, H. M.,** Mitomycin C: interstitial pneumonitis and haemolytic-uraemic syndrome, *Neth. J. Med.,* 26, 159, 1983.
61. **Verwey, J., Boven, E., Van der Meulen, J., and Pinedo, H. M.,** Recovery from mitomycin C-induced hemolytic uremic syndrome, *Cancer,* 54, 2878, 1984.
62. **Bradner, W. T., Rose, W. C., Schurig, J. E., Schlein, A., and Huftalen, J. B.,** Antitumor activity and toxicity in animals of RR-150 (7-cystaminomitosane) a new mitomycin derivative, *Cancer Res.,* 44, 5619, 1984.
63. **Sakurai, Y.,** M-83: a new derivative of mitomycin, *Cancer Treat. Symp.,* 1, 29, 1983.
64. **Bradner, W. T., Schurig, J. E., Schlein, A., Rose, W. C., Farwell, A. R., Florczyk, J. B., Huftalen, J. B., and Kaneko, T.,** Experimental antitumor activity and toxicity of BMY-26606, a potentially less myelosuppressive mitomycin derivative, *Cancer J.,* 1, 118, 1986.
65. **Bradner, W. T., Remers, S. A., and Vyas, D. M.,** Structure-activity comparison of mitomycin C and mitomycin A analogues, *Anticancer Res.,* 9, 1095, 1989.
66. **Asanuma, F.,** Antitumor activity and pharmacokinetics of 7-*N*-(*p*-hydroxyphenyl)mitomycin C in human tumor xenografts transplanted into nude mice, *J. Antibiot.,* 38, 401, 1985.
67. **Bleyer, W. A.,** Methotrexate: clinical pharmacology, current status and therapeutic guidelines, *Cancer Treat. Rev.,* 4, 87, 1977.
68. **McGuire, J. J., Hsieh, P., Franco, C. T., and Piper, J. R.,** Folylpolyglutamate synthetase inhibition and cytotoxic effects of methotrexate analogs containing 2,w-diaminoalkanoic acids, *Biochem. Pharmacol.,* 35, 2607, 1986.
69. **O'Dwyer, P. J., Shoemaker, D. D., Plowman, J., Cradock, J., Grillo-Lopez, A., and Leyland-Jones, B.,** Trimetrexate: a new antifol entering clinical trials, *Invest. New Drugs,* 3, 71, 1985.
70. **Lin, J. T., Cashmore, A. R., Baker, M., Dreyer, R. N., Ernstoff, M., Marsh, J. C., Bertino, J. R., Whitfield, L. R., Delap, R., and Grillo-Lopez, A.,** Phase I studies with trimetrexate: clinical pharmacology, analytical methodology, and pharmacokinetics, *Cancer Res.,* 47, 609, 1987.

71. **Schmid, F. A., Sirotnak, F. M., Otter, G. M., and DeGraw, J. I.,** New folate analogs of the 10-diaza-aminopterin series: markedly increased antitumor activity of the 10-ethyl analog compared to the parent compound and methotrexate against some human tumor xenografts in nude mice, *Cancer Treat. Rep.*, 69, 551, 1985.
72. **Braakhuis, B. J. M., Leyva, A., Schoevers, E. J., Boerrigter, G. H., Schornagel, J. H., and Snow, G. B.,** Lack of effect of methotrexate on human head and neck tumours transplanted in athymic nude mice, *Acta Otolaryngol. (Stockholm.)*, 99, 208, 1985.
73. **Brown, D. H., Braakhuis, B. J. M., Van Dongen, G. A. M. S., Van Walsum, M., Bagnay, M., and Snow, G. B.,** Activity of the folate analog 10-ethyl,10-deaza-aminopterin (10-EdAM) against human head and neck cancer xenografts, *Anticancer Res.*, 9, 1549, 1989.
74. **Wallace, R. E., Lindh, D., and Durr, F. E.,** Arabinofuranosyl-5-azacytosine: activity against human tumors in athymic mice, *Cancer Chemother. Pharmacol.*, 25, 117, 1989.
75. **Rivard, G. E., Momparler, R. L., Demers, J., Beiwit, P., Raymond, R., Lin, K., and Momparler, L. F.,** Phase I study on 5-aza-2′-deoxycytidine in children with acute leukemia, *Leuk. Res.*, 5, 453, 1981.
76. **Van Groeningen, C. J., Leyva, A., O'Brien, A. M. P., Gall, H. E., and Pinedo, H. M.,** Phase I and pharmacokinetic study of 5-aza-2′-deoxycytidine (NSC 127716) in cancer patients, *Cancer Res.*, 46, 4831, 1986.
77. **Braakhuis, B. J. M., Van Dongen, G. A. M. S., Peters, G. J., Van Walsum, M., Leyva, A., and Snow, G. B.,** Chemotherapy of human head and neck cancer xenografts with 5-aza-2′-deoxycytidine, in *5-Aza-2′-deoxycytidine: Preclinical and Clinical Studies*, Momparler, R. L. and De Vos, D., Eds., Pharmachemie, Haarlem, N.L., 1990, 67.
78. **Boven, E., Schipper, H., Erkelens, C. A. M., Hatty, S. A., and Pinedo, H. M.,** Gemcitabine (2′,2′-difluoro-deoxycytidine), a new cytidine analog with schedule- rather than dose-related activity in human tumor xenografts of ovarian cancer and soft tissue sarcoma, *Eur. Soc. Med. Oncol.*, abstr., 1990.
79. **Peters, G. J., Laurensse, E., Leyva, A., Lankelma, J., and Pinedo, H. M.,** Sensitivity of human, murine, and rat cells to 5-fluorouracil and 5′-deoxy-5-fluorouridine in relation to drug-metabolizing enzymes, *Cancer Res.*, 46, 20, 1986.
80. **Chang, A. E., Schneider, P. D., Sugarbaker, P. H., Simpson, C., Culnane, M., and Steinberg, S. M.,** A prospective randomized trial of regional versus systemic continuous 5-fluorodeoxyuridine chemotherapy in the treatment of colorectal liver metastases, *Ann. Surg.*, 206, 685, 1987.
81. **Hohn, D. C., Stagg, R. J., Friedman, M. A., Hannigan, J. F., Rayner, A., Ignoffo, R. J., Acord, P., and Lewis, B. J.,** A randomized trial of continuous intravenous versus hepatic intraarterial floxuridine in patients with colorectal cancer metastatic to the liver: the Northern California Oncology Group trial, *J. Clin. Oncol.*, 7, 1646, 1989.
82. **Al-Sarraf, M.,** Clinical trials with fluorinated pyrimidines in patients with head and neck cancer, *Invest. New Drugs*, 7, 71, 1989.
83. **Valdivieso, M., Bodey, G. P., Gottlieb, J. A., and Freireich, E. J.,** Clinical evaluation of ftorafur (pyrimidine-deoxyribose N_1-2′-furanidyl-5-fluorouracil), *Cancer Res.*, 36, 1821, 1976.
84. **Fujimoto, S., Wang, Y., Inoue, K., and Ogawa, M.,** Antitumor activity of a new fluoropyrimidine derivative 5′-deoxy-5-fluorouridine, against murine and human experimental tumors, *Jpn. J. Cancer Res. (Gann)*, 76, 644, 1985.
85. **Peters, G. J., Braakhuis, B. J. M., De Bruijn, E. A., Laurensse, E. J., Van Walsum, M., and Pinedo, H. M.,** Enhanced therapeutic efficacy of 5′-deoxy-5-fluorouridine in 5-fluorouracil resistant head and neck tumours in relation to 5-fluorouracil metabolising enzymes, *Br. J. Cancer*, 59, 327, 1989.
86. **Anai, H., Maehara, Y, Kusumoto, H., Kusumoto, T., and Sugimachi, K.,** Sensitivity test for 5-fluorouracil and its analogues, 1-(2-tetrahydrofuryl)-5-fluorouracil, uracil/1-(2-tetrahydrofuryl)-5-fluorouracil (4:1) and 1-hexacarbamoyl-5-fluorouracil, using the subrenal capsule assay, *Oncology*, 45, 144, 1988.
87. **Foster, B. J., Harding, B. J., Leyland-Jones, B., and Hoth, D.,** Hexamethylmelamine: a critical review of an active drug, *Cancer Treat. Rev.*, 13, 197, 1986.
88. **Foster, B. J., Clagett-Carr, K., Marsoni, S., Simon, R., and Leyland-Jones, B.,** Role of hexamethylmelamine in the treatment of ovarian cancer: where is the needle in the haystack?, *Cancer Treat. Rep.*, 70, 1003, 1986.
89. **Goldberg, R. S., Griffin, J. P., McSherry, J. W., and Krakoff, I. H.,** Phase I study of pentamethylmelamine, *Cancer Treat. Rep.*, 64, 1319, 1980.
90. **Muindi, J. R. F., Newell, D. W., Smith, I. E., and Harrap, K. R.,** Penta-methylmelamine (PMM): phase I clinical and pharmacokinetic studies, *Br. J. Cancer*, 47, 27, 1983.
91. **Connors, T. A., Cumber, A. J., Ross, W. C. J., Clarke, S. A., and Mitchley, B. C. V.,** Regression of human lung tumors xenografts induced by water-soluble analogs of hexamethylmelamine, *Cancer Treat. Rep.*, 61, 927, 1977.
92. **Rutty, C. J., Newell, D. R., Muindi, J. R. F., and Harrap, K. R.,** The comparative pharmacokinetics of pentamethylmelamine in man, rat and mouse, *Cancer Chemother. Pharmacol.*, 8, 105, 1982.
93. **Boven, E., Nauta, M. M., Schlüper, H. M. M., Erkelens, C. A. M., and Pinedo, H. M.,** Superior efficacy of trimelamol to hexamethylmelamine in human ovarian cancer xenografts, *Cancer Chemother. Pharmacol.*, 18, 124, 1986.

Chapter 15

INVESTIGATIONAL AGENTS

Ghanem Atassi

TABLE OF CONTENTS

I. INTRODUCTION

Since the observation in the 1950s that nitrogen mustard could induce responses in patients with refractory lymphoma, an extensive research program initiated by the National Cancer Institute in the U.S. and, in 1970, joined by the European Organization for Research on Treatment of Cancer (EORTC) in collaboration with pharmaceutical companies and chemical industries led to the discovery of many active chemotherapeutic agents. At present, there are approximately 40 anticancer drugs in clinical use which were selected by murine tumor models.[1] However, it is now apparent that this methodology merely identified compounds with a similar level of activity and with a continuous degree of moderate to severe toxicity.[2] As a matter of fact, the models and the methodology have been repeatedly reviewed[2-4] and a massive effort was made to introduce new approaches to improve the predictive correlation of data from chemotherapy experiments in tumor-bearing animals or from *in vitro* tests to clinical responses in cancer patients.[5-13]

In spite of these efforts, there is still no prospective study demonstrating the predictive value of any preclinical model in use for anticancer drug development. However, retrospective attempts to correlate published data with clinical results indicated that human tumor xenografts in nude mice respond to such drugs in a way that would be expected on the basis of clinical experience.[14-17] The increasing interest in the nude mouse-human tumor model for drug testing is derived in part from the dissatisfaction with transplantable murine tumors which apparently display a spectrum of drug sensitivities that is not always related to the drug sensitivity of human tumors. However, even the use of human tumor xenografts will lead to dissatisfaction if the relevance of the methodology adopted and the criteria chosen to evaluate drug efficacy have to be questioned.

This chapter aims to (1) review the antitumor activity of various chemical structures recognized as investigational anticancer drugs, which have shown antitumor activity in human tumor xenograft systems, and (2) reach, in the light of this review, some conclusions which hopefully will improve the selection of true positive anticancer drugs.

The methodology used in the xenograft studies to be described is of utmost importance for the interpretation of the "treatment" results. Nevertheless, for the purpose of this review, only the most relevant features of the methodology will be mentioned here, whereas details should be retrieved from the original paper(s).

II. INVESTIGATIONAL AGENTS

A. NORLEUCINE DERIVATIVES

DON or 6-diazo-5-oxo-L-norleucine and azotomycin or *N*(*N*-glutamyl-6-diazo-5-oxo-norleucinyl)-6-diazo-5-oxo-norleucine are structurally related antitumor antibiotics (Figure 1). Those drugs were tested against four human tumor xenografts.[18] Mice carried s.c. tumors and drugs were administered i.p. on an every-4-d schedule. The efficacy of drugs was assessed on the basis of the tumor size of treated mice as compared to that of the control; the optimal T/C% corresponds to the lowest relative tumor weight of the treated group (T) as a percentage of the relative tumor weight of the control group (C) calculated at the optimal dose. The tumor designated CX-I appeared to be refractory to both drugs (T/C% > 50), CX-2 responded to treatment with either drug (0 < T/C% < 10), and MX-l and LX-l were very sensitive to both drugs at 100 mg/kg (T/C% < 10). Clinical trials with DON and azotomycin generated results that were not too promising.[19-22]

B. THYMIDINE

The finding that an excess of thymidine could inhibit the growth of human melanomas transplanted to nude mice[23] encouraged some investigators to develop a system enabling

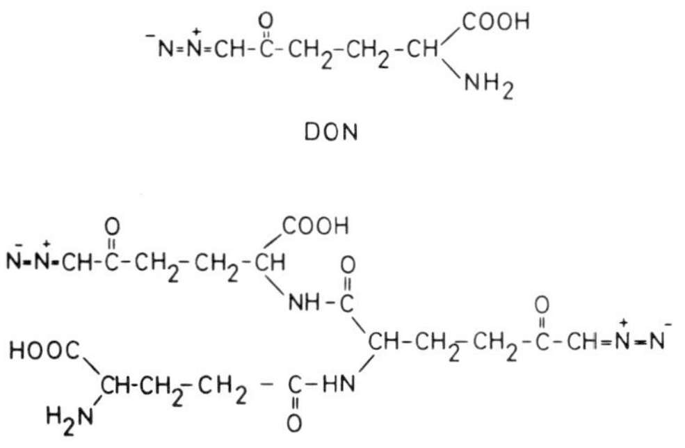

FIGURE 1. Structures of DON and azotomycin.

continuous infusion of the drug into these animals. Infusion of thymidine (444 to 888 mg/kg/h administered continuously for 96 to 140 h) induced total tumor regressions of xenografts of 1 teratocarcinoma, 5 different melanomas, and 1 lung cancer.[24]

In a second study, a melanoma (T242), a small cell lung cancer (T293), a colon cancer (T183), and a duct carcinoma of the breast (Tl12) were inoculated s.c. and treated with thymidine (1 g/kg/d) s.c. for 72 h each week for 3 weeks.[25] A significant effect was found only for the melanoma xenograft. However, there were no complete tumor regressions, not even when melanomas of very small initial volume were treated. These data indicated that thymidine administered at the same dose used by Lee et al.[24] was insufficient to rid the mice of even a small tumor burden. Exposure to thymidine in the schedule described produced significant toxicity consisting of myelosuppression.[25]

In a third study, the therapeutic effect of thymidine was tested against two different heterotransplanted lymphomas.[26] Doses were given at 90% of the LD_{10} at the time tumors measured 0.5 tol cm^3. These doses caused a weight loss between 3 and 10% in treated animals. Thymidine, as well as cyclophosphamide used as the positive control, was given by the i.p. route. Both heterotransplanted lymphomas were sensitive to cyclophosphamide given as a single dose of 200 mg/kg. In contrast, thymidine was ineffective at a dose of 500 mg/kg every day × 10 against these lymphomas.

C. ALKYL LYSOPHOSPHOLIPIDS

Alkyl lysophospholipids are synthetic analogs of the naturally occurring 2-lysophosphatidylcholine and represent a new class of antimetabolite antitumor agents. They inhibit or prevent the growth of syngeneic transplantable tumors in mice and rats.[27] The effects of the racemic 1-octadecyl-2-methyl-sn-glycero-3-phosphocholine (Et-l8-OCH_3) on xenotransplanted human tumors in nude mice was investigated.[28] To exclude hormonal influence on the growth of the gynecological tumors, 6-week-old male mice were castrated before the start of the experiment. Treatment with alkyl lysophospholipids was started in an ovarian cancer xenograft 3 d after transplantation of the tumor. Complete tumor regressions were observed in several mice following i. v. treatment in various schedules for 3 weeks. Moreover, Et-18-OCH_3 had no side effects after this prolonged administration.[28]

D. DIANHYDROGALACTITOL

Dianhydrogalactitol (Figure 2) belongs to those drugs which have been selected for clinical trials on the basis of activity against a large panel of murine tumors.[29] The drug entered phase I-II trials, where its activity against solid tumors appeared to be disappointing.

In the treatment of well-established s. c. xenografts of a diffuse histiocytic lymphoma and a

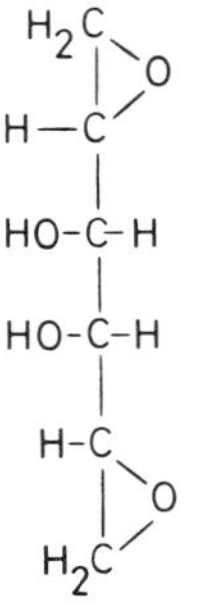

FIGURE 2. Structure of dianhydrogalactitol.

lymphoblastic lymphoma in nude mice, the i.p. administration for 5 consecutive days showed antitumor effects against both tumors.[26] The therapeutic effect of the drug was as significant as the effect of cyclophosphamide on the diffuse histiocytic lymphoma.

The efficacy of dianhydrogalactitol studied as a single agent against solid tumors was rather disappointing; no responses occurred in the treatment of 2 brain tumors,[30] 1 response was observed in 12 different xenografts of stomach cancer,[31] and no responses in 4 xenografts of colon cancer.[32]

E. α-DIFLUOROMETHYLORNITHINE (DFMO)

DFMO is a specific enzyme-activated, irreversible inhibitor of ornithine decarboxylase.[33] DFMO is able to suppress the appearance of several transplanted animal tumors.[34,35] The relationship between the tumor burden and host toxicity for the efficacy of the drug was examined in a small cell lung cancer xenograft.[36] When DFMO was given to animals for 1 week prior to inoculation of tumor cells, none of the 16 animals developed tumors after 20 weeks, compared to 44 of 47 animals (94%)which developed tumors in the control group. In the late DFMO-treatment group, tumors measured 1.5 cm^3 and complete arrest of tumor growth was evident after 4 weeks. The survival of the treated animals was significantly increased. This study suggests that DFMO could play a role in the treatment of small cell lung cancer and be specifically effective in the setting of low tumor burden. No clinical study has been published testing this hypothesis. Otherwise, clinical trials have suggested therapeutic effects only in patients with acute leukemia.[37]

F. DIAZIQUONE (AZQ)

AZQ (Figure 3) is a lipid-soluble synthetic benzoquinone which showed impressive efficacy against standard animal brain tumor models.[38] In early clinical trials, the drug demonstrated interesting activity against brain tumors and showed modest toxicity.[39]

The activity of AZQ against a series of human brain tumor lines grown in nude mice has also been investigated.[40,41] In 7 different brain tumor xenografts, the median growth delay giving the drug i.p. daily for 5 d ranged between 2.3 and 16.4 d (at 75% of the LD_{10} dose). Growth delays using a single-dose schedule were in the range of 6.2 to 13 d. There were few significant tumor regressions in any of the xenografts. Exceptions were the N-456 xenograft in which 5 regressions among 10 tumors were observed and the N-518 xenograft with 3 regressions out of 8 tumors.[40]

AZQ showed superior antitumor effects as compared to conventional agents in a sophisticated model for retinoblastoma transplanted into the anterior chamber of the nude mouse eye, of potential clinical benefit.[42]

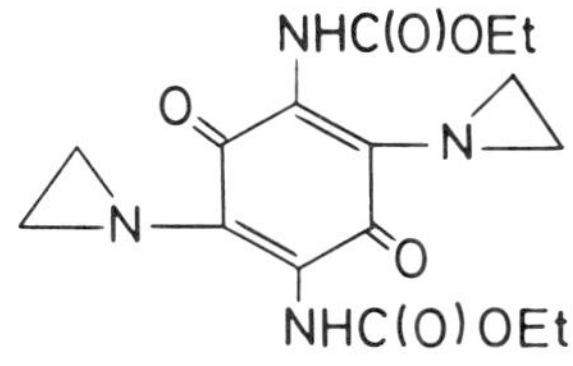

FIGURE 3. Structure of diaziquone (AZQ).

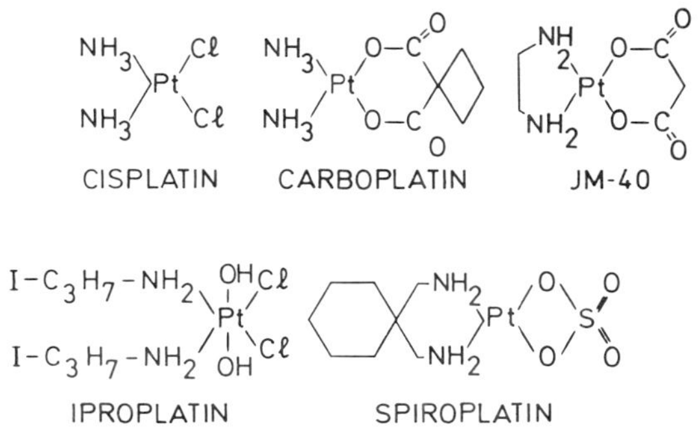

FIGURE 4. Structures of platinum complexes.

G. PLATINUM ANALOGS

In an attempt to assess the predictive value of two human ovarian cancer lines grown in nude mice, the efficacy of new platinum analogs under clinical investigation was compared to that of cisplatin.[43] Besides cisplatin, the following platinum analogs were studied at equitoxic doses: carboplatin, JM-40, iproplatin, and spiroplatin (Figure 4). In the tumor line designated MRI-H-207, complete remissions were obtained with cisplatin, carboplatin, iproplatin, and JM-40. In mice carrying the Pe ovarian cancer xenograft, the drugs showed moderate activity for cisplatin and carboplatin and minimal activity for iproplatin and JM-40. Spiroplatin was the least effective drug. These data correlated well with recent findings from clinical studies with these drugs in advanced ovarian cancer.[44] Additional *in vivo* studies with platinum analogs have been described in Chapter 14.

H. METALLOCENES

Organometallic compounds of the metallocene dihalide type (Figure 5) have been shown to possess antitumor properties against experimental tumors.[35] Therefore, the analysis of the antitumor effects of two titanocene dihalides, $(C_5H_5)_2$ TiX_2 with X = Cl or Br, against human tumors in nude mice was carried out.[46] The responses of a human colon cancer xenograft (S90) to administration of titanocene dihalides were compared with the effect of cisplatin. The drugs cisplatin, titanocene dichloride (Tc), and titanocene dibromide (Tb) were administered at equitoxic doses i. p. every 2 d × 5, and the best results obtained were a T/C of 73% for cisplatin, of 31% for Tc, and 28% for Tb. The failure of cisplatin to induce antitumor effects in colorectal cancer agrees with clinical findings.[47]

In another study, the influence of single doses of diethoxy-(I-phenyl-1,3-butanedionato) titanium IV (DBT) on the antitumor activity and toxicity in a human breast (M3a), colorectal (S90), and lung cancer line (L261) grown as xenografts was reported.[48] Minimal antitumor

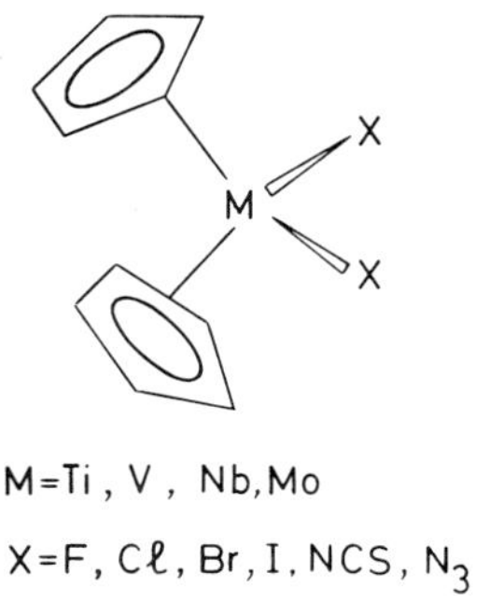

FIGURE 5. Structures of metallocene dihalides.

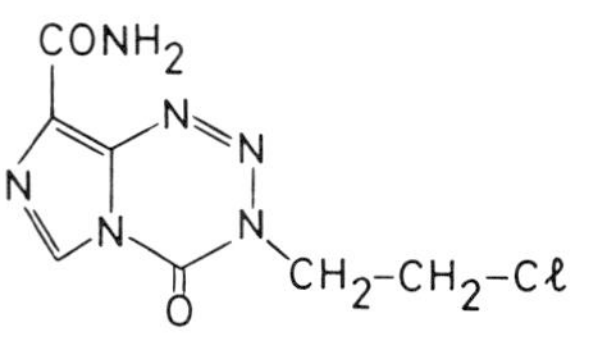

FIGURE 6. Structure of mitozolomide.

activity of DBT was shown against the human lung tumor only at a toxic dose which induced early deaths and weight loss. No activity was reported against the breast and colorectal cancer xenograft.[48]

I. MITOZOLOMIDE

Mitozolomide is a new imidazotetrazine (Figure 6) with promising effects in murine tumor systems. Therefore, investigations were extended to human tumor xenografts in order to establish the differential sensitivity of the various tumor types.[49] The human tumors used were a melanoma tumor line (LOX), a small cell lung cancer line (H-249), an osteosarcoma (TPX), and a colon cancer (WIDR) xenograft. Mitozolomide and CCNU were clearly more active than cisplatin in the LOX melanoma s. c. xenograft. In the mitozolomide-treated animals, the tumors disappeared and did not regrow during the observation period (52 d).[50] For the lung cancer H-249 grown s.c., the tumors of the mitozolomide-treated mice disappeared almost completely. In contrast, doxorubicin, cisplatin, and etoposide had only minor effects on tumor growth.[50] After two doses of mitozolomide, the s.c. implants of the osteosarcoma TPX started to decrease in size. In some of these animals, the tumors finally disappeared without regrowth, whereas, in additional mice, the tumors were still decreasing in size upon termination of the experiment. Abrin had a statistically significant growth-inhibitory effect, whereas doxorubicin had no effect on the osteosarcoma line. The results for the WIDR colon cancer line were as follows. 5-Fluorouracil was the most effective drug tested. However, mitozolomide and abrin also showed clear inhibition of growth. If the human tumor xenografts used in these experiments are indeed better representative than murine tumors, significant clinical responses could be anticipated. However, for mitozolomide, no clinical activity was reported in subsequent clinical phase II evaluation.

J. NEW ANTHRACYCLINE DERIVATIVES

A variety of anthracycline derivatives have been developed over the past years. The *in vivo* testing of new doxorubicin analogs is extensive and has been described in Chapter 14.

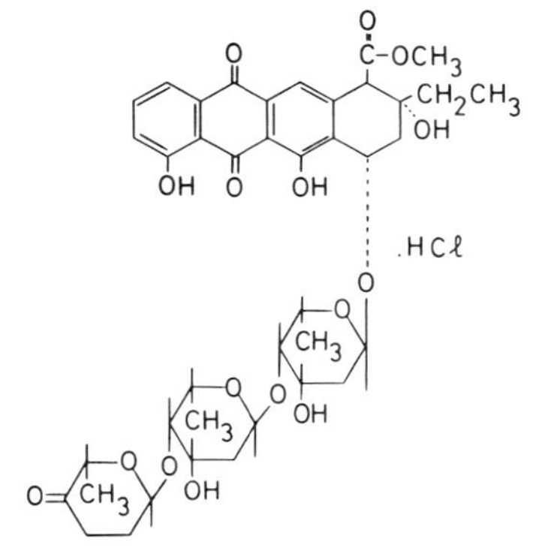

FIGURE 7. Structure of aclacinomycin hydrochloride.

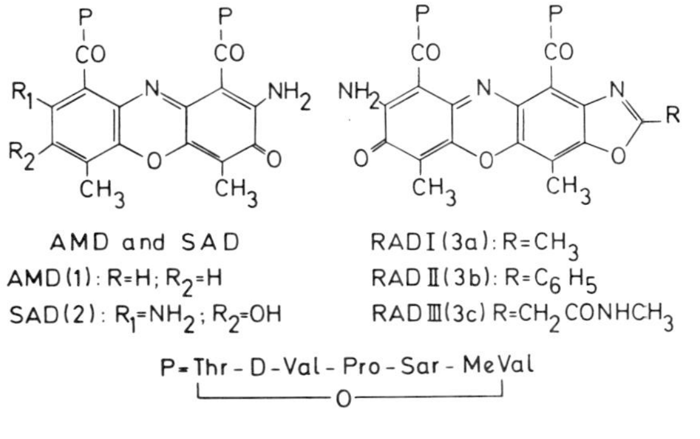

FIGURE 8. Structures of actinomycin D analogs.

Another anthracycline antibiotic, aclacinomycin A (Figure 7), has been found active in a wide spectrum of murine tumors as well as in some human tumor xenografts.[51,52] The painful peritonitis associated with i.p. doxorubicin administration in patients prompted the evaluation of aclacinomycin A as an alternative to doxorubicin for this treatment approach. Aclacinomycin A has been tested i.p. in a nude mouse i.p. xenograft model for human ovarian cancer (OVCAR-3).[53] Aclacinomycin A was able to increase the median survival time to 54 d, as compared to 37 d in untreated control mice.[54] No early deaths were recorded in treated mice, which could be attributed to drug toxicity. It is worth mentioning that aclacinomycin A has been shown to have clinical efficacy in phase I trials in ovarian cancer.

K. ACTINOMYCIN D ANALOGS

Considerable efforts have been put into the search for actinomycin D analogs (Figure 8) having a broader range of antitumor activity and reduced host toxicity. The antitumor activity of three "reverse" analogs, RAD II, RAD III, and SAD, against CX-1, MX-1, and LX-1 human tumor xenografts are summarized.[55] Of the three analogs, the tumors responded best to the treatment with RAD III. The tumors also responded fairly well to SAD treatment, but at a dose of 60 mg/kg only. RAD II showed marginal but positive activity. Actinomycin D and doxorubicin were ineffective in this tumor at the doses tested. None of these drugs were active against the MX-1 and LX-1 xenografts at the doses applied.

L. NEW FOLATE ANALOGS OF THE 10-DEAZA-AMINOPTERIN SERIES

Superior antitumor activity was reported for 10-ethyl-10-deaza-aminopterin as compared to methotrexate.[56] Results of limited studies carried out in three human tumor xenografts in nude mice are as follows.[57] Against the MX-1 breast cancer xenograft, methotrexate was essentially inactive and l0-deaza-aminopterin was only minimally active. However, the 10-ethyl derivative of 10-deaza-aminopterin showed substantial antitumor activity which even resulted in one of the mice free of tumor on day 26. Against the LX-1 lung cancer xenograft, the overall antitumor activity was similar to that obtained against the MX-1 tumor, but no animals were found to be tumor-free following treatment with 10-ethyl-10-deaza-aminopterin. Treatment of the CX-1 colon cancer xenograft with methotrexate or 10-deaza-aminopterin was essentially without effect.

M. AMPLIGEN

Double-stranded RNAs are inducers of different molecular forms of human interferon (IFN), as well as activators of the IFN-associated intracellular mediators.[58] Ampligen, a mismatched analog of $rI_n.rC_n$, designated $rI_n.r(C_{12},U)_n$, was shown to have retained the ability to induce an antiproliferative state in animal test systems, while being very inefficient in triggering various other responses such as fever, antibody formation, and bone marrow-, liver-, or kidney-toxicities, all of which have been associated with IFN-inducers or αIFN therapy.[59] In order to demonstrate that ampligen is superior to IFN in the treatment of cancer, a human renal cell cancer (786-O) xenograft model was used.[60,61] Nude mice implanted with the line 786-O will die from tumor cell proliferation and the production of a substance producing a hypercalcemic state.[61] IFN treatment did not alter the process of tumor cell proliferation, whereas ampligen therapy resulted in a statistically significant prolongation of life.[62] Median survival time of control and IFN-treated mice was 21 weeks; while at 32 weeks after transplantation of the tumor, 90% of the mice were alive in the ampligen-treated group.[63]

N. PROSTAGLANDIN A

Prostaglandin A (PGA) inhibits the growth of rodent and human tumor cells *in vitro*.[64] This observation led to the investigation of PGA in the possible inhibition of the growth of human melanoma xenografts.[65] MIRW5 cells were injected intracutaneously into the area of the upper abdomen and PGA was injected s.c. into the lower abdomen at day 6 after tumor cell inoculation. MIRW5 tumors treated for 5 consecutive days with PGA at a dose of 20 mg/kg/d showed signicifant retardation in tumor growth.[66] The PGA treatment did not result in weight loss, skin toxicity, or diarrhea. Also, PGA doses of 20 to 40 mg/kg/d for 8 consecutive days induced growth inhibition without side effects. Another group of melanoma-bearing mice was treated with PGA 200 mg/kg for 2 d on day 6 after implantation, which resulted in a larger decrease in tumor growth. However, an inflammatory response and diarrhea were observed with complete recovery within 10 d.

O. TAXOL

Taxol, a complex taxane derivative isolated from the bark of the *Taxus brevifolia*, promotes the formation of abnormal bundles of microtubules.[67] Preclinical studies have demonstrated that taxol A (Figure 9) has antitumor activity in the L1210, P388, and P1534 leukemia systems.[68] The activity of this drug against human tumor xenografts has also been investigated.[69-71] The treatment of a breast cancer xenograft with taxol starting on day 30 after implantation produced total regression of tumors for 20 d.[70] Treatment of a tongue squamous cell carcinoma xenograft starting 7 d after transplantation resulted in a tumor growth delay of 21 d (significant as compared to control). The treatment of a bronchial adenocarcinoma 32 d after the tumor transplant also resulted in a significant tumor growth delay (34 d). No signs of toxicity were observed except s.c. ulceration at the site of the s.c. injection.

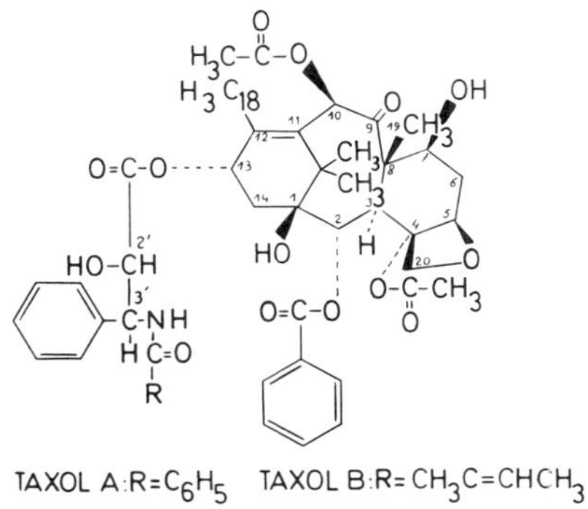

FIGURE 9. Structures of taxol derivatives.

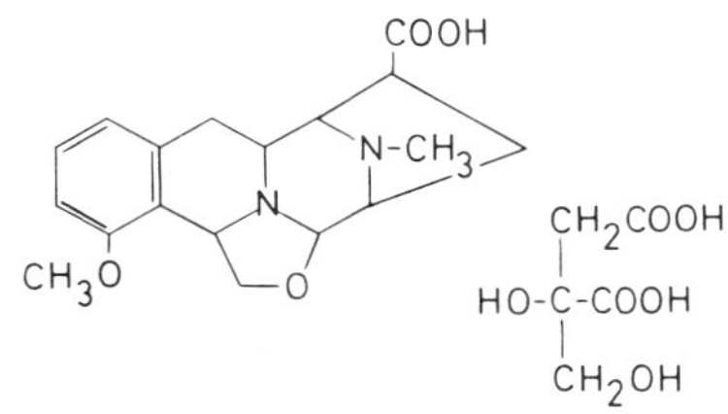

FIGURE 10. Structure of quinocarmycin citrate.

The antitumor activity of taxol A in an ovarian tumor was shown where the drug induced a growth delay of 33 d and a significant reduction in the tumor volume and weight.[71] Under the same conditions, taxol B appeared to be less active or inactive as compared to taxol A.

P. QUINOCARMYCIN CITRATE

Quinocarmycin citrate (QCM) is a novel antibiotic produced by *Streptomyces meranovinaceus*. This agent consists of the novel skeleton 8,11-iminoazepino-isoquinoline (Figure 10), and revealed antitumor activity against some experimental tumors.[72] The antitumor activity of QCM against 11 human tumor xenografts was evaluated.[73] One breast cancer (MX-1), one colon cancer (Co-4), seven gastric cancer (St-4, St-15, St-40, SC-2-JCK, SC-6-JCK, Exp-4, and H-111), and two small cell lung cancer (Lu-24 and Lu-130) xenografts were used. The antitumor activity of QCM following i.p. treatment was evaluated in terms of the lowest T/C during the experiment. With the exception of St-4 and St-15, QCM in different schedules appeared to be active in all models. Lu-130 tumors even disappeared following QCM treatment. No clinical data are yet available to assess the predictive value of this study.

Q. SUGAR-CONTAINING NITROSOUREA DERIVATIVES

CGP 6809 (ethyl-6-deoxy-3,5-di-*0*-methyl-6-(3-methyl-3-nitrosoureido)-x-D-glucofuranoside) (Figure 11) is a new methylnitrosourea-sugar derivative that lacks the strongly alkylating 2-chloroethyl group at the N′-position that is characteristic for BCNU, CCNU, MeCCNU, and chlorozotocin. For CGP 6809, a high antitumor activity has been demonstrated in transplantable tumors in mice and rats.[74] Promising antitumor activity was also observed against 10 human tumor xenografts in nude mice. In three xenografts, greater than 50% reduction of the inital tumor size was obtained and, in another xenograft, the drug prevented tumor growth. Regressions were observed in very rapidly, as well as slowly, growing tumors.[75] Clinical studies should demonstrate whether there is a good correlation between these experimental data and clinical responses.

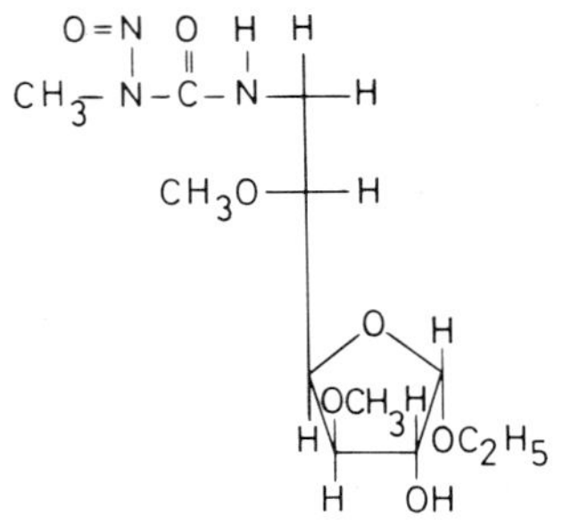

FIGURE 11. Structure of CGP6809.

III. DISCUSSION AND CONCLUSIONS

In this chapter, the chemotherapeutic effects of some investigational drugs against human tumor xenografts have been reviewed. In general, the response to antitumor agents of human tumors transplanted to nude mice has been observed to be similar to that in the clinic.[15,76-79] However, for some investigational drugs, the correlation between experimental data generated with xenografts and clinical data was very weak to nonexistent. This was the case for DON, azotomycin, DFMO, and mitozolomide.[19-22,36,39] Kubota et al.[17,79] found that the correlation between the nude mouse system and clinical reports was not excellent for colon cancer and small cell lung cancer, while this correlation was better for stomach and moderate for breast cancer. The tendency appears for human tumor xenografts to overestimate the activity of most standard drugs studied as a single agent.[16]

The reason for the different chemosensitivity profiles of some tumor types in nude mice may be explained by several factors. When the xenograft is originated from a cultured cell line, it tends to be sensitive to most of the drugs tested.[17] Indeed, when human tumor cells have been cultured *in vitro* for several years, any mutation might occur of relevance to its chemosensitivity. Another important factor is the question whether xenografts are representative for the clinical disease. It is well known that the overall take rate of the primary transplantation is less than 50%.[80] This might suggest an incomplete adaptation to nude mice of some human tumor cells with consequences for the pattern of sensitivity to drugs of some tumor types in nude mice.[17] However, the chemosensitivity profile from the original patients seems to be preserved in human tumor xenografts originated *in vivo* and is generally consistent throughout serial transfer.[81,82] This finding suggests that human tumor xenografts in nude mice may be excellent tools to supply a large amount of tumor cells *in vivo* for continuous and repeated experiments. However, one should be aware of the variables which may alter the validity of data obtained with the nude mouse-human tumor model and the experimental design should therefore reflect the treatment conditions in the clinic as much as possible.

1. For experimental therapy, such as with investigational agents, several aspects should be taken under consideration to increase the predictive value of human tumor xenografts (see also Chapter 21) The total body burden of tumor cells at the time of drug delivery should be high enough to reflect advanced cancer in the clinic, which means that the antitumor activity should be evaluated against well-established tumors. Alkyl lysophospholipids were active when administered 3 d after tumor implantation.[28] No evidence is available that those derivatives would show the same activity if administered when tumors have reached 600 mm^3. In the reports used for this chapter, the volume of tumors at the time of treatment initiation varied from 2 to 6 mm in diameter and occasionally reached 10 mm

in diameter.[48] When reported in cubic millimeters (mm^3), the tumor volume at the initiation of treatment varied between 50 and 200 mm^3 and occasionally measured 600 mm^3.[46]

2. The dose of the drug administered should be taken into consideration. While it is possible to study doses up to the LD_{10} in nude mice, equitoxic doses cannot be used in patients. Non-toxic doses effective in reducing tumor growth are the most suitable, while doses equal to or exceeding the LD_{10} — even when eliciting a 100% tumor growth inhibition — are meaningless.
3. The schedule and route of administration are factors of major importance. Repeated administration every 5 or 7 d may allow for sustained plasma drug levels during a relatively long period. The i.v. and the oral routes are desirable. The i.p. route is suitable to treat s.c. xenografts, while the i.p. route used to treat i.p. implants will favor the selection of drugs which poorly cross the physiological barriers[53] unless the experiment aims to select a drug for treatment of i.p. metastases. Unfortunately, administration of the drug in drinking water[36] will not favor the reproducibility of experimental data.
4. The criteria to evaluate antitumor activity in the laboratory should be comparable to those in use in the clinic. In this respect, tumor regression is of more importance in antitumor activity than merely tumor growth delay.
5. Obviously, each institute is using its own methodology and techniques to test investational agents, which may explain the conflicting results obtained between laboratories.[24,25] Techniques applied to transplant tumors are critical and may influence the composition of tumor cell populations. Because of this, the xenograft used in one laboratory may not have the same biological characteristics as a presumably identical xenograft grown in another laboratory. The mode of transplantation can affect the cellular heterogeneity and expression of phenotypic characteristics. Commonly, tumors are serially transplanted via tumor fragments. This repeated transfer of fragments may lead to the decrease of phenotypic heterogeneity.
6. Lack of uniformity in animal strains, animal number (treated and control groups), in animal husbandry, environmental conditions, and animal care may also bring about discrepancies in experimental results. The health status of nude mice and microbial infection of tumor cells adversely affect the outcome and the reproducibility of the screening of new drugs. It is desirable to use the same strain of mice from one experiment to another and in laboratories in case of a close collaboration. The number of tumor-bearing mice should not be below six for test groups to ensure statistical significance.
7. In order to assess the efficacy of investigational agents against human tumor xenografts of a particular tumor type, biological characteristics varying between the lines and the number of xenografts as a representation of the clinical disease should be sufficient. In most reports, the number of xenografts for a particular tumor type were often very low, except in the studies by Fiebig et al.[75] and Kubota et al.[17]
8. It is of major importance to identify a positive and a negative control drug for each human tumor line and to include these drug controls while investigating a new compound. Such controls will identify intrinsic test variations and experiments that are technical failures.

Finally, it is reasonable to say that a good evaluation of the predictive capacity of the human tumor xenograft model in the detection of new antitumor agents cannot be carried out correctly because clinical data are rarely available for drugs inactive in experimental models. However, it might be possible to improve the predictive value if the most relevant experimental conditions are chosen from the variables discussed. In addition, it is desirable to further investigate the host differences (e.g., drug metabolism and secretion) which may occasionally be the reason for the differences in tumor responses to a drug.

REFERENCES

1. **Geran, R. I., Greenberg, N. H., MacDonald, M. M., Schumacher, A. M., and Abbott, B. J.,** Protocols for screening chemical agents and natural products against animal tumors and other biological systems, *Cancer Chemother. Rep.,* 3, 1, 1972.
2. **Goldin, A. and Venditti, J.,** The NCI screen and its implications for clinical evaluation, *Rec. Results Cancer Res.,* 56, 207, 1978.
3. **Schabel, F. M., Griswold, D. P., Corbett, T. H., Laster, W. R., Mayo, J. G., and Lloyd, H. H.,** Testing therapeutic hypotheses in mice and man: observations on the therapeutic activity against advanced solid tumors of mice treated with anticancer drugs that have demonstrated a potential clinical utility for treatment of advanced solid tumors of man, in *Cancer Drug Development. Part B. Methods in Cancer Research,* Busch, H. and DeVita, V., Eds., Academic Press, New York, 1979, 3.
4. **Venditti, J. M.,** The model's dilemma, in *Design of Models for Testing Cancer Therapeutic Agents,* Fidler, I. J. and White, R. J., Eds., Van Nostrand Reinhold, New York, 1982, 80.
5. **Atassi, G. and Staquet, M.,** The clinical predictive value of the mouse screening methods, in *Anticancer Drug Development,* Hilgard, P. and Hellmann, K., Eds., J. R. Prous Publishers, Barcelona, 1983, 24.
6. **Atassi, G.,** Do we need new chemosensitive experimental models?, *Eur. J. Cancer Clin. Oncol.,* 20, 1217, 1984.
7. **Atassi, G., Dumont, P., and Fournier, J.,** New methods for determining tumor sensitivity, *Eur. J. Cancer Clin. Oncol.,* 21, 1299, 1985.
8. **Dumont, P., Van der Esch, E. P., Jabri, M., Lejeune, F., and Atassi, G.,** Chemosensitivity of human melanoma xenografts in immunocompetent mice and its histological evaluation, *Int. J. Cancer,* 33, 447, 1984.
9. **Abrams, J., Jacobovitz, D., Dumont, P., Semal, P., Mommen, P., Klastersky, J., and Atassi, G.,** Subrenal capsule assay of fresh human tumors: problems and pitfalls, *Eur. J. Cancer Clin. Oncol.,* 22, 1387, 1986.
10. **Hamburger, A. W. and Salmon, S. E.,** Primary bioassay of human tumor stem cells, *Science,* 197, 461, 1977.
11. **Shoemaker, R. H.,** New approaches to antitumor drug screening: the human tumor colony forming assay, *Cancer Treat. Rep.,* 70, 9, 1986.
12. **Weisenthal, L. M. and Lippman, M. E.,** Clonogenic and non-clonogenic *in vitro* chemosensitivity assays, *Cancer Treat. Rep.,* 63, 615, 1985.
13. **Kiss, R., de Launoit, Y., Gras, S., Thibaut, S., and Atassi, G.,** A new assay to evaluate cell growth and drug sensitivity in culture using a cell image processor, *Anticancer Res.,* 8, 765, 1988.
14. **Steel, G. G., Courtenay, V. D., and Peckham, M. J.,** The response to chemotherapy of a variety of human tumour xenografts, *Br. J. Cancer,* 47, 1, 1983.
15. **Giovanella, B. C., Stehlin, J. S., Shepard, R. C., and Williams, L. J.,** Correlation between response to chemotherapy of human tumors in patients and in nude mice, *Cancer,* 52, 1146, 1983.
16. **Mattern, J., Bak, M., Hahn, E. W., and Volm, M.,** Human tumor xenografts as model for drug testing, *Cancer Metast. Rev.,* 7, 263, 1988.
17. **Kubota, T., Ishibiki, K., and Abe, O.,** The clinical usefulness of human xenografts in nude mice, in *Prediction of Response to Cancer Therapy,* Liss, A. R., Ed., Alan R. Liss, New York, 1988, 213.
18. **Ovejera, A. A., Houchens, D. P., Catane, R., Sheridan, M. A., and Muggia, F. M.,** Efficacy of 6-diazo-5-oxo-L-norleucine and *N*(*N*- -glytamyl-6-diazo-5-oxo-norleucinyl)-6-diazo-5-oxo-neurleucine against experimental tumors in conventional and nude mice, *Cancer Res.,* 39, 3220, 1979.
19. **Ansfield, F. J.,** Phase I study of azotomycin (NSC-56654), *Cancer Chemother. Rep.,* 46, 37, 1965.
20. **Magill, G. B., Myers, W. P. L., Reilly, H. C., Putnam, R. C., Magill, J. W., Sykes, M. P., Escher, G. C., Karnofsky, D. A., and Burchenal, J. H.,** Pharmacological and initial therapeutic observations on 6-diazo-5-oxo-L-norleucine (DON) in human neoplastic disease, *Cancer,* 10, 1138, 1957.
21. Veterans Administration Cancer Chemotherapy Study Group. A clinical study of the comparative effect of nitrogen mustard and DON in patients with bronchogenic carcinoma, Hodgkin's disease lymphosarcoma and melanoma, *J. Natl. Cancer Inst.,* 22, 433, 1959.
22. **Weiss, A. J., Ramirez, G., Crage, T., Strawitz, J., Goldman, L., and Downing, V.,** Phase II study of azotomycin (NSC-56654), *Cancer Chemother. Rep.,* 52, 611, 1968.
23. **Lee, S. S., Giovanella, B. C., and Stehlin, J. S.,** Effect of excess thymidine on the growth of human melanoma cells transplanted in thymus deficient nude mice, *Cancer Lett.,* 3, 209, 1977.
24. **Lee, S. S., Giovanella, B. C., Stehlin, J. S., and Brunn, J. C.,** Regression of human tumors established in nude mice after continuous infusion of thymidine, *Cancer Res.,* 39, 2928, 1979.
25. **Howell, S. B., Jenkins, R. S., and Streifel, J.,** Activity of thymidine as a chemotherapeutic agent against human tumor xenografts in nude mice, *Cancer Res.,* 39, 3875, 1979.
26. **Sordillo, P. P., Helson, C., Lesser, M., and Helson, L.,** Effect of phase I and II chemotherapeutic agents against human lymphomas heterotransplanted in nude mice, *Oncology,* 40, 15, 1983.
27. **Tarnowski, G. S., Mountain, J. M., Stock, C. C., Munder, P.G., Weltzien, H.U., and Westphal, O.,** Effect of lysolecithin and analogs on mouse ascites tumors, *Cancer Res.,* 38, 339, 1978.

28. **Runge, M. H., Andreesen, R., Pfleiderer, A., and Munder, P. G.,** Destruction of human solid tumor by alkyl lysophospholipids, *J. Natl. Cancer Inst.*, 64, 1301, 1980.
29. **Németh, L., Institóris, L., Somfai, S., Gál, F., Pályi, I., Sugár, J., Csuka, O., Szentirmay, Z.,** and **Kellner, B.,** Pharmacologic and antitumor effects of 1,2:5,6-dianhydrogalactitol (NSC-132313), *Cancer Chemother. Rep.*, 56, 593, 1972.
30. **Houchens, D. P., Ovejera, A. A., Riblet, S. M., and Slagel, D. E.,** Human brain tumor xenografts in nude mice as a chemotherapy model, *Eur. J. Cancer Clin. Oncol.*, 19, 799, 1983.
31. **Fiebig, H. H., Widmer, K. H., Fiedler, L., Wittekind, C., and Löhr, G. W.,** Development and characterization of 51 human tumor models for large bowel, stomach and esophageal cancers, *Dig. Surg.*, 1, 225, 1984.
32. **Kopper, L., Lapis, K., and Hegedus, C.,** Chemotherapeutic sensitivity of human colorectal tumor xenografts, *Oncology*, 37 (Suppl. 1), 42, 1980.
33. **Metcalf, B. W., Bey, P., Danzin, C., Jung, M. J., Casara, P., and Vevert, J. P.,** Catalytic irreversible inhibition of mammalian ornithine decarboxylase by substrate and product analogues, *J. Am. Chem. Soc.*, 100, 2551, 1978.
34. **Kellen, J. A., Chan, A. W. L., and Mirarian, A.,** Effect of α-difluoromethyl-ornithine on the transplantable rat hepatoma 5123, 1-1, *Res. Commun. Chem. Path. Pharmacol.*, 30, 377, 1980.
35. **Prakash, N. J., Schechter, P. J., Grove, J., and Koch-Weser, J.,** Effect of α-difluoromethylornithine, an enzyme-activated irreversible inhibitor of ornithine decarboxylase, on L1210 leukemia in mice, *Cancer Res.*, 38, 3059, 1978.
36. **Luk, G. D., Abeloff, M. D., Griffin, C. A., and Baylin, S. B.,** Successful treatment with DL-α-difluoromethylornithine in established human small cell variant lung carcinoma implants in athymic mice, *Cancer Res.*, 43, 4239, 1983.
37. **Siimes, J., Seppänen, P., Alhonen-Hongisto, L., and Jänne, J.,** Synergistic action of two polyamine antimetabolites leads to a rapid therapeutic response in childhood leukemia, *Int. J. Cancer*, 28, 567, 1981.
38. **Khan, A. H. and Driscoll, J.,** Potential central nervous system antitumor agents. Aziridinylbenzaquinones.1, *J. Med. Chem.*, 19, 313, 1976.
39. **Schilsky, R. L., Kelley J. A., Ihde, D. C., Howser, D. M., Cordes, R. S., and Young, R. C.,** Phase I trial and pharmacokinetics of aziridinylbenzoquinone (NSC 182986) in humans, *Cancer Res.*, 42, 1582, 1982.
40. **Schold, S. C., Friedman, H. S., Bjornsson, T. D., and Bigner, D. D.,** Treatment of human glioma and medulloblastoma tumor lines in athymic mice with diaziquone and diaziquone-based combinations, *Cancer Res.*, 44, 2352, 1984.
41. **Schold, S. C., Bullard, D. E., Bigner, S. H., Jones, T. R., and Bigner, D. D.,** Growth, morphology and serial transplantation of anaplastic human gliomas in athymic mice, *J. Neuro-Oncol.*, 1, 5, 1983.
42. **White, L., Szirth, B. C., and Benedict, W. F.,** Evaluation of response to chemotherapy in retinoblastoma heterotransplanted to the eyes of nude mice, *Cancer Chemother. Pharmacol.*, 23, 63, 1989.
43. **Boven E., Van der Vijgh, W. J. F., Nauta, M. M., Schlüper, H. M. M., and Pinedo, H. M.,** Comparative activity and distribution studies of five platinum analogues in nude mice bearing human ovarian carcinoma xenografts, *Cancer Res.*, 45, 86, 1985.
44. **Evans, B. D., Raju, K. S., Calvert, A. H., Harland, S. J., and Wiltshaw, E.,** Phase II study of JM8, a new platinum analog, in advanced ovarian carcinoma, *Cancer Treat. Rep.*, 67, 997, 1983.
45. **Köpf-Maier, P. and Köpf, H.,** Antitumor metallocenes: New developments and toxicologic features, *Anticancer Res.*, 6, 227, 1986.
46. **Köpf-Maier, P., Moormann, A., and Köpf, H.,** Activity of titanocene dihalides against a human colon carcinoma heterotransplanted to athymic mice, *Eur. J. Cancer Clin. Oncol.*, 21, 853, 1985.
47. **Kovach, J. S., Moertel, C. C., Schutt, A. J., Reitemeier, R. G., and Hahn, R. G.,** Phase II study of *cis*-diammine-dichloroplatinum (NSC 119875) in advanced carcinoma of the large bowel, *Cancer Chemother. Rep., (Part I)*, 57, 357, 1973.
48. **Mattern, J., Keppler, B., and Volm, M.,** Preclinical evaluation of diethoxy-(l-phenyl-l,3-butanedionato) titanium (IV) in human tumour xenografts, *Arzneim. Forsch./Drug Res.*, 34 (II) 1289, 1984.
49. **Fodstad, Ø., Aamdal, S., Pihl, A., and Boyd, M. R.,** Activity of mitozolomide (NSC 353451), a new imidazotetrazine, against xenografts from human melanomas, sarcomas, and lung and colon carcinomas, *Cancer Res.*, 45, 1778, 1985.
50. **Fodstad, Ø., Aass, N., and Pihl, A.,** Assessment of tumour growth and of response to chemotherapy of human melanomas in athymic, nude mice, *Br. J. Cancer*, 41(Suppl. IV), 146, 1980.
51. **Hori, S., Shirai, M., Hirano, S., Oki, T., Invi, T., Tsukagoshi, S., Ishizuka, M., Takeuchi, T., and Umezawa, H.,** Antitumor activity of new anthracycline antibiotics, aclacinomycin A and its analogs, and their toxicity, *Gann*, 68, 685, 1977.
52. **Oki, T., Takeuchi, T., Oka, S., and Umezawa, H.,** New anthracycline antibiotic aclacinomycin A: experimental studies and correlation with clinical trials, *Rec. Results Cancer Res.*, 76, 21, 1981.
53. **Hamilton, T. C., Young, R. C., Louie, K. C., Behrens, B. C., McKoy, W. M., Grotzinger, K. R., and Ozols, R. F.,** Characterization of a xenograft model of human ovarian carcinoma which produces ascites and intraabdominal carcinomatosis, *Cancer Res.*, 44, 5286, 1984.

54. **Louie, K. C., Hamilton, T. C., Behrens, B. C., Grotzinger, K. R., McKoy, W. M., Myers, C. E., Young, R. C., and Ozols, R. F.,** Preclinical evaluation of aclacinomycin A for the intraperitoneal treatment of human ovarian carcinoma, *Cancer Chemother. Pharmacol.,* 18, 153, 1986.
55. **Sengupta, S. K., Kogan, Y., Kelly, C., and Szabo, J.,** New actinomycin D analogues as superior chemotherapeutic agents against primary and advanced colon tumors and colon xenografts in nude mice, *J. Med. Chem.,* 31, 768, 1988.
56. **Sirotnak, F. M., DeGraw, J. I., Schmid, F. A., Goutas, L. J., and Moccio, D. M.,** New folate analogs of the 10-deaza-aminopterin series. Further evidence for markedly increased antitumor efficacy compared with methotrexate in ascitic and solid tumors models, *Cancer Chemother. Pharmacol.,* 12, 26, 1984.
57. **Schmid, F. A., Sirotnak, F. M. Otter, G. M. and DeGraw, J. I.,** New folate analogs of 10-deaza-aminopterin series: markedly increased antitumor activity of the 10-ethyl analog compared to the parent compound and methotrexate against some human tumor xenografts in nude mice, *Cancer Treat. Rep.,* 69, 551, 1985.
58. **Greene, J. J., Ts'o, P. O. P., Strayer, D. R., and Carter, W. A.,** Therapeutic applications of double-stranded RNAs, in *Handbook for Experimental Pharmacology,* Came, P. E. and Carter, W. A., Eds., Springer-Verlag, Berlin, 1984, 535.
59. **Robinson, R. A., DeVita, V. T., Levy, H. B., Baron, S., Hubbard, S. P., and Levine A. S.,** A phase I-II trial of multiple-dose polyriboinosinic-polyribocytidylic acid in patients with leukemia or solid tumors, *J. Natl. Cancer Inst.,* 57, 599, 1976.
60. **Carter, W. A., O'Mally, J., Beeson, M., Cunnington, P., Kelvin, A., Vere-Hodge, A., Alderfer, J. L., and Ts'o, P. O. P.,** An integrated and comparative study of the antiviral effects and other biological properties of the polyinosinic-polycytidylic acid duplex and its mismatched analogues. III. Chronic effects, and immunological features, *Mol. Pharmacol.,* 12, 440, 1976.
61. **Strewler, G. J., Williams, R. D., and Nissenson, R. A.,** Human renal carcinoma cells produced hypercalcemia in the nude mouse and a novel protein recognized by parathyroid hormone receptors, *J. Clin. Invest.,* 71, 769, 1983.
62. **Hubbell, H. R., Kvalnes-Krick, K., Carter, W. A., and Strayer, D. R.,** Antiproliferative and immunomodulatory actions of β-interferon and double-stranded RNA, individually and in combination, on human bladder tumor xenografts in nude mice, *Cancer Res.,* 45, 2481, 1985.
63. **Carter, W. A., Strayer, D. R., Hubbell, H. R., and Brodsky, I.,** Preclinical studies with ampligen (mismatched double-stranded RNA), *J. Biol. Resp. Modif.,* 4, 495, 1985.
64. **Bregman, M. D. and Meyskens, F. L.,** Inhibition of human malignant melanoma colony-forming cells *in vitro* by prostaglandin A, *Cancer Res.,* 43, 1642, 1983.
65. **Bregman, M. D., Funk, C., and Fukushima, M.,** Inhibition of human melanoma growth by prostaglandin A, D and J analogues, *Cancer Res.,* 46, 2740, 1986.
66. **Levin, N., Queen, L., Chalom, A. A., and Daniels, L. J.,** Animal model of intracutaneous melanoma, *J. Invest. Dermatol.,* 78, 191, 1982.
67. **Schiff, P. B., Fant, J., and Horwitz, S. B.,** Promotion of microtubule assembly *in vitro* by taxol, *Nature, (London),* 22, 665, 1979.
68. **Wani, M. C., Taylor, H. L., Wall, M. E., Coggon, P., and McPhail, A. T.,** Plant antitumor agents. VI. The isolation and structure of taxol, a novel antileukemic and antitumor agent from taxus brevifolia, *J. Am. Chem. Soc.,* 93, 2325, 1971.
69. **Jacrot, M., Riondel, J., Picot, F., Leroux, D., Mouriquand, C., Beriel, H., and Potier, P.,** Action du taxol vis-à-vis de tumeurs humaines transplantées sur des souris athymiques, *C.R. Seances Acad. Sci. Paris,* III, 297, 597, 1983.
70. **Riondel, J., Jacrot, M., Picot, F., Beriel, H., Mouriquand, C., and Potier, P.,** Therapeutic response to taxol of six human tumors xenografted into nude mice, *Cancer Chemother. Pharmacol.,* 17, 137, 1986.
71. **Riondel, J., Jacrot, M., Nissou, M. F., Picot, F., Beriel, H., Mouriquand, C., and Potier, P.,** Antineoplastic activity of two taxol derivatives on an ovarian tumor xenografted into nude mice, *Anticancer Res.,* 8, 387, 1988.
72. **Fujimoto, K., Oka, T., and Morimoto, M.,** Antitumor activity of a novel antitumor antibiotic, quinocarmycin citrate (KW2152), *Cancer Res.,* 47, 1516, 1987.
73. **Inoue, S., Kubota, T., Ohishi, T., Kuzuoka, M., Oka, S., Shimoyama, Y., Kikuyama, S., Ishibiki, K., and Abe, O.,** Antitumor activity of quinocarmycin citrate (KW-2152) against human tumor xenografts serially transplanted into nude mice, *Keio J. Med.,* 37, 355, 1988.
74. **Schieweck, K., Stanek, J., Kanter, P. M., Schmidt-Ruppin, K. H., Müller, M., and Matter, A.,** CGP6809, a sugar-containing nitrosourea derivative: pharmacological and physiochemical properties, *Cancer Chemother. Pharmacol.,* 23, 341, 1988.
75. **Fiebig, H. H., Widmer, K. H., Winterhalter, B. R., and Löhr, G. W.,** CGP6809 — a new nitrosoureido-sugar derivative with activity in human tumor xenografts, *Cancer Chemother. Pharmacol.,* 23, 337, 1989.
76. **Polvsen, C. O. and Jacobsen, G. K.,** Chemotherapy of a human malignant melanoma transplanted in the nude mouse, *Cancer Res.,* 35, 2790, 1975.

77. **Osieka, R., Houchens, D. P., Goldin, A., and Johnson, R. K.,** Chemotherapy of human colon cancer xenografts in athymic nude mice, *Cancer,* 40, 2640, 1977.
78. **Fujita, M., Hayata, S., and Taguchi T.,** Relationship of chemotherapy on human cancer xenografts in nude mice to clinical response in donor patient, *J. Surg. Oncol.,* 15, 211, 1980.
79. **Kubota, T., Shimosato, Y., and Nagai, K.,** Experimental chemotherapy of carcinoma of the human stomach and colon serially transplanted into nude mice, *Gann,* 69, 299, 1978.
80. **Shimosato, Y., Kameya, T., Nagai, K., Hirohashi, S., Koide, T., Hayashi, H., and Nomura, T.,** Transplantation of human tumors in nude mice, *J. Natl. Cancer Inst.,* 56, 1251, 1976.
81. **Kubota, T., Kabouchi, K., Koh, J., Enomoto, K., Ishibiki, K., and Abe, O.,** Human breast carcinoma (MCF-7) serially transplanted into nude mice, *Jpn. J. Surg.,* 13, 381, 1983.
82. **Kubota, T., Nakada, M., Tsuyuki, K., Inada, T., Asanuma, F., Ishibiki, K., and Abe, O.,** Cell kinetics and chemosensitivity of human carcinomas serially transplanted into nude mice, *Jpn. J. Cancer Res., (Gann),* 77, 502, 1986.

Chapter 16

HORMONES

G. J. van Steenbrugge

TABLE OF CONTENTS

I. INTRODUCTION

Nude mice are of great potential usefulness to endocrinologists because they provide a congenitally immunosuppressed host in which transplantable endocrine tumor lines can be developed originating from human tumor tissue. Generally, the transplantability of human endocrine tumors is limited. Most, if not any, of the hormone-responsive tumor types, such as breast, prostate, and endometrial cancer, belong to the group of tumors with a moderate to poor take rate in the nude mouse. It is tempting to speculate that the relatively poor transplantability can at least partially be attributed to the impaired endocrine functions of the nude mouse. Determination of the endocrine milieu, that is, estimation of the circulating levels of (steroidal) hormones, is a mandatory step for further studies on the behavior of hormone-responsive tumors as xenografts in these animals. This also applies to the comparison between endocrine profiles in nude mice and in man. A good similarity in hormone levels obviously determines the validity of the nude mouse as a model for studies on the hormonal control of hormone-dependent growth in human tumor xenografts. In general, the suitability of an established tumor model depends on the extent to which its characteristics reflect the properties of that particular cancer in the clinical situation. For hormone-responsive cancers, additional criteria should be fulfilled, such as the retention of hormone receptors when originally present and prolonged responsiveness of the tumor tissue upon hormonal manipulation. Once these criteria have been met, such experimental tumor systems in nude mice allow detailed analysis of the mechanism of action of hormones in their (tumor) target tissue and of the effects of different forms of endocrine therapy in a physiological environment.

II. ENDOCRINE FUNCTIONS IN NUDE MICE

A. REPRODUCTIVE FUNCTIONS

Congenitally athymic nude (nu/nu) female mice are known to have severe deficiencies in the reproductive system, including reduced ovarian weight, increased follicular atresia, and decreased fertility, as compared to their heterozygous littermates.[1,2] Interaction between the thymus gland and the reproductive system has been proposed for several years.[3,4] In general, nude mice weigh less than their heterozygous littermates, as do most of their endocrine organs. Rebar et al.[5] showed that nude mice have significantly reduced circulating levels of both FSH and LH. These data strongly suggest that the decreased gonadotropin concentrations result in the reproductive abnormalities observed in nude mice. Weinstein[6] demonstrated that the ovaries of female nudes are normal and that the malfunction resulted from insufficient gonadotropin secretion from the pituitary, but his results also suggested that these animals suffer from impairment of hypothalamic control of the pituitary. It was also shown that the reduction of gonadotropin concentrations in nude mice was completely prevented in females and partially prevented in males by thymic transplantation on the first day of life.[7] The reduced concentrations of gonadotropins in nude mice appear to be causally related to the absence of the thymus and the data suggest that the thymus, directly or indirectly, is necessary for the development of normal function of the hypothalamic-pituitary-gonadal axis in mice. However, nude mice are fertile[8] and have a normal development of sexual dimorphism, as shown by Davidson et al.[9] They also demonstrated that the abnormal endocrine and reproductive functions of nude mice observed by other investigators[1,2] may not entirely result from thymic aplasia; when nude mice were reared and maintained in a specific pathogen-free (SPF) environment, these abnormalities were significantly reduced. Evidently, this finding relates to environmental stress to which nude mice are known to be very susceptible.

TABLE 1
Serum Steroid Levels in BALB/c Nude Mice and Their Immunocompetent Littermates

	E_2	T	DHT	A	P
BALB/c +/+ males	<40	22 ± 11.88	0.66 ± 0.18	1.0 ± 0.28	14.4 ± 6.48
BALB/c +/+ females	<40	0.77 ± 0.14	0.49 ± 0.13	0.75 ± 0.21	9.8 ± 4.41
BALB/c nu/nu males	<40	10.3 ± 3.09	0.50 ± 0.14	0.71 ± 0.20	17.6 ± 7.92
BALB/c nu/nu castrated males	<40	0.73 ± 0.06	0.49 ± 0.13	0.79 ± 0.22	6.9 ± 2.90
BALB/c nu/nu females	<40	0.91 ± 0.11	0.49 ± 0.13	0.83 ± 0.23	3.7 ± 1.67
BALB/c nu/nu castrated females	<40	0.82 ± 0.07	0.42 ± 0.11	0.87 ± 0.24	3.4 ± 1.53

Note: Mice at the age of 8 weeks. E_2: estradiol (pmol/l ± S.D.); T: testosterone; DHT: dihydrotestosterone; A: androstenedione; P: progesterone (nmol/l ± S.D.).

B. STEROIDOGENESIS

Determination of the endocrine profile in nude mice and a comparison with that in man are obligatory steps for valid application of the nude mouse model in studies on the influences of hormones on human tumor tissue. In a number of studies, a comparison has been made for serum steroid levels between nude mice and their heterozygous littermates. Generally, it was shown that nudes have a normal steroidogenesis, but decreased serum steroid levels as compared to normal immunocompetent mice.[10-12] This was attributed mainly to the previously described hypothalamic lesions in these animals,[6] which consequently impair gonadotropin secretion by the pituitary gland.[5]

Particularly in the first several days of life, nude mice have significantly reduced circulating levels of both FSH and LH.[4] Data on gonadotropin concentrations in blood and in the pituitary of nude mice differ greatly and depend on the time of blood sampling (e.g., cycling female mice), the strain of mice investigated, and the environmental conditions, as well as on the method used for measurement of gonadotropin levels.[5-7,12,13]

In a large-scale study by Brünner et al.[10] using 8-week-old BALB/c nude mice, a comparison was made between serum steroid levels found in intact and endocrine ablated nude mice of both sexes and in their normal homozygous littermates; the most relevant data are shown in Table 1. In all groups of mice, the levels of androstenedione were low, as were the levels (<5 nmol/l) of dehydroepiandrosterone (not included in the table). Observations made by Van Weerden et al.[14] confirmed that circulating levels of the adrenal androgens, androstenedione and dehydroepiandrosterone (respectively 0.9 and 0.7 nmol/l), were relatively low in comparison to those found in men. These low levels seemed to be related to a 17-hydroxylase deficiency in the adrenal glands of the mice.

With respect to circulating steroid levels, it is noteworthy that, contrary to man, the mouse plasma does not contain detectable amounts of sex hormone-binding globulin (SHBG).[11] Consequently, in the mouse the fraction of circulating free steroids is much higher than in man.

1. Female Mice

The serum level of estradiol in female nude mice, as in all other groups of animals investigated by Brünner et al.,[10] was below the detection limit (40 pmol/l) of the assay used. These levels are in agreement with values reported by our group,[11,15] but are far below those published by Seibert et al.[16] and by Davidson et al.[9] who measured serum estradiol levels of the order of 100 to 150 pmol/l. The last investigators claimed that the normal and relatively high steroid levels in their nude mice could be ascribed to the SPF conditions. However, the results reported by Brünner et al.[10] were also obtained in SPF mice kept under sterile conditions. Particularly for studies on transplantable human breast cancers and, to a lesser extent, on endometrial cancers in nude mice, knowledge on the circulating estradiol levels is essential for the interpretation of the results. In female nude mice, the level of progesterone was significantly lower than in normal littermates, whereas no difference was observed with regard to the levels of serum testosterone, dihydrotestosterone, and androstenedione.[10]

In order to determine the relevance of female nude mice for the clinical situation of female patients, a comparison was made between serum steroid concentrations in both species. Serum androstenedione and, more importantly, estradiol in female mice appeared to be significantly lower than the levels in women, whereas serum testosterone and dihydrotestostereone levels in female mice were similar to the levels in both pre- and post-menopausal women. Progesterone levels were found to be higher than the levels reported in women. For more information on steroid levels in different species, including man and mice, the literature survey of Overpeck et al.[17] is recommended.

2. Male Mice

In male mice, the levels of testosterone and androstenedione were found to be significantly lower in nude mice than in immunocompetent littermates, whereas the levels of dihydrotestosterone and progesterone were similar in the two types of mice.[10] We observed a wide range of circulating levels of testosterone in intact male nude mice, varying from less than 1.0 nmol/l to levels exceeding 90 nmol/l (Figure 1).[11] Although these data were obtained in tumor-bearing nude mice which were maintained under different circumstances, the same variability was observed in mice of similar age, housed separately, and sampled at the same time of the day. This variability in testosterone levels is not unusual in males; Bartke et al.[18] observed a 30-fold difference in plasma testosterone levels in normal male mice housed and bled under identical circumstances. It was concluded that, in male (nude) mice, testosterone is released in a pulsatile fashion, as is also described for the rat.[19] In order to improve the endocrine circumstances in male nude mice for the acceptance and growth of androgen dependent tumors (i. e., prostate cancers), constant release implants with testosterone were successfully applied in this host.[11,19] Such implants are an important tool for hormonal manipulation in tumor bearing nude mice.

Compared to the levels in man, the serum concentrations of testosterone, dihydrotestosterone, and androstenedione in male nude mice were substantially lower, whereas serum progesterone was higher than the levels described for human males.[10]

With respect to the wide-spread use of LHRH-analogs to obtain a chemical castration in the man, it is noteworthy that mice can be completely or partially insensitive to the antitesticular effects of such compounds.[20,21] This phenomenon was also demonstrated using an LHRH-antagonist in the male nude mouse.[22] Recently, Redding and Schally[23] obtained a complete suppression of plasma testosterone levels in male nude mice by application of a sustained release formulation of an LHRH-antagonist. They thus proved that nude mice can be valuable for studies on the effects of LHRH analogs on tumor biology.

III. ENDOCRINE-DEPENDENT TUMORS

A. TRANSPLANTABILITY

A survey of the literature on human tumor transplantation to nude mice reveals that tumors of the colon, skin, and lung are generally easy to establish as serial lines, whereas hormone-dependent or hormone-secreting tumors, as well as lymphomas, present more difficulty.[24,25] The hormone-dependent tumors must be provided with the proper trophic hormone stimulation in the host, a requirement that, in certain cases, may be impossible to meet. For hormone-secreting tumors, on the other hand, it can be even more difficult to provide the proper metabolic milieu *in vivo*. Many hormone-secreting cells are not only by themselves dependent on other trophic

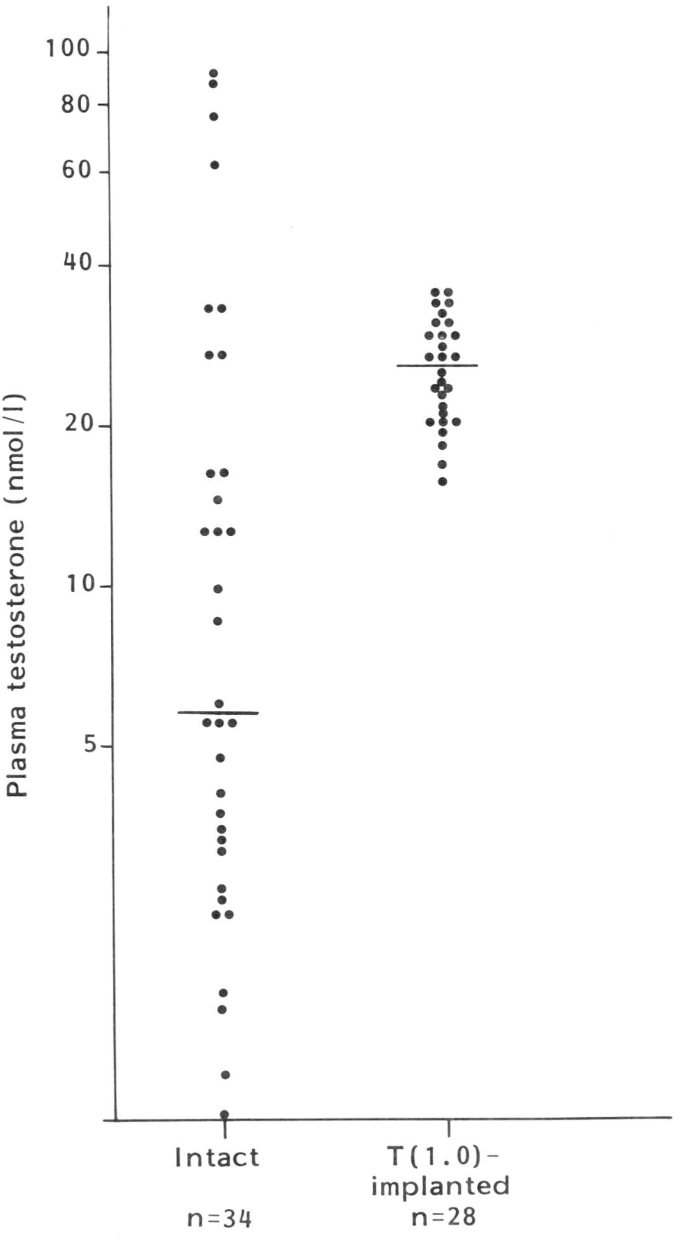

FIGURE 1. Plasma levels of testosterone (T) in intact and in T-implanted PC-82 tumor-bearing male nude mice.

stimuli, but also lack a physiological feedback inhibiting the secretion of tumor-specific hormone, which might cause death of the host. Therefore, it can be predicted that the human tumor lines established from endocrine tissues originate from either target tissues, hormone-secreting tumors that have lost all or most of the secretory function, or tumors with hormonal products, even when secreted in large amounts, do not have a lethal effect on the nude mouse host.

Human tumors in which hormonal treatment may be of importance and successfully established in nude mice include breast,[26-29] prostate,[11,30-33] and endometrial cancers,[34-37] but also some ovarian,[38] laryngeal,[39] pancreatic,[40] colorectal,[41,42] gastric,[43] and even lung tumor models[44] have been described to respond to hormonal treatment. Serial transplantability should be used as a main criterion for a certain tumor to be considered as an established model in nude mice. In this respect, it is generally recognized that breast, prostate, and endometrial cancers have a relatively low take rate, particularly when hormonal responsiveness is used as another main criterium for these tumors. Among the models established from tumors clinically known to be endocrine dependent, cell lines are also described with a completely hormone-independent pattern of growth.

For the three tumor types with clearly proven endocrine dependence, overall take rates are described as varying from less than 5% for prostate,[45-47] and 10 to 15% for breast[29,48,49] and endometrial[36,50] cancers. In these figures, transplantable tumor lines are included which either were directly established in nude mice or were initially cultured *in vitro,* and subsequently inoculated into nude mice. There is a tendency towards a higher take rate for established *in vitro* cell lines than for tumor samples directly taken from the patient, exemplified by Ozello et al.[51] for breast cancer cell lines.

B. HORMONE RESPONSIVENESS

Upon review of the literature on endocrine-dependent tumors, it appears that concepts like hormone responsiveness, dependency, and sensitivity are not uniformely used. On the one hand, responsiveness is used as a general term to indicate whether or not the growth of a certain tumor is affected by endocrine manipulation. This, for example, applies for hormonal as well as anti-hormonal treatment.[29] On the other hand, the tumor itself is assumed to consist of a number of phenotypically distinct cell clones with different degrees of hormone sensitivity. As a result of studies with the Dunning rat prostate tumor system, Isaacs and Coffey[52] described three possible cell types in prostate cancer: androgen-dependent, androgen-sensitive, and androgen-independent cells. Based on these findings with an androgen target tissue, it can be defined more generally that hormone-dependent cells chronically require a critical level of hormone stimulation for their continued growth. In contrast to the hormone-dependent cells which die without adequate hormonal stimulation, hormone-sensitive cells grow faster in the presence of sufficient levels of the trophic hormone and can still continue to grow without hormone. Hormone-independent cells neither die nor slow down their continuous growth following hormone ablation; that is, the growth of such cells is completely autonomous.

The extent to which hormones are involved in growth and function of human tumor xenografts varies from completely dependent, as shown for some of the established prostate and breast cancer models, to a certain degree of hormone sensitivity, as described for some endometrial, ovarian, colorectal, and other cancers. Examples of breast cancer xenografts that are completely dependent on estrogens are the MCF-7, ZR-75, and BR-10 tumors which do not grow in oophorectomized nude mice, but which develop in estrogenized animals.[29] Similarly, the androgen-dependent prostate cancers, PC-82, PC-EW, HONDA, and TEN-12, cannot be grown in female or castrated male mice, whereas they readily develop when sufficient levels of circulating testosterone are maintained in these animals.[30-33] Table 2 contains information on various tumor types established as a model in nude mice and on the hormone(s) by which these tumors are influenced.

C. STEROID HORMONE RECEPTORS

Hormone receptors are specific proteins found in hormone target tissues. The interaction of a hormone with its receptor is an early event for the initiation of hormonal responses.[53] The presence of hormone receptors in a certain type of (tumor) tissue is assumed to be indicative for the responsiveness of such a tissue to the specific hormone. Determination of (steroid) hormone

TABLE 2
Transplantable Hormone-Responsive Tumor Lines Established in Nude Mice

Tissue type	Hormone(s) involved	Dependent (D)/ sensitive (S)[1]	Ref.
Breast	Estrogens/ progestins	D and/or S[2]	26–29,48,49,51,57, 59–61,64,65,67
Prostate	Androgens	D	11,30–33,47
Endometrium	Estrogens/ progestins	S	34–37,50,62,63
Ovary	Estrogens/ androgens	S	38
Larynx	Estrogens	S	39
Colorectum	Estrogens/ androgens	S	41,42
Pancreas	Androgens	S	40
Brain (meningioma)	Progestins	S	54

[1] Hormone dependency or sensitivity.
[2] Both estrogen dependency and sensitivity have been reported; progestin involvement is restricted to hormone sensitivity.

receptors in human tumor lines grown in nude mice should be part of their characterization. Significant amounts of high-affinity binding proteins were detected in breast and endometrial cancers (estrogen and progesterone receptors),[28,29,34,49] in laryngeal tumors (estrogen receptors),[39] meningiomas (progesterone receptors),[54] prostate (androgen receptors),[30-33] and lung tumors (glucocorticoid receptors) grown as xenografts.[43] The majority of the receptor-containing tumor models appeared to be hormone responsive. Some exceptions have been described, such as a hepatoma xenograft expressing high amounts of androgen receptors, but without response to androgen withdrawal therapy.[55] Steroid receptors in human tumor models may be well preserved during serial transfer, which is in contrast to rodent models such as mammary tumors where loss of estrogen receptors is a common observation.[56]

IV. HORMONAL MANIPULATION OF ENDOCRINE DEPENDENT TUMORS

A. TYPES OF ENDOCRINE THERAPY

Similar to the clinical situation for endocrine treatment of hormone-dependent tumors in nude mice, four categories can be distinguished: (1) ablative endocrine therapy in which hormone-producing organs are removed; (2) inhibitive endocrine treatment in which synthesis of steroids is blocked chemically; (3) additive endocrine treatment in which high doses of steroids are administered; and (4) competitive endocrine treatment utilizing agents that bind to steroid hormone receptors. As most experience with endocrine therapy in nude mice has been obtained with transplantable breast, prostate, and (to a lesser extent) endometrial cancers, this chapter will focus mainly on these three types of tumors.

1. Ablative Endocrine Therapy

Upon ablative treatment (oophorectomy, orchiectomy, adrenalectomy, or hypophysectomy), the circulating levels of endogeneously produced hormones are reduced. In the nude mouse, oophorectomy (via lateral incisions in the flanks) and orchiectomy (via the scrotal route) are easily performed. This type of treatment is commonly used for initially checking the endocrine dependence of tumor models once established, and for control of this characterisitic during subsequent transfer of hormone responsive tumor lines. The changes reported in serum steroid hormones following oophorectomy[10,16] and orchiectomy[10,19] correspond to those observed after similar endocrine ablations in man. Adrenalectomy is a difficult procedure and glucocorticoid replacement is necessary; moreover, it is associated with relatively high mortality. Hypophysectomy in nude mice is technically feasible,[27] but data on the effects on serum hormone levels are not available.

2. Inhibitive Endocrine Treatment

The blockade of steroid hormone synthesis, either at the peripheral or at the target site, will result in a reduction of the levels of the specifically inhibited hormone. The number of studies with this type of treatment in nude mice is limited. Fukutomi et al.[57] applied an aromatase (converting androgens to estrogens) inhibitor to breast cancer xenografts. The treatment led to a reduction of serum estradiol, a decrease in mouse uterine weight, and a (partial) tumor growth inhibition, which is consistent with clinical experience with this drug. Administration of a 5α-reductase inhibitor (blocking the conversion of testosterone to dihydrotestosterone) to nude mice bearing androgen-dependent human prostate cancers resulted in a castration-like effect on this tumor type.[58]

3. Additive Endocrine Therapy

This type of endocrine treatment includes the administration of additional, usually pharmacological, concentrations of steroids such as estrogens, progestins, or androgens. High doses of estradiol, applied via Silastic implants resulting in circulating E_2 levels of 4 to 5 nmol/l, administered to male nude mice were shown to suppress the circulating concentration of testosterone to (near) castrate level. As a consequence, the growth of the PC-82 tumor, an androgen-dependent prostate cancer in nude mice, was effectively inhibited.[15] Whether these high estrogen levels also directly affect the growth of this experimental prostate tumor could not clearly be proven. With respect to additive endocrine therapy in breast cancer xenografts, it is noteworthy that pharmacological doses, in the range of those mentioned above (plasma levels up to 10 nmol/l), were stimulatory for the estrogen-dependent MCF-7 cell line, and inhibited the growth of the T-61 mammary tumor cell line.[28,29,59] The fact that both cell lines contain a significant number of estrogen receptors suggests that the underlying mechanisms for these opposite reponses are still poorly understood.

4. Competitive Endocrine Treatment

Competitive endocrine treatment utilizes agents that bind to hormone receptors, thereby blocking the biological activity of the native hormone. This treatment type (i. e., the application of anti-hormones) has been studied to a limited extent in nude mice and mainly in breast cancer xenografts. The effects of the anti-estrogen, tamoxifen, and, to a lesser extent, LY156758 and toremifene were studied in transplants of breast tumors[28,60,61] and endometrial cancers.[62,63] In two studies of Brünner et al.[61,64] in breast cancer xenografts, the pharmacokinetics of tamoxifen, the most widely applied anti-estrogen, was also described. Recently, Schneider et al.[65] reported on the successful application of a progesterone antagonist (onapristone) to the T61 breast tumor cell line grown in nude mice.

B. ROUTES OF HORMONE ADMINISTRATION

Additive endocrine therapy comprises the administration of exogenous hormones, either at a physiological or at a pharmacological dose. For hormonal manipulation experiments in endocrine-dependent xenografts, various routes of administration have been described: the single or frequent injection of hormonal solutions; the use of time release pellets; steroid-containing silicone tubing (Silastic implants), or Alzet minipumps; and the administration of hormonal compounds by gavage, that is, by intubation (p.o.). The methods most often used will be discussed in this chapter. Whatever route is selected for hormonal manipulation experiments, monitoring of circulating levels of the hormone of interest is an obligatory step to validate this therapeutic approach.

1. Injection

Administration of steroids has been performed either by injection of a single depot or by frequent injection of solutions or suspensions. In a study by Brünner et al.[66] using a breast cancer model, 17β-estradiol (E_2) was administered as a single i. m. depot. The injected E_2 at doses of 0.001 to 1.0 mg resulted in peak plasma concentrations at day 1 ranging from levels that would be physiological (0.7 n*M*) to pharmacological (35 n*M*) in premenopausal women. With a gradual decline, these levels lasted for periods up to 3 weeks. Other investigators administered steroids as i. m. injection (E_2-dipropionate; dissolved in sesame oil) once a week,[67] as daily s. c. injection (E_2 dissolved in a mixture of saline, polysorbate, carboxymethylcellulose, and benzylalcohol),[38] or as daily i.p. injection (E_2 dissolved in saline with Myrj 53).[41] However, in none of these studies was the measurement of circulating steroid levels included. Particularly when constant hormone levels are necessary, the use of slow release pellets or Silastic implants are preferable to the frequent injection of oil suspensions, a method that is associated with fluctuating steroid concentrations in the blood caused by the rapid absorption of the steroid hormone from the site of injection.[68]

2. Slow Release Pellets

In a study by Shafie and Grantham,[27] constant and physiological plasma levels of E_2 were obtained in nude mice by s.c. implantation (in the interscapular region) of pellets containing the steroid and cholesterol as carrier. In the same study, prolactin was administered to mice bearing breast cancer xenografts by using Alzet minipumps. Many investigators have used slow release pellets consisting of a cholesterol/methylcellulose/lactose matrix. These pellets are designed to provide a controlled release of the relevant compound (e.g., hormone) for a period of 60 or 90 d, and they are easily placed in the interscapular region of (nude) mice by means of a trocar. With pellets containing 0.05, 0.5, and 5 mg of E_2, applied to nude mice bearing an endometrial cancer cell line, plasma levels of 25, 250, and 2500 pg/ml, respectively, could be reached.[35] In estrogen-dependent breast tumors in nude mice, such as the MCF-7 cell line, a standard pellet of 0.5 mg E_2 had to be applied for optimal growth.[27,28]

3. Silastic Implants

Another reliable method for long-term administration of steroids to nude mice, is the use of silicone tubing ("Silastic" implants) filled with crystalline steroid.[69] Such devices can be implanted s. c. into nude mice, and steroids are released into the blood of the animals at a constant rate, providing a stable hormonal milieu for long periods (more than 3 months) of time. A disadvantage of the method is the fact that the implants are not commercially available and, therefore, have to be made by hand via a rather time-consuming procedure.[19] Testosterone (T) filled silastic implants were successfully applied to the PC-82 model, a hormone-dependent cell line transplantable to nude mice.[11,15] By the use of a standard 1.0-cm T-implant, fairly constant levels of plasma-T were reached, in contrast to extremely fluctuating levels of circulating T in intact male nude mice (Figure 1). This significantly improved the take and growth rate of the PC-82 tumor.[11] In addition, the implants can be easily installed and removed, which facilitates hormonal manipulation experiments with endocrine-dependent tumors.[11,15,61] Maximum blood levels of the substituted hormone were shown to be reached at 4 h after installation of the implants.[11]

The release rate of T from the Silastic implants appeared to be too high to obtain low near-castrate levels of circulating T in nude mice.[14] To circumvent this problem, Silastic capsules of 0.6 cm length were packed with cholesterol mixed with different proportions of T. Accordingly, castrated male or female nude mice substituted with these implants showed plasma-T levels

TABLE 3
Plasma Concentration of Estradiol (E_2) and Testosterone in Control, E_2-Implanted, or Castrated Male Nude Mice

Group	N	Plasma-E_2, pmol/l			Plasma-T nmol/l		
Intact controls	11	42	±	15	11.1	±	9.8
E_2-implant (0.5 cm)	9	946	±	211	2.5	±	1.0[1,2]
E_2-implant (2 × 2.0 cm)	8	4442	±	962	1.0	±	0.2[1,2]
Castration	8	40	±	13	0.4	±	0.3[1]

Note Blood of the mice was collected at 7 d after implantation; values expressed as mean ± S.D.

[1] Significantly different from control group ($p < 0.001$).
[2] Significantly different from castrated group ($p < 0.001$).

varying from 0.2 (castrate level) up to 20 nmol/l (the normal physiological level in men). In a study by Van Weerden et al.,[70] the critical level of androgen for growth response of the PC-82 tumor could be assessed with these low-dose implants. In another study with the PC-82 tumor, relatively low and high circulating levels of plasma-E_2 were reached by the application of a single E_2-implant of 0.5 cm and a double implant of 2.0 cm length (Table 3), respectively.[15] PC-82 is an androgen-dependent tumor; concomitant measurements of plasma-T levels were obligatory in order to draw conclusions from the results obtained. In Table 3, it is indicated that even with the high (pharmacological) dose of E_2 applied, the plasma-T levels in the male mice were still significantly higher than castrate level.

C. EXPERIMENTAL DESIGN AND EVALUATION OF EFFECTS

For the study on hormonal influences on tumor cells by use of human tumor xenografts in nude mice, two main concepts can be distinguished: (1) the direct approach, considering both the take rate and growth of a tumor transplant, and (2) the indirect approach in which established tumors in the host are treated. Van Weerden et al.,[70] studying the androgen responsiveness of the PC-82 prostate tumor model, have described both concepts. In the direct approach, nude mice are implanted with tumor tissue and simultaneously substituted with hormone, or a comparison is made between take and growth of a hormone-responsive tumor in intact and in hormone-deprived animals. Generally, in such a protocol, the tumor transplant is directly exposed to the hormonal milieu. With the indirect approach, the possible effect on the take of the growth-stimulating hormone(s) of interest is circumvented by transplanting the tumor under standard (hormonal) conditions prior to treatment. Once tumors have been developed, treatment is started, and a possible influence of the hormone(s) on the exponential growth can be determined.

The techniques available for studies on the effect of endocrine treatment in human tumor xenografts in nude mice can be divided into *in situ* assays and excision assays. *In situ* assays include measurement of regression rate, quantification of tumor shrinkage, calculation of regrowth delay, and cell kinetic assays on cells obtained from fine needle tumor aspirates. Excision assays include studies with tracer isotopes, estimation of cell survival by the *in vitro* clonogenic assay, and morphological studies.

The most frequently used method to evaluate the response of xenografts to endocrine therapy is the calculation of changes in tumor growth, by measuring tumor nodules with calipers, determined as a function of time after treatment. Here, reference is made to the majority of reports dealing with the subject on (endocrine) treatment of experimental tumors. Conventional cell kinetic methods (e.g., labeling with tracer isotopes and histological determination of the mitotic index) have also been used to evaluate the effect of hormonal treatment.[28,67] Sequential fine-needle aspirations for flow cytometric DNA analysis have been successfully applied by

Brünner et al.[59,66] This assay was used to evaluate the effect of oophorectomy as well as treatment with estradiol or tamoxifen on the cell cycle distribution of human breast cancer in nude mice. Recently, the Ki-67 monoclonal antibody has become available, which is directed against a proliferation-associated nuclear antigen. This antibody was shown to be a very suitable marker for monitoring the response of the PC-82 prostate tumor to hormonal manipulation, that is, androgen withdrawal and resubstitution.[47,71] Another approach to assess this proliferative activity of human prostate cancer tissue in nude mice is the BrdU anti-BrdU immunohistochemical technique.[47,72]

V. CONCLUDING REMARKS

In the present evaluation of hormones in the nude mouse-human tumor model, it was demonstrated that the animal is suitable for studies of hormone-tumor interactions in hormone-responsive human cancers. Pharmacological monitoring of endocrine treatment in nude mice is necessary to document the serum concentrations required, rather than to assume that levels are equivalent to those reported in man.[10,11,15,19] The available information on experiments with human tumor xenografts has shown that a variety of endocrine therapeutic approaches in use in the clinic are feasible for use in the nude mouse model. Ablative and inhibitive endocrine treatment have been shown to induce changes in hormone concentrations comparable to those observed in patients following similar treatments.

It must be emphasized that potential differences in the host with respect to studies on hormones (i.e., its hormonal status) and also hormonal pharmacokinetics and metabolism, immunological reactions, and other factors, preclude direct extrapolation of results obtained from the nude mouse to patients. For example, in some studies with estrogenized nude mice bearing estrogen-negative endometrial or laryngeal tumors, it has been suggested that the observed growth enhancement was caused by host-related rather than tumor-related factors.[36,39] The fact that nude mice have normal or even increased activities of natural killer (NK) cells,[73] and that estradiol is a strong inhibitor of NK cell activity,[74] might explain such observations. For testing this particular aspect, Jordan et al.[75] have considered the use of the beige (NK cell-deficient) mice. Another host-related factor that should not be ignored is the observation that hormones affect the secretion of growth factors. For instance, androgens have been shown to stimulate the secretion of epidermal growth factor from the submandibular gland in the mouse.[76] The involvement of growth factors in growth and function of experimental human tumor cell lines *in vitro*, as well as *in vivo*, has been clearly demonstrated.[77]

Finally, although xenografts and cell lines grown as tumors in nude mice respond appropriately to hormonal manipulation as expected from their hormone receptor status, it is unlikely that this model will prove useful to predict endocrine dependence in the individual patient. Particularly, the low take rate and slow growth of tumor samples of endocrine-related tumors, the long period of observation required, and the possible selection of subpopulations upon growth in the host make the nude mouse model impractical for direct patient-related research. Nevertheless, as an *in vivo* model, the nude mouse will continue to contribute to further information on the biology of hormone-responsive human cancers.

REFERENCES

1. **Besedovsky, H. O. and Sorkin, E.,** Thymus involvement in sexual maturation, *Nature (London)*, 249, 356, 1974.
2. **Lintern-Moore, S. and Pantelouris, E. M.,** Ovarian development in the athymic nude mice. I. The size and composition of the follicle population, *Mech. Ageing Dev.*, 4, 385, 1975.
3. **Flanagan, S. P.,** "Nude": a new hairless gene with pleiotropic effects in the mouse, *Genet. Res.*, 8, 295, 1966.
4. **Nishizuka, Y. and Sakakura, I.,** Thymus and reproduction: sex-linked dysgenesis of the gonad after neonatal thymectomy, *Science*, 166, 753, 1969.
5. **Rebar, R. W., Morandini, I. C., Erickson, G. F., and Petze, J. E.,** The hormonal basis of reproductive defects in athymic mice: diminished gonadotropin concentrations in prepubertal females, *Endocrinology*, 108, 120, 1981.
6. **Weinstein, Y.,** Impairment of the hypothalamo-pituitary-ovarian axis of the athymic "nude" mouse, *Mech. Ageing Dev.*, 8, 63, 1978.
7. **Rebar, R. W., Morandini, I. C., Benirschke, K., and Petze, J. E.,** Reduced gonadotropins in athymic mice: prevention by thymus transplantation, *Endocrinology*, 107, 2130, 1980.
8. **Poiley, W. M., Overeja, A. A., Otis, A. P., and Reeder, C. R.,** Reproductive behavior of athymic nude (*nu/nu*-BALB/c/A/BOM cr) mice in a variety of environments, in *Proc. 1st Int. Workshop on Nude Mice*, Rygaard, J. and Povlsen, C. O., Eds., Gustav Fischer Verlag, Stuttgart, 1975, 189.
9. **Davidson, K. A., Holland, J. M., Hall, J. W., and Gipson, L. C.,** Endocrine morphology and reproductive function in athymic nude mice, in *Proc. 3rd Int. Workshop on Nude Mice*, Reed, N., Ed., Fischer, New York, 1982, 317.
10. **Brünner, N., Svenstrup, B., Spang-Thomson, M., Benett, P., Nielsen, A., and Nielsen, J.,** Serum steroid levels in intact and endocrine ablated BALB/c nude mice and their intact littermates, *J. Steroid Biochem.*, 25, 429, 1986.
11. **van Steenbrugge, G. J., van Dongen, J. J. W., Reuvers, P. J., de Jong, F. H., and Schroeder, F. H.,** Transplantable human prostatic carcinoma (PC-82). I. Hormone-dependence and the concentration of androgens in plasma and tumor tissue, *Prostate*, 11, 195, 1987.
12. **Rebar, R. W., Morandini, I. C., Petze, J. E., and Erickson, G. F.,** Hormonal basis of reproductive defects in athymic mice: reduced gonadotropins and testosterone in males, *Biol. Reprod.*, 27, 1267, 1982.
13. **Ultee-van Gessel, A. M., van Steenbrugge, G. J., Leemborg, F. G., Schroeder, F. H., and de Jong, F. H.,** *In vivo* administration of an LHRH-antagonist to male mice: effects on pituitary gonadotropin secretion *in vitro*, *Acta Endocrinol. (Copenhagen)*, 120, 308, 1989.
14. **van Weerden, W. M., van Steenbrugge, G. J., van Kreuningen, A., Moerings, E. P. C. M., de Jong, F. H., and Schröder, F. H.,** Effects of low testosterone levels and of adrenal androgens on growth of prostate tumor models in nude mice, *J. Steroid Biochem.*, (accepted for publication).
15. **van Steenbrugge, G. J., Groen, M., van Kreuningen, A., de Jong, F. H., Gallee, M. W. P., and Schröder, F. H.,** Transplantable human prostatic carcinoma (PC-82) in athymic nude mice. III. Effects of estrogens on the growth of the tumor, *Prostate*, 12, 157, 1988.
16. **Seibert, K., Shafie, S. M., Triche, T. J., Wang-Peng, J. J., O'Brien, S. J., Toney, J. H., Huff, K. K., and Lippman, M. E.,** Clonal variation of MCF-7 breast cancer cells *in vitro* and in athymic nude mice, *Cancer Res.*, 43, 2223, 1983.
17. **Overpeck, J. G., Colson, S. H., Hohmann, J. R., Applestine, M. S., and Reilly, J. F.,** Concentrations of circulating steroids in normal prepubertal and adult male and female humans, chimpanzees, rhesus monkeys, rats, mice, and hamsters: a literature survey, *J. Toxicol. Environ. Health*, 4, 785, 1987.
18. **Bartke, A., Steele, R. E., Musto, N., and Caldwell, B. V.,** Fluctuations in plasma testosterone levels in adult rats and mice, *Endocrinology*, 92, 1223, 1973.
19. **van Steenbrugge, G. J., Groen, M., de Jong, F. H., and Schroeder, F. H.,** The use of steroid-containing silastic implants in male nude mice: plasma hormone levels and the effect of implantation on the weights of the ventral prostate and seminal vesicles, *Prostate*, 5, 639, 1984.
20. **Bex, F. J., Corbin, A., and France, E.,** Resistance of the mouse to the antifertility effects of LHRH agonists, *Life Sci.*, 30, 1263, 1982.
21. **Wang, N. G., Sunderam, K., Pavlou, S., Rivier, J., Vale, W., and Bardin, C. W.,** Mice are insensitive to the antitesticular effects of luteinizing hormone-releasing hormone agonists, *Endocrinology*, 112, 331, 1983.
22. **van Steenbrugge, G. J., Ultee-van Gessel, A. M., Groen, M., de Jong, F. H., and Schroeder, F. H.,** Administration of an LHRH-antagonist to male mice: effects on *in vivo* secretion of hormones and on the growth of a transplantable human prostatic carcinoma, *Life Sci.*, 40, 1335, 1987.
23. **Redding, T.W. and Schally, A.V.,** Inhibition of the pituitary-gonadal axis in nude male mice by continuous administration of LH-RH agonists and antagonists, *J. Endocrinol.*, (in press).

24. **Sharkey, F.E. and Fogh, J.,** Considerations in the use of nude mice for cancer research, *Cancer Metast. Rev.*, 3, 341, 1984.
25. **Reid, L.M. and Shin, S.,** Transplantation of heterologous endocrine tumor cells in nude mice, in *The Nude Mouse in Experimental and Clinical Research*, Fogh, J. and Giovanella, B. C., Eds., Academic Press, New York, 1978, 313.
26. **Rae-Venter, B. and Reid, L. M.,** Growth of human breast carcinomas in nude mice and subsequent establishment in tissue culture, *Cancer Res.*, 40, 95, 1980.
27. **Shafie, S. M. and Grantham, F. H.,** Role of hormones in the growth and regression of human breast cancer cells (MCF-7) transplanted into athymic nude mice, *J. Natl. Cancer Inst.*, 67, 51, 1981.
28. **Osborne, C. K., Hobbs, K., and Clark, G. M.,** Effect of estrogens and antiestrogens on growth of human breast cancer cells in athymic nude mice, *Cancer Res.*, 45, 584, 1985.
29. **Brünner, N., Osborne, C. K., and Spang-Thomsen, M.,** Endocrine therapy of human breast cancer grown in nude mice, *Breast Cancer Res. Treat.*, 10, 229, 1987.
30. **Hoehn, W., Schroeder, F. H., Riemann, J. F., Joebsis, A. C., and Hermanek, P.,** Human prostatic adenocarcinoma: some characteristics of a serially transplantable line in nude mice (PC-82) *Prostate*, 1, 95, 1980.
31. **Hoehn, W., Wagner, M., Riemann, J. F., Williams, E., Walther, R., and Schrueffer, R.,** Prostatic adenocarcinoma PC-EW, a new human tumor line transplantable in nude mice, *Prostate*, 5, 445, 1984.
32. **Ito, Y. Z., Mashimo, S., Nakazato, Y., and Takikawa, H.,** Hormone dependency of a serially transplantable human prostatic cancer (HONDA) in nude mice, *Cancer Res.*, 45, 5058, 1985.
33. **Harper, M. E., Sibley, P. E. C., Rowlands, A., Buttifaut, C., Beacock, C., and Griffith, K.,** Hormonal modulation of the growth of a new transplantable prostatic cell line in athymic nude mice, *Urol. Res.*, 14, 156 (abstract), 1986.
34. **Satyaswaroop, P. G., Zaino, R. J., and Mortel, R.,** Human endometrial adenocarcinoma transplanted into nude mice: growth regulation by estradiol, *Science*, 219, 58, 1983.
35. **Satyaswaroop, P. G., Zaino, R., Clarke, C. L., and Mortel, R.,** Nude mouse system in the study of tumor biology, treatment strategies and progesterone receptor physiology in human endometrial carcinoma, *J. Steroid Biochem.*, 27, 431, 1987.
36. **Friedl, A., Gottardis, M. M., Pink, J., Buchler, D. A., and Jordan, V. C.,** Enhanced growth of an estrogen receptor-negative endometrial adenocarcinoma by estradiol in athymic mice, *Cancer Res.*, 49, 4758, 1989.
37. **Tseng, L., Tseng, J. K., Mann, W. J., Chumas, J. C., Stone, M. L., Mazella, J., Sun, B., Amalfitano, T. G., and Wallach, R. C.,** Endocrine aspects of human uterine sarcoma: a preliminary study, *Am. J. Obstet. Gynecol.*, 155, 95, 1986.
38. **Sawada, M., Terada, N., Wada, A., Mori, Y., Yamasaki, M., Saga, T., and Endo, K.,** Estrogen- and androgen-responsive growth of human ovarian adenocarcinoma heterotransplanted into nude mice, *Int. J. Cancer*, 45, 359, 1990.
39. **Somers, K., Koenig, M., and Schechter, G. L.,** Growth of head and neck squamous cell carcinoma in nude mice: potentiation of laryngeal carcinoma by 17B-estradiol, *J. Natl. Cancer Inst.*, 80, 688, 1988.
40. **Greenway, B., Duke, D., Pym, B., Iqbal, M. J., Johnson, J., and Williams, R.,** The control of human pancreatic adenocarcinoma xenografts in nude mice by hormonal therapy, *Br. J. Surg.*, 69, 595, 1982.
41. **Izbicki, J. R., Schmitz, R., Hoppen, H. O., Izbicki, W., and Troidl, H.,** Effects of steroid hormone therapy on primarily xenotransplanted human colorectal adenocarcinomas, *J. Cancer Res. Clin. Oncol.*, 108, 345, 1984.
42. **Stebbings, W. S. L., Vinson, G. P., Farthing, M. J. G., Balkwill, F., and Wood, R. F. M.,** Effect of steroid hormones on human colorectal adenocarcinoma xenografts, of known steroid-receptor status, in nude mice, *J. Cancer Res. Clin. Oncol.*, 115, 439, 1989.
43. **Watson, S., Durrant, L., and Morris, D.,** Gastrin: growth enhancing effects of human gastrin and colonic tumour cells, *Br. J. Cancer*, 59, 554, 1989.
44. **Mattern, J., Klinga, K., Runnebaum, B., and Volm, M.,** Influence of hormone therapy on human lung tumors transplanted into nude mice, *Oncology*, 42, 388, 1985.
45. **Gittes, R. F.,** The nude mouse — its use as tumor-bearing model of the prostate, in *Models for Prostate Cancer, Prog. Clin. Biol. Res.*, Vol. 37, Murphy, G. P., Ed., Alan. R. Liss, New York, 1980, 31.
46. **Reid, L. M., Minato, N., and Rojkind, M.,** Human prostatic cells in culture and in conditioned animals, in *Male Accessory Sex Glands*, Spring-Mills, E. and Hafez, E. S. E., Eds., Elsevier, North-Holland Biomedical Press, 1980, 617.
47. **Steenbrugge, G. J.,** *Transplantable Human Prostate Cancer (PC-82) in Athymic Nude Mice: A Model for the Study of Androgen-Regulated Tumor Growth*, Ph.D. thesis, Erasmus University, Rotterdam, 1988.
48. **Bailey, M. J., Gazet, J. C., and Peckham, M. J.,** Human breast cancer xenografts in immune-suppressed mice, *Br. J. Cancer*, 42, 524, 1980.
49. **Bastert, G., Fortmeyer, H. P., Eichholz, H., Michel, R. Th., Huck, R., and Schmidt-Matthiesen, H.,** Human breast cancer in thymusaplastic nude mice, in *Thymusaplastic Nude Mice and Rats in Clinical Oncology*, Bastert, G. B. A., Fortmeyer, H. P., and Schmidt-Matthiesen, H., Eds., Gustav Fischer, Stuttgart, 1981, 157.

50. **Satyaswaroop, P. G., Zaino, R., and Mortel, R.,** A nude mouse model for studying estradiol-regulated growth of human endometrial carcinoma, in *Proc. 4th Int. Workshop on Immune-Deficient Animals in Exp. Res.*, Sordat B., Ed., Karger, Basel, 1984, 268.
51. **Ozello, L. and Sordat, M.,** Behavior of tumors produced by transplantation of human mammary cell lines in athymic nude mice, *Eur. J. Cancer,* 16, 553, 1980.
52. **Isaacs, J. T. and Coffey, D. S.,** Adaptation versus selection as the mechanism responsible for the relapse of prostatic cancer to androgen ablation therapy as studied in the Dunning R-3327-H adenocarcinoma, *Cancer Res.,* 41, 5070, 1981.
53. **Liao, S., Mezzetti, G., and Chen, C.,** Androgen receptor and early biochemical responses, in *The Cell Nucleus,* Vol. 7, Busch, H., Ed., Academic Press, New York, 1979, 201.
54. **Olson, J. J., Beck, D. W., Schlechte, J. A., and Loh, P. M.,** Effect of the antiprogesterone RU-38486 on meningioma implanted into nude mice, *J. Neurosurg.,* 66, 584, 1987.
55. **Dunk, A. A., Kyprianou, N., Davies, P., and Thomas, H. C.,** Human hepatocellular carcinoma tumor xenografts: their androgen-receptor status and growth responses to castration, *Digest. Dis. Sci.,* 4, 472, 1988.
56. **Sluyser, M. and van Nie, R.,** Estrogen receptor content and hormone-responsive growth of mouse mammary tumors, *Cancer Res.,* 34, 3253, 1974.
57. **Fukutomi, T., Kubota, T., Ikeda, T., Isobe, Y., Kikuyima, S., Shimada, A., Nakamura, A., Nishiumi, T., Enomoto, K., Ihibiki, K., and Aba, O.,** Experimental combined hormone therapy on human breast carcinomas serially transplanted into nude mice, *Jpn. J. Cancer Res. (Gann),* 77, 92, 1986.
58. **Andriole, G. L., Rittmaster, R. S., Loriaux, D. L., Kish, M. L., and Linehan, W. M.,** The effect of 4MA, a potent inhibitor of 5alpha-reductase on the growth of human, androgen-responsive genitourinary tumors grown in athymic nude mice, *Prostate,* 10, 189, 1987.
59. **Brünner, N., Spang-Thomsen, M., Vindeløv, L., and Nielsen, A.,** Effect of 17B-estradiol on growth curves and flow cytometric DNA distribution of two human breast carcinomas grown in nude mice, *Br. J. Cancer,* 47, 641, 1983.
60. **Brünner, N., Spang-Thomsen, M., Vindeløv, L., Wolff, J., and Engelholm, S. A.,** Effect of tamoxifen on the receptor-positive T61 and receptor-negative T60 human breast carcinomas grown in nude mice, *Eur. J. Cancer Clin. Oncol.,* 21, 1349, 1985.
61. **Robinson, S. P. and Craig Jordan, V.,** Antiestrogen action of toremifene on hormone-dependent, -independent, and heterogeneous breast tumor growth in the athymic mouse, *Cancer Res.,* 49, 1758, 1989.
62. **Gottardis, M. M., Ricchio M. E., Satyaswaroop, P. G., and Jordan, V. C.,** Effect of steroidal and nonsteroidal antiestrogens on the growth of a tamoxifen-stimulated human endometrial carcinoma (Enca101) in athymic mice, *Cancer Res.,* 50, 3189, 1990.
63. **Zaino, R. J., Satyaswaroop, P. G., and Mortel, R.,** Hormonal therapy of endometrial adenocarcinoma in a nude mouse model, *Cancer Res.,* 45, 539, 1985.
64. **Brünner, N., Spang-Thomsen, M., Henriksen, U., Rasmussen, S., Svenstrup, B., Bennet, P., Wolff, J., and Engelholm, S. A.,** Application of endocrine therapy in the nude mouse model, in *Immune Deficient Animals in Biomedical Research,* Rygaard et al., Eds., Karger, Basel, 1987, 340.
65. **Schneider, M. R.,** Potency and mechanism of action of progesterone antagonists in experimental mammary tumors, *J. Cancer Res. Clin. Oncol.,* 116 (Suppl.), 912, 1990.
66. **Brünner, N., Spang-Thomsen, M., Vindeløv, L., Engelholm, S. A., and Svenstrup, B.,** Dose-dependent effect of 17β-estradiol determined by growth curves and flow cytometric DNA analysis of a human breast carcinoma (T61) grown in nude mice, *Exp. Cell Biol.,* 53, 220, 1985.
67. **Kubota, T., Kubouchi, K., Tsuyuki, K., Nakada, M., Asanuma, F., Koh, J., Ishibiki, K, Abe, O., and Kawamura, E.,** Experimental endocrine therapy of human breast carcinomas transplanted into nude mice, in *Proc. 4th Int. Workshop on Immune-Deficient Animals in Exp. Res.,* Sordat, B., Ed., Karger, Basel, 1984, 385.
68. **James, K. C., Nicholls, P. J., and Roberts, M.,** Biological half-lifes of (4-^{14}C) testosterone and some of its esters after injection into rats, *J. Pharmacol.,* 21, 24, 1969.
69. **Dziuk, P. J. and Cook, B.,** Passage of steroids through silicone rubber, *Endocrinology,* 78, 208, 1966.
70. **van Weerden, W. M., van Steenbrugge, G. J., van Kreuningen, A., Moerings, E. P. C. M., de Jong, F. H., and Schröder, F. H.,** Assessment of the critical level of androgen for growth response of transplantable human prostatic carcinoma (PC-82) in nude mice, *J. Urology,* (accepted for publication), 1990.
71. **Gallee, M. P. W., van Steenbrugge, G. J., ten Kate, F. J. W., Schroeder, F. H., and van der Kwast, Th. H.,** Determination of proliferative fraction of a transplantable hormone-dependent human prostatic carcinoma (PC-82) by the monoclonal antibody Ki-67; potential application for hormone therapy monitoring, *J. Natl. Cancer Inst.,* 79, 1333, 1987.
72. **van Weerden, W. M., van Steenbrugge, G. J., Moerings, E. P. C. M., Vermeij, M., and Schroeder, F. H.,** Proliferative activity of human prostatic cancer tissue in athymic nude mice assessed by the Ki-67 and BrdU monoclonal antibody, *J. Cancer Res. Clin. Oncol.,* 116 (Suppl.), 462, 1990.

73. **Habu, S., Fukui, H., Kasai, M., Nagai, Y., Okumura, K., and Tamaoki, N.,** *In vivo* effects of anti-sialo CMI.1. Reduction of NK activity and enhancement of transplanted tumor growth in nude mice, *J. Immunol.*, 127, 34, 1981.
74. **Seaman, W. E., Blackman, M. A., Gindhart, T. D., Roubinian, J. R., Loeb, J. M., and Talal, N.,** β-estradiol reduces natural killer cells in mice, *J. Immunol.*, 121, 2193, 1978.
75. **Jordan, V. C., Gottardis, M. M., Robinson, S. P., and Friedl, A.,** Immune-deficient animals to study "hormone-dependent" breast and endometrial cancer, *J. Steroid Biochem.*, 34, 169, 1989.
76. **Byyny, R. L., Orth, D. N., Cohen, S., and Doyne, E. S.,** Epidermal growth factor: effects of androgens and adrenergic agents, *Endocrinology*, 95, 776, 1974.
77. **Dickson, R. B., McManaway, M. E., and Lippman, M. E.,** Estrogen-induced factors of breast cancer cells partially replace estrogen to promote tumor growth, *Science*, 232, 1540, 1986.

Chapter 17

MONOCLONAL ANTIBODIES

Hidde J. Haisma

TABLE OF CONTENTS

I. INTRODUCTION

The development of the hybridoma technology by Kohler and Milstein[1] for the production of monoclonal antibodies created a new tool for the diagnosis and therapy of cancer.[2] For the determination of *in vivo* usefulness of (mostly murine) monoclonal antibodies directed against human tumor-associated antigens, relevant tumor systems are needed. Human tumors transplanted to nude mice provide such a model. In previous years, the model has been used successfully in analyzing the efficacy of radiolabeled monoclonal antibodies to localize in tumor lesions. Moreover, conjugates of monoclonal antibodies with radionuclides, drugs, toxins, or enzymes have been evaluated in the nude mouse-human tumor model for therapeutic effectiveness before their introduction into the clinic.

In this chapter, some aspects of monoclonal antibodies for localization and therapy in human tumor xenografts will be discussed (Table 1). Xenografts are established either directly from patient's tumor tissue or from *in vitro* cell lines. Almost all studies with monoclonal antibodies have been performed in animals with tumors inoculated s.c. in the flanks of the mice and, therefore, the results and the problems of other sites of tumor growth will not be addressed.

II. ANTIGEN EXPRESSION AND TUMOR CHARACTERISTICS

A. ANTIGEN EXPRESSION

The validity of the nude mouse-human tumor model for the preclinical evaluation of monoclonal antibodies depends on many factors, of which the growth characteristics and relevant antigen expression are the most important.

Only a few studies have addressed the question if xenografts maintain the antigenic phenotype of the original tumors. Ripamonti et al.[3] studied the MCF-7, HT-29, and SW626 cell lines. A dramatic decrease in the level of antigen expression of ovarian cancer-associated antigens, and frequently a drop below the detection limit, was observed when the *in vitro* cultured cells were implanted i. p. into nude mice. Friedlander et al.[4] and Ward et al.[5] found no differences in the antigenic expression of 12 and 8 human epithelial ovarian tumors, respectively, when established as s.c. xenografts in nude mice. In the latter study, the expression of one antigen was lost for one cell line cultured *in vitro* upon establishment as a xenograft. From these studies, it is apparent that in most instances human tumor xenografts resemble the patients' tumors as far as antigen expression is concerned, but antigen expression has to be confirmed in the xenograft tissue.

B. SITE OF ANTIGEN

Although antibodies directed against cell surface antigens have been used almost exclusively for tumor targeting, some reports deal with the localization of antibodies directed against intracellular antigens. Welt et al.[6] used a monoclonal antibody recognizing a pigmentation-associated glycoprotein in melanoma cells which specifically localized in pigmented melanoma transplants and persisted in the xenografts for up to 8 to 10 weeks. Dairkee et al.[7] demonstrated selective localization of a monoclonal antibody against keratin in fresh malignant breast cancer specimens after passive infusion. Epstein et al.[8] introduced a new method of tumor localization and therapy based on the loss of membrane integrity in degenerating cells. An antibody, TNT-1, directed against an intranuclear antigen was demonstrated to bind to xenografts that contain necrotic areas. A human IgM antibody, 16.88, reactive with an altered form of cytokeratin showed localization in a human colon cancer xenograft model,[9] as well as in 9 of 12 patients with colorectal cancer.[10]

C. QUANTITY OF ANTIGEN

The quantity of antigen in the xenografts seems to be important for the localization of monoclonal antibodies in the tumors. In a study of breast cancer xenografts, Mushakis et al.[11]

TABLE 1
Factors Involved in Tumor Localization with Monoclonal Antibodies or Conjugates

Antigen- or tumor-related
Quantity of antigen
Cell surface or intracellular antigen
Circulating antigen
Vascularization of tumor
Tumor size
Antibody-related
Antibody specificity
Antibody class
Size of antibody, fragment, conjugate
Antibody dose
Immunoreactivity after conjugation

found uptake of an iodinated anti-carcinoembryonic antigen (CEA) antibody only in those xenografts with a high amount of CEA. Similar results were obtained by Hedin et al.[12] and Philben et al.[13] A correlation between the number of epidermal growth factor (EGF) receptors per cell and antibody targeting was found by Goldenberg et al.[14]: 2×10^6 receptors per cell gave good tumor targeting, 3×10^5 receptors per cell resulted in lower localization, whereas a cell line with 5×10^3 receptors per cell showed no preferential uptake of the antibody. In clinical studies, monoclonal antibodies show preferential uptake in antigen-positive tumor sites. A direct correlation has been demonstrated between the results of immunoscintigraphy and those of the immunohistochemical analysis of the tumor tissue.[15]

In vitro exposure of cancer cell lines to either α- or γ-interferon can increase the expression of some tumor-associated antigens by enhancing both the number of antigen molecules per cell and the number of positive cells.[16,17] Antibody uptake in tumors may therefore be improved by pretreatment of the animals with interferons. Administration of human interferon-α to nude mice bearing human colon cancer xenografts effectively increased the antigen expression and augmented the localization of a radiolabeled monoclonal antibody to the tumor site.[18] *In vivo* treatment with recombinant human γ-interferon enhanced or induced the expression of HLA-DR antigens and a tumor-associated antigen in breast cancer xenografts in nude mice. Subsequently, a radioiodinated monoclonal antibody directed against HLA-DR was demonstrated to specifically localize in the tumors of these animals, but not in tumors of non-pretreated mice.[19] These preclinical data have led to preliminary studies in patients, and treatment with interferon-α increased the uptake of an antibody in melanoma lesions.[20]

D. CIRCULATING ANTIGEN

Many monoclonal antibodies react with antigens that are shed into the circulation by tumor cells. Consequently, circulating antigen can bind to antibodies injected i.v., thus inactivating the antibody by formation of immune complexes.[21] Monoclonal antibodies directed against CEA are widely used in clinical and animal studies. Circulating CEA from human tumor xenografts has been shown to induce the removal of anti-CEA antibodies from the vascular compartment by clearance of complexes in the liver and the spleen. The net result was a decreased tumor uptake of the antibody.[22,23] In clinical studies, however, circulating antigen may also complex with the antibody used for radioimmunodetection,[21] but such complexes have neither led to rapid clearance nor side effects, or prevented successful imaging. In general, patients with high serum titers of the antigen have a higher chance for positive tumor detection.[24]

E. SITE OF ANTIBODY LOCALIZATION IN TUMORS

For antibody-mediated therapy, the antibody should specifically target to malignant cells and not to the stromal cells present in the tumors. Antibody administered i.v. has to cross the capillary wall to be able to bind to tumor cells. Large molecules move through the vascular endothelium by diffusion and pinocytosis. In the case of tumor capillaries, the pores may be much larger than in normal capillaries. This alone may result in antibody localization in tumors. The capacity of antibodies to penetrate into tumor tissue has been investigated using radiolabeled antibodies and autoradiography. Localization of antibody in the tumor stroma[25,26] or tumor stroma interface,[27] with little penetration into the tumor mass, has been reported. Heterogeneous localization of antibodies in the tumor has been described by other groups.[28,29] In these studies, only a few antibody-positive cells could be detected to have bound the antibody. In another approach, Ong et al.[30] found a relatively homogeneous penetration into the tissue and binding of antibody to the surface of tumor cells. The latter group used biotinylated antibody and immunohistochemistry instead of radiolabeled antibody and autoradiography, which may explain the differences between the studies.

F. SIZE OF TUMORS

As tumors expand, there can be a relative reduction in blood flow because the number of vessels in tumors are proportionally lower as the tumor enlarges.[31] The reduced blood supply means a decrease in antibody delivery, which can result in diminished antibody uptake in larger tumors. Findings have been inconsistent in the relation of the tumor size to the percentage of antibody localization in experimental tumors. Hedin et al.[12] did not find such a correlation, but Mann et al.[32] reported that tumor uptake was well correlated with tumor size, with larger xenografts producing better images. In contrast, better tumor localization in smaller xenografts was found by Philben et al.[13] using anti-CEA antibodies. Tumors in patients are generally much larger than xenografts. The relative uptake in smaller lesions is shown to be higher than in larger tumors, because of which a higher radiation dose may be expected in case of radioimmuno-therapy.[33] The clinical consequence of the little specific uptake of antibodies in larger tumors may be that antibody-mediated therapy can only be successful in an adjuvant setting.

III. ANTIBODY CHARACTERISTICS

The biodistribution and the pharmacokinetics of monoclonal antibodies depend on factors related to the specificity of the antibody, the (sub)class of the antibody and its size, that is, whole immunoglobulin, $F(ab')_2$ or Fab.

A. ANTIBODY SPECIFICITY

No single antibody has been reported that reacts with tumor-specific rather than tumor-associated antigens. The exceptions are the anti-idiotype antibodies reactive with a single B cell clone. Most monoclonal antibodies currently used for cancer therapy cross-react with some normal (human) tissues. Generally, in patients, treatment with monoclonal antibodies has not led to the destruction of normal tissues. Toxicity, if present, is related to the immunogenicity of the antibody [IgE response or human anti-mouse antibodies (HAMA)] or to the conjugated drug or radionuclide; for example, an immune response against a conjugated toxin or bone marrow suppression caused by irradiation.

B. IMMUNOREACTIVITY OF ANTIBODY

The immunoreactivity of antibody preparations (i.e., the antigen binding capacity) may be reduced by damage of the antibody during *in vivo* or *in vitro* production, purification storage, or conjugation procedures. Loss of immunoreactivity by the labeling step will result in diminished *in vivo* localization of tumors. High levels of radioiodination[34] or high substitution of antibodies with chelating groups to allow attachment of radiometals[35] has led to preparations with reduced immunoreactivity and tumor localizing properties. In a study by Pimm et al.,[36] the antigen-binding capacity of the antibody appeared to decrease by mixing the preparation with immunologically inert immunoglobulin, and a progressive decline in localization of human colon cancer xenografts was noticed.

In clinical studies, the quality of antibody preparations is usually controlled so that the immunoreactivity is kept constant. No direct comparisons have been made between 'good' and 'bad' antibody preparations in patients. The studies with human tumor xenografts, however, indicate that the quality of the antibody after conjugation is a very important determinant in antibody localization in tumors.

C. ANTIBODY PHARMACOKINETICS

Murine monoclonal IgG (mol wt 150,000 Da) antibodies in mice have a half-life of 3 to 10 d and IgM (mol wt 900,000 Da) antibodies have a half-life of approximately 3 d. Antibodies consist of an antigen-binding portion (Fab) and an Fc region, the latter being responsible for immunological effector functions like fixation of complement or binding to Fc receptors on macrophages. The proteolytic enzyme pepsin cleaves off most of the Fc region, leaving 2 Fab fragments bound together in a divalent structure referred to as the $F(ab')_2$ fragment (mol wt 100,000 Da). The enzyme papain removes a larger portion of the Fc region, resulting in 2 monovalent Fab fragments (mol wt 50,000 Da) and an intact Fc fragment. The half-life of $F(ab')_2$ in mice is 3 to 6 h and of Fab, around 2 h. Catabolism of IgG and $F(ab')_2$ occurs in the liver and the spleen, whereas Fab is catabolized principally in the kidney.[37,38]

In patients, murine antibodies have longer half-lives than in mice. Half-lives are approximately 40 h for IgM, 28 h for IgG, and 26 h and 1.6 h for $F(ab')_2$ and Fab fragments of IgG, respectively.[39] Homologous antibodies are catabolized at a slower rate than murine antibodies: for IgG the half-life is 21 d, and for IgM it is 5 d.[39]

D. ROUTE OF ADMINISTRATION

In attempts to increase the uptake of antibody into tumors, regional delivery of the antibody is an elegant approach. In patients, monoclonal antibodies have been administered into the lymphatics or the pleural, pericardial, and peritoneal cavities. The clearance from these sites is usually much slower than from the circulation, and preferential retention of antibody in the tumor may be achieved.[39,40]

In mice, radiolabeled antibodies transfer very rapidly from the peritoneal cavity to the bloodpool. After 2 h, more than 50% of the injected dose is present in the blood. Favorable tumor localization may be obtained in small tumor nodules of i.p. grown xenografts after i.p. injection of antibody.[41]

E. ANTIBODY UPTAKE IN TUMORS

Significant improvement of localization in human tumor xenografts has been reported for $F(ab')_2$ fragments, relative to either the parent IgG or the Fab fragment.[42-45] Wahl et al.[43] have compared an anti-CEA IgG antibody and its $F(ab')_2$ or Fab fragments in the detection of human colon cancer xenografts. $F(ab')_2$ fragments gave better images at an earlier time point than either Fab or whole IgG. Buchegger et al.[44] compared four monoclonal antibodies and their fragments directed to CEA in a colon cancer nude mouse model. All four antibodies were selected for high affinity to CEA and low cross-reactivity to other antigens. In each case, $F(ab')_2$ proved to be superior to whole antibody with respect to better tumor to tissue ratios. Absolute uptake of antibody varied between 1 and 20% of the injected dose. Uptake of $F(ab')_2$ in all studies was less than that of whole IgG.

In clinical studies, the tumor concentration of antibody after i.v. injection varies from 0.1 to 0.001% injected dose/gram and is highest for IgG, lowest for Fab, and intermediate for $F(ab')_2$, which correlates with their half-lives in the blood.[46,47] The low tumor uptake in patients remains the greatest limitation for the use of antibodies in a clinical setting. Moreover, the difference in antibody uptake between human tumor xenografts and patient's tumors causes difficulty in the extrapolation of preclinical data to the situation in patients.

F. ANTIBODY DOSE

An increase in the amount of antibody administered to the mice leads to a relatively reduced non-specific uptake of antibody in the reticuloendothelial system, probably due to the saturation of binding sites. Rogers et al.[48] studied the dose-dependent localization of $F(ab')_2$ fragments of an anti-CEA antibody in human colon cancer xenografts. Optimal localization was found with a dose of 58 μg of antibody. Increasing the dose of antibody to 200 to 400 μg led to reduced localization. In contrast, Sharkey et al.[49] found a linear relationship between the concentration of antibody in the tumor and the antibody dose (0.1 to 1.0 mg) in hamsters bearing human tumor xenografts in the cheeck pouch. The discrepancy between these two studies might be explained by the difference in size of the animals used and to different saturation levels for antibody in the xenografts.

In patients, it has been demonstrated that higher doses of murine antibody (>20 mg) results in an increased serum half-life, a relative decrease in non-specific uptake, and a higher uptake in tumor lesions. The improvement of tumor detection is thought to be caused by the saturation of non-specific binding sites, especially in the liver, which leads to a greater availability of antibody for the target antigen.[39]

G. HUMAN OR HUMANIZED ANTIBODIES

In clinical trials, the major problem with repeated injections of murine monoclonal antibodies is the development of a human anti-mouse antibody (HAMA) response. The HAMA may enhance antibody clearance and thereby impair the therapeutic effectiveness. HAMA occurs less frequently with single low doses of antibody fragments, but repeated doses of antibody will result in HAMA in most patients. Immunosuppressive agents are of limited benefit for the prevention of HAMA, thus making the production of human or humanized antibodies a necessary approach to enable monoclonal antibody-mediated therapy.[39]

Chimeric antibodies with human constant regions and murine variable regions may greatly reduce the problems of HAMA. In a study by Colcher et al.,[50] a chimeric antibody was compared with the native antibody, B72.3, in human colon cancer xenografts. The chimeric antibody showed a more rapid clearance from the circulation and showed less absolute tumor localization. In a study of McCabe et al.,[51] a human IgM monoclonal antibody was shown to localize in xenografts of human colon cancer, but uptake was low because of fast clearance of the antibody. These limited studies indicate that, in mice, human or humanized antibodies clear more rapidly from the circulation than homologous antibodies. Therefore, the nude mouse model may be a limitation for evaluating these antibodies.

IV. ANTIBODY-RADIONUCLIDE CONJUGATES FOR TUMOR LOCALIZATION

Radiolabeled antibodies hold considerable promise as a means of tumor detection by external scintigraphy. Initial work by Pressman and Korngold[52] in the 1950s with radiolabeled polyclonal antisera in rats has shown immunological tumor targeting. A series of radionuclides are available for imaging and the choice is based on the nature of rays emitted, the half-life, the availability, and the antibody coupling characteristics (Table 2).

The nude mouse-human tumor model has been used to evaluate the localization properties of antibodies conjugated with different radionuclides by a variety of techniques. Quantitative imaging in mice is almost impossible because of the small size of the animal and the relative large spatial resolution of the gamma-cameras used. Therefore, quantification is usually performed by dissecting the animals and measuring the distribution of the antibody by counting selected tissues at several time points.

A. IODINE

Iodine radionuclides can be easily coupled to monoclonal antibodies by electrophilic

TABLE 2
Radionuclides in Use for Tumor Localization or Therapy

Isotope	Half-life	Decay mode	Imaging	Therapy	Advantage	Disadvantage
^{99m}Tc	6 h	γ	Exellent	No	Cheap; availability	Short $T_{1/2}$; Renal uptake
^{111}In	3 d	γ	Exellent	No		Liver uptake
^{123}I	13 h	γ	Exellent	No	Availability	
^{131}I	8 d	β and γ	Not good	Yes	Images; cheap	
^{90}Y	8 d	β	No	Yes	Chelate chem.	Bone uptake
^{186}Re	3.5 d	β	Yes	Yes	Tc chemistry	Availability
^{211}At	7 h	α	No	Yes	Iodine chem.	

substitution of hydrogen atoms in tyrosine molecules.[53] A major disadvantage of these isotopes is that they are dehalogenated *in vivo,* which results in stomach and thyroid uptake of radioiodine. The antithyroid drug propylthiouracil, which inhibits the deiodination of thyroxine, also inhibits the dehalogenation of monoclonal antibodies, thereby increasing the half-life threefold.[54] Indirect iodination methods have also been shown to decrease the dehalogenation of antibodies, which resulted in an increased tumor uptake of the antibodies.[55,56]

B. INDIUM

^{111}In is generally coupled to antibodies by a chelate of diethylenetriamine pentaacetic acid (DTPA) which has been conjugated to the protein. The interaction between indium and protein-bound DTPA is ionic and is the cause of the slow separation of the radionuclide from the chelate *in vivo,* whereafter it accumulates in the reticuloendothelial system. Liver accumulation of ^{111}In is the major problem when ^{111}In-labeled antibodies are used for tumor localization studies.

The introduction of a diester linkage between the antibody and an EDTA analog showed a reduction in blood activity and an increase in the tumor-to-blood ratio in nude mice carrying a B cell lymphoma. However, tumor-to-liver and tumor-to-kidney ratios had not improved when the diester conjugate was compared with the control conjugate.[57] Another approach to reduce the uptake of ^{111}In in the liver is the addition of a metabolizable linker between the antibody and the DTPA chelate, thereby accelerating the cleavage of the chelate from the antibody. When the distribution of ^{111}In-DTPA linked to a diester and labeled to an anti-melanoma monoclonal antibody was compared to a peptide-linked conjugate, tumor to blood, liver, and kidney ratios were substantially improved with the diester conjugate. The whole body biological half-life of the diester conjugate was three times shorter than for the peptide conjugate. As a result, earlier and better images of the tumor xenografts were obtained.[58]

C. TECHNETIUM

^{99m}Tc is the most widely used radionuclide in nuclear medicine. ^{99m}Tc is coupled to antibodies either by chelates, ligands (N_2S_2) or bound to thiol groups of the antibody. Disadvantages of the ligand exchange method are aggregate formation and the relative high temperature (50°C) necessary for labeling. Immunoreactive antibody can be obtained and ^{99m}Tc ligand exchange labeled anti-melanoma antibody 9.2.27 showed a higher concentration in tumor xenografts than the same antibody labeled with ^{125}I.[59]

D. CLINICAL STUDIES

Tumor localization in the nude mouse-human tumor model has preceded radioimmunolocalization in patients and has pointed out ways to improve the diagnostic results. In a recent review, Larson[46] summarized 61 clinical studies with radiolabeled monoclonal antibodies. The majority of these trials were performed with ^{131}I-labeled antibodies (40) followed by ^{111}In (14), ^{123}I (10), and ^{99m}Tc (3). Sensitivity ranged from 38 to 100% with an average of 70%. ^{111}In showed

TABLE 3
Immunoconjugates for Therapy

Radionuclides
^{90}Y, ^{131}I, ^{186}Re, and ^{211}Astatine
Toxins
Abrin, Ricin, Gelonin, Diphtheria toxin, Pseudomonas Exotoxin
Drugs
Daunomycin, doxorubicin, methotrexate, and vinblastine
Enzyme-prodrug systems
Alkaline phosphatase and carboxypeptidase

better results than ^{131}I, if liver uptake of ^{111}In did not complicate the detection of the lesions. Sensitivity of ^{123}I-labeled antibody could be increased by introducing computerized tomography. The indirect iodinated antibodies and antibodies labeled with ^{111}In by metabolizable linkers have not yet been evaluated in the clinic.

V. THERAPY

A. UNCONJUGATED ANTIBODIES

In nude mice bearing xenografts, therapy with unlabeled monoclonal antibodies may be very effective. Macrophages seem to be responsible for cell killing because complement depletion did not abolish the efficacy of the monoclonal antibody. The role of T and natural killer (NK) cells as effector cells could also be excluded.[60] Murine monoclonal antibodies from the IgG2a and IgG3 isotype were most effective, whereas IgG1 and IgG2b showed limited tumor growth inhibition.[61-63]

Because of the importance of growth factors for the proliferation of tumor cells, monoclonal antibodies directed against growth factor receptors are investigated for therapy. Treatment of xenografts of non-small cell lung cancer with an antibody directed against the EGF receptor,[64] or xenografts of small cell lung cancer with an antibody against bombesin,[65] has resulted in inhibition of tumor growth.

Clinical trials with unlabeled murine monoclonal antibodies have demonstrated only minor responses.[60] Therefore, it appears that the nude mouse model is not clinically relevant for determining the efficacy of unmodified antibody therapy. The limited success of this approach in patients has led to the production of antibodies conjugated to a variety of potentially toxic compounds in an effort to increase their therapeutic effectiveness (Table 3).

B. RADIONUCLIDES

The principle advantage of using radionuclide-labeled antibodies rather than drug or toxin conjugates is that the radioconjugates do not need to internalize to mediate cell killing, or bind to every cell to be therapeutically effective. This latter property is also the greatest disadvantage because of the potential toxicity to adjacent normal tissues. Some radionuclides in use for radioimmunotherapy are listed in Table 2. Rofstad[66] reviewed the usefulness of the nude mouse-human tumor model for radiotherapeutic research. He concluded that the xenografts retain the intrinsic radiosensitivity and may therefore be employed in preclinical predictive studies.

Numerous reports on animal radioimmunotherapy describe either an incomplete cytotoxic effect[54,67,68] or a temporary antitumor response.[69,70] Selection of tumor clones that lack the relevant antigen expression may be one reason for the regrowth of tumors after initial effective radioimmunotherapy.[71] A single, high dose of radiolabeled IgG appears to be more effective than multiple fractions of smaller doses.[71,72] Blumental et al.[73] compared the efficacy of IgG with that of $F(ab')_2$ fragments of two anti-CEA antibodies. Fractionated high doses of $F(ab')_2$

fragments were equally effective, but less toxic than IgG. Toxicity included a loss in body weight of 10 to 20% and a drop in peripheral white blood cell counts. Higher doses of $F(ab')_2$ fragment could be given because of the faster clearance as compared to IgG. Effective radiation doses in human tumor xenografts ranged from 700[54] to 5000[73] cGy. Most of these values are considerably lower than clinically effective external radiation doses. The small size and the rapid growth of the xenografts may be the cause of a relatively greater sensitivity to irradiation.

The encouraging results from animal radioimmunotherapeutic studies have incited clinical trials. Responses have already been shown for ^{131}I-labeled antibodies in patients with hepatoma, Hodgkin's disease, non-Hodgkin's lymphoma, chronic lymphocytic leukemia, and ovarian cancer, but patients have not been cured.[33] Up to 500 mCi of ^{131}I has been given, which resulted in estimated radiation doses to the tumor of 800 to 4000 cGy. Toxicity, as predicted from analogous animal experiments, was limited to bone marrow suppression.

C. TOXINS

Antibody-toxin conjugates or immunotoxins[74] combine the specificity of antibodies with the potent lethality of plant or bacterial toxins. Theoretically, one toxin molecule is sufficient for cell death because the toxins catalytically inhibit protein synthesis. Toxins and conventional cytostatic agents have different mechanisms of action. In particular, toxins will kill non-proliferating cells, whereas most cytostatics interfere with metabolically active or dividing cells. FitzGerald and Pastan[75] have reviewed the use of immunotoxins.

Most targets for immunotoxin therapy will be cell-surface antigens or receptors. For these conjugates to be effective, the antibody must bind to the tumor cell and then be internalized into the cytoplasm. A major disadvantage of immunotoxins is that only cells will be killed that express the relevant antigen. Therefore, for effective therapy, the antigen or receptor has to be expressed on every tumor cell.

Antitumor effects of immunotoxins in nude mice bearing xenografts have been shown by several groups. Weil-Hillman et al.[76] used an immunotoxin consisting of an anti-T cell leukemia antibody and ricin. Established solid tumors of T cell leukemia (CEM cell line) in nude mice regressed after intratumoral injection of the conjugates. Antigen specificity could only be achieved after systemic administration of the immunotoxin together with lactose to block binding of ricin to antigen negative cells. FitzGerald et al.[77] used an immunotoxin composed of an antibody against the human transferrin receptor coupled to the ricin A chain. Dose-dependent inhibition of growth of ovarian tumors (NIH:OVCAR-3) was shown. Moreover, treatment specificity was proven by the blocking of the antitumor effect by excess antibody. The same group demonstrated growth inhibition in the ovarian cancer model with an immunotoxin prepared with Pseudomonas exotoxin.[78] Similar studies on diphteria toxin A chain coupled to an antibody against a melanoma-associated antigen revealed growth inhibition of melanoma xenografts.[79] Anti-EGF receptor antibody conjugates prepared with gelonin, a ribosome inactivating glycoprotein with the same potency as ricin A chain, were tested in nude mice carrying EGF receptor-hyperproducing squamous cell carcinoma xenografts. Daily i.p. injections for 5 consecutive days with the conjugate showed inhibition of tumor growth for about 7 d without side effects. Unconjugated mixtures of antibody and gelonin were ineffective.[80]

To date, two clinical studies with ricin-immunotoxins have been reported: a therapeutic study in chronic lymphocytic leukemia[81] and another in melanoma.[82] Antitumor effects have been minimal. Side effects included minor signs of neurotoxicity which were not predicted from animal studies. Prominent immunogenicity of the conjugates was observed, which limited their repeated administration in patients.

D. CYTOSTATIC AGENTS

Immunoconjugates composed by antibodies and cytostatic agents have the advantage that their antitumor activity and toxicity are well-defined in the clinic as well as in animal models.

By targeting the drug to the tumor via the antibody, the toxicity of the drugs to normal tissues should be reduced. Disadvantages of antibody-drug conjugates are similar to those of immunotoxins, and include the fact that they will be effective only towards cells expressing the relevant antigen. Also, the conjugates have to be internalized. Other disadvantages are that drugs are less potent to toxins in causing cell death and most cytostatics are only effective against dividing cells.

In order to obtain a sufficiently high drug concentration in the tumor, the cytostatic agents are usually coupled to antibodies at ratios ranging from 5 to 20 molecules per molecule antibody. Major problems associated with the production of monoclonal antibody-drug conjugates include: poor yields, formation of high molecular weight aggregates, and inactivation of the antibody and/or the drug. Conjugates with daunomycin,[83] doxorubicin,[84] methotrexate,[85,86] and vindesine[87] have shown enhanced growth inhibition of human tumor xenografts in nude mice as compared to treatment with the drugs alone.

An alternative way of drug targeting has been explored by Corvalan et al.[88] using a hybrid-hybrid monoclonal antibody with specificity to both the tumor associated antigen (CEA) and the cytostatic agent (vinblastine). In this model, the drug is delivered to the tumor through the antigen binding site of the bispecific antibody. The hybrid-hybrid monoclonal antibody localized specifically in the CEA-expressing tumor and targeted vinblastine to the tumor. Treatment with the vinca alkaloid hybrid-hybrid monoclonal antibody was more effective in inhibiting tumor growth than with vinblastine given as a single agent.

Clinical trials with antibody-drug conjugates have been limited to investigations on the possible side effects of the conjugates and their potential tumor localization. Responses to treatment have not yet been reported.[89] The development of more potent drugs, linkers, or delivery systems such as liposomes[90] may provide the basis for successful therapy in patients.

E. ENZYMES

In the approach of antibody-enzyme conjugates, antibodies are used to deliver enzymes to tumor cells. The enzymes will then convert relatively non-toxic prodrugs, which are administered after the conjugates have bound to the antigens, into active cytotoxic agents. Ideally, the prodrug should be inert and have a long biological half-life. The drug released should be readily diffusible and enter the cells immediately, or be toxic at the membrane level. The application of an anticarcinoma antibody-alkaline phosphate conjugate in combination with the prodrug etoposide phosphate was presented by Senter et al.[91] The alkaline phosphatase converts etoposide phosphate into etoposide, which is more than 100-fold more toxic than the prodrug *in vitro*. Despite the presence of alkaline phosphatase in many tissues, a profound antitumor response was observed in xenografts treated with the conjugate followed by etoposide phosphate. Similar results were obtained in the same model with mitomycin C phosphate as the prodrug.[92] Carboxypeptidase G2-anti-hCG conjugates, in combination with the prodrug *para*-N-*bis* (2-chloroethyl) aminobenzoic acid, were used by Bagshawe et al.[93] to treat nude mice bearing human choriocarcinoma xenografts. Carboxypeptidase G2 was chosen as the enzyme because it has no known equivalent activity in man or animals. Marked inhibition of tumor growth was shown in animals treated with the conjugate and prodrug, whereas the prodrug alone or the cytotoxic agent were not active in this model.

Prodrugs sequentially administered with enzyme-antibody conjugates avoid the obstacles for immunoconjugates to be effective, such as the internalization of the antigen-antibody complex and the problem of heterogeneity in antigen expression. The preparation of prodrugs of clinically effective drugs may offer a promising approach for tumor site-selective therapy mediated through antibody-enzyme conjugates and they should therefore be promptly investigated for their effectiveness in patients.

VI. CONCLUSIONS

Human tumor xenografts generally retain the antigenic profile of the original patients' tumors. The nude mouse-human tumor model has been useful for predicting the effectiveness of monoclonal antibody conjugates in patients. Many concepts, such as the targeting to intracellular antigens vs. membrane-associated sites, the use of $F(ab')_2$ fragments vs. whole immunoglobulins, high doses of antibody, or the effect of interferons on antigen expression to enhance tumor localization, have been evaluated in the nude mouse model and effects were predictable for those in patients. For the selection of radionuclides and the various methods of conjugation for radioimmunolocalization or radioimmunotherapy, the nude mouse has shown to be of great value.

The extent of tumor detection and the competence of therapy with radiolabeled monoclonal antibodies in human tumor xenografts are overestimated when compared to the human situation: tumor uptake of antibodies is 100- to 1000-fold lower in patients than in mice. Furthermore, the use of mouse antibodies in a murine system does not predict for the behavior of such antibodies in patients, especially when multiple injections are needed which may elicit an immune response in patients, but not in the animals. In analogy, the effects of human or humanized monoclonal antibodies in the mouse bearing human tumor xenografts may be difficult to translate to the clinical situation.

REFERENCES

1. **Kohler G. and Milstein, C.,** Continuous cultures of fused cells secreting antibody of predefined specificity, *Nature (London)*, 256, 495, 1975.
2. **Haisma, H. J., Goedemans, W., and Hilgers, J.,** Tumor imaging with monoclonal antibodies, in *Tumor Immunology-Mechanisms, Diagnosis, Therapy,* Den Otter, W. and Ruitenberg, E. J., Eds., Elsevier Science Publishers, New York, 1987, 171.
3. **Ripamonti, M., Canevari, S., Menard, S., Mezzanzanica, D., Miotti, S., Orlandi, R., Tagliabue, E., and Colnaghi, M. I.,** Human carcinoma cell lines xenografted in athymic mice: biological and antigenic characteristics of an intraabdominal model, *Cancer Immunol. Immunother.*, 24, 13, 1987.
4. **Friedlander, M. L., Russel, P., Taylor, I. W., and Tattersall, M. H. N.,** Ovarian tumor xenografts in the study of the biology of human epithelial ovarian cancer, *Br. J. Cancer*, 51, 319, 1985.
5. **Ward, B. G., Wallace, K., Shepherd, J. H., and Balkwill, F. R.,** Intraperitoneal xenografts of human epithelial ovarian cancer in nude mice, *Cancer Res.*, 47, 2662, 1987.
6. **Welt, S., Mattes, M. J., Grando, R., Thomson, T. M., Leonard, R. W., Zanzonico, R. W., Bigler, R. E., Yeh, S., Oettgen, H. F., and Old, L. J.,** Monoclonal antibody to an intracellular antigen images human melanoma transplants in nu/nu mice, *Proc. Natl. Acad. Sci. U.S.A.*, 84, 4200, 1987.
7. **Dairkee, S. and Hackett, A. J.,** Internal antigens accessible in breast cancer: implications for tumor targeting, *J. Natl. Cancer Inst.*, 80, 1216, 1988.
8. **Epstein, A. L., Chen, F. M., and Taylor, C. R.,** A novel method for the detection of necrotic lesions in human cancers, *Cancer Res.*, 48, 5842, 1988.
9. **Haisma, H. J., Kessel, M. A. P., Silva, C., van Muijen, M., Roos, J. C., Bril, H., Martens, H. J. M., McCabe, R., and Boven, E.,** Human IgM monoclonal antibody 16.88: pharmacokinetics and distribution in mouse and man, *Br. J. Cancer*, 62, 40, 1990.
10. **Steis, R. G., Carrasquillo, J. A., McCabe, R., Bookman, M. A., Reynolds, J. C., Larson, S. M., Smith, J. W., Clark, J. W., Dailey, V., Del Vecchio, S., Shuke, N., Pinsky, C. M., Urba, W. J., Haspel, M., Perentesis, P., Paris, B., Longo, D. L., and Hanna, M. G.,** Toxicity, immunogenicity, and tumor radioimmunodetecting ability of two human monoclonal antibodies in patients with metastatic colorectal carcinoma, *J. Clin. Oncol.*, 8, 476, 1990.
11. **Moshakis, V., McIlhinney, R. A. J., Raghavan, D., and Neville, A. M.,** Localization of human tumor xenografts after i.v. administration of radiolabeled monoclonal antibody, *Br. J. Cancer*, 44, 91, 1981.

12. **Hedin, A., Wahren, B., and Hammerstrom, S.,** Tumor localization of CEA-containing human tumors in nude mice by means of monoclonal anti-CEA antibodies, *Int. J. Cancer*, 30, 547, 1982.
13. **Philben, V. J., Jakowatz, J. G., Beatty, B. G., Vlahos, W. G., Paxton, R. J., Williams, L. E., Shiveley, J. E., and Beatty, J. D.,** The effect of tumor CEA content and tumor size on tissue uptake of indium 111-labeled anti-CEA monoclonal antibody, *Cancer*, 57, 571, 1986.
14. **Goldenberg, A., Masui, H., Divgi, C., Kamrath, H., Pentlow, K., and Mendelsohn, J.,** Imaging of human tumor xenografts with an indium-111 labeled anti-epidermal growth factor receptor monoclonal antibody, *J. Natl. Cancer Inst.*, 81, 1616, 1989.
15. **Abdel-Nabi, H. H., Schwartz, A. N., Goldfogel, G., Ortman, J. A., Matsuoka, D. M., Unger, M. W., and Wechter, D. G.,** Colorectal tumors: scintigraphy with In-111 anti-CEA monoclonal antibody and correlation with surgical, histopathologic and immunohistochemical findings, *Radiology*, 166, 747, 1988.
16. **Giacomini, P., Aguzzi, A., Pestka, S., Fisher, P. B., and Ferrone, S.,** Modulation by recombinant DNA leukocyte (α) and fibroblast (β) interferons of the expression and shedding of HLA- and tumor-associated antigens by human melanoma cells, *J. Immunol.*, 133, 1649, 1984.
17. **Greiner, J. W., Hand, P. H., Noguchi, P., Fischer, P. B., Pestka, S., and Schlom, J.,** Enhanced expression of surface tumor-associated antigens on human breast and colon tumor cells after recombinant leukocyte a-interferon treatment, *Cancer Res.*, 44, 3208, 1984.
18. **Greiner, J. W., Guadagni, F., Noguchi, P., Pestka, S., Colcher, D., Fisher, P. B., and Schlom, J.,** Recombinant interferon enhances monoclonal antibody-targeting of carcinoma lesions *in vivo*, *Science*, 235, 895, 1987.
19. **Rowlinson, G., Balkwill, F., Snook, D., Hooker, G., and Epenetos, A.,** Enhancement by γ-interferon of *in vivo* tumor radiolocalization by a monoclonal antibody against HLA-DR antigen, *Cancer Res.*, 46, 6413, 1986.
20. **Rosenblum, M. G., Lamki, L. M., Murray, J. L., Carlo, D. J., and Gutterman, J. U.,** Interferon-induced changes in pharmacokinetics and tumor uptake of 111-In labeled antimelanoma antibody 96.5 in melanoma patients, *J. Natl. Cancer Inst.*, 80, 160, 1988.
21. **Haisma, H. J., Battaile, A., Stradtman, E. W., Knapp, R. C., and Zurawski, V. R.,** Antibody-antigen complex formation following injection of OC125 monoclonal antibody in patients with ovarian cancer, *Int. J. Cancer*, 40, 758, 1987.
22. **Hagan, P. L., Halpern, S. E., Chen, A., Krishnan, L., Frincke, J., Bartholomew, R. M., David, G. S., and Carlo, D.,** *In vivo* kinetics of radiolabeled monoclonal anti-CEA antibodies in animal models, *J. Nucl. Med.*, 26, 1418, 1985.
23. **Pimm, M. V., Durrant, L. G., and Baldwin, R. W.,** Influence of circulating antigen on the biodistribution and tumor localization of radiolabeled monoclonal antibody in a human tumor:nude mouse xenograft model, *Eur. J. Cancer Clin. Oncol.*, 25, 9, 1325, 1989.
24. **Goldenberg, D. M., Goldenberg, H., Sharkey, R. M., Lee, R. E., Higgenbotham, E., Horowitz, J. A., Hall, T. C., Pinsky, C. M., and Hansen, H. J.,** Imaging of colorectal carcinoma with radiolabeled antibodies, *Semin. Nucl. Med.*, 19, 262, 1989.
25. **Moshakis, V., Ormerod, M. G., Westwood, J. H., Imrie, S., and Neville, A. M.,** The site of binding of anti-CEA antibodies to tumor CEA *in vivo*: an immunocytochemical and autoradiographic approach, *Br. J. Cancer*, 46, 18, 1982.
26. **Lewis, J. C. M., Boxer, G. M., Sharle, R., and Bagshawe, K. D.,** The comparative distribution of monoclonal antibodies to CEA in colorectal xenografts, *Tumor Biol.*, 5, 255, 1984.
27. **Pimm, M. V. and Baldwin, R. W.,** Localization of an anti-tumor monoclonal antibody in human tumor xenografts: kinetic and quantitative studies with the 791T/36 antibody, in *Monoclonal Antibodies for Cancer Detection and Therapy*, Baldwin, R. W. and Byers, V. S., Eds., Academic Press, New York, 1985, 97.
28. **Fand, I., Sharkey, R. M., McNally, W. P., Bril, A. B., Som, P., Yamamoto, K., Primus, F. J., and Goldenberg, D. M.,** Quantitative whole body autoradiography of radiolabeled antibody distribution in a xenografted human cancer model, *Cancer Res.*, 46, 271, 1986.
29. **Jones, P. L., Gallagher, B. M., and Sands, H.,** Autoradiograhic analysis of monoclonal antibody distribution in human colon and breast tumor xenografts, *Cancer Immunol. Immunother.*, 22, 139, 1986.
30. **Ong, L. I. and Mattes, M. J.,** Penetration and binding of monoclonal antibodies in experimental human solid tumors grown in mice, *Cancer Res.*, 49, 4264, 1989.
31. **Guillino, P. and Grantham, F.,** Studies on the exchange of fluids between host and tumor. The blood flow of hepatomas and other tumors in mice and rats, *J. Natl. Cancer Inst.*, 27, 1465, 1961.
32. **Mann, B., Cohen, M., Saxton, R. E., Morton, D. L., Benedict, W. F., Korn, E. L., Spolter, L., Graham, L. S., Chang, C. C., and Burk, M. W.,** Imaging of human tumor xenografts in nude mice with radiolabeled antibodies, *Cancer*, 54, 1319, 1984.
33. **Goldenberg, D. M.,** Future role of radiolabeled monoclonal antibodies in oncological diagnosis and therapy, *Semin. Nucl. Med.*, 4, 332, 1989.

34. **Matzku, S., Kirchgessner, H., Dippold, W. G., and Bruggen, J.,** Immunoreactivity of monoclonal anti-melanoma antibodies in relation to the amount of radioactive iodine substituted to the antibody molecule, *Eur. J. Nucl. Med.*, 11, 260, 1985.
35. **Sakahara, H., Endo, K., Nakashima, T., Koizumi, M., Ohta, H., Torizuka, K., Furukawa, T., Ohmono, Y., Yokoyama, A., Okada, K., Yoshida, O., and Nishi, S.,** Effect of DTPA conjugation on the antigen binding activity and biodistribution of monoclonal antibodies against α-fetoprotein, *J. Nucl. Med.*, 26, 750, 1985.
36. **Pimm, M. V. and Baldwin, R. W.,** Comparative tumor localization properties of radiolabeled monoclonal antibody preparations of defined immunoreactivities, *Eur. J. Nucl. Med.*, 13, 348, 1987.
37. **Covell, D. G., Barbet, J., Holton, O. D., Black, C. D. V., Parker, R. J., and Weinstein, J. N.,** Pharmacokinetics of monoclonal immunoglobulin G1, $F(ab')_2$ and Fab′ in mice, *Cancer Res.*, 46, 3969, 1986.
38. **Holton, O. D., Black, C. D. V., Parker, R. J., Covell, D. G., Barbet, J., Sieber, S., Talley, M. J., and Weinstein, J. N.,** Biodistribution of monoclonal IgG1, $F(ab')_2$ and Fab′ in mice after intravenous injection, *J. Immunol.*, 139, 3041, 1987.
39. **Zuckier, L. S., Rodriguez, L. D., and Scharff, M. D.,** Immunologic and pharmacologic concepts of monoclonal antibodies, *Semin. Nucl. Med.*, 19, 166, 1989.
40. **Haisma, H. J., Moseley, K. R., Battaile, A., Griffiths, T. C., and Knapp, R. C.,** Distribution and pharmacokinetics of radiolabeled monoclonal antibody OC125 after intravenous and intraperitoneal administration in gynecologic tumors, *Am. J. Obstet. Gynecol.*, 159, 843, 1988.
41. **Moseley, K. R., Battaile, A., Knapp, R. C., and Haisma, H. J.,** Localization of radiolabeled $F(ab')_2$ fragments of monoclonal antibodies in nude mice bearing intraperitoneally growing human ovarian cancer xenografts, *Int. J. Cancer*, 42, 368, 1988.
42. **Colcher, D., Zalutski, M., Kaplan, W., Kufe, D., Austin, F., and Schlom, J.,** Radiolocalization of human mammary tumors in athymic mice by monoclonal antibody, *Cancer Res.*, 40, 736, 1983.
43. **Wahl, R. L., Parker, C. W., and Philpott, G. W.,** Improved imaging and tumor localization with monoclonal $F(ab')_2$, *J. Nucl. Med.*, 24, 316, 1983.
44. **Buchegger, F., Haskell, C. M., Schreyer, M., Scazziga, B. R., Randin, S., Carrel, S., and Mach, J. P.,** Radiolabeled fragments of monoclonal antibodies against carcinoembryonic antigen for localization of human colon carcinoma grafted in nude mice, *J. Exp. Med.*, 158, 413, 1983.
45. **Colapinto, E. V., Humphrey, P. A., Zalutski, M. R., Groothuis, D., Friedman, H. S., de Tribolet, N., Carrel, S., and Bigner, D. D.,** Comparative localization of murine monoclonal antibody Me1-14 $F(ab')_2$ fragment and whole IgG2a in human glioma xenografts, *Cancer Res.*, 48, 5701, 1988.
46. **Larson, S. M.,** Clinical radioimmunodetection, 1978-1988: overview and suggestions for standarization of clinical trials, *Cancer Res.*, 50, 892s, 1990.
47. **Larson, S. M., Brown, J. P., Wright, P. W., Carrasquillo, J. A., Hellström, I., and Hellström, K. E.,** Imaging of melanoma with I-131-labeled monoclonal antibodies, *J. Nucl. Med.*, 24, 123, 1983.
48. **Rogers, G. T., Harwood, P. J., Pedley, R. B., Boden, J., and Bagshawe, K. D.,** Dose-dependent localization and potential for therapy of $F(ab')_2$ fragments against CEA studied in a human tumor xenograft model, *Br. J. Cancer*, 54, 341, 1986.
49. **Sharkey, R. M., Primus, F. J., and Goldenberg, D. M.,** Antibody protein dose and radioimmunodetection of GW-39 human colon tumor xenografts, *Int. J. Cancer*, 39, 611, 1987.
50. **Colcher, D., Milenic, D., Roselli, M., Raubitschek, A., Yarranton, G., King, D., Adair, J., Whittle, N., Bodmer, M., and Schlom, J.,** Characterization and biodistribution of recombinant and recombinant/chimeric constructs of monoclonal antibody B72.3, *Cancer Res.*, 49, 1738, 1989.
51. **McCabe, R. P., Peters, L. C., Haspel, M. V., Pomato, N., Carrasquillo, J. A., and Hanna, M. G.,** Preclinical studies on the pharmacokinetic properties of human monoclonal antibodies to colorectal cancer and their use for detection of tumors, *Cancer Res.*, 48, 4348, 1988.
52. **Pressman, D. and Korngold, L.,** The *in vivo* localization of anti-Wagner osteogenic sarcoma antibodies, *Cancer*, 6, 7619, 1953.
53. **Haisma, H. J., Hilgers, J., and Zurawski, V. R.,** Iodination of monoclonal antibodies for diagnosis and radiotherapy using a convenient one vial method, *J. Nucl. Med.*, 27, 1890, 1986.
54. **Zalcberg, J. R., Thompson, C. H., Lichtenstein, M., and McKenzie, I. F. C.,** Tumor immunotherapy in the mouse with the use of ^{131}I-labeled monoclonal antibodies, *J. Natl. Cancer Inst.*, 72, 697, 1984.
55. **Wilburg, D. S., Hadley, S. W., Hylarides, M. D., Abrams, P. G., Beumier, P. A., Morgan, A. C., Reno, J. M., and Fritzberg, A. R.,** Development of a stable radioiodinating reagent to label monoclonal antibodies for radiotherapy of cancer, *J. Nucl. Med.*, 30, 216, 1989.
56. **Zalutski, M. R., Noska, M. A., Colapinto, E. V., Garg, P. K., and Bigner, D. D.,** Enhanced tumor localization and *in vivo* stability of a monoclonal antibody radioiodinated using radioiodinated *N*-succinimidyl 3(tri-*n*-butylstannyl) benzoate, *Cancer Res.*, 49, 5543, 1989.

57. **Haseman, M. K., Googwin, D. A., Meares, C. F., Kaminski, M. S., Wensel, T. G., McCall, M. J., and Levy, R.,** Metabolizable ^{111}In chelate conjugated antiidiotype monoclonal antibody for radioimmunodetection of lymphoma in mice, *Eur. J. Nucl. Med.*, 12, 455, 1986.
58. **Paik, C. H., Yokoyama, K., Reynolds, J. C., Quadri, S. M., Min, C. Y., Shin, S. Y., Maloney, P. J., Larson, S. M., and Reba, R. C.,** Reduced background activities by introduction of a diester linkage between antibody and a chelate in radioimmunodetection of tumor, *J. Nucl. Med.*, 30, 1693, 1989.
59. **Fritzberg, A. R., Abrams, P. A., Beaumier, P. L., Kasina, S., Morgan, A. C., Rao, T. N., Reno, J. M., Sanderson, J. A., Srinivasan, A., Wilbur, D. S., and Vanderheyden, J.,** Specific and stable labeling of antibodies with technetium-99m with a diamide dithiolate chelating agent, *Proc. Natl. Acad. Sci. U.S.A.*, 85, 4025, 1988.
60. **Foon, K. A.,** Biological response modifiers: the new immunotherapy, *Cancer Res.*, 49, 1621, 1989.
61. **Bernstein, I. D., Tam, M. R., and Nowinski, R. C.,** Mouse leukemia: therapy with monoclonal antibodies against a thymus differentiation antigen, *Science*, 207, 68, 1980.
62. **Herlyn, D. and Koprowski, H.,** IgG2a monoclonal antibodies inhibit human tumor growth through interaction with effector cells, *Proc. Natl. Acad. Sci. U.S.A.*, 79, 4761, 1982.
63. **Seto, M., Takahashi, T., Nakamura, S., Matsudaira, Y., and Nishizuka, Y.,** *In vivo* antitumor effects of monoclonal antibodies with different immunoglobulin classes, *Cancer Res.*, 43, 4768, 1983.
64. **Masui, H., Kawamoto, T., Sato, J. D., Wolf, B., Sato, G., and Mendelsohn, J.,** Growth inhibition of human tumor cells in athymic mice by anti-epidermal growth factor receptor monoclonal antibodies, *Cancer Res.*, 44, 1000, 1984.
65. **Cuttitta, F., Carney, D. N., Mulshine, J., Moody, T. W., Fedorko, J., Fischler, A., and Minna, J. D.,** Bombesin-like peptides can function as autocrine growth factors in human small cell lung cancer, *Nature (London)*, 316, 823, 1985.
66. **Rofstad, R.,** Human tumor xenografts in radiotherapeutic research, *Radiother. Oncol.*, 3, 35, 1985.
67. **Goldenberg, D. M., Gaffar, S. A., Bennet, J., and Beach, L.,** Experimental radioimmunotherapy of a xenografted human colonic tumor (GW-39) producing carcinoembryonic antigen, *Cancer Res.*, 41, 4354, 1981.
68. **Esteban, J. M., Schlom, J., Mornex, F., and Colcher, D.,** Radioimmunotherapy of athymic mice bearing human colon carcinomas with monoclonal antibody B72.3, *Eur. J. Cancer Clin. Oncol.*, 23, 643, 1987.
69. **Jones, D. H., Goldman, A., Gordon, I., Pritchard, J., Gregory, B. J., and Kenshead, J. T.,** Therapeutic application of a radiolabeled monoclonal antibody in nude mice xenografted with human neuroblastoma: tumoricidal effects and distribution studies, *Int. J. Cancer*, 35, 715, 1985.
70. **Cheung, N. K., Landmeier, B., and Neely, J.,** Complete tumor ablation with iodine-131 radiolabeled disialoganglioside GD2-specific monoclonal antibody against neuroblastoma xenografted in nude mice, *J. Natl. Cancer Inst.*, 77, 739, 1986.
71. **Esteban, J. M., Hyams, D. M., Beatty, B. G., Wanek, P., and Beatty, J. D.,** Effect of Yttrium-90 labeled anti-carcinoembryonic antigen monoclonal antibody on the morphology and phenotype of human tumors grown as peritoneal carcinomatosis in athymic mice, *Cancer*, 63, 1343, 1989.
72. **Hyams, D. M., Esteban, J. M., Beatty, B. G., Wanek, P. M., and Beatty, J. D.,** Prolongation of survival of nude mice bearing human colon cancer, *Arch. Surg.*, 124, 175, 1989.
73. **Blumental, R. D., Sharkey, R. M., Kashi, R., and Goldenberg, D. M.,** Comparison of therapeutic efficacy and host toxicity of two different 131-I labeled antibodies and their fragments in the GW-39 colonic cancer xenograft model, *Int. J. Cancer*, 44, 292, 1989.
74. **Blythman, H. E., Casellas, P., Gros, O., Jansen, F. K., Paolucci, F., Pau, B., and Vidal, H.,** Immunotoxins: hybrid molecules of monoclonal and a toxin subunit specificcally kill tumor cells, *Nature (London)*, 290, 145, 1981.
75. **FitzGerald, D. and Pastan, I.,** Targeted toxin therapy for the treatment of cancer, *J. Natl. Cancer Inst.*, 81, 1455, 1989.
76. **Weil-Hillman, G., Runge, W., Jansen, F. K., and Vallera, A. A.,** Cytotoxic effect of anti-Mr 67000 protein immunotoxins on human solid tumors in a nude mouse model, *Cancer Res.*, 45, 1328, 1985.
77. **FitzGerald, D. J., Bjorn, M. J., Ferris, R. J., Winkelhake, J. L., Frankel, A. E., Hamilton, T. C., Ozols, R. F., Willingham, M. C., and Pastan, I.,** Antitumor activity of an immunotoxin in an nude mouse model of human ovarian cancer, *Cancer Res.*, 47, 1407, 1987.
78. **Willingham, M. C., FitzGerald, D. J., and Pastan, I.,** Pseudomonas exotoxin coupled to a monoclonal antibody against ovarian cancer inhibits the growth of human ovarian cancer cells in a mouse model, *Proc. Natl. Acad. Sci. U.S.A.*, 84, 2474, 1987.
79. **Bumol, T. F., Wang, Q. C., Reisfeld, R. A., and Kaplan, N. O.,** Monoclonal antibody and antibody-toxin conjugate to a cell surface proteoglycan of melanoma cells suppress *in vivo* tumor growth, *Proc. Natl. Acad. Sci. U.S.A.*, 80, 529, 1983.
80. **Hirota, N., Ueda, M., Ozawa, S., Abe, O., and Shimizu, N.,** Suppression of an epidermal growth dactor receptor-hyperproducing tumor by an immunotoxin conjugate of gelonin and a monoclonal anti-epidermal growth factor receptor antibody, *Cancer Res.*, 49, 7106, 1989.

81. **Laurent, G., Pris, J., and Farcet, J. P.,** Effects of therapy with T101-ricin A chain immunotoxin in two leukemia patients, *Blood,* 67, 1680, 1986.
82. **Spitler, L. E., Del Rio, M., and Khentigan, A.,** Therapy of patients with malignant melanoma using a monoclonal antimelanoma antibody ricin A chain immunotoxin, *Cancer Res.,* 47, 1717, 1987.
83. **Aron, R. and Sela, M.,** *In vitro* and *in vivo* efficacy of conjugates of daunomycin with anti-tumor antibodies, *Immunol. Rev.,* 62, 5, 1982.
84. **Dillman, R. O., Shawler, D. L., and Johnson, D. E.,** Preclinical trials with combinations of T101-monoclonal antibody and doxorubicin, *Cancer Res.,* 46, 4886, 1986.
85. **Deguchi, T., Chu, T. M., Leong, S. S., Horoszewicz, J. S., and Lee, C.,** Effect of methotrexate-monoclonal anti-prostatic acid phosphatase antibody conjugate on human prostate tumor, *Cancer Res.,* 46, 3751, 1986.
86. **Kanellos, J., Pietersz, G. A., and McKenzie, I. F. C.,** Studies of methotrexate monoclonal antibodies conjugates for immunotherapy, *J. Natl. Cancer Inst.,* 75, 319, 1985.
87. **Rowland, G. F., Simmonds, R. G., Gore, V. A., Marsden, C. H., and Smith, W.,** Drug localization and growth inhibition studies of vindesin-monoclonal anti-CEA conjugates in a human tumor xenograft, *Cancer Immunol. Immunother.,* 19, 1, 1986.
88. **Corvalan, J. R. F., Smith, W., and Gore, V. A.,** Tumor therapy with vinca alkaloids targeted by a hybrid-hybrid monoclonal antibody recognizing both CEA and vinca alkaloids, *Int. J. Cancer,* 2 (Suppl.), 22, 1988.
89. **Elias, D. J., Hirschowitz, L., Kline, L. E., Kroener, J. F., Dillman, R. O., Walker, L. E., Robb, J. A., and Timms, R. M.,** Phase I clinical comperative study of monoclonal antibody KS1/4 and KS1/4-methotrexate immunoconjugate in patients with non-small cell lung carcinoma, *Cancer Res.,* 50, 4145, 1990.
90. **Weinstein, J. N. and Leserman, L. D.,** Liposomes as drug carriers in cancer chemotherapy, *Pharmacol. Ther.,* 24, 207, 1989.
91. **Senter, P. D., Saulnier, M. G., Schreiber, G. J., Hirschberg, D. L., Brown, J. P., Hellström, I., and Hellström, K. E.,** Anti-tumor effects of antibody-alkaline phosphatase conjugates in combination with etoposide phosphate, *Proc. Natl. Acad. Sci. U.S.A.,* 85, 4842, 1988.
92. **Senter, P. D., Schreiber, G. J., Hirschberg, D. L., Ashe, S. A., Hellström, K. E., and Hellström, I.,** Enhancement of the *in vitro* and *in vivo* antitumor activities of phosphorylated mitomycin C and etoposide derivatives by monoclonal antibody-alkaline phosphatase conjugates, *Cancer Res.,* 49, 5789, 1989.
93. **Bagshawe, K. D., Springer, C. J., Searle, F., Antoniw, P., Sharma, S. K., Melton, R. G., and Sherwood, R. F.,** A cytotoxic agent can be generated selectively at cancer sites, *Br. J. Cancer,* 58, 700, 1988.

Chapter 18

CYTOKINES

S. T. A. Malik and F. R. Balkwill

TABLE OF CONTENTS

I. INTRODUCTION

The role of cytokines in the therapy and the biology of cancer is a subject of increasing interest. Several factors have contributed to this interest, particularly the availability of recombinant cytokines such as the colony stimulating factors, tumor necrosis factor, the interferons and interleukins for clinical trials, and detailed laboratory studies. Coupled with the use of these proteins as therapeutic agents, there is an increasing awareness of their roles in modifying both normal and tumor cell behavior and metabolism. The bewildering array of *in vitro* effects of cytokines has made it important to look at overall *in vivo* effects in tumor models, and to try and extrapolate some of the results to the therapy and understanding of the biology of human cancers.[1]

When studying cytokine actions in nude mouse xenograft models, it is important to appreciate the limitations and interpretability of the results obtained, which can be complicated by the pleiotropic actions of cytokines and their species specificity. However, concomitant studies to investigate the role of T cell-mediated immunity, and comparison of *in vitro* and *in vivo* effects on tumor cells can yield information relevant to the antitumor effects of cytokines in man. As cytokine actions are not restricted to cytostatic/cytotoxic effects on tumor cells, but include other actions such as growth stimulation, effects on differentiation of tumor cells, and effects on host stromal and immune cell polulations, speculations can be made on the role of endogenously produced cytokines in determining tumor behavior *in vivo*. Although many cytokines are produced by T cells, there is no evidence that nude mice lack the ability to generate most cytokines, and it is imperative that factors that lead to endogenous production of cytokines, e.g., infections, are carefully monitored in experiments assessing the effect of exogenous cytokines in tumor systems. Successful tumor transplantation in nude mice may be under the influence of cytokines produced by the host. For example; pretreatment of mice with anti-interferon (IFN) antibodies was shown to enhance the tumor take rate of human prostate cancer.[2] Observations such as these are in themselves relevant to the use of cytokines as therapeutic agents in man, and point out one of the potential artifacts of the nude mouse system, that is, the tumor populations that do grow in nude mice may be selected for resistance, or indeed sensitivity, to endogenous cytokines. There is experimental support for this from a recent study which showed that a murine fibrosarcoma cell line, highly sensitive to tumor necrosis factor (TNF) *in vitro*, rapidly became TNF resistant when injected s.c. into mice.[3]

With the known cross-species specificity of many cytokines, their effects on the host and subsequent tumor behavior must be borne in mind. The excellent study of Kopf-Maier and Jackel[4] illustrates the complexity of host-tumor interactions in a nude mouse xenograft model. Analysis of the relative amounts of host (mouse) and colorectal tumor tissue (human) at various times in s.c. xenografts revealed that 16 to 24 h after transplantation, the number of tumor cells had reduced markedly. Host infiltrating cells became the dominant population of cells between 2 and 8 d. Tumor cells began to proliferate on day 6 and became cytokinetically stable between days 14 and 30, coinciding with a decrease in host cells, and maximal tumor growth rate. At 30 d after transplantation, degeneration of both tumor and host cells was observed. Clearly, the administration of a cytokine could lead to the modulation of a number of parameters, and its effect may be determined by the time at which it is introduced in the xenograft model.

In this chapter, studies of individual cytokines will be discussed, highlighting some of the aspects pointed out above.

II. THE INTERFERONS

A. BACKGROUND

The three major classes of IFN, IFN-α and IFN-β, and IFN-γ, have been noted to have antiproliferative effects against a variety of tumor cell lines *in vitro*.[5] Of the three types, IFN-α is currently the only cytokine that has an established role in the therapy of human cancer, particularly in the treatment of hairy cell leukemia, chronic myeloid leukemia, and in solid tumors such as carcinoid tumors and renal cell cancer.[6] IFNs have been investigated in a variety of syngeneic murine tumor models over the last 2 decades, and the work done by Gresser[7] in this field has laid the foundations for the use of IFNs in cancer therapy. Predictably, antitumor activity has been described against virus-induced murine neoplasms, spontaneously developing tumors in mice, and in mice injected with murine tumor cell lines. In most of these models, an optimal antitumor effect was noted if the tumor load was low and IFN preparations were given daily.[8]

The use of IFNs in the nude mouse-human tumor model poses one major problem — the species specificity of IFNs. Thus, human IFNs, while inducing 2′-5′A synthetase activity in human tumor xenografts, do not affect the levels of this enzyme in mouse tissues or enhance NK cell activity in the mouse. Mouse IFNs do not induce 2′-5′A synthetase activity in human tissues.[9,10] This makes it difficult to assess the role of host antitumor mechanisms that may be activated by IFNs (e.g., NK cells and antitumor macrophages), although these problems can be partially overcome by adoptive transfer of syngeneic or xenogeneic cells activated *in vitro* by cytokines (see Section IV on interleukin-2). The species specificity of IFNs also leads to the exclusion of the assessment of other potential avenues of host-tumor interactions, such as the induction of class II MHC antigens [11] and tumor specific antigens (e.g., CA125[12]) by IFNs, which may theoretically increase the potential for immunological attack on the tumor. However, useful information can be gained on the importance of the direct antitumor effects of IFNs, and how these can be modified by varying the route of administration and dose scheduling. Parallel experiments with murine IFNs may indicate whether indirect effects of IFNs may also be involved.

There have been numerous reports on the antitumor effects of IFNs against human tumor xenografts in nude mice. A selection of these is shown in Tables 1 and 2, and the ensuing discussion will concentrate on some of the major findings of these studies.

B. TYPE OF TUMOR, TUMOR LOAD, AND TUMOR PROLIFERATIVE STATE

Although no single study has evaluated the effects of IFNs on a large sample of different tumors, consistent antitumor effects of IFNs have been noted particularly in renal cell, breast, ovarian and colorectal cancers, and osteosarcomas transplanted to nude mice. Not surprisingly, the antitumor effect is more noticable when IFN therapy is begun shortly after tumor injection; for example, in the therapy of bone sarcomas.[15] Studies such as this suggest that the IFNs are most effective against low tumor burdens and best used clinically as adjuvant or neo-adjuvant therapies. An increasing number of studies have shown that antiproliferative effects are also evident in mice bearing "established" tumors. It has been suggested that they are more effective against slower growing xenografts, such as breast cancer, than against more aggressive tumors such as some fibrosarcomas.[28] In general, histological examination of human tumor xenografts inhibited in growth by human IFNs shows few changes. The tumors usually appear to be miniatures of the control tumors, although some exceptions have been reported. Intratumoral therapy of an s.c. glioblastoma multiform xenograft with human IFN-β led to growth inhibition.[29] Histological examination of IFN-treated tumors showed an increase of round and multinucleate cells in tumor areas, focal necrosis, and increased collagen deposition. The authors pointed out that IFN therapy led to an increase of polyploid cells in the tumors, a change associated with a better clinical prognosis. Brosjö et al.[19,30] reported that IFN-α therapy of some osteosarcoma xenografts not only inhibited their growth, but also led to the formation of normal bone in tumor areas. Although this indicated that IFN-α led to differentiation of the tumor into normal bone, recent studies suggest that the normal bone is of mouse origin, that is, a bone-inducing substance was produced by the action of IFN on the tumor.[31] However, it is not known if this was a specific action of IFN or it was related to the growth inhibition *per se*.

TABLE 1
Studies of α-Interferons in Nude Mouse Human Tumor Xenograft Models

Interferon	Dose	Route	Xenograft	Antitumor effect	Ref.
HLI	5×10^6 μ/kg	s.c.	Fibrosarcoma	No effect	13
HLI	2×10^4 μ/d	s.c.	Breast cancer	Growth inhibition	14
IFN α-2 HLI	2×10^4 μ/d	s.c.	Bone sarcoma fragments	Only effective before or shortly after implantation	15
HLI	1×10^5 μ/d	s.c. and i.t.	Renal cell cancer HeLa cells Nasopharyngeal cancer Breast cancer Laryngeal cancer	Only effective against renal cell cancer	16
HLI	2×10^4 μ/d	s.c.	Gliomas	No antitumor effect	17
IFN-α	1×10^4 μ/d	s.c.	HeLa cell	No antitumor effect	18
IFN-α	10^5–10^6 μ/d	s.c.	Osteosarcoma	Growth arrest in 5 of 11 xenografts and normal bone formation in some	19
HLI	5×10^4 μ/d	s.c.	Osteosarcoma	Growth inhibition of early or late tumors	20
HLI	2×10^5 μ/d	s.c.	Colorectal, breast ovarian cancer	Effective tumor stasis in 9 of 12 xenografts	21

Note: s.c., subcutaneous; i.p., intraperitoneal; i.t., intratumoral; HLI, human lymphoblastoid interferon.

TABLE 2
Studies of β and γ-Interferons in Nude Mouse Human Tumor Xenograft Models

Interferon	Dose	Route	Xenograft	Antitumor effect	Ref.
IFN-β	$1\text{-}6 \times 10^5$ μ/kg	s.c. and i.t.	Melanoma	i.t. IFN better: tumor stasis	22
IFN-β	2×10^5 μ/d	i.p.	Transitional cell cancer	Only effective on early tumors	23
rHu-γ	$2\text{-}3.3 \times 10^6$ μ/d	i.v. and i.t.	Melanoma	No significant effect	24
rHu-γ	2×10^5 μ/d	i.p./s.c./i.t.	Colorectal Breast Ovarian cancers	No significant effect	21
rHu-γ	2×10^5 μ/injection	Intrasplenic	Intrasplenic colon cancer cell lines	Only effective in one cell line	25
rHu-γ	5×10^4 μ/d	i.p.	Intraperitoneal ovarian cancer	Marked antitumor activity in 2 of 3 xenografts	26
rHu-γ	2×10^5 μ/d	s.c.	Osteosarcoma	Tumor stasis in 2 of 3 xenografts	27

Note: s.c., subcutaneous; i.p., intraperitoneal; i.t., intratumoral; IFN, interferon; rHu-γ, recombinant human interferon gamma.

C. DOSE SCHEDULING AND ROUTE

The results of several studies indicate that the optimal antitumor effects of IFNs are related to the dose of IFN administered. For example, this was noted in studies on the effect of human leucocyte IFN and IFN-α, respectively, against breast cancer and osteosarcoma xenografts.[9,30] In addition, daily and prolonged administration of IFN was found to be optimal, and at least 2 weeks of therapy elapsed before the antitumor effect was evident in the breast cancer xenograft study. The dose dependence of the direct effect of IFNs on tumors is also demonstrated by the increased antiproliferative activity of intratumoral IFN-β, as compared to systemic administration, despite similar plasma levels.[22] The effective doses of IFN-α in different studies vary from 1×10^4 U/mouse/d to 8×10^6 U/mouse/d. The effective dose of 2×10^5 U/mouse/d against breast cancer xenografts[9] corresponds to 20×10^6 U/m^2/d in man, and is likely to be poorly tolerated by most patients over a prolonged period. Based on the data from nude mouse studies, a possible reason for the inefficacy of IFNs as single agents in most solid tumors in man may be related to the fact that continued and sustained administration of high doses of IFNs are impractical in cancer patients because of toxicity. Therefore, innovative approaches may be required to try and increase the selective delivery of IFNs to tumors in man.

D. THE COMPARATIVE ACTIVITY OF DIFFERENT INTERFERONS

One of the first studies to look into the differential effects of subtypes of IFN-α showed that there was a marked difference in the antitumor activity of subtypes IFN-αA and IFN-αD in breast cancer models.[32] A follow-up study confirmed the differential activity of six pure IFN preparations against a breast and a colorectal cancer xenograft. The most active IFNs were of the alpha group, whereas IFN-β and IFNγ were largely inactive.[32] Human IFN-α was also compared to human recombinant IFN-γ against 12 xenografts of different tumors (breast, bowel, and ovary), and the superior antitumor effect of IFN-α was again demonstrated.[21] IFN-γ was also inactive, except at very high doses, against melanoma cell lines injected into nude mice.[24] These studies, like others,[21] pointed out the discrepancy between the *in vitro* growth inhibitory effect of IFN-γ and its lack of antitumor activity *in vivo*. However, in experimental lung metastasis models using colorectal cancer and melanoma cell lines,[33,34] IFN-γ had undoubtedly antitumor effects. This suggests that the antitumor activity of IFN-γ may be site dependent. In i.p. ovarian cancer xenograft models, IFN-γ was active against two of three xenografts.[26] One of these models was completely resistant to IFN-γ, while mice treated with intraperitoneal IFN-α were cured (unpublished observations). In a study of the comparative efficacy of alpha and recombinant-gamma IFNs in the treatment of three osteosarcoma xenografts, IFN-γ was as effective as IFN-α in two tumors, whereas IFN-α was effective against all three.[27] Comparison of the antitumor activity of natural IFN-γ with recombinant IFN-γ in this study showed that the latter was a more effective antitumor agent and that this was related to the different pharmacokinetics of the two IFN-γ preparations.

E. MECHANISMS OF ANTITUMOR ACTIVITY

In view of the species specificity of IFNs, it is clear that human IFNs have direct antitumor effects on some human tumor xenografts in nude mice. However, IFNs have a wide variety of cell regulatory effects, not only on cells of the immune system, but on other non-hemopoietic cell populations.[5] In syngeneic murine models, the indirect antitumor effects of IFNs are well documented.[35] For example, IFN-β was shown to induce hemorrhagic necrosis in a murine tumor model.[36] The assessment of such indirect effects is of importance and is one of the major drawbacks of studying IFN actions in the nude mouse model. That indirect effects may play an important role is indicated by studies on the antitumor effect of murine IFNs against human xenografts.[37,38] In both studies, human tumors were inhibited in growth, but the mechanism of the antitumor effect of murine IFN was not related to the activation of NK cells. Probably, the antitumor effect was caused by some unknown modulation of the host-tumor relationship.

Macrophage activation by IFN-γ may be responsible for some of its antitumor effects. Adoptive transfer studies of human monocytes activated *in vitro* by IFN-γ demonstrated antitumor effects.[39] Experiments on a possible synergism of 5-fluorouracil and murine IFN-γ against colorectal cancer cell lines were also suggestive for macrophage activation as an important component in the antitumor action of IFN-γ.[25] In the Meth A sarcoma model in nude mice, the antitumor effect of mouse IFN was greatly diminished, suggesting that in immunogenic tumors T cell-mediated immunity is of importance.[40] T cell effects are clearly not assessable in the nude mouse-human tumor model, but are of questionable relevance for the antitumor effects of IFNs in man because of the low immunogenicity of most human cancers.

F. SYNERGISTIC EFFECTS

Several investigators have shown that the antitumor effects of IFNs can be enhanced by other treatment modalities, particularly cytokines and cytotoxic drugs. Hyperthermia and irradiation can also potentiate the growth inhibitory effect of IFNs.[18,41]

1. Combinations of Cytokines

Additive antitumor effects of combinations of either IFN-α or IFN-β and IFN-γ against human tumor xenografts have been noted against s.c., i.p., and metastatic human tumors in nude mice,[33,42,43] but not by other investigators.[32] Whether true synergy occurred is difficult to say because IFNs administered as single agents were also active against the tumors studied and isobologram analysis was not performed. Cure of tumors was rare, as is known for single IFN therapy. No conclusions can be drawn from these studies on the potential toxicity of such combinations in man.

IFN-γ potentiation of the antitumor activity of TNF against a variety of cell lines has been noted *in vitro*.[44,45] Balkwill et al.[26] have described the results of i.p. therapy of three i.p. human ovarian cancer xenografts with recombinant human TNF, recombinant human IFN-γ or the two combined. Additive antitumor activity was noted in two of the three xenografts, and a third xenograft, resistant to both rhuIFN-γ and rhuTNF as single agents, was cured by the combination. Other investigators have reported similar results in i.p. ovarian cancer models.[46,47] The use of murine IFN-γ in combination with TNF, while also showing an enhanced antitumor effect as compared to either cytokine alone, can however result in marked toxicity in tumor-bearing animals (unpublished observations) This underlines the need for caution in extrapolating the therapeutic index of results obtained in mice with species-specific and human cytokines to man. Synergistic interactions with IFN-α and interleukin-2 have been reported in murine tumor models;[48] but to our knowledge, this has not been assessed in the nude mouse-human tumor model.

2. Cytotoxic Drugs

There is a wealth of preclinical data showing that the combination of IFNs with cytotoxic drugs can lead to enhanced antitumor activity. A selection of these studies is summarized in Table 3. IFNs have been shown to affect drug metabolism, although this did not explain the synergy between IFNs and cyclophosphamide and doxorubicin in inhibiting the growth of a breast cancer xenograft.[55] The mechanisms underlying these additive effects are likely to provide some intriguing insights, not only into the mechanisms of IFN action, but also into those of cytotoxic drugs. Thus, IFN-α was shown to enhance the uptake of doxorubicin into tumor cells *in vitro*, and this correlated with the additive antitumor effects of these agents *in vivo*.[56] IFN-α may therefore have potential as an agent to counteract the effects of the overexpression of the multidrug resistant phenotype, which is associated with accelerated efflux of doxorubicin from tumor cells. Several studies have noted the synergy between the antitumor action of the fluoropyrimidines, particularily 5-fluorouracil (5-FU) and IFNs *in vitro* and *in vivo* (see Table 3). The synergy between 5-FU and IFN-α was reversed by addition of exogenous thymidine,

TABLE 3
Positive Interactions between Human Interferons and Cytotoxic Drugs in Human Tumor Xenograft Models

Drug	IFN	Schedule	Tumor	Effect	Ref.
DOX	α	Concurrent	Breast cancer	Tumor stasis or regression and increased survival	49
DOX	γ	Concurrent	Colon cancer	Reduced tumor growth and inhibition of metastasis	50
CY	α	Concurrent and sequential	Breast cancer	Tumor regression and increased survival	49
CDDP	α	Concurrent	Mesothelioma	Tumor stasis	51
CDDP	α	Concurrent	Non-small cell lung cancer	Reduced tumor growth rate	52
MITO C	α	Concurrent	Mesothelioma	Tumor stasis	18
5-FU	β	Concurrent (alternate days)	HeLa	Reduced tumor growth rate	18
5-FU	γ	Concurrent for 5 d then IFN-γ for 5 more days	Colon cancer	Reduced tumor growth in spleen; inhibition of liver metastasis	25
5-FU	α	IFN-α × 3 weekly, 11 doses; 5-FU × 1 weekly 3 doses starting after 3 IFN injections	Colon cancer	Spleen tumors failed to develop after intrasplenic injection	53
5-FU	α	IFN-α daily 5-FU once weekly	Colon cancer	Reduced tumor growth rate	54

Note: DOX, doxorubicin; CY, cyclophosphamide; CDDP, cisplatin; MITO C, mitomycin C; 5-FU, f-fluorouracil.

suggesting an inhibitory effect of IFN on thymidylate synthetase activity in tumor cells.[57] Cells that become resistant to 5-FU often express high levels of thymidylate synthetase, and IFN may be able to reverse the drug-resistant phenotype by blocking this compensatory increase in enzyme activity. IFNs may also inhibit the membrane transport of extracellular thymidine.[58]

The potential importance of preclinical studies in nude mice is reflected by a recent clinical trial of IFN-α in combination with 5-FU.[59] A response rate of 76% in previously untreated patients with metastatic colorectal cancer was achieved (IFN-α alone has not been shown to have any effect in metastatic colorectal cancer, and the response rate with 5-FU alone is generally in the region of 7 to 20%) It may now be possible to explore some of the other potentially additive effects of IFNs with cytotoxic drugs in a variety of poorly chemosensitive tumors. Further studies in nude mouse models may be able to identify effective IFN-drug combinations in these tumors.

III. TUMOR NECROSIS FACTOR

Tumor necrosis factor (TNF or TNF-α) was first identified as a protein that was induced in the serum of endotoxin-treated mice and caused hemorrhagic necrosis of syngeneic, chemically induced tumors.[60] The TNF gene has since been cloned in several species,[61,62] and the recombinant TNF made available has been actively investigated in clinical trials and experimental tumor models. In the syngeneic Meth A sarcoma model, TNF caused hemorrhagic necrosis of the tumor, but this effect is greatly decreased in nude mice, suggesting a requirement for T cells for the predominantly vascular damage.[63] This has been confirmed by adoptive transfer experiments.[64] Similarly, the TNF-induced red cell influx in the murine SAI fibrosarcoma is less pronounced in T cell-deficient mice.[65] Comparisons of the effects in normal mice with nude mice suggest an important role of T cells in mediating the toxic effect of TNF on tumor vasculature *in vivo*. However, the murine tumors studied are known to be immunogenic and the role of T cell-mediated immunity against human tumors which are poorly immunogenic is uncertain.

The antitumor effect of TNF against human tumor xenografts has been reported by several groups.[66-69] Most investigators have looked at its effect on s.c. xenografts and have found that the antitumor effect is dose related. Intratumoral TNF is the most effective route of administration. Balkwill et al.[66] showed that intratumoral, but not i.p., TNF was effective against a selection of s.c. breast, colon, and ovarian cancer xenografts. It would therefore appear that local/locoregional administration of TNF would be the most effective means for clinical application. There are several reports on the use of i.p. TNF in i.p. tumor models in nude mice.[26,46,47]

In our laboratory, in two of three i.p. xenografts TNF was able to prolong the survival of mice, whereas the addition of IFN-γ had a synergistic effect.[26] Detailed investigations have revealed that the effect of i.p. TNF therapy alone in these models is complex.[70] Although TNF administration was shown to eradicate ascitic tumor cells, detailed post-mortem studies revealed a concomitant promotion of tumor adhesions and invasion of the peritoneal suface. This appearance was strikingly similar to metastatic ovarian cancer in patients, and suggested that endogenous TNF production may influence the pattern of ovarian cancer metastasis formation. We have compared the metastatic behavior of Chinese hamster ovary cells (CHO) transfected with the human TNF gene[71] with that of CHO cells transfected with only the expression vector.[72] Nude mice injected with CHO/TNF cells died earlier of ascites (15 d) as compared to those injected with CHO/NEO cells which died later (30 d) and had multiple intra-abdominal solid tumors. Detailed histological examination of the peritoneum and liver at 3 to 14 days showed that the CHO/TNF cells invaded the peritoneum and liver surface, while the CHO/NEO cells did not. This was confirmed by studying the localization of radiolabeled cells in nude mouse tissues following i.p. injection, and it was also revealed that the CHO/TNF cells metastasized to the lungs. The enhanced metastasis formation of TNF transfectants was specifically inhibited by anti-TNF antibodies. The potential clinical implications of these findings have been underlined by the discovery of TNF mRNA and protein in pathological specimens of human ovarian cancer by using *in situ* hybridization and immunohistochemical techniques in our laboratory.[73] Other investigators have studied the metabolic changes in nude mice injected with TNF transfectants and have shown a potential role of tumor-produced TNF in cancer cachexia, anemia, and hypercalcemia.[71,74] The information gained from the behavior of TNF transfectants and their effects on the host underlines the potential of the nude mouse model in looking into the effects of other tumor-produced factors *in vivo*.

IV. INTERLEUKIN-2 AND LYMPHOKINE-ACTIVATED KILLER CELLS

A great deal of interest has been generated by the demonstration that lymphoid cells can be activated by interleukin-2 (IL-2) to lyse a wider range of tumor cells than NK cells.[75] This activity is termed "lymphokine-activated-killer" (LAK) cell activity and is probably mediated by diverse cell populations including T cells, NK cells, and macrophages. In syngeneic murine tumor models, concomitant infusion of *in vitro* IL-2 activated lymphoid cells with IL-2 led to regression of tumors, while IL-2 alone or LAK cells alone were not as effective.[76] Clinical trials based on these studies of IL-2 alone, or in combination with LAK cells, generated *in vitro* by incubation with IL-2 have been conducted, and significant responses have been documented, particularly against renal cell cancer and melanoma.[77,78]

It is notable that LAK cell activity can be induced in nude mice by infusions of human recombinant IL-2, confirming that LAK activity can be mediated by non-T cell populations. It is also possible to adoptively transfer xenogeneic LAK cells into nude mice and compare their relative efficacy to *in vivo* induced mouse LAK cells following injection of IL-2 alone. Ortaldo et al.[79] described the adoptive cellular immunotherapy of the human ovarian cancer cell line NIH:OVCAR-3 implanted i.p. into nude mice. Mice were injected i.p. with tumor cells on day 0 and therapy began the next day with recombinant human IL-2 alone, *in vitro* IL-2 activated human peripheral blood mononuclear cells alone, or both simultaneously. The results showed that IL-2 alone had no antitumor activity, but that weekly injections of IL-2 plus LAK cells for 4 weeks prolonged the median survival of mice from 35 to 60 d. Further experiments to characterize the subpopulations of human peripheral blood mononuclear cells mediating LAK cell activity showed that IL-2 activated large granular lymphocytes and T cells were equally effective in mediating antitumor activity. Similar results were reported by Marincola et al.[80] in nude mice bearing intrapancreatic human pancreatic cancer xenografts, again using human peripheral blood lymphocytes activated *in vitro* with IL-2. In the group of mice treated with IL-2 alone, deaths due to toxicity consisting of fluid retention with ascites and pleural effusions were reported in 37%of mice. These side effects are similar to those reported in IL-2/LAK trials in man.[77] Interestingly, this form of toxicity was not noted in the IL-2/LAK group, the major cause of early mortality in this group being pulmonary embolism at the time of injection of LAK cells. IL-2 alone or LAK cells alone again had no therapeutic effect, but the combination prolonged the survival of mice bearing pancreatic tumor implants. A significant antitumor effect on s.c. pancreatic tumor implants was noted in parallel with the IL-2/LAK combination. Shimizu et al.[81] investigated IL-2/LAK treatment of s.c. human endometrial cancer xenografts, but did not find an antitumor effect. The addition of lentinan (a polysaccharide with T cell-activating properties isolated from an edible Japanese mushroom) to the IL-2/LAK regimen led to a significant antitumor effect. These three investigations suggest that, for antitumor activity, the concomitant injection of IL-2 to adoptively transferred LAK cells is required and the effects can be enhanced by other immunomodulating agents. However, recent data from clinical trials have shown that prolonged continuous infusion of IL-2 alone can also result in *in vivo* activation of host antitumor mechanisms.[78] This is also suggested by an ingenious experiment reported by Bubenik et al.[82] The human IL-2 gene was transfected into a fibroblast cell line, and IL-2 secreting clones were isolated. IL-2 transfectants, or control fibroblasts, were injected peritumorally into nude mice bearing s.c. HeLa cell xenografts. The growth of xenografts was significantly inhibited by the IL-2 transfectants, but not the control fibroblasts. Histological examination of the regressing tumors showed intensive infiltration by mononuclear cells of the peritumoral stroma and necrosis of the tumor. The nature of the cell populations in the peritumoral areas was not fully defined, but this experiment demonstrated conclusively that LAK activity can be generated in nude mice *in vivo*. The role of endogenous IL-2 production in restricting tumor development *in vivo* was demonstrated by the study of Saito et al.[83] Pretreatment of nude mice with IL-2 inhibited the tumor take rate of the human hepatoma cell line HCC-H, and this effect could be abolished by administration of anti-asialo GM1 antiserum, suggesting that augmented NK cell activity mediated this effect. In a rare report on direct antiproliferative effects of IL-2 on tumor cells, the hormone-dependent MCF-7 breast cancer cell line was shown to be inhibited in growth *in vitro* as well as *in vivo* in nude mice.[84]

V. OTHER CYTOKINES

There have been limited studies on the effects of other cytokines on human tumor xenografts, such as interleukin-1 (IL-1) IL-6, transforming growth factor-β (TGF-β), and the colony stimulating factors. IL-1 (α and β) is cytostatic for some tumor cell lines *in vitro*, and has antitumor activity in syngeneic tumor models.[85-87] In a study on the effects of IL-1α against two murine tumors in both euthymic and nude mice, IL-1α led to a hemorrhagic response, indicating that a non-T cell dependent mechanism was operative in this model.[88] We have tested the effect of i.p. human recombinant IL-1β on the three i.p. human ovarian cancer models in nude mice, and compared the effect with recombinant TNF (unpublished data). Our results show that IL-1β markedly enhanced the peritoneal adhesion and implantation of two of these ascitic tumors without effect on survival, while in a TNF-resistant tumor, IL-1β had a marked antitumor effect. These results may indicate that like TNF, IL-1 could be of influence on the metastatic behavior of human cancers. In another study of human squamous cell cancer lines injected into nude mice, metastatic properties were not related to different levels of IL-1 production, although it was suggested that cell lines producing IL-1 showed higher growth rates.[89] A role for tumor-derived IL-1 in the hypercalcemia of malignancy has been suggested in squamous cell cancers.[90,91] For example, a squamous cell thyroid cancer derived from a patient with hypercalcemia and leucocytosis induced similar metabolic changes in nude mice. Other factors associated with hypercalcemia and bone resorption derived from nude mouse studies include TNF, TGF-α, and TGF-β.[74,92] TGF-β has been reported to inhibit the growth and increase the differentiation of a human lung adenocarcinoma in nude mice.[93]

Human cancers have been shown to produce colony stimulating factors in nude mouse experiments,[94] and such models have been exploited for the study of the T cell dependency on the development of bone marrow lineages *in vivo*.[95] Antitumor effects of any of the colony stimulating factors have not yet been reported in nude mouse models.

VI. CONCLUSIONS

The study of cytokine effects in the nude mouse-human tumor model, like other experimental tumor models, has limitations, some of which have been discussed in this chapter. However, if the data gained from such experiments is integrated with information from *in vitro* studies, adoptive transfer of cytokine-activated host cells and the results obtained from syngeneic tumor models, issues of relevance to the therapy and biology of human cancer, can be addressed under controlled experimental conditions.

Human tumor xenograft studies have identified the differential actions of IFN-α and IFN-γ and have also shown that commonly occurring human tumors can be directly inhibited in growth by IFNs. Furthermore, the preclinical data obtained in nude mouse models with IFN-cytotoxic drug combinations is now being supported by the results of clinical trials.

The availability of human tumor cell lines that produce cytokines, either constitutively or after transfection with the relevant cytokine genes, makes the nude mouse model an important vehicle to study the role of tumor-derived cytokines in determining tumor behavior *in vivo*. Such studies are already giving new insights into the role of cytokines, not only as therapeutic agents, but as autocrine growth factors and modulators of tumor metastases.

REFERENCES

1. **Balkwill, F. R.,** *Cytokines in Cancer Therapy*, Oxford Medical Publications, 1989.
2. **Reid, L. M., Minato, N., Gresser, I., Holland, J., Kadish, A., and Bloom, B. R. ,** Influence of anti-mouse interferon serum on the growth and metastasis of tumour cells persistently infected with virus and of human prostatic tumours in nude mice, *Proc. Natl. Acad. Sci. U.S.A.*, 78, 1171, 1981.
3. **Lattime, E. C. and Stutman, O.,** Tumor growth *in vivo* selects for resistance to tumor necrosis factor, *J. Immunol.*, 143, 4317, 1989.

4. **Kopf-Maier, P. and Jackel, M.,** Proliferation behaviour of xenografted tumours: a flow cytometric study, *AntiCancer Res.,* 8, 1355, 1989.
5. **Strander, H.,** Interferon treatment of human neoplasia, *Adv. Cancer Res.,* 46, 1986.
6. **Kelly, S., Malik, S. T. A., and Balkwill, F. R.,** Cytokine therapy, in *The science of cancer treatment,* Waring, M. J. and Ponder, B. A. J., Eds., Kluwer Academic Publishers, Dordrect, 1990, 127.
7. **Gresser, I.,** in *Interferon and Cancer,* Ed. Sikora, K., Plenum, New York, pp 65, 1983.
8. **Balkwill, F. R.,** Animal models for investigating antitumor effects of interferon, *Meth. Enzymol.,* 119, 649, 1986.
9. **Balkwill, R. R., Moodie, E. M., Freedman, V., and Fantes, K. H.,** Human interferon inhibits the growth of established human breast tumours in the nude mouse, *Int. J. Cancer,* 30, 231, 1982.
10. **Riviere, Y. and Hovanessian, A. G.,** Direct action of interferons and inducers of interferon on tumour cells in athymic nude mice, *Cancer Res.,* 43, 4596, 1989.
11. **Balkwill, F. R., Stevens, M. H., Griffin, D. B., Thomas, J. A., and Bodmer, J.,** Interferon gamma regulates HLA-D expression on solid tumours *in vivo, Eur. J. Cancer Clin. Oncol.,* 23, 101, 1987.
12. **Marth, C., Fuith, L. C., Bock, G., Daxenbichler, G., and Dapunt, O.,** Modulation of ovarian carcinoma tumour marker CA-125 by γ-interferon, *Cancer Res.,* 49, 6538, 1989.
13. **DeClerq, E., Georgiadis, J., Edy, V. G., and Sobis, M. H.,** Effect of human and mouse interferon and of polyriboinosinic acid, polyribocytidilic acid on the growth of human fibrosarcoma and melanoma tumours in nuder mice, *Eur. J. Cancer Clin. Oncol.,* 14, 1273, 1978.
14. **Backwill, F. B., Taylor-Papadimitriou, J., Fantes, K. H., and Sebesteny, A.,** Human interferon can inhibit the growth of human breast cancer xenografts in nude mice, *Eur. J. Cancer, Clin. Oncol.,* 16, 569, 1980.
15. **Hofmann, V., Groscurth, P., Morant, R., Cserhati, M., Honegger, H. P., and von Hochstetter, A.,** Effects of leukocyte interferon (E. coli) on human bone sarcoma growth *in vitro* and in the nude mouse, *Eur. J. Cancer Clin. Oncol.,* 21, 859, 1985.
16. **Yamaoka, T., Takada, H., Yanagi, Y., Kataoka, T., and Sakurai, Y.,** The antitumor effects of human lymphoblastoid interferon on human renal cell carcinoma in athymic nude mice, *Cancer Chemother. Pharmacol.,* 14, 184, 1985.
17. **Bradley, N. J., Darling, J. L., Oktar, N., Bloom, H. J. G., Thomas, D. G. T., and Davies, A. J. S.,** The failure of human leucocyte interferon to influence the growth of human glioma cell populations, *in vitro* and *in vivo* studies, *Br. J. Cancer,* 48, 819, 1983.
18. **Namba, M., Yamamoto, S., Tanaka, H., Kanamori, T., Nobuhara, M., and Kimoto, T.,** *In vitro* and *in vivo* studies on potentiation of cytotoxic effects of anticancer drugs or cobalt 60 gamma rays by interferon on human neoplastic cells, *Cancer,* 54, 2262, 1985.
19. **Brosjö, O., Bauer, H. C. F., Brostrom, L., Nillson, O. S., Reinholt, F. P., and Tribukait, B.,** Growth inhibition of human osteosarcomas in nude mice by interferon-α. Significance of dose and tumour differentiation, *Cancer Res.,* 47, 258, 1987.
20. **Masuda, S., Fukuma, H., and Beppu, Y.,** Antitumour effect of human leucocyte interferon on osteosaracoma transplanted into nude mice, *Eur. J. Cancer Clin. Oncol.,* 19, 1521, 1983.
21. **Balkwill, F. R., Griffin, D. B., and Lee, A. E.,** Interferons alpha and gamma differ in their ability to cause tumour stasis and regression *in vivo, Eur. J. Cancer Clin. Oncol.,* 25, 1481, 1989.
22. **Ida, N., Uenishi, N., Kajita, A., and Satoh, Y.,** Antitumor effect of human fibroblast interferon on the growth of human melanoma cells implanted in nude mice, *Gann,* 73, 952, 1982.
23. **Hubbell, H. R., Kvalnes-Krick, K., Carter, W. A., and Strayer, D. R.,** Antiproliferative and immunomodulatory actions of beta-interferon and double-stranded RNA, individually and in combination, on human bladder tumor xenografts in nude mice, *Cancer Res.,* 45, 2481, 1985.
24. **Trotta, P. P. and Harrison, S. D.,** Evaluation of the antitumor activity of recombinant human gamma-interferon employing human melanoma xenografts in athymic nude mice, *Cancer Res.,* 47, 5437, 1987.
25. **Morikawa, K., Fan, D., Denkins, Y. M., Levin, B., Gutterman, J. U., Walker, S. M., and Fidler, I. J.,** Mechanisms of combined effects of gamma-interferon and 5-fluorouracil on human colon cancers implanted into nude mice, *Cancer Res.,* 49, 799, 1989.
26. **Balkwill, F. R., Ward, B. G., Moodie, E., and Fiers, W.,** Therapeutic potential of tumor necrosis factor-alpha and gamma-interferon in experimental human ovarian cancer, *Cancer Res.,* 47, 4755, 1989.
27. **Brosjö, O., Bauer, H. C. F., Nillson, O. S., Halverson, D., Reinholt, F. P., and Tribukait, B.,** Effect of human interferon alpha and interferon gamma on growth, histolohy, and DNA content of human osteosarcomas in nude mice, *J. Interferon Res.,* 9, 475, 1989.
28. **De Clercq, E., Georgiades, J., Edys, V. G., and Sobis, H.,** Effect of human and mouse interferon and of polyriboinosinic acidand polyribocytidylic acid on the growth of human fibrosaracoma and melanoma tumors in nude mice, *Eur. J. Cancer Clin. Oncol.,* 11, 1273, 1982.
29. **Tanaka, N., Nagao, S., Togho, A., Sekiguchi, F., Kohno, M., Ogawa, H., Matsui., T., and Matsutani, M.,** Effects of human fibroblast interferon on human gliomas transplanted into nude mice, *Gann,* 74, 308, 1983.

30. **Brosjö, O., Bauer, H. C. F., Brostrom, L., Nilsonne, U., Nillson, O. S., Reinholt, F. P., Strander, H., and Tribukait, B.,** Influence of human a-interferon on four osteosarcoma xenografts in nude mice, *Cancer Res.*, 45, 5598, 1985.
31. **Forster, S., Triffitt, J. T., Bauer, H. C. F., Brosjö, O., Nilsson, O. S., Smith, R., and Sykes, B.,** Interferon-inhibited human osteosarcoma xenografts induce host bone in nude mice, *J. Bone Mineral Res.*, 3, 199, 1983.
32. **Balkwill, F. R., Goldstein, L., and Stebbing, N.,** Differential action of six human interferons against two human carcinomas grwoing in nude mice, *Int. J. Cancer*, 35, 613, 1985.
33. **Kondo, H., Tanaka, N., Naomoto, Y., and Orita, K.,** Antitumor effect of recombinant human interferon-beta and interferon-gamma in combination against human colon cancer cell line *in vitro* and in nude mice, *Jpn. J. Cancer Res.*, 78, 1258, 1987.
34. **Ramani, P. and Balkwill, F. R.,** Human interferons inhibit experimental metasases of a human melanoma cell line in nude mice, *Br. J. Cancer* , 58, 350, 1988.
35. **Berladelli, F., Gresser, I., Maury, C., and Manoury, M. T.,** Antitumour effects of interferon in mice injected with interferon sensitive and interferon resistant Friend leukaemia cells, *Int. J. Cancer*, 30, 821, 1982.
36. **Dvorak, H. F., and Gresser, I.,** Microvascular injury in the pathogenesis if interferon-induced necrosis of subcutaneous tumors in mice, *J. Natl. Cancer Inst.*, 81, 497, 1989.
37. **Uenishi, N., Ida, N., Kajita, A., and Usuki, K.,** Host-mediated antitumor effect of interferon in nude mice transplanted with human tumor, *Gann*, 74, 730, 1983.
38. **Balkwill, F. R. and Projetti, E.,** Effects of mouse interferon on human tumour xenografts in the nude mouse host, *Int. J. Cancer*, 38, 375, 1986.
39. **Dumont, S., Hartmann, D., Poindron, P., Oberling, F., Faradji, A., and Bartholeyns, J.,** Control of the antitumoral activity of human marcrophages produced in large amounts in view of adoptive transfer, *Eur. J. Cancer Clin. Oncol.*, 24, 1691, 1988.
40. **Kataoka, T., Oh-haski, F., Sakurai, Y., Usuki, K., and Ida, N.,** Relative contribution of antiproliferative and host immunity-associated activity of mouse interferon in murine tumor therapy, *Cancer Res.*, 4, 5661, 1984.
41. **Onishi, T., Machida, T., Mori, Y., Iizuka, N., Masuda, F., Mochizuki, S., Tsukamoto, H., and Harada, N.,** Hyperthermia with simultaneous administration of interferon using established human renal carcinoma heterotransplanted in nude mice, *Br. J. Urol.*, 6, 227, 1989.
42. **Naomoto, Y., Tanaka, N., Kondo, H., and Orita, K.,** Synergistic effect of recombinant human interferon-beta and -gamma on human colon cancer transplanted into nude mice, *Gan To Kagaku Ryoho*, 13, 3155, 1986.
43. **Ozzello, L., Habif, D. V., De-Rosa, C. N., and Cantell, K.,** Treatment of human breast cancer xenografts using natural interferons-alpha and -gamma injected singly or in combination, *J. Interferon Res.*, 8, 679, 1987.
44. **Fransen, L., Van Der Heyden, J., Ruysschaert, R., and Fiers, W.,** Recombinant tumour necrosis factor, its effect and its syenrgism with interferon gamma on a variety of norman and transformed human cell lines, *Eur. J. Cancer Clin. Oncol.*, 22, 419, 1986.
45. **Dealtry, G. B., Naylor, M. S., Fiers, W., and Balkwill, F. R.,** The effect of recombinant human tumour necrosis factor on growth and macromolecular synthesis of human epithelial cells, *Exp. Cell Res.*, 170, 428, 1985.
46. **Rubin, S. C., Hoskins, W. J., Markman, M., Feferici, M., Lloyd, K. O., and Lewis, J. L.,** Synergistic activity of tumor necrosis factor and interferon in a nude mouse model of human ovarian cancer, *Gynecol. Oncol.*, 34, 353, 1989.
47. **Manetta, A., Podczaski, E., Zaino, R. J., and Satyaswaroop, P. G.,** Therapeutic effect of recombinant human tumor necrosis factor in ovarian carcinoma xenograft in nude mice, *Gynecol. Oncol.*, 34, 360, 1989.
48. **Brunda, M. J., Bellantoni, D., and Sullich, V.,** *In vivo* antitumour activity of combinations of interferon alpha and interleukin-2 in a murine model. Correlation of efficacy with the induction of cytotoxic cells resembling natural killer cells, *Int. J. Cancer*, 40,367, 1987.
49. **Balkwill, F. R. and Moodie, E. M.,** Positive interactions between human interferon and cyclophosphamide or adriamycin in a human tumor model system, *Cancer Res.*, 44, 904, 1984.
50. **Kondo, H., Tanaka, N., and Orita, K.,** Antitumor effect of human recombinant interferon-gamma alone and in combination with adriamycin on a human colon cancer cell line (RPMI 4788) *in vitro* and in nude mice, *Jpn. J. Cancer Res.*, 78, 1400, 1987.
51. **Sklarin, N. T., Chanian, A. P., Feuer, E. J., Lahman, L. A., Szrajer, L., and Holland, J. F.,** Augmentation of activity of *cis*-diamminedichloroplatinum (II) and mitomycin C by interferon in human malignant mesothelioma xenografts in nude mice, *Cancer Res.*, 48, 64, 1988.
52. **Carmichael J., Fergusson, P. J., Wolf, R. C., Balkwill, F. R., and Smyth, J. F.,** Augmentation of cytotoxicity of chemotherapy by human a-interferons in human non-small cell lung cancer xenografts, *Cancer Res.*, 46, 4916, 1986.
53. **Birsic, W., D'Oro, L., Charoensiri, S., and Katoh, A.,** The combined effect of interferon and 5-FU on tumor-cell metastasis in the nude mouse, *Dis. Colon Rectum*, 32, 340, 1989.
54. **Nishiyama, M., Takagami, S., Kirkhara, Y., Saeki, T., Niimi, K., Kim, R., Jinushi, K., Toge, T., Niimoto, M., and Hattori, T.,** Combined effects of interferon alpha A/D with flouropyrimidine derivatives in the subrenal capsule assay, *Gan To Kagaku Ryoho*, 15, 2285, 1988.

55. **Balkwill, F. R., Mowshowitz, S., Seilman, S. S., Moodie, E. M., Griffin, D. B., Fantes, K. H., and Wolf, R. C.,** Positive interactions between interferon and chemotherapy due to direct tumor action rather than effects on host drug-metabolizing enzymes, *Cancer Res.*, 44, 5249, 1984.
56. **Yoneda, K., Yamamoto, T., and Osaki, T.,** Influence of interferon on adriamycin uptake of cultured tumour cells, *Int. J. Cancer*, 44, 483, 1989.
57. **Elias, L. and Crissman, H. A.,** Interferon effects upon the adenocarcinoma 38 and HL-60 cell lines, antiproliferative responses and synergistic interactions with halogenated pyrimidine antimetabolites, *Cancer Res.*, 48, 4868, 1988.
58. **Gewert, D. R., Shah, S., and Clemens, M. J.,** Inhibition of cell division by interferons, changes in the transport and intracellular metabolism of thymidine in human lymphoblastoid (Daudi) cells, *Eur. J. Biochem.*, 116, 487, 1981.
59. **Wadler, S., Schwartz, E. L., Goldman, M., Lyver, A., Rader, M., Zimmerman, M., Itri,., Weinberg, V., and Wiernik, P. H.,** Flourouracil and recombinant alpha-2a-interferon. An active regiman against advanced colorectal carcinoma, *J. Clin. Oncol.*, 7, 1769, 1989.
60. **Carswell, E. A., Old, L. J., Kassel, R. J., Green, S., Fiore, N., and Williamson, B.,** An endotoxin-induced serum factor that causes necrosis of tumours, *Proc. Natl. Acad. Sci. U.S.A.*, 72, 3666, 1975.
61. **Pennica, D., Nedwin, G. E., Hayflick, J. S., Seeburg, P. H., Derynck, R., Palladino, M. A., Kohur, W. J., Aggarwal, B. B., and Goeddel, D. V.,** Human tumour necrosis factor, precursor structure, expression and homology to lymphotoxin, *Nature (London)*, 312, 724, 1984.
62. **Goeddel, D. V., Aggarwal, B. B., Gray, P. W., Leung. D. W., Nedwin, G. E., Palladino, M. A., Patton, J. S., Pennica, D., Shepard, H. M., Sugarman, B. J., and Wong, G. H. W.,** Tumour necrosis factors, gene structure and biological activities, *Cold Spring Harbor Symposia on Quantative Biology*, 51, 597, 1986.
63. **Haranaka, K., Satomi, N., and Sakurai, A.,** Antitumor activity of murine tumour necrosis factor (TNF) against transplanted murine tumours and heterontransplanted human tumours in nude mice, *Int. J. Cancer*, 34, 263, 1984.
64. **Palladino, M. A.J. R., Refaat Shalaby, M., Kramer, S. M., Ferraiolo, B. L., Baughman, R. A., Deleo, A. B., Crase, D., Marafino, B., Aggarwal, B. B., and Figari, I. S.,** Characterization of the antitumor activities of human tumour necrosis factor alpha and the comparison with other cytokines, induction of tumour-specific immunity, *J. Immunol.*, 138, 4023, 1987.
65. **Havell, E. A., Fiers, W., and North, R. J.,** The antitumour function of tumour necrosis factor (TNF). I. Therapeutic action of TNF against an established murine sarcoma is indirect, immunologically dependent, and limited by severe toxicity, *J. Exp. Med.*, 167, 1067, 1988.
66. **Balkwill, F. R., Lee, A., Adlam, G., Moodie, E., Thomas, A., Tavernier, J., and Fiers, W.,** Human tumour xenografts treated with tumour necrosis factor alone or in combinations with interferons, *Cancer Res.*, 46, 3990, 1986.
67. **Creasey, A. A., Reynolds, M. T., and Laird, W.,** Cures and partial regression of murine and human tumours by recombinant human tumour necrosis factor, *Cancer Res.*, 46, 5687, 1986.
68. **Haranaka, K., Satomi, N., Sakurai, A., and Haranaka, R.,** Antitumour effects of tumour necrosis factor, cytotoxic or necrotising activity and its mechanism,.*Ciba Found. Symp.*, 131, 140, 1985.
69. **Nosoh, Y., Toge., Nishiyama, M., Yamaguchi, M., Hirabayashi, N., Niimoto, M., and Hattori, T.,** Antitumour activity of recombinant human tumour necrosis factor against human tumour xenografts transplanted into nude mice, *Jpn. J. Surg.*, 17, 51, 1987.
70. **Malik, S. T. A.,Griffin, D. B., Fiers, W., and Balkwill, F.** R., Paradoxical effects of tumour necrosis factor in experimental ovarian cancer, *Int. J. Cancer*, 44, 918, 1989.
71. **Oliff, A., Defeo-Jones, Boyer, M., Martinez, D., Kiefer, D., Vuocolo, G., Wolfe, A., and Socher, S. H.,** Tumours secreting human TNF/cachectin induce cachexia in mice, *Cell*, 50, 555, 1987.
72. **Malik, S. T. A., Naylor, M. S., Oliff, O., and Balkwill, F. R.,** Enhanced metastatic properties of chinese hamster ovary cells secreting tumor necrosis factor, (submitted for publication).
73. **Naylor, M. S., Malik, S. T. A., Stamp, G., Jobling, T., and Balkwill, F. R.,** *In situ* detection of tumour necrosis factor in human ovarian cancer, (submitted for publication).
74. **Johnson, R. A., Boyce, B. F., Mundy, G. R., and Roodman, G. D.,** Tumours producing human tumour necrosis factor induce hypercalcaemia and osteoclastic bone resorption in nude mice, *Endocrinology*, 124, 1424, 1989.
75. **Grimm, E. A.,** Human lymphokine-activated killer cells (LAKcells) as a potential immunotherapeutic modality, *B. B. A.*, 865, 267, 1986.
76. **Rosenberg, S. A.,** Adoptive immunotherapy of cancer using lymphokine activated killer cells and recombinant interleukin-2, in *Important Advances in Oncology*, DeVita, V. T., Hellman, S., and Rosenberg, S. A., Eds., J. B. Lippincott, Philadelphia, 1986, 55.
77. **Rosenberg, S. A., Lotze, M. T., Muul, L. M., Chang. A. E., Avis, F. P., Leitman, S., Lineham, W. M., Robertson, C. N., Lee, R. E., Rubin, J. T., Seipp, C., Simpson, C. G., and White, D. E.,** A progress report on the treatment of 157 patients with advanced cancer using lymphokine-activated killer cells and interleukin-2 or high dose interleukin-2 alone, *N. Engl. J. Med.*, 316, 889, 1987.

78. **West, W. H., Tauer, K. W., and Yanneli, J. R.,** Constant infusion recombinant interleukin-2 in adoptive cellular therapy of advanced cancer, *N. Engl. J. Med.,* 316, 898, 1987.
79. **Ortaldo, J. R., Porter, H. R., Miller, P., Stevenson, H. C., Ozols, R. F., and Hamilton, T. C.,** Adoptive cellular immunotherapy of human ovarian carcinoma xenografts in nude mice, *Cancer Res.,* 46, 4414, 1986.
80. **Marincola, F. M., DaPozzo, L. F., Howick, J. C., Drucker, B, J., Hough, K. L., and Holder, W. D.,** Prolonged survival of nude mice bearing human pancreatic cancer and treated with lymphokine-activated killer cells and interleukin-2, *Current Surg.,* 46, 304, 1989.
81. **Shimizu, H., Inoue, M., and Tanizawa, O.,** Adoptive cellular immunotherapy to the endometrial carcinoma cell line xenografts in nude mice, *Gynecol. Oncol.,* 34, 195, 1989.
82. **Bubenik, J., Voitenok, N. N., Kieler, J., Prassolov, V. S., Chumakov, P. M., Bubenikova, D., Simova, J., and Jandlova, T.,** Local administration of cells containing an inserted IL-2 gene and preducing IL-2 inhibits growth of human tumours in nu/nu mice, *Immunol. Lett.,* 19, 279, 1988.
83. **Saito, H., Morizane, T., Inagaki, Y., Kumagai, N., Watanabe, T., Nakamura, T., Satoh, I., Sawaguchi, K., Iwabuchi, N., Tsuchimoto, and Tsuchiya,** Decrease of transplantability by the immunopotentiators, OK-432 and interleukin-2. Experiments on a human hepatoma cell line in nude mice, *Eur. J. Cancer Clin. Oncol.,* 25, 79, 1989.
84. **Pacoitti, G. F., and Tamarkin, L.,** Interleukin-2 differentially affects the proliferation of a hormone-dependent and a hormone-independent human breast cancer cell line *in vitro* and *in vivo, AntiCancer Res.,* 8, 1233, 1988.
85. **Onozaki, K., Matsushima, K., Aggarwal, B. B., and Oppenheim, J. J.,** Human interleukin 1 is a cytocidal factor for several tumor cell lines, *J. Immunol.,* 135, 3962, 1985.
86. **Tsai, S. and Gaffney, E. V.,** Inhibition of cell proliferation by interleukin 1 derived from monocytic leukemia cells, *Cancer Res.,* 46, 1471, 1986.
87. **Nakamura, S., Nakata, K., Kashmoto, S., Yoshida, H., and Yamada, M.,** Antitumour effect of recombinant human interleukin 1 alpha against murine syngeneic tumors, *J. Cancer Res.,* 77, 767, 1986.
88. **Braunschweiger, P. G., Johnson, C. S., Kumar, N., Ord, V., and Furmanski, P.,** Antitumor effects of recombinant human interleukin 1 alpha in RIF-1 and Panc02 solid tumors, *Cancer Res.,* 48, 6011, 1988.
89. **Price, J. E., Sauder, D. N., and Fidler, I. J.,** Tumorigenicity and metastatic behavior in nude mice of two human squamous cell carcinoma lines that differ in production of the cytokine ETAF/IL-1, *J. Invest. Dermatol.,* 91, 258, 1987.
90. **Sato, K., Fuji, Y., Ono, M., Nomura, H., and Shizume, K.,** Production of interleukin 1 alpha-like factor and colony-stimulating factor by a squamous cell carcinoma of the thyroid (T3M-5) derived from a patient with hypercalcemia and leukocytosis, *Cancer Res.,* 47, 6474, 1987.
91. **Kobayaski, H., Endo, K., Nishimura, K., Kasagi, K., Yamamoto, I., Konishi, J , Abe, M., and Shimizu, Y.,** A case report of thyoid papillary cancer that manifested leukocytosis and hypercalcemia after radiotherapy in bone metastasis, *Gan. No. Rinsho.,* 35, 55, 1989.
92. **Linkhart, T. A., Mohan, S., Jennings, J. C., and Baylink, D. J.,** Copurification of osteolytic and transforming growth factor beta activities produced by human lung tumor cells associated with humoral hypercalcemia of malignancy, *Cancer Res.,* 49, 271, 1989.
93. **Twardzik, D. R., Ranchalis, J. E., McPherson, J. M., Ogawa, Y., Gentry, L., Purchio, A., Plata, E., and Todaro, G. J.,** Inhitition and promotion of differentiated like phenotype of a human lung carcinoma in athymic mice by natural and recombinant forms of transforming growth factor-beta *J. Natl Cancer Inst.,* 81, 1182, 1989.
94. **Ishiwata, I., Ishiwaata, C., Nozawa, S., Iizuka, R., and Ishikawa, H.,** Change in leukocyte count in peripheral blood of nude mouse bearing transplanted gynecological malignant tumors, *Nippon Sanka Fujinka Gakkai Zasski.,*40, 1767, 1988.
95. **Lepault, F., Fache, M. P., and Frindel, E.,** Thymic dependency of the humoral regulation of CFU-s proliferation in mice bearing a CSF-producing tumor, *Int. J. Cell Cloning.,* 6, 252, 1988.

Chapter 19

DIFFERENTIATION INDUCERS

Guus A. M. S. van Dongen and Boudewijn J. M. Braakhuis

TABLE OF CONTENTS

I. INTRODUCTION

The major objective of cancer chemotherapy is tumor cell death. However, many common tumor types respond poorly to the present drugs. Furthermore, toxicity limits the optimal use of these agents in the majority of patients. There is thus an urgent need for new effective drugs with less toxic side effects. A novel approach of increasing interest is the use of so-called "differentiation inducers". The effect of such drugs may involve the stimulation of tumor cells to differentiate to a more benign cell type. The mechanism of action is to enhance the maturation of tumor stem cells to such an extent that terminal differentiation ensues with a finite number of tumor cell divisions. It has been validated in *in vitro* and *in vivo* systems that tumor cells can be converted to mature end-stage cells by differentiation inducers.[1-12] Recently, the first clinical trials on such compounds have been conducted.[13-19] The design of these treatment protocols was based solely on data derived from *in vitro* experiments. The resulting clinical observations suffer from the inability to distinguish between maturation and cytotoxicity, which makes it difficult to evaluate the underlying mechanisms of therapeutic effects of these drugs. Furthermore, the requirements for optimal differentiation induction *in vivo* are generally unknown, and additional fundamental information on the application of maturation therapy is needed before the merits of this approach can be established. Thus far, the nude mouse-human tumor model has been used infrequently for studies on differentiation inducers, though it may be a favorable system for development of maturation therapy. This chapter deals with advantages as well as limitations of the nude mouse model with respect to maturating agents and describes methods for monitoring differentiation and proliferation *in vivo*.

II. DIFFERENTIATION CAPACITY OF TUMOR CELLS

Neoplasms are generally initiated by transformation of a single normal cell. During the development of a neoplasm, new characteristics are acquired, but the genotype is largely stable during transformation and progression. Neoplastic formation can therefore be described as a misprogramming in the expression of genes relevant for normal processes like tissue development, tissue renewal, and wound healing.[20] It has been proposed that the functional lesion in a neoplastic cell is the diminished capacity to differentiate.[1] Nevertheless, it is generally assumed that neoplastic cells retain many of their genes responsible for differentiation, but that in these cells the process of proliferation in some way is favored at the expense of the process of differentiation.[4] Mechanisms have been proposed by which the neoplastic cell, under normal homeostatic conditions, remains frozen in its proliferation mode. These mechanisms include an increased sensitivity to and/or endogenous production of so-called "proliferation factors" probably related to oncogene products, or a decreased responsiveness to or availability of differentiation factors.[20,21] This concept indicates possibilities to alter the proliferation program by environmental influences.

It has been validated in *in vitro* and *in vivo* experimental systems, as well as in patients, that malignant cells can be converted to benign cells without proliferation capacity. First, there is the spontaneous maturation of a proportion of tumor cells within tumors. In a nice study of Pierce and Wallace,[23] it was demonstrated that undifferentiated cells became terminally differentiated as determined with ^{3}H-thymidine administered to rats bearing squamous cell carcinoma and by following the label in time throughout the tumor by autoradiography. However, in most tumors, the percentage of cells differentiating to mature end-stage cells is low. Second, clinicians have reported "spontaneous" maturation of neuroblastoma,[24,25] testicular cancer,[26] and acute myeloid leukemia.[27] A variety of treatment modalities were used in these patients, including chemotherapy, radiotherapy, as well as administration of Coley's toxin. The extent to which these treatments contributed to the phenomenon of differentiation is not known. Third, tumor cells can revert to normal when placed in a proper embryonic field. This phenomenon was described by

Brinster[28] who injected embryonal carcinoma cells into the blastocyst cavity of mice. Normal cells arose which were responsive to physiological stimuli regulating proliferation and differentiation. Fourth, a number of agents are able to induce malignant cells to differentiate both *in vitro* and *in vivo*. Most of them have only been tested in hematological neoplasms where they induce cells to mature.[5,6] The concept of differentiation induction also applies for *in vitro* cultured cell lines derived from solid tumors including neuroblastoma,[29] colon cancer,[30] lung cancer,[31] melanoma,[32] and choriocarcinoma.[33] However, as became clear, many *in vitro* cultured cell lines derived from solid tumors show alteration of differentiation characteristics without undergoing terminal differentiation. Recently, induction of differentiation has also been demonstrated in primary cultures of patients' tumors.[34]

Many excellent reviews deal with differentiation induction as a potential therapeutic approach in the treatment of hematological and solid neoplasms.[1-11] Some authors have classified the compounds that have a differentiation-inducing capacity. Pierce and Speers[11] have made a valid division into three categories: (1) embryonic inducers, (2) naturally occurring substances, and (3) chemical agents. The first division includes examples causing neoplastic cells to differentiate into normal cells when placed in a normal embryonic environment. Induction may be due to diffusible factors, mediated by cell-cell contacts or both. The second category of naturally occurring substances, or physiological substances as indicated by others, comprises growth factors, hormones, cytokines, and vitamins.[5,6] The third group contains chemicals with the potential to induce differentiation and it encompasses a broad spectrum of agents. Among these are fatty acids, polar solvents, vitamin analogs, DNA-intercalators, antimetabolites, phorbol esters, proteases, and toxins.

III. TUMOR-BEARING NUDE MICE AND DIFFERENTIATION INDUCERS: GENERAL CONSIDERATIONS

Studies on the processes of proliferation and differentiation in neoplasms require the establishment of reliable models for continuous supply of material with reproducible morphological and biological characteristics of the original tumor. Tumor cells cultured *in vitro* as a monolayer seem to be an unfavorable model since culturing conditions are far from physiological and not permissive for major modulation of differentiation. The major advantage of human xenografts is their use as a model providing three dimensional growth of grafted tissue with the potential for tissue-specific organization and differentiation. This in turn permits cytodifferentiation when this is dependent on such organization. The retention of histochemical and biochemical properties, as well as the capacity to respond to physiological stimuli, have been studied for normal or malignant tissues propagated either *in vitro*, as a cell line, or *in vivo* in the nude mouse, as a xenograft. Extensive analysis of the *in vitro* vs. *in vivo* behavior of normal keratinocytes and cells derived from squamous cell carcinoma revealed that human skin keratinocytes tend to lose many of the differentiation characteristics upon culturing *in vitro*.[35] In these cells, morphological and ultrastructural changes have been observed, as well as an alteration of the expression of biochemical differentiation markers, such as involucrin, filaggrin, and keratins.[36-41] Also, squamous epithelia of other origin with individual distinctive features like cornea, esophagus, or conjunctiva convert to an almost similar unspecific "culture type" pattern of morphological differentiation and expression of biochemical differentiation markers.[36,41] A similar phenomenon was observed for the neoplastic counterpart, squamous cell carcinoma cell lines,[42,43] and for other specialized normal epithelia such as breast,[44] intestine,[45,46] bronchus,[47] and endometrium.[48] However, when some of these normal and malignant cultured cells were injected into nude mice, they reexpressed the individual organization and keratinization patterns, indicating that both genetic and environmental factors control overt differentiation.[35,36,39,41-43,46] In addition, nonmalignant pathological keratinocytes, such as those isolated from nonbullous congenital ichthyosiform erythroderma lesions were shown to lose their disease-characteristic features upon culturing *in vitro,* but expression recurred after grafting of these cultured cells to nude mice.[49]

In elegant experiments with dermal-epidermal recombinants, Fusenig et al.[35,50,51] showed that interaction of keratinocytes with mesenchyme is a prerequisite for a complete expression of epidermal tissue architecture and keratinization. Extracellular matrix constituents like collagen, as well as diffusible factors from the mesenchyme, seem to be regulating factors. Also, studies on organogenesis during embryonic development in normal systems demonstrate the complexity of this process. In reviews of Pierce and Speers,[11] Gurdon,[52] and of Freshney,[5] data on normal and abnormal regulatory control mechanisms of the process of differentiation in various embryonic and adult tissues are summarized. As illustrated in these reviews, cell-cell interactions, extracellular matrix components, and/or diffusible factors seem to be required for normal organotypic differentiation besides the fact that cells should be competent to respond to these environmental signals.

It seems obvious that a system in which tumor cells are interacting with a biological framework that mimics the environment of the original tumor provides the most relevant model for studying differentiation inducers.[53,54] The fact that a variety of human tumors including basal cell carcinoma,[55] colon cancer,[46] pancreatic cancer,[56] renal cell cancer,[57] squamous cell carcinoma,[58] and testicular embryonal carcinoma[59] maintain their morphological characteristics upon transplantation to nude mice indicates that the regulatory control mechanism of the process of differentiation in this new host seems comparable to that in man. In addition, tumor functions may well be retained after grafting to nude mice, as illustrated by reports on the formation of specialized products like mucin,[60] casein,[61] enzymes,[56,62,63] hormones,[64,65] and so on, and the capacity to respond to the physiological stimuli like hormones[56,63] and growth factors.[66,67] In contrast, some tumor types have been described with frequently or incidentally altered architecture and differentiation immediately upon grafting to the nude mouse[68–70] or during repeated transfer in this new host.[71-73] The new host environment may cause selection of certain malignant cell clones or induce various changes in the proliferation of neoplastic and stromal subpopulations. With regard to the latter, it has to be considered that the stroma within human tumor xenografts is of mouse origin. Furthermore, phenotypic changes following transplantation can also be acquired, as documented for the expression of tumor markers in primary human esophageal cancers and their xenografts in a comparative study of Burg-Kurland et al.[70] Phenotypic changes may be induced by local stromal factors differing from the microenvironment of the organ of origin, as indicated by studies on the analysis of tumor characteristics in relation to various sites of implantation.[62,74] For instance, the expression of plasminogen activator in colon cancer, which is considered to be associated with tumorous invasiveness, was much higher when cells were inoculated in the gut of nude mice than when injected s.c.[62] The importance of the locoregional environment in modulating architectural and cytological tissue features was also demonstrated by Mackenzie et al.[74] With heterotransplants of epithelium and mesenchyme from various body locations, they demonstrated the modulating role of mesenchyme in epidermal differentiation, which resulted in regionally different patterns of keratin expression and morphology.

For studies on tumor types in which the processes of proliferation and differentiation are not obviously associated with 3-D growth, such as hematological neoplasms, the nude mouse system may be of value for the evaluation of the role of the microenvironment. In this respect, mouse tissues are known to synthesize naturally occurring substances that affect differentiation of these malignant cells in culture. These factors are considered to be T cell independent.[75] Host influences became clear when human leukemia cells from the established KG-1, KG-1a, and HL-60 cell lines were inoculated s.c. into newborn nude mice.[76] Transplanted cells revealed enzymatic changes compatible with induced cell differentiation. These observations cannot be totally explained by the selection of subpopulations *in vivo* and it was concluded that expression of some new differentiation characteristics was evidently acquired. Data of Okabe et al.[77] clearly illustrate that a system in which tumor cells interact with a biological environment may provide the most relevant model for studying fundamental aspects of growth regulation and for testing potential differentiation inducers. They used mouse myeloid leukemia cells which appeared to be resistant to differentiation into macrophages and granulocytes *in vitro* upon treatment with differentiation inducers, as well as *in vivo* when maintained in diffusion chambers in syngeneic mice and exposed to endogenous inducers. However, when the cells had been pretreated *in vitro* with actinomycin D, differentiation was found under *in vivo* conditions, indicating that the drug had sensitized the resistant cells to endogenous inducers.

IV. ASSAYS TO MONITOR DIFFERENTIATION *IN VIVO*

A fundamental problem when studying differentiation inducers *in vivo* is the linkage between reduced proliferation and induced cellular differentiation, and the distinction of this latter process from the cytotoxic action of the agent. In *in vitro* culture systems, it has been demonstrated that the concentration of drugs required to induce differentiation may be marginally lower than those which cause cytoxicity.[78,79] Therefore, one may expect that both effects occur simultaneously when tumors are exposed to these drugs *in vivo*. The only technique described until now, which permits an independent estimate of the degree of cytotoxicity and the commitment to differentiation *in vivo*, has been presented by Schwartz et al.[78] The *in vivo* differentiating activity of the anthracycline antibiotics marcellomycin, aclacinomycin, and doxorubicin was studied in HL-60 cells grown in agar in diffusion chambers implanted in the murine peritoneal cavity. This procedure permitted the independent determination of the cloning efficiency as well as the number of differentiated cells within individual colonies.

Adequate analytical methods should be available for potential differentiation inducers to be able to investigate the interrelation of proliferation and differentiation at the cellular level. Evaluation of maturation requires markers that are indicative for differentiated cells. Simultaneously, tumor growth characteristics and/or the proliferative behavior of individual tumor cells should be assessed.

Since for each tumor type a number of distinct differentiation markers have been recognized, description of analytical approaches to assess maturation will only be performed in general terms in this section. Among the criteria used to confirm maturation are morphological features at the cellular as well as the tissue level, physiological functions such as locomotive activity and phagocytosis, enzyme activity, production and excretion of products like hormones, mucins, and casein, binding capacity for specific biological products like growth factors or lectins, expression of genes related with differentiation, and the expression of antigens associated with differentiation.

Morphological examination of possible maturation includes light microscopy and electron microscopy, frequently combined with immunohistochemical techniques. Light microscopy may reveal alteration of the general cellular organization and the presence of multicellular organized structures, such as horn pearls in squamous cell carcinoma, domes in breast and lung cancer, duct-like or glandular structures in colon and pancreatic cancer, and tubule-like structures in rhabdomyosarcoma. Also, cytodifferentiation may be assessed by light and electron microscopy. Differentiated rhabdomyosarcoma cells are marked by multiple nuclei and the abundance of orientated filaments in the cytoplasm. Differentiated cells of squamous cell carcinoma contain a cornified envelope underneath the plasma membrane and have a high expression of tonofilaments and desmosomes. Acinar cells of human salivary glands contain numerous secretory granules which are involved in amylase excretion, while goblet cells in lung have specialized morphological features associated with mucin secretion and so on. Morphological criteria were used to demonstrate differentiation induction *in vivo* in xenografts of salivary cancer treated with dibutyryl cyclic AMP[80] and 12-*O*-tetradecanoyl-phorbol-13-acetate,[81] of glioma treated with glioma maturation factor,[66] of squamous cell carcinoma treated with the polar solvent *N,N*-dimethylformamide,[82] of lung cancer treated with transforming growth factor-β,[67] of rhabdomyosarcoma upon irradiation,[83] of teratocarcinoma cells treated with retinoids,[84] and also in leukemic cells grown *in vivo* treated with a variety of differentiation inducing agents.[78,85] In addition, morphological criteria were handled to assess preservation of differentiated features in these xenografts.

A limitation in the use of morphological criteria to assess the extent of maturation is the objective quantification. For maturation induction, it should be proven conclusively that the relative increase in differentiated appearance is actually due to accelerated maturation and not the result of a selective pressure by a growth inhibitory effect. This means that tissue sampling has to be conducted in a proper way. First of all, because of variations in microscopical appearance between individual control and treated tumors, several tumors from each group should be examined. Secondly, because heterogeneous cell populations might be present within a tumor, sections should be taken from multiple levels of all tissues available.

Quantification of differentiation markers can be facilitated by performing immunohistochemical techniques. A nice example has been described by Takizawa et al.,[83] who evaluated the effect of irradiation in relation to dose on the degree of rhabdomyoblastic differentiation, as estimated by the extent of immunostaining for myoglobin, creatine phosphokinase-MM, and desmin. Per treatment schedule, they selected several tumors responding to irradiation that had subsequently regrown. Tumors were excised when they reached pretreatment volume and were processed for immunohistochemical examination. The number of stained cells counted in ten randomly selected microscopical fields were compared for treated and control tumor tissues. Based on numerically distinct differences between the irradiated and control tumors, it was concluded that the expression of at least one of the examined differentiation markers had been induced by irradiation.

For immunohistochemical examination of differentiation markers in xenograft tissue, the use of mouse monoclonal antibodies in indirect staining techniques (e.g., monoclonal antibody binding followed by a secondary reaction step with peroxidase conjugated rabbit anti-mouse IgG antibody) meets some restrictions. It is known that the presence of endogenous mouse immunoglobulins within the tissue sections may lead to extensive background staining. This problem can be overcome by the application of biotinylated antibodies in an avidin-biotin peroxidase technique.

Flow cytometry is another technical approach for estimation of the percentage of cells expressing a distinct antigen associated with differentiation. For this purpose, tumors have to be dispersed into single cells by the use of enzymatic digestion. In our laboratory, we have used a two-step digestion procedure of collagenase plus hyaluronidase, followed by trypsin plus dithiothreitol to disperse squamous cell carcinoma xenografts which had been treated with the differentiation inducer *N,N*-dimethylformamide (DMF) *in vivo*.[82] The collagenase fraction mainly consisted of cells of murine origin, while the trypsin fraction contained both mouse and human cells. Two-color (green vs. red) dual parameter flow cytometric measurement was performed for simultaneous assessment of DNA content and antigen expression in individual cells derived from control or DMF-treated tumors. Human tumor cells and normal mouse cells, which mainly consisted of fibroblasts and blood cells, could be distinguished in two ways; that is, on the basis of their DNA content and by their reactivity pattern obtained with a broad-spectrum cytokeratin antiserum recognizing human cancer cells only. When binding was performed with a monoclonal antibody recognizing the differentiation marker cytokeratin 10, none of the cells derived from control tumors showed a positive reaction, while for DMF-treated tumors 20% of the cells expressed this marker. Drawbacks of this procedure are the possible loss of cellular antigens after enzymatic digestion and the undissolved parts of the tissue because of which it is unclear whether released cells are representative for the whole tumor.

Apart from cellular expression of differentiation, some analytical approaches include the entire tumor tissue. In such methods, tumor extracts have been prepared to measure hormones, enzymes, or other markers related to differentiation. For example, Sekiya et al.[65] have determined human chorionic gonadotrophin or α-fetoprotein levels in sera of tumor-bearing nude mice and in tumor extracts, to prove *in vivo* differentiation of embryonal carcinoma cells into choriocarcinoma or yolk sac carcinoma cells, respectively.

As mentioned earlier, stromal elements within xenografts are considered to be of mouse origin. Comparative studies on the expression of differentiation markers may be hampered by variability in the proportions of human and murine tissue in tumor extracts. Therefore, Cajot et al.,[62] in studies on the modulation of plasminogen activator activity by host tissue in colon cancer xenografts, developed a method for estimation of the proportion of human tissue. They determined the electrophoretic lactate dehydrogenase (LDH) isoenzyme pattern, and the staining intensity of human LDH-2 was expressed as a percentage of the total enzyme staining intensity. This ratio was used to adjust enzyme activities in tumor extracts as a reflection of the true proportion of human tissue (values in mIU human extract protein per milliliter).

The distinction of human tissue from mouse tissue may also be important for a proper interpretation of drug effects, as illustrated by Brosjö and Forster.[86,87] In a first paper, Brosjö et al.[86] describe the antitumor effect of interferon-α on human osteosarcoma transplants in nude mice. Histological examination of responding tumors revealed that xenografts had mineralized and were partly replaced by normally appearing bone and bone marrow, suggestive for differentiation. However, upon further analysis, it became clear that interferon had induced the osteosarcoma cells to produce a bone-inductive agent that interacted with host cells leading to normal bone tissue formation.[87] By applying species-specific anticollagen antibodies, they showed that normal bone was of mouse origin, whereas the residual mineralized tissue was of human origin.

V. ASSAYS TO MONITOR PROLIFERATION *IN VIVO*

The most important aspect of maturation induction is that a better differentiated phenotype in tumors results in loss of growth potential. In most studies on differentiation inducers tested in tumor-bearing nude mice, growth delay has been used as a parameter to express antitumor effects. Tumor size was determined with a slide caliper to calculate the volume at regular time intervals after transplantation. Analytical methods for direct correlation of differentiation changes with loss of proliferative potential at the cellular level have not yet been applied. However, novel techniques enable studies on the kinetic behavior of specific cell populations. The proliferative behavior of a subpopulation with distinct differentiated features can be analyzed by methods providing simultaneous information on cell cycle traverse and specific antigen expression. One of the conventional methods is based on the *in vivo* injection of ^{3}H-thymidine, followed by autoradiography of paraffin-embedded or frozen tumor tissue sections to detect proliferating cells. This method can be combined with immunocytochemical staining of cellular differentiation antigens. A similar approach is provided by bromodeoxyuridine (BrdUrd), a pyrimidine analog incorporated into the DNA of proliferating cells, which can be immunocytochemically detected after *in vivo* administration to mice.[88,89] An illustration is offered by Verstijnen et al.[88] who used double-labeling immunocytochemical techniques for the demonstration of BrdUrd incorporation and the expression of differentiation markers in intestinal cells. The BrdUrd method has clear advantages because neither time-consuming autoradiographic techniques nor radiolabeled DNA precursors are needed. These aspects make the BrdUrd method a feasible approach for clinical studies.

Most recently, some monoclonal antibodies have been developed that identify antigens associated with proliferation.[90-94] One of the best-documented antibodies is Ki-67, which recognizes a nuclear antigen in proliferating normal and malignant cells of various mammalian species during G1, S, G2, and M phases of the cell cycle, but not during G0 and early G1 phases.[90,95] Some studies indicate that Ki-67 may be a reliable tool for monitoring the proliferative fraction of tumor cells during growth manipulation.[96] Again, simultaneous assessment of such proliferation and differentiation markers is possible when double-labeling techniques are being applied. An obvious limitation of all methods described above, which is inherent to *in vivo* experimentation, is the inability to determine whether growth inhibitory effects on individual cells are reversible or not.

In spite of the various assays available, it still may be difficult to elucidate the mechanism of action of so-called "differentiation inducers". Conceptually, tumor growth inhibition upon treatment with differentiation inducers is considered to be a direct result of maturating effects of the drug. However, literature data indicate that other mechanisms of drug action may explain the phenomenon observed. For example, Twardzik et al.[67] have described concomitant inhibition of growth and induction of differentiation in human lung cancer xenografts upon treatment with transforming growth factor-β (TGF-β). In addition, they observed a diminished vascularization in TGF-β-treated tumors which may have a causal relation with growth inhibition and differentiation induction. The enhancement of tumor antigenicity, followed by an antitumor response of the host immune system, has also been recognized as a possible means by which differentiating agents may cause growth inhibition.[2]

VI. ADMINISTRATION SCHEDULES AND METHODS FOR DIFFERENTIATION INDUCERS

In the process of differentiation, two specific stages can be distinguished. First, the cells become committed to terminal differentiation. At this stage, cells still express their undifferentiated phenotype, but have acquired the capacity to differentiate even in the absence of the inducing agent. During the second stage, committed cells continue to divide into a finite number of cells and to express differentiation features, processes which are modulated by environmental factors. For some *in vitro* cultured cell lines, the period of drug exposure needed for commitment of cells to differentiate has been established. Fibach et al.[97] showed that only a small proportion of murine erythroleukemia cells was committed after exposure to hexamethylene bisacetamide for 10 to 12 h, while more than 90% was committed when this time period was extended to 48 to 60 h. Also, for differentiation induction of HL-60 cells by retinoic acid, the percentage of mature cells appeared directly related to drug exposure time.[98] Comparable data are not available for cell lines derived from solid tumors, because these cells do not undergo terminal differentiation under conventional culturing conditions as monolayer on plastic.

Optimal use of differentiation inducers *in vivo* is assumed to require a continuous exposure of tumor cells to an effective drug concentration for a prolonged period of time. Several attempts have been made to meet these conditions.[80-82,84,86,99,100] However, when these studies were guided by pharmacokinetic investigations, it became clear that for some drugs the assumed requirements for optimal therapeutic efficacy were not fulfilled. Pharmacokinetic considerations revealed that an optimal treatment might be attained by continuous infusion, but due to technical difficulties this hypothesis has never been tested in tumor-bearing nude mice. While most of the conventional procedures for prolonged continuous infusion are cumbersome, new applications have been presented to achieve more or less constant plasma levels of a given drug. Kozielsky et al.[101] developed an infusion apparatus which can be disconnected in between the treatment cycles, resulting in minimal stress. Another method to achieve constant drug levels in serum is the use of osmotic minipumps. These pumps are self-powered (based on osmotic principles), can be implanted in mice either i.p. or s.c., and enable continuous delivery of agents. Nishimura et al.[102] described the application of these pumps for *in vivo* administration of r-interleukin-2 (r-IL-2), a drug with a short half-life *in vivo*. Slow release of r-IL-2 by osmotic minipumps caused a significant prolongation of serum levels. Claessen et al.[103] used the same system for prolonged local delivery of chemotherapeutics in guinea pigs. Minipumps available at the moment are designed to release their content within a period of 7 d. For testing of low toxic drugs like differentiation inducers at or near maximum tolerated dose, the contents of these pumps may appear to be insufficient.

Another interesting approach to deliver drugs for a prolonged period of time has been described by Otaka et al.[104] Liposomes were conjugated with a specific monoclonal antibody for selective targeting of the differentiation inducer sodium butyrate to human colon cancer xenografts in nude mice. Butyrate-encapsulated antibody conjugated liposomes can be considered as a system for local, slow delivery of drug. Liposomes were administered repeatedly (3 or 4 times) every 4th or 5th day. In the group of antibody-conjugated liposomes, but not in the group treated with liposomes alone or treated with liposomes bound to non-specific mouse IgG, significant growth inhibition was observed. Inhibitory effects were associated with changes in cell morphology and increases in membrane-bound alkaline phosphatase and gamma-glutamyltranspeptidase, indicating occurrence of butyrate-induced differentiation.

VII. CONCLUSIONS

Data presented in this chapter indicate that the nude mouse provides the best human tumor model as yet available for reproduction of the regulatory mechanisms of tumor cell differentiation in man. Therefore, it also appears to be the most relevant model for testing differentiation-inducing agents. This holds true especially for studies on maturating events in solid tumors. Furthermore, for tumor types in which proliferation and differentiation are not obviously associated with 3-D growth, such as hematological neoplasms, the nude mouse model may add information on the role of the microenvironment in these processes. The human-tumor-bearing nude mice model is also an advantageous preclinical tool, because important *in vivo* aspects of differentiation inducers like the pharmacokinetics, the therapeutic index, as well as the host-mediated modulation of differentiation can be determined.

Techniques for simultaneous monitoring of the processes of differentiation and proliferation in xenografts grown in nude mice are presently available. Nevertheless, a major problem of studies with differentiation inducers *in vivo,* remains the distinction of maturating activity from cytotoxic effects. For the design of optimal treatment schedules, the nude mouse shows clear restrictions, especially when protocols for continuous long-term drug delivery are being considered. New administration procedures have been developed, but their value in studies on differentiation inducing agents has yet to be proven.

REFERENCES

1. **Sachs, L.,** Control of growth and normal differentiation in leukemic cells. Regulation of the developmental program and restoration of the normal phenotype in myeloid leukemia, *J. Cell. Physiol.,* 1 (Suppl.), 151, 1982.
2. **Spremulli, E. N. and Dexter, D. L.,** Polar solvents: a novel class of antineoplastic agents, *J. Clin. Oncol.,* 2, 227, 1984.
3. **Sartorelli, A. C.,** Malignant cell differentiation as a potential therapeutic approach, *Br. J. Cancer,* 52, 293, 1985.
4. **Bloch, A.,** Induced cell differentiation in cancer therapy, *Cancer Treat. Rep.,* 68, 199, 1984.
5. **Freshney, R. I.,** Induction of differentiation in neoplastic cells, *Anticancer Res.,* 5, 111, 1985.
6. **Reiss, M., Gamba-Vitalo, C., and Sartorelli, A. C.,** Induction of tumor cell differentiation as a therapeutic approach: preclinical models for hematopoietic and solid neoplasms, *Cancer Treat. Rep.,* 70, 201, 1986.
7. **Marks, P. A. and Rifkind, R. A.,** Differentiation modifiers, *Cancer,* 54, 2766, 1984.
8. **Gabrilove, J. L.,** Differentiation factors, *Semin. Oncol.,* 13, 228, 1986.
9. **Rifkind, R. A.,** Differentiation inducers: freeing cancer cells to mature, *Adv. Oncol.,* 2, 3, 1985.
10. **Seifter, E. J., Levitt, M. L., and Kramer, B. S.,** An outlier theory of cancer curability: tumor cell differentiation as a therapeutic goal, *Am. J. Med.,* 83, 757, 1987.

11. **Pierce, G. B. and Speers, W. C.,** Tumors as caricatures of the process of tissue renewal: prospects for therapy by directing differentiation, *Cancer Res.*, 48, 1996, 1988.
12. **Sachs, L.,** Cell differentiation and bypassing of genetic defects in the suppression of malignancy, *Cancer Res.*, 47, 1981, 1987.
13. **Egorin, M. J., Sigman, L. M., VanEcho, D. A., Forrest, A., Whitacre, M. Y., and Aisner, J.,** Phase I clinical and pharmacokinetic study of hexamethylene bisacetamide (NSC 95580) administered as a five-day continuous infusion, *Cancer Res.*, 47, 617, 1987.
14. **Rowinski, E. K., Ettinger, D. S., Grochow, L. B., Brundett, R. B., Cates, A. E., and Donehower, R. C.,** Phase I and pharmacologic study of hexamethylene bisacitamide in patients with advanced cancer, *J. Clin. Oncol.*, 4, 1835, 1986.
15. **Young, C. W., Fanucchi, M. P., Walsh, T. D., Blatzer, L., Yaldaei, S., Stevens, Y. W., Gordon, C., Tong, W., Rifkind, R. A., and Marks, P. A.,** Phase I trial and clinical pharmacological evaluation of hexamethylene bisacetamide by ten-day continuous intravenous infusion at twenty-eight-day intervals, *Cancer Res.*, 48, 7304, 1988.
16. **Gold, E. J., Bosl, G. J., and Itri, L. M.,** Phase II trial of 13-*cis* retinoic acid in patients with advanced nonseminomatous germ cell tumors, *Cancer Treat. Rep.*, 68, 1287, 1984.
17. **Lippman, S. M., Kessler, J. F., and Meyskens, F. L.,** Retinoids as preventive and therapeutic anticancer agents, *Cancer Treat. Rep.*, 71, 493, 1987.
18. **Clagett-Carr, K., Sarosy, G., Plowman, J., Hoth, D. F., and Leyland-Jones, B.,** *N*-Methylformamide: cytotoxic, radiosensitizer, or chemosensitizer, *J. Clin. Oncol.*, 6, 906, 1988.
19. **Cheson, B. D. and Simon, R.,** Low-dose ara-C in acute non-lymphocytic leukemia and myelodysplastic syndromes: a review of 20 years experience, *Semin. Oncol.*, 14 (Suppl. 1), 126, 1987.
20. **Markert, C. L.,** Neoplasia: a decrease of cell differentiation, *Cancer Res.*, 28, 1908, 1968.
21. **Sporn, M. B., Roberts, A. B., and Driscoll, J. S.,** Principles of cancer biology: growth factors and differentiation, in *Cancer: Principles and Practice of Oncology*, 2nd ed., Hellman, S., Rosenberg, S. A., and DeVita, V., Jr., Eds., J. B. Lippincott, Philadelphia, 1985, 49.
22. **Goustin, A. S., Leof, E. B., Shipley, G. D., and Moses, H. L.,** Growth factors and cancer, *Cancer Res.*, 46, 1015, 1986.
23. **Pierce, G. B. and Wallace, C.,** Differentiation of malignant to benign cells, *Cancer Res.*, 31, 127, 1971.
24. **McLaughlin, J. E. and Urich, H.,** Maturating neuroblastoma and ganglioneuroblastoma: a study of four cases with long survival, *J. Pathol.*, 121, 19, 1977.
25. **Dyke, P. C. and Mulkey, D. S. B.,** Maturation of ganglioneuroblastoma to ganglioneuroma, *Cancer*, 20, 1343, 1967.
26. **Hong, W. K., Wittes, R. E., Hajdy, S. T., Cvitkovi, E., Whitmore, W. F., and Golbey, R. B.,** The evolution of mature teratoma from malignant tumours, *Cancer*, 40, 2987, 1977.
27. **Novogrodsky, A., Dvir, A., Ravid, A., Shkolnik, T., Stenzel, K. H., Rubin, A. L., and Zaizov, R.,** Effect of polar organic compounds on leukemic cells: butyrate-induced partial remission of acute myelogenous leukemia in a child, *Cancer*, 51, 9, 1983.
28. **Brinster, R. L.,** Effect of cells transferred into the mouse blastocyst on subsequent development, *J. Exp. Med.*, 140, 1049, 1974.
29. **Schubert, D. and Jacob, F.,** 5-Bromodeoxyuridine-induced differentiation of a neuroblastoma, *Proc. Natl. Acad. Sci. U.S.A.*, 67, 247, 1970.
30. **Dexter, D. L. and Hager, J. C.,** Maturation induction of tumor cells using a human colon carcinoma model, *Cancer*, 45, 1178, 1980.
31. **Tralka, T. S. and Rabson, A. S.,** Cilia formation in cultures of human lung cancer cells treated with dimethylsulfoxide, *J. Natl. Cancer Inst.*, 57, 1383, 1976.
32. **Huberman, E., Heckman, C., and Langerbach, R.,** Stimulation of differentiated functions in human melanoma cells by tumor promoting agents and dimethylsulfoxide, *Cancer Res.*, 39, 2618, 1979.
33. **Friedman, S. J. and Skehan, P.,** Morphological differentiation of human choriocarcinoma cells, *Cancer Res.*, 39, 1960, 1979.
34. **Schroy, P. C., Carnright, K., Winaver, S. J., and Friedman, E. A.,** Heterogeneous responses of human colon carcinomas to hexamethylene bisacetamide, *Cancer Res.*, 48, 5487, 1988.
35. **Breitkreutz, D., Bohnert, A., Herzmann, E., Bowden, P. E., Boukamp, P., and Fusenig, N. E.,** Differentiation specific functions in cultured and transplanted mouse keratinocytes: environmental influences on ultrastructure and keratin expression, *Differentiation*, 26, 154, 1984.
36. **Sun, T. T. and Green, H.,** Cultured epithelial cells of cornea, conjunctiva and skin: absence of marked intrinsic divergence of their differentiated status, *Nature (London)*, 269, 489, 1977.
37. **Fuchs, E. and Green, H.,** The expression of keratin in epidermis and cultured epidermal cells, *Cell*, 15, 887, 1978.
38. **Watt, F. M., Boukamp, P., Hornung, J., and Fusenig, N. E.,** Effect of growth environment on spatial expression of involucrin by human epidermal keratinocytes, *Arch. Dermatol. Res.*, 279, 335, 1987.

39. **Bohnert, A., Hornung, J., Mackenzie, I. C., and Fusenig, N. E.,** Epithelial-mesenchymal interactions control basement membrane production and differentiation in cultured and transplanted mouse keratinocytes, *Cell Tissue Res.,* 244, 413, 1986.
40. **Dale, B. A., Gown, A. M., Fleckman, P., Kimball, J. R., and Resing, K. A.,** Characterization of two monoclonal antibodies to human epidermal keratohyalin: reactivity with filaggrin and related proteins, *J. Invest. Dermatol.,* 88, 306, 1987.
41. **Doran, T. I., Vidrich, A., and Sun, T. T.,** Intrinsic and extrinsic regulation of the differentiation of skin, corneal and esophageal epithelial cells, *Cell,* 22, 17, 1980.
42. **Fusenig, N. E., Breitkreutz, D., Dzarlieva, R. T., Boukamp, P., Bohnert, A., and Tilgen, W.,** Growth and differentiation characteristics of transformed keratinocytes from mouse and human skin *in vitro* and *in vivo, J. Invest. Dermatol.,* 81, 168s, 1983.
43. **Boukamp, P., Rupniak, H. T. R., and Fusenig, N. E.,** Environmental modulation of the expression of differentiation and malignancy in six human squamous cell carcinoma cell lines, *Cancer Res.,* 45, 5582, 1985.
44. **Easty, G. G., Easty, D. M., Monaghan, P., Ormerod, M. G., and Neville, A. M.,** Preparation and identification of human breast epithelial cells in culture, *Int. J. Cancer,* 26, 577, 1980.
45. **Kedinger, M., Simon-Assmann, P., Alexandre, E., and Haffen, K.,** Importance of a fibroblastic support for *in vitro* differentiation of intestinal endodermal cells and for their response to glucocorticoids, *Cell Diff.,* 20, 171, 1987.
46. **Richman, P. I. and Bodmer, W. F.,** Control of differentiation in human colorectal carcinoma cell lines: epithelial-mesenchymal interactions, *J. Pathol.,* 156, 197, 1988.
47. **Barrett, L. A., McDowell, E. M., Frank, A. L., Harris, C. C., and Trump, B. F.,** Long-term organ culture of human bronchial epithelium, *Cancer Res.,* 36, 1003, 1976.
48. **Sananes, N., Weiller, S., Baulieu, E. E., and Le Goascogne,** *In vitro* decidualisation of rat endometrial cells, *Endocrinology,* 103, 86, 1978.
49. **Haftek, M., Thivolet, J., Thomas, J., Joubaud, A., and Faure, M.,** Re-expression of disease-characteristic features of non-bullous congenital ichthyosiform erythroderma (CIE) after grafting of the pathological keratinocyte cultures to athymic mice, *J. Cut. Pathol.,* 16, 1, 1989.
50. **Mackenzie, I. C. and Fusenig, N. E.,** Regeneration of organized epithelial structure, *J. Invest. Dermatol.,* 81, 189s, 1983.
51. **Hornung, J., Bohnert, A., Phan-Than, L., Krieg, T., and Fusenig, N. E.,** Basement membrane formation by malignant mouse keratinocyte cell lines in organotypic culture and transplants: correlation with degree of morphologic differentiation, *J. Cancer Res. Clin. Oncol.,* 113, 325, 1987.
52. **Gurdon, J. B.,** Embryonic induction — molecular aspects, *Development,* 89, 385, 1987.
53. **Giovanella, B. C. and Fogh, J.,** The nude mouse in cancer research, *Adv. Cancer Res.,* 44, 69, 1985.
54. **Shimosato, Y., Kameya, T., and Hirohashi, S.,** Growth, morphology, and function of xenotransplanted human tumors, in *Pathology Annual,* Vol. 14, Sommers, S. and Rosen, P., Eds., Appleton Century Crofts, New York, 1979, 215.
55. **Stamp, G. W. H., Quaba, A., Braithwaite, A., and Wright, N. A.,** Basal cell carcinoma xenografts in nude mice: studies on epithelial differentiation and stromal relationships, *J. Pathol.,* 156, 213, 1988.
56. **Horoszewicz, J. S., Leong, S. S., Kawinski, E., Karr, J. P., Rosenthal, H., Ming Chu, T., Mirand, E. A., and Murphy, G. P.,** LNCaP model of human prostatic carcinoma, *Cancer Res.,* 43, 1809, 1983.
57. **Knöfel, W. T., Otto, U., Baisch, H., and Klöppel, G.,** Stability of human renal cell carcinomas during long term serial transplantation into nude mice: histopathology, nuclear grade, mitotic rate, and DNA content in thirty tumors, *Cancer Res.,* 47, 221, 1987.
58. **Braakhuis, B. J. M., Sneeuwloper, G., and Snow, G. B.,** The potential of the nude mouse xenograft model for the study of head and neck cancer, *Arch. Oto-rhino-laryngol.,* 239, 69, 1983.
59. **Walt, H. and Hedinger, Chr. E.,** Differentiation of human testicular embryonal carcinoma and teratocarcinoma grown in nude mice and soft-agar cultures, *Cell Differentiation,* 15, 81, 1984.
60. **Hollande, E., Trocheris de St-Front, V., Louet-Hermitte, P., Bara, J., Péquignot, J., Estival, A., and Clemente, F.,** Differentiation features of human pancreatic tumor cells maintained in nude mice and in culture: immunocytochemical and ultrastructural studies, *Int. J. Cancer,* 34, 177, 1984.
61. **Dubois, J. D., O'Hare, M. J., Monaghan, P., Bartek, J., Norris, R., and Gusterson, B. A.,** Human breast epithelial xenografts: an immunocytochemical and ultrastructural study of differentiation and lactogenic response, *Differentiation,* 35, 72, 1987.
62. **Cajot, J. F., Sordat, B., and Bachmann, F,** Human primary colon carcinomas xenografted into nude mice. II. Modulation of tumor plasminogen activator activity by the host tissue environment, *J. Natl. Cancer Inst.,* 77, 1099, 1986.
63. **Ito, Y. Z., Mashimo, S., Nakazoto, Y., and Takikawa, H.,** Hormone dependency of a serially transplantable human prostatic cancer (HONDA) in nude mice, *Cancer Res.,* 45, 5058, 1985.
64. **Casper, J., Schmoll, H. J., Schnaidt, U., and Fonatsch, C.,** Cell lines of human germinal cancer, *Int. J. Andrology,* 10, 105, 1987.

65. **Sekiya, S., Ishige, K., Yamazawa, K., Kawata, M., Inaba, N., Nagao, K., and Takamizawa, H.,** The capacity for differentiation and production of human chorionic gonadotropin and α-fetoprotein in human embryonal carcinoma cell lines, *Int. J. Gyncecol. Pathol.*, 7, 373, 1988.
66. **Lim, R., Hicklin, D. J., Ryken, T. C., Han, X. M., Liu, K. N., Miller, J. F., and Baggenstoss, B. A.,** Suppression of glioma growth *in vitro* and *in vivo* by glia maturation factor, *Cancer Res.*, 46, 5241, 1986.
67. **Twardzik, D. R., Ranchalis, J. E., McPherson, J. M., Ogawa, Y., Gentry, L., Purchio, A., Plata, E., and Todaro, G. J.,** Inhibition and promotion of differentiated-like phenotype of a human lung carcinoma in athymic mice by natural and recombinant forms of transforming growth factor-β, *J. Natl. Cancer Inst.*, 81, 1182, 1989.
68. **Löning, T. H. and Mackenzie, I. C.,** Immunohistochemical studies of basal cell carcinomas transplanted into nude mice, *Arch. Dermatol. Res.*, 279, 37, 1986.
69. **Jacobsen, P. F., Parker, T. J., and Papadimitriou, J. M.,** Ependymal differentiation of a glioblastoma multiforme after heterotransplantation into nude mice, *J. Natl. Cancer Inst.*, 79, 771, 1987.
70. **Burg-Kurland, C. L., Purnell, D. M., Combs, J. W., Valerio, M. G., Harris, C. C., and Trump, B. F.,** Immunocytochemical evaluation of primary human esophageal carcinomas and their xenografts for keratin, β-chorionic gonadotropin, placental lactogen, α-fetoprotein, carcinoembryonic antigen, and nonspecific cross-reacting antigen, *Cancer Res.*, 46, 5730, 1986.
71. **Horvath, G., Alm, P., Tropé, C., and Fernö, M.,** Changes in tumor morphology in one human endometrial carcinosarcoma heterotransplanted into nude mice, *In Vivo*, 3, 113, 1989.
72. **Wennerberg, J.,** Changes in growth pattern of human squamous cell carcinomas of the head and neck during serial passages in nude mice, *Int. J. Cancer*, 33, 245, 1984.
73. **Pollack, A., Block, N. L., Stover, B. T., and Irvin, G. L.,** Tumor progression in serial passages of the Dumming R3327-G rat prostatic adenocarcinoma: growth rate response to endocrine manipulation, *Cancer Res.*, 45, 1052, 1985.
74. **Mackenzie I. C. and Hill, M. W.,** Connective tissue influences on patterns of epithelial architecture and keratinization in skin and oral mucosa of the adult mouse, *Cell Tissue Res.*, 235, 551, 1984.
75. **Metcalf, D.,** Sources and biology of regulatory factors active on mouse myeloid leukemic cells, *J. Cell. Physiol.*, 1 (Suppl.), 175, 1982.
76. **Machado, E. A., Gerard, D. A., Lozzio, C. B., Lozzio, B. B., Mitchell, J. R., and Golde, D. W.,** Proliferation and differentiation of human myeloid leukemic cells in immunodeficient mice: electron microscopy and cytochemistry, *Blood*, 63, 1015, 1984.
77. **Okabe, J., Honma, Y., Hayashi, M., and Hozumi, M.,** Actinomycin D restores *in vivo* sensitivity to differentiation induction of non-differentiating mouse myeloid leukemia cells, *Int. J. Cancer*, 24, 87, 1979.
78. **Schwartz, E. L., Brown, B. J., Nierenburg, M., Marsh, J. C., and Sartorelli, A. C.,** Evaluation of some anthracycline antibiotics in an *in vivo* model for studying drug-induced human leukemia cell differentiation, *Cancer Res.*, 43, 2725, 1983.
79. **Langdon, S. P. and Hickman, J. A.,** Correlation between the molecular weight and potency of polar compounds which induce the differentiation of HL-60 human promyelocytic leukemia cells, *Cancer Res.*, 47, 140, 1987.
80. **Hayashi, Y., Yanagawa, T., Azuma, M., Yura, Y., Yoshida, H., and Sato, M.,** Induction of cells with a myoepithelial cell phenotype by treatment with dibutyryl cyclic AMP in human salivary adenocarcinoma cells grown in athymic nude mice, *Virchows Arch. B*, 50, 1, 1985.
81. **Hayashi, Y., Yoshida, H., Nagamine, S., Yanagawa, T., Yura, Y., Azuma, M., and Sato, M.,** Induction of cells with acinar cell phenotype including presence of intracellular amylase: treatment with 12-*O*-tetradecanoyl-phorbol-13-acetate in a neoplastic human salivary intercalated duct cell line grown in athymic nude mice, *Cancer*, 60, 1000, 1987.
82. **Van Dongen, G. A. M. S., Braakhuis, B. J. M., Leyva, A., Hendriks, H. R., Kipp, B. B. A., Bagnay, M., and Snow, G. B.,** Anti-tumor and differentiation-inducing activity of *N,N*-dimethylformamide (DMF) in head-and-neck cancer xenografts, *Int. J. Cancer*, 43, 285, 1989.
83. **Takizawa, T., Matsui, T., Maeda, Y., Okabe, S., Mochizuki, M., Tanaka, A., Kawaguchi, K., Fukayama, M., Funata, N., Koike, M., and Mastuda, T.,** *Lab. Invest.*, 60, 22, 1989.
84. **Strickland, S. and Sawey, M. J.,** Studies on the effect of retinoids on the differentiation of teratocarcinoma stem cells *in vitro* and *in vivo*, *Develop. Biol.*, 78, 76, 1980.
85. **Ishikawa, Y.,** Further studies on the differentiation of a cell line of myeloid leukemia, *J. Cell. Physiol.*, 76, 175, 1970.
86. **Brosjö, O., Bauer, H. C. F., Broström, L. A., Nilsson, O. S., Reinholt, F. P., and Tribukait, B.,** Growth inhibition of human osteosarcomas in nude mice by human interferon-α significance of dose and tumor differentiation, *Cancer Res.*, 47, 258, 1987.
87. **Forster, S., Triffitt, J. T., Bauer, H. C. F., Brosjö O., Nilsson, O. S., Smith, R., and Sykes, B.,** Interferon-inhibited human osteosarcoma xenografts induce host bone in nude mice, *J. Bone Mineral Res.*, 3, 199, 1988.
88. **Verstijnen, C. P. H. J., Ten Kate, J., Arends, J. W., Schutte, B., and Bosman, F. T.,** Xenografting of normal colonic mucosa in athymic nude mice, *J. Pathol.*, 155, 77, 1988.

89. **Schutte, B., Reynders, M. M. J., Bosman, F. T., and Blijham, G. H.,** Studies with anti-bromodeoxyuridine antibodies. II. Simultaneous immunocytochemical detection of antigen expression and DNA synthesis by *in vivo* labeling of mouse intestinal mucosa, *J. Histochem. Cytochem.*, 35, 371, 1987.
90. **Gerdes, J., Schwab, U., Lemke, H., and Stein, H.,** Production of a mouse monoclonal antibody reactive with a human nuclear antigen associated with cell proliferation, *Int. J. Cancer,* 31, 13, 1983.
91. **Ogata, K., Kurki, P., Celis, J. E., Nakamura, R. M., and Tan, E. M.,** Monoclonal antibodies to a nuclear protein (PCNA/cyclin) associated with DNA replication, *Exp. Cell Res.*, 168, 475, 1987.
92. **Black, A., Freeman, J. W., Zhou, J. W., and Busch, H.,** Novel cell cycle related nuclear proteins found in rat and human cells with monoclonal antibodies, *Cancer Res.*, 47, 3266, 1987.
93. **Chatterjee, A., Freeman, J. W., and Busch, H.,** Identification and partial characterization of Mr 105,000 nucleolar antigen associated with cell proliferation, *Cancer Res.*, 47, 6329, 1987.
94. **Yankulov, K. Y., Hadjiolova, K. V., Kounec, K. V., Getov, C., and Hadjiolov, A. A.,** The use of a monoclonal antibody against a proliferating cell nuclear matrix antigen in the study of solid human tumors, *Int. J. Cancer,* 43, 800, 1989.
95. **Gerdes, J., Lemke, H., Baisch, H., Wasker, H. H., Schwab, U., and Stein, H.,** Cell cycle analysis of a cell proliferation-associated human nuclear antigen defined by the monoclonal antibody Ki-67, *J. Immunol.*, 133, 1710, 1984.
96. **Gallee, M. P. W., van Steenbrugge, G. J., Ten Kate, F. J. W., Schroeder, F. H., and Van der Kwast, T. H.,** Determination of the proliferative fraction of a transplantable, hormone-dependent, human prostatic carcinoma (PC-82) by monoclonal antibody Ki-67: potential application for hormone therapy monitoring, *J. Natl. Cancer Inst.*, 79, 1333, 1987.
97. **Fibach, E., Reuben, R. C., Rifkind, R. A., and Marks, P. A.,** Effect of hexamethylene bisacetamide on the commitment to differentiation of murine erythroleukemia cells, *Cancer Res.*, 37, 440, 1977.
98. **Breitman, T. S., Selonick, S. E., and Collins, S. J.,** Induction of differentiation of the human promyelocytic leukemia cell line (HL-60) by retinoic acid, *Proc. Natl. Acad. Sci. U.S.A.*, 77, 2936, 1980.
99. **Dexter, D. L., Spremulli, E. N., Matook, G. M., Diamond, I., and Calabresi, P.,** Inhibition of the growth of human colon cancer xenografts by polar solvents, *Cancer Res.*, 42, 5018, 1982.
100. **Chang, J. H. T. and Prasad, K. N.,** Differentiation of mouse neuroblastoma cells *in vitro* and *in vivo* induced by cyclic adenosine monophosphate (cAMP) *J. Ped. Surg.*, 11, 847, 1976.
101. **Kozielski, T., Lockshin, A., Giovanella, B. C., and Stehlin, J. S.,** A method for prolonged, repeated infusion of nude mice, *Lab. Animal Sci.*, 36, 696, 1986.
102. **Nishimura, T., Uchiyama, Y., Yagi, H., and Hashimoto, Y.,** Administration of slowly released recombinant interleukin 2: augmentation of the efficacy of adoptive immunotherapy with lymphokine-activated killer (LAK) cells, *J. Immunol. Meth.*, 91, 21 1986.
103. **Claessen, A. M. E., Valster, H., Bril, H., Meyer, S., and Scheper, R. J.,** Cell-mediated immunity is enhanced by cytostatic drugs continuously released at the site of antigenic stimulation, *Cancer Immunol. Immunother.*, 28, 131, 1989.
104. **Otaka, M., Singhal, A., and Hakomori, S.,** Antibody-mediated targeting of differentiation inducers to tumor cells: inhibition of colonic cancer cell growth *in vitro* and *in vivo*, *Biochem. Biophys. Res. Commun.*, 158, 202, 1989.

Chapter 20

LIMITATIONS FOR STUDIES IN HUMAN TUMOR BIOLOGY

Øystein Fodstad

TABLE OF CONTENTS

I. INTRODUCTION

Human tumor xenografts in nude mice are regarded as the best model available for the study of human tumor biology. The experiments can be performed under reproducible and controlled conditions, and practically the same tumor tissue can be examined by any method and drug available at present and in the future. Xenografts thus provide a unique and renewable source of tumor material for biological studies. In previous chapters, a broad range of areas is discussed in which the usefulness of xenografts has been proven, with emphasis on their possibilities and advantages. This chapter will focus on some inherent limitations of the nude mouse-human tumor model, particularly those disadvantages that have received most attention in the literature. However, a discussion of limitations and drawbacks cannot be complete without balancing these with some of the more positive aspects of the model. Examples of ways to reduce the implications of the limitations will be presented, and the prospects of using the model in new areas of oncology research will be outlined.

II. STRUCTURAL AND PHYSIOLOGICAL CHARACTERISTICS OF XENOGRAFTS

A. MURINE STROMA AND BLOOD SUPPLY

The relevance of studies in the biology of human tumors, when grown as xenografts in nude mice, is based on the assumption that the interaction between the tumor cells and the host environment is closely similar to that in patients. One striking difference between xenografts and their parent tumors is, however, that the blood vessels and stromal cells in the xenografts are of murine origin. It would, therefore, not be surprising if this had significant impact on important characteristics of the human cells in the tumor.

A tumor's blood supply is generally regarded as one of the main limiting factors for tumor expansion and growth. In an s.c. xenograft, the blood supply is provided by the blood vessels which enter the tumor from the peripheral fibrous pseudocapsule. In some xenografts, central necrosis is already observed in rather small (<8 to 10 mm) tumors, suggesting an insufficient blood supply, which is possibly due to poor neovascularization.[1] In other cases, however, such necrosis can be a very late event, occurring first when the diameter is more than 15 mm.

Growth-regulating signal molecules should be present and reach the tumor cells in xenografts equally well as in the human situation. In the cases where human tumors are difficult to establish in nude mice, part of the cause may be related to inadequate supply of factors needed to promote growth in the local environment. Moreover, it is known that some endothelial cell receptors may be different between mouse and man and, in the cases where it is necessary that factors of influence on tumor growth bind to such receptors, tumor take may also be affected.

Part of the tumor stroma is formed by intercellular substance produced by the tumor cells, for example, osteoid production in osteosarcomas. Whether such production is regulated by inherent properties of the tumor cells, or as a result of the environment in which the cells grow, may be of importance for the relevance of the model. In several cases of osteosarcomas studied in our laboratory, osteoid production in s.c. grown osteosarcomas has apparently been a normal and stable feature of each tumor line.[2]

B. IMMUNOLOGICAL FACTORS

The nu/nu mouse has provided an important tool for the elucidation of the different mechanisms of the immune system participating in the "immune surveillance" of tumor cells.[3] The lack of functional T cells permits the growth of most human tumors, but it has been shown that T cell-like activity may be induced[4] that probably can affect tumor take. Whereas the B cell function in nude mice is relatively unaffected,[5] natural killer (NK) cell activity is high.[6]

Moreover, the activity of the macrophages,[7] particularly in the peritoneal cavity and in the lungs, may be different from that in man. It is conceivable that such differences may interfere with different types of biological studies in nude mice. This is clearly possible in chemotherapy experiments with drugs that affect the immune system.

C. HORMONAL AND METABOLIC FACTORS

Hormones, various enzymes, and other physiological host factors may also be of influence on the growth and the biological characteristics of the human tumor cells in xenografts.

Hormone dependency of different human tumor types, such as breast cancer and prostate cancer, has been confirmed in xenografts.[8-11] In some cases of breast cancer, the addition of estrogens is necessary, even to female mice, in order to support tumor growth; whereas androgen supply may permit the propagation of some prostate tumors.[12] Hormone levels and the metabolism of hormones[13] or anti-hormones may differ between mice and man. Kitamura et al.[14] suggested that the immunological status of the host could affect the hormone dependency of an undifferentiated carcinoma. Kondo et al.[15] showed that cachexia appearing in melanoma-bearing nude mice was caused by remarkably increased levels of serum sialyltransferase and decreased liver catalase activity. Moreover, enzyme activity in the liver and the kidney has been found to differ between nude Swiss and regular Swiss mice.[16]

D. HOST INFECTIONS

The immune deficiency makes nude mice prone to attract various types of infections,[12,17] several of which may affect the immune system, as well as enzyme and possible hormone activity in the various tissues. This hazard is not fully elucidated, but experimentation in non-healthy animals is not recommended. When the symptoms caused by the infections are easily detectable, precautions can be taken. However, it is well known that minor infections giving few and diffuse symptoms can be harbored in a nude mouse colony for a long time without being recognized by the investigator, or without any action taken to prevent its significant implications. It cannot be excluded that inconsistency in results obtained in different laboratories, or in the same laboratory at different times, may have been caused by problems of this category.

III. TUMOR TRANSPANTABILITY AND GROWTH

A. REPRESENTATIVITY OF TUMOR PANELS

The take rate of human tumors in nude mice differs significantly between tumor types,[12,18] a topic that is discussed in Chapter 6. Breast, prostate, and testicular cancers are particularly difficult to establish as permanent lines, and very few cases of leukemia and lymphoma xenografts have been described. When only a small fraction of the tumor biopsies transplanted to nude mice shows progressive growth to an extent permitting biological studies, it is questionable whether the resulting xenografts are representative for their respective tumor types. With the known heterogeneity in tumor biology[19-21] and in response to therapy, even in tumors of similar morphology, it is not likely that results obtained in one or a few xenografts can be representative for a particular tumor type on the whole. Furthermore, even for those tumor types that are relatively easy to establish as xenografts, not all biopsies in nude mice will grow, because of which an important clinical subgroup will be represented only.

Most studies of xenografts have been performed in xenografts grown s.c.,[12] a site that usually bears no resemblance with the situation in patients. In spite of this, the biology of the s.c. xenografts seems to be surprisingly similar to that of their parent tumors.[7,12,22-25] Arguments have been given for growing the human cells at a clinically more relevant site, that is, by "orthotopic" implantation. Naito et al.[26] have shown that a renal cell cancer gave a much higher number of metastases when the tumor cells were injected into the kidney than into other tissues. Similarly,

McLemore et al.[27] found a higher take rate for lung cancer by the intrabronchial route than by s.c. injection of the cells. It is not surprizing that the microenvironment surrounding the implanted cells decisively influences cell proliferation.[1] However, this issue has to be further elucidated before drawing any definite conclusions on the value of orthotopic transplantations. Lymphomas and brain tumors have been established as xenografts by intracerebral injection.[12] Although this site is advantageous because it is immunologically privileged, the procedure is not optimal for detailed biological studies.

One prerequisite for studying human tumor biology in xenografts is that the tumor growth rate is comparable to that observed in the clinic.[28] The conclusion of some studies addressing this question is that xenografts have a shorter tumor volume-doubling time than their clinical counterparts.[29] Such comparisons are difficult, as tumors with diameters as small as in xenografts are rarely amenable to investigation in patients. In our opinion, the observed differences in growth characteristics and size between clinical and implanted tumors do not seem to form a major limitation for the relevance of biological studies in nude mice.

B. SELECTION OF TUMOR SUBPOPULATIONS

The heterogeneity of human tumors[19-21] may constitute an obstacle to obtain xenografts that are representative for the whole tumor tissue. Thus, implantation of small tissue cubes, or cell suspensions prepared from a minor portion of the patient's tumor, provides possibilities for selecting subpopulations that share characteristics with just a minority of the cells in the tumor of origin. It is well known that during tumor progression the diversity in cell characteristics increases, and the possibility for selecting subpopulations of tumor cells therefore increases with time. In spite of the possibility of such a selection, very little evidence has been provided for this to be a major problem in the use of xenografts for biological studies.

Isolation of subpopulations of tumor cells from xenografts has been reported, and is usually performed through *in vitro* cultivation of the cells.[20] This procedure may, by itself, also induce epigenetic alterations of cell phenotypes that incorrectly may be regarded as a selection of preexisting cell subpopulations. In this respect, we failed several times in the intentional selection of subpopulations of cells from heterogeneous tumor lines grown *in vivo* (unpublished). However, we have seen that by continuous transfer of a human osteosarcoma, its ability for osteoid production and expression of an osteosarcoma-associated antigen was permanently lost. Thus, it may have been possible that a particular cell subpopulation was selected by *in vivo* transfer which had a growth advantage in the environment provided by the nude host. Another possibility would be that the interaction between the tumor cells and the host environment had induced the change in tumor cell characteristics. As another example, on two different occasions we performed selection experiments with FEMX-I melanoma cells taken from extrapulmonary metastases appearing after i.v. injection by reinjecting these cells i.v. into new recipient mice. After the first 5 to 6 generations of metastases, the tumor cell characteristics unexpectedly changed into a more differentiated phenotype characterized by increased melanin production and a reduced growth rate (manuscript in preparation). When such cells were transferred s.c. to other mice, the tumor reverted to the original phenotype after two passage generations, and thus illustrate the effect of the local growth conditions on the tumor cells.

C. INDUCTION OF MURINE TUMORS

A series of investigators working with human tumor xenografts have observed the emergence of malignant murine tumors. The majority of these show fibrosarcoma-like morphology, but lymphosarcomas and lymphomas have also been reported.

The "spontaneously" arising tumors may be divided in at least two categories:

1. Tumors that develop without any direct association with the site of growth of the human xenografts. These seem to be of lymphoreticular origin and contain mixtures of null, B, and T cells.[30-32] Baird et al.[30] demonstrated that the lymphoma development was preceded

FIGURE 1. Lungs of a nude mouse 41 d after injection of 1×10^6 cells of the LBX human small cell lung cancer line. The grape-like appearance is caused by numerous murine tumor nodules.

by the secretion of a murine leukemia virus glycoprotein antigen into the blood stream. The relationship between the growth of the xenografts and the development of these tumors is not clear. Sharkey and Fogh[33] showed that the incidence of spontaneous tumors in nude mice was similar to that previously reported for regular mice of the same background.

2. The most commonly induced murine tumors are sarcomas that develop in close association with the human tumor implant.[34-40] The majority of these have a fibrosarcoma-like appearance. Generally, the human tumor cells are completely overgrown by the mouse sarcoma cells.[34] In some cases, the two populations of malignant cells may coexist in the tumor, and a selection of pure murine tumor cells can be obtained in tissue culture.[36,37] In some cases, the murine sarcomas have been transplantable, not only in nude mice, but also in regular thymus-bearing animals.[36] Our group reported a case where attempts to cultivate a human melanoma xenograft *in vitro* always resulted in the emergence of malignant mouse cells.[41] Tumors did form after the injection of these cells s.c. into thymus-bearing mice; but after reaching a size of about 10 mm, the tumors regressed and were not transplantable into other animals. In another melanoma xenograft, the fraction of murine, apparently malignant, cells increased gradually during subsequent transfer to represent up to 80% of the cells in the tumor tissue. Some generations later, however, the murine tumor cells spontaneously disappeared, and the line now contains human malignant cells only (unpublished). In other cases, we have observed a rapid development of murine tumors in the lungs of nude mice upon i.v. injection of various types of human tumor cells (Figure 1). These tumors did not seem to originate from any particular cell type in the lung.

The mechanisms underlying the transformation of the murine host cells are not known. It has been suggested that transformation occurs more frequently with small cell lung cancer cells, and that the human cells may produce growth factors that are involved in this transformation process.[37] During the latency period from cell inoculation to the development of a growing xenograft, most of the tumor cells will die. One could speculate that at the time of death of these cells, some genetic information could be transferred to the host cells and induce their transformation. In a mouse carrying a human nonsmall cell lung cancer of mixed morphology, we observed the development of a murine tumor about three weeks after chemotherapy had resulted in complete regression of the xenograft (unpublished). The murine tumor was examined for the presence of human DNA, but no Alu-sequences were detected.

It is necessary to routinely check the tumors for the coexistence of murine transformed cells, or the possibility of such cells to overgrow the human cells. Morphological examination may not necessarily give definite answers, and the cells should also be studied by screening isoenzyme profiles, chromosome patterns, and possibly the binding of antibodies recognizing human tumor-associated antigens.

IV. METASTASIS STUDIES

The apparent lack of spontaneous metastasis formation from s.c. tumor xenografts in nude mice represents a major limitation for studying this important aspect of tumor biology.[7,12,24] However, to start with Giovanella et al.[42] in 1973, scattered observations of such metastases have been reported. In several of the cases where metastases have been described, tumor cell suspensions were injected to establish the primary s.c. tumor.[42,43] With this procedure, the possibility exists that some of the cells have been injected into some of the small blood vessels, thereby enhancing the probability of artificial tumor growth in other organs. In our own area of work, we observed several cases of the development of tumor foci, particularly in the pancreas. When tissue cubes of the same tumors were transplanted, spontaneous metastases only occurred in extremely rare, single cases. This is in accordance with the experience of Sharkey and Fogh[12] who observed metastases in only 14 of 1045 nude mice carrying 11 different tumor lines. Interestingly, Sordat et al.[44] have demonstrated that spontaneous metastases are much more frequent if the tumor tissue is transplanted to newborn nude mice.

A variety of factors control metastasis formation. The intrinsic ability of a tumor to invade adjacent tissues and to disseminate tumors is one of the most important single factors. The fact that most studies on xenografts are performed in the tumors growing s.c. does most certainly contribute to the low frequency of metastasis formation. In the patients, tumors are usually located in organs from which metastatic spread occurs more easily. Also, in contrast to the situation in man, the "primary" tumor xenografts in the mice are usually not removed by surgery or radiotherapy, and it is therefore possible that the animals are not given sufficient time to develop manifest metastases during the period of observation.[1] Since tumors grown at different locations in the same animal may influence the growth of each other, it is also possible that a large s.c. tumor may suppress the development of overt metastases in other areas. Consistent with this hypothesis, Shoemaker et al.[45] found that when lung tissue from mice carrying s.c. LOX melanoma xenografts was transplanted s.c. to recipient mice, LOX tumors developed, in spite of the lack of detectable lung metastases in the donor animals (unpublished).

Several models for human tumor metastasis in nude mice have recently been reported.[7,26,46-53] In most cases, these models involve inoculation of tumor cells by different routes, such as i.v., i.p., intrasplenic, intra-arterial, and intracardial injections. Inoculation into the mouse footpad has also given regional as well as distant metastases. By a variety of manipulations, many tumor models have now been obtained that are relevant for the clinical situation, as metastasis formation is observed in the tissues typical for the respective tumor type in patients

(Chapter 10). We have established several models in nude mice, as well as in nude rats, where the tissue-preferential growth of the metastatic cells corresponds closely to the patterns in the original patients.[48-52] That such metastases do retain important characteristics of the parent tumors is exemplified by our finding that several human tumors metastasizing to the bone/bone marrow of mice and rats display the typical lytic or sclerotic lesions observed in the clinic, as demonstrated by bone scintigraphy (unpublished).

V. STABILITY OF TUMOR CHARACTERISTICS

One major argument against the validity of biological studies in human tumor xenografts has been the changes observed in tumor characteristics during consecutive transfer.[29,54-57] It is a common experience that xenografts frequently increase their growth rates during the first three to five passage generations, possibly reflecting an adaptation to the new environment; but thereafter, the growth is usually stable for many years of continuous transfer.[12,17,28,58-66]

As already mentioned, questions have been raised with regard to the short tumor volume-doubling times of the xenografts compared to those assessed in patients.[29] In addition, variations in the morphology of tumors in different generations as xenografts, and differences between the morphology of the tissue in mice as compared to that of the original tumor in the patient, have also led to concern about the stability of the human tumor lines.

The concept of tumor heterogeneity involves the view that a biologically and morphologically heterogenous tumor may contain distinct subpopulations of tumor cells.[21] If the morphological heterogeneity sometimes seen in xenografts represents different and stable subpopulations, transplantation of small pieces of these tumors should lead to a wide variety of sublines of the original tumor. However, this does not seem to be the case. Thus, human tumor lines can usually be serially transplanted without loss of their biological and biochemical characteristics.[11,12,59,62-64,67] Nevertheless, examples of tumor sublines with different growth characteristics, morphological patterns, and sensitivity to drugs have been reported.

We have observed that two human melanomas with stable growth patterns for some passage generations spontaneously increased their growth rates, until they had volume-doubling times of approximately one half or one third of those that were initially calculated (Figure 2). In one of the cases, we could demonstrate that the change occurred simultaneously with an increased fraction of cells in S phase and a reduced cell loss.[53] The tumor morphology did not change, but the sensitivity to some cytostatic agents increased with the increased growth rate. However, no evidence could be obtained for a possible selection of a subpopulation of tumor cells.

Recently, Horvath et al.[56] showed that in the third passage of a human endometrial carcinosarcoma, two independent tumor lines developed; one line was a poorly differentiated adenocarcinoma without sarcoma components, whereas the other line was still a carcinosarcoma. In a study on soft tissue sarcomas and their xenografts, Donhuijsen et al.[57] showed an instability in several characteristics. Of particular interest was their finding that, upon serial transplantation, the different xenografts showed either stability, decrease, increase, or alteration in characteristics, such as the degree of differentiation, the cellularity, the number of mitosis, and the DNA index. Some characteristics changed from one passage to another, possibly influenced by the local growth conditions, which may differ between animals and various stages of growth. This phenomenon is consistent with our findings that successive passages show irregular changes among themselves, as observed in sarcomas, melanomas, and lung cancers (unpublished). It is conceivable that interactions between tumor and host cells may occasionally lead to transient alterations in characteristics, whereas in other cases, more permanent changes can be induced. Alterations in important tumor variables should be recognized as early as possible, and the implication of these changes has to be considered for correct interpretation of experimental results.

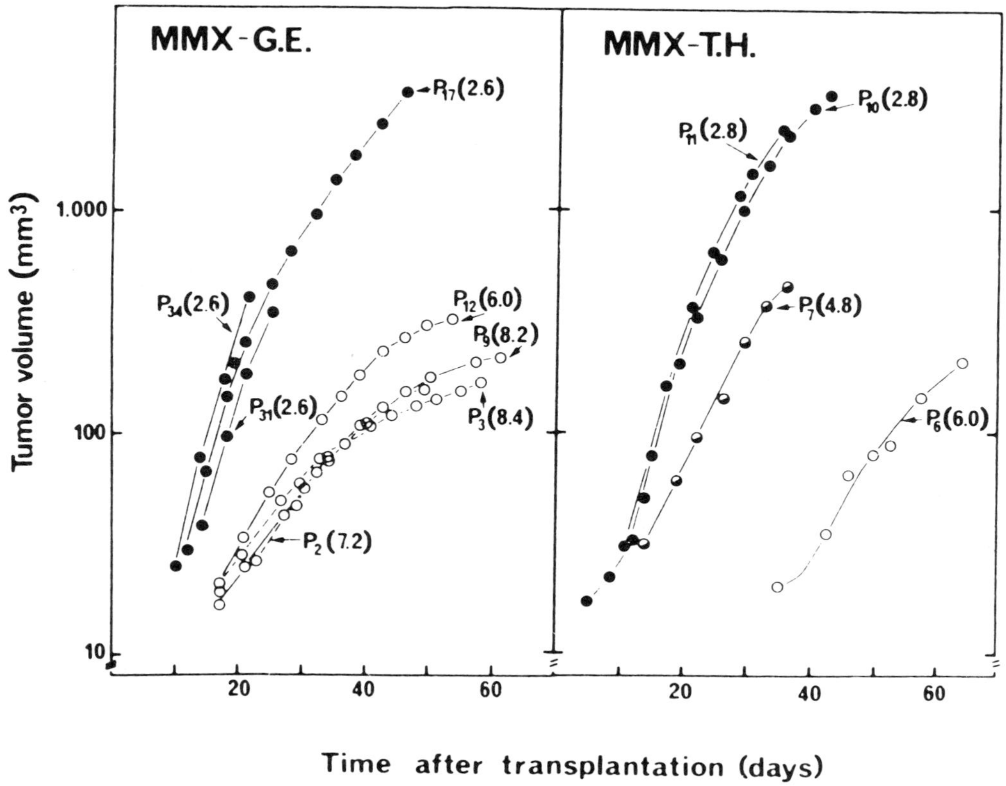

FIGURE 2. Growth at different passages of two transplanted melanomas. Numbers in parentheses give the time, in days, required to double mean tumor volume (30 to 60 mm^3).

The observed trend towards a more malignant phenotype in xenografts might reflect evolutionary differences between primary tumors and metastases in the human host.[12] We have recently followed a case of renal cell cancer in which the primary tumor of the patient did not contain papillary structures, a subsequent metastasis showed a mixture of solid and papillary tissue, whereas a xenograft established from this metastasis had pronounced papillary structures (unpublished). Therefore, it may be conceivable that xenografts also provide possibilities to study these aspects of tumor biology.

VI. MONOCLONAL ANTIBODIES AND CYTOKINES

Xenografts have been used over 10 years in experiments with monoclonal antibodies[68-73] with the aim to improve the diagnosis and the therapy of human cancer. These aspects are discussed in Chapter 17 of this book, and this section will be restricted to the notice that antigen expression is highly dependent on the conditions under which the tumor cells are grown.[73] Moreover, in order to investigate the homing of antibodies or their conjugates *in vivo,* the vascularity of the tumor and the pharmacokinetics of the antibodies are of decisive importance. Since most monoclonal antibodies against tumor-associated antigens are of murine origin, several of the difficulties faced with these molecules in man are of lesser magnitude in the murine hosts. However, the nude mouse-human tumor model may be used to obtain important and relevant data, if these facts are taken into consideration.

In recent years, a number of new biologically active molecules have been introduced in oncology research and the therapy of cancer, that is, interferons, interleukins, tumor necrosis factor, and a variety of colony stimulating factors (Chapter 18). The investigation of these compounds in nude mice is limited by their complex mechanisms of action. Thus, their primary effect on the host cells will indirectly affect the growth and differentiation of implanted tumor cells. Because of species-dependent specificity, the usefulness of some cytokines in the nude mouse model is questionable. Exceptions exist, such as the recent models in the multiple immunodeficient mice for studies in human leukemia and the engraftment and growth of human hematopoietic stem cells.[74-76] In these models, it should be possible to study the effect of various biologically active human factors on normal human cells, as well as the associated direct or indirect implications for the tumor cells.

VII. XENOGRAFTS IN MOLECULAR BIOLOGY

Tumorigenicity in nude mice is regarded as the most stringent criterion for successful malignant transformation of non-tumorigenic cells.[77] The introduction of oncogenes into mouse fibroblasts resulting in their transformation led to numerous studies where s.c. tumor formation in nude mice is being used as definite evidence for malignancy.[77-79] The validity of the test may depend on the type of recipient cells used for the oncogene transformation experiments. It should also be kept in mind that not all human tumors can be established as progressively growing xenografts in nude mice.

In the last few years there has been growing interest in the use of xenografts for studies on oncogene activation, rearrangement, and expression in the different types of tumors.[78-81] In most cases, a good correlation has been reported between the findings in the xenografts and their clinical counterparts. Since some oncogenes code for growth factor receptors,[78,80,82] the xenografts may also be used for studies on the activity of growth factors and for attempts to block the effect of such factors.

Recently, the significance of tumor suppressor genes in tumorigenesis and tumor progression has been demonstrated.[77,83-85] Inactivation of the retinoblastoma (Rb) gene is sufficient to induce malignancy in retinal cells, and this gene is also involved in the development of several other tumor types such as sarcomas,[86,87] small cell lung cancer,[88] and breast cancer.[89] The p53 gene seems to be involved in colon cancer and lung cancer tumorigenesis.[84] Suppressor genes have been studied in *in vitro* cell lines, but xenografts may be a better source of human tumor cells to investigate the factors involved in tumor development. Interestingly, the group of Lee have recently shown that retinoblastoma,[90] osteosarcoma, and prostate[91] cancer cells with tumorigenicity in nude mice lose this ability after reintroduction of a functional Rb gene into the tumor cells.

Some reports have described either a positive or an inverse correlation between the expression of recently identified genes in tumor cells[92,93] and the ability of the cells to form metastases. For the further analysis of the role of such genes, tumors that produce metastases in nude mice and rats may prove to be of particular value.

VIII. CONCLUSIONS

Xenografts in nude mice provide the best available model for studies on human tumor biology *in vivo*. The usefulness of the model is, however, limited by several factors inherent to either the host, the tumor, or the interaction between the two. Therefore, the various limitations of human tumor xenografts must be taken into consideration in the design of the experiments. Low transplantability of some tumor types seems to be the main disadvantage, as it may affect the general validity of the data obtained. The defective resistance of nude mice against infections may result in health problems seriously affecting the experiments. Another drawback has been the low frequency of metastases observed from s.c. xenografts. In the last few years, however, several relevant models in nude mice for human tumor metastasis have been described. Although the biological characteristics of the xenografts seem to reflect those of their parent tumors, some characteristics may change during serial transplantation, and measures should be taken to monitor and detect such alterations. The use of xenografts in diagnostic and therapeutic experiments involving monoclonal antibodies has provided valuable information, but the model underestimates some of the problems faced in man. Because of the species differences, the value of the nude mouse in studies on the effects of cytokines is limited; but for certain purposes, interesting possibilities have recently come available. In the area of molecular biology, xenografts have already shown their validity and usefulness, and this may be a promising field to obtain an increased understanding of mechanisms involved in tumorigenesis and metastasis formation.

REFERENCES

1. **Sordat, B. and Wang, W. R.,** Human colorectal tumor xenografts in nude mice: expression of malignancy, *Behring Inst. Mitt.*, 74, 291, 1984.
2. **Fodstad, Ø., Brøgger, A., Bruland, Ø., Solheim, Ø. P., Nesland, J. M., and Pihl, A.,** Characteristics of a cell line established from a patient with multiple osteosarcoma, appearing 13 years after treatment for bilateral retinoblastoma, *Int. J. Cancer*, 38, 33, 1986.
3. **Rygaard, J.,** *Thymus and Self. Immunobiology of the Mouse Mutant Nude*, FADL, Copenhagen, 1973.
4. **MacDonald, H. R.,** Phenotypic and functional characteristics of "T"-like cells in nude mice, *Exp. Cell Biol.*, 50, 305, 1982.
5. **Kindred, B.,** Nude mice in immunology, *Prog. Allergy*, 26, 137, 1979.
6. **Fodstad, Ø., Hansen, C. T., Cannon, G. B., Statham, C. N., Lichtenstein, G., and Boyd, M. R.,** Lack of correlation between natural killer activity and tumor growth control in nude mice with different immune defects, *Cancer Res.*, 44, 4403, 1984.
7. **Fidler, I. J.,** Rationale and methods for the use of nude mice to study the biology and therapy of human cancer metastasis, *Cancer Met. Rev.*, 5, 29, 1986.
8. **Clarke, R., Brünner, N., Katzenellenbogen, B. S., Thompson, E. W., Norman, M. J., Koppi, C., Paik, S., Lippman, M. E., and Dickson, R. B.,** Progression of human breast cancer cells from hormone-dependent to hormone-independent growth both *in vitro* and *in vivo*, *Proc. Natl. Acad. Sci. U.S.A.*, 86, 3649, 1989.
9. **Brünner, N., Bronzert, D., Vindeløv, L. L., Rygaard, K., Spang-Thomsen, M., and Lippman, M. E.,** Effect on growth and cell cycle kinetics of estradiol and tamoxifen on MCF-7 human breast cancer cells grown *in vitro* and in nude mice, *Cancer Res.*, 49, 1515, 1989.
10. **Osborne, C. K., Hobbs, K., and Clark, G. M.,** Effect of estrogens and antiestrogens on growth of human breast cancer cells in athymic nude mice, *Cancer Res.*, 45, 584, 1985.
11. **Kameya, T., Shimosato, Y., Hayashi, H., and Tsumuraya, M.,** Growth and differentiation of hormone-producing human tumors in nude mice, in *Proc. 2nd Int. Workshop on Nude Mice*, Nomura, T., Ohsawa, N., Tamaoki, N., and Fujiwara, K., Eds., University of Tokyo Press, Tokyo, 1977, 405.
12. **Sharkey, F. E. and Fogh, J.,** Considerations in the use of nude mice for cancer research, *Cancer Met. Rev.*, 3, 341, 1984.
13. **Robinson, S. P., Langan-Fahey, S. M., and Jordan, V. C.,** Implications of tamoxifen metabolism in the athymic mouse for the study of antitumor effects upon human breast cancer xenografts, *Eur. J. Cancer Clin. Oncol.*, 25, 1769, 1989.
14. **Kitamura, Y., Uchida, N., Hayata, I., Yamaguchi, K., Okamoto, S., Narita, N., and Matsumoto, K.,** Effect of serial passage in female nude athymic mice on androgen dependency of Shionogi carcinoma 115, *Cancer Res.*, 40, 4781, 1980.
15. **Kondo, Y., Sato, K., Ueyama, Y., and Ohsawa, N.,** Serum sialyltransferase and liver catalase activity in cachectic nude mice bearing a human malignant melanoma, *Cancer Res.*, 41, 2912, 1981.
16. **Freudenthal, R. I., Leber, A. P., Emmerling, D. C., Kerchner, G. A., and Ovejera, A. A.,** Comparison of the drug metabolizing enzymes in the liver and kidneys from homozygous nude Swiss, heterozygous normal Swiss, homozygous normal Swiss and DBA/2 mice, *Res. Commun. Chem. Pathol. Pharmacol.*, 15, 267, 1976.

17. **Fodstad, Ø.**, Representativity of xenografts for clinical cancer. Tumor and host characteristics as variables of tumor take rate, in *Human Tumour Xenografts in Anticancer Drug Development*, Winograd, B., Pinedo, H. M., and Peckham, M. J., Eds., Springer-Verlag, Heidelberg, 1988, 15.
18. **Giovanella, B. C., Stehlin, J. S., Fogh, J., and Sharkey, F. E.**, Serial transplantation of human malignant tumors and their use in experimental chemotherapy, in *The Nude Mouse in Experimental and Clinical Research*, Fogh, J. and Giovanella, B., Eds., Academic Press, London, 1978, 163.
19. **Sirachy, J.**, An approach to the problem of heterogeneity of human tumour-cell populations, *Br. J. Cancer*, 39, 570, 1979.
20. **Dexter, D. L., Spremulli, E. N., Fligiel, Z., Barbosa, J. A., Vogel, R., VanVoorhees, A., and Calabresi, P.**, Heterogeneity of cancer cells from a single human colon carcinoma, *Am. J. Med.*, 71, 949, 1981.
21. **Fidler, I. J. and Hart, I. R.**, The origin of metastatic heterogeneity in tumors, *Eur. J. Cancer Clin. Oncol.*, 17, 487, 1981.
22. **Giovanella, B. C., Stehlin, J. S., Shepard, R. C., and Williams, L. J.**, Correlation between response to chemotherapy of human tumors in patients and in nude mice, *Cancer*, 52, 1146, 1983.
23. **Shimosato, Y., Kameya, T., Nagai, K., Hiroshashi, S., Koide, T., Hayashi, H., and Nomura, T.**, Transplantation of human tumors in nude mice, *J. Natl. Cancer Inst.*, 56, 1251, 1976.
24. **Giovanella, B. C. and Fogh, J.**, Present and future trends in investigations with the nude mouse as a recipient of human tumor transplants, in *The Nude Mouse in Experimental and Clinical Research*, Fogh, J. and Giovanella, B. C., Eds., Academic Press, New York, 1978, 281.
25. **Gould, V. E., Fodstad, Ø., Memoli, V. A., Warren, W. H., Dordi, L. E., and Johannessen, J. V.**, Neuroendocrine cells and neuroendocrine neoplasms of the skin, in *Evolution and Tumor Pathology of the Neuroendocrine System*, Falkner, S., Håkansson, R., and Sundler, F., Eds., Elsevier, Amsterdam, 1984, 545.
26. **Naito, S., von Eschenbach, A. C., Giavazzi, R., and Fidler, I. J.**, Growth and metastasis of tumor cells implanted into different organs of nude mice, *Cancer Res.*, 46, 4109, 1986.
27. **McLemore, T. L., Blacker, P. C., Gregg, M., Jessee, S. E., Alley, M. C., Abbott, B. J., Shoemaker, R. H., Litterst, C. C., Hubbard, W. C., Brennan, R. N., Fine, D. L., Eggleston, J. C., Mayo, J. G., and Boyd, M. R.**, Intrabronchial implantation: a method for the orthotopic propagation of human lung tumors in athymic nude mice, *Cancer Res.*, 47, 5132, 1987.
28. **Pickard, R. G., Cobb, L. M., and Steel, G. G.**, The growth kinetics of xenografts of human colorectal tumours in immune deprived mice, *Br. J. Cancer*, 31, 36, 1975.
29. **Mattern, J., Wayss, K., Haag, D., Toomes, H., and Volm, M.**, Different growth rates of lung tumors in man and their xenografts in nude mice, *Eur. J. Cancer Clin. Oncol.*, 46, 289, 1980.
30. **Baird, S. M., Beattie, G. M., Lannom, R. A., Lipsick, J. S., Jensen, F. C., and Kaplan, N. O.**, Induction of lymphoma in antigenically stimulated athymic mice, *Cancer Res.*, 42, 198, 1982.
31. **Beattie, G. M., Baird, S. M., Lipsick, J. S., Lannom, R. A., and Kaplan, N. O.**, Induction of T- and B-lymphocyte responses in antigenically stimulated athymic mice, *Cancer Res.*, 41, 2322, 1981.
32. **Tralka, T. S., Costa, J., Sindelar, W. F., Rabson, A., Gullino, P. M., Henson, E., Chu, E. W., and O'Connor, G. T.**, Spontaneous lymphosarcoma arising in a nude mouse: characterization *in vivo* and *in vitro*, *J. Natl. Cancer Inst.*, 58, 977, 1977.
33. **Sharkey, F. E. and Fogh, J.**, Incidence and pathological features of spontaneous tumors in athymic nude mice, *Cancer Res.*, 39, 833, 1979.
34. **Beattie, G. M., Knowles, A. F., Jensen, F. C., Baird, S. M., and Kaplan, N. O.**, Induction of sarcomas in athymic mice, *Proc. Natl. Acad. Sci. U.S.A.*, 79, 3033, 1982.
35. **Goldenberg, D. M. and Pavia, R. A.**, Malignant potential of murine stromal cells after transplantation of human tumors into nude mice, *Science*, 212, 65, 1981.
36. **Goldenberg, D. M. and Pavia, R. A.**, *In vivo* horizontal oncogenesis by a human tumor in nude mice, *Proc. Natl. Acad. Sci. U.S.A.*, 79, 2389, 1982.
37. **Soranzo, C., Ingrosso, A., Pratesi, G., Lombardi, L., Pilotti, S., and Zunino, F.**, Malignant transformation of host cells by a human small cell lung cancer xenografted into nude mice, *Anticancer Res.*, 9, 361, 1989.
38. **Sparrow, S., Jones, M., Billington, S., and Stace, B.**, The *in vivo* malignant transformation of mouse fibroblasts in the presence of human tumour xenografts, *Br. J. Cancer*, 53, 793, 1986.
39. **Gupta, V., Rajaraman, S., Gadson, P., and Costanzi, J. J.**, Primary transfection as a mechanism for transformation of host cells by human tumor cells implanted in nude mice, *Cancer Res.*, 47, 5194, 1987.
40. **Costa, J., Yee, C., and Rabson, A.**, *In vivo* transformation of mesenchymal murine cells in immuno-suppressed animals bearing human xenografts, in *Immune-deficient Animals*, Sordat, B., Ed., Karger, Basel, 1984, 230.
41. **Tveit, K. M. and Pihl, A.**, Do cell lines *in vitro* reflect the properties of the tumours of origin? A study of lines derived from human melanoma xenografts, *Br. J. Cancer*, 44, 775, 1981.
42. **Giovanella, B. C., Yim, S. O., Morgan, A. C., Stehlin, J. S., and Williams, L. J.**, Brief communication: metastases of human melanomas transplanted in "nude" mice, *J. Natl. Cancer Inst.*, 50, 1051, 1973.
43. **Giovanella, B. C., Stehlin, J. S., and Williams, L. J.**, Heterotransplantation of human malignant tumors in "nude" thymusless mice. II. Malignant tumors induced by injection of cell cultures derived from human solid tumors, *J. Natl. Cancer Inst.*, 52, 921, 1974.

44. **Sordat, B., Merenda, C., and Carrel, S.,** Invasive growth and dissemination of human solid tumors and malignant cell lines grafted subcutaneously to newborn nude mice, in *Proc. 2nd Int. Workshop on Nude Mice,* Nomura, T., Ohsawa, N., Tamaoki, N., Fujiwara, K., Eds., University of Tokyo Press, Tokyo, 1977, 313.
45. **Shoemaker, R. H., McLemore, T. L., Abbott, B. J., Fine, D. L., Gorelik, E., Mayo, J. G., Fodstad, Ø., and Boyd, M. R.,** Human tumour xenograft models for use with an *in vitro*-based, disease-oriented antitumour drug screening program, in *Human Tumour Xenografts in Anticancer Drug Development,* Winograd, B., Pinedo, H. M., and Peckham, M. J., Eds., Springer-Verlag, Heidelberg, 1988, 115.
46. **Giavazzi, R., Campbell, D. E., Jessup, J. M., Cleary, K., and Fidler, I. J.,** Metastatic behaviour of tumor cells isolated from primary and metastatic human colorectal carcinomas implanted into different sites in nude mice, *Cancer Res.,* 46, 1928, 1986.
47. **Arguello, F., Baggs, R. B., Johnstone, L., Duerst, R. E., and Frantz, C. N.,** Role of bone marrow stroma in bone metastasis, *Proc. Am. Assoc. Cancer Res.,* 31, 64, 1990.
48. **Fodstad, Ø., Aamdal, S., McMenamin, M., Nesland, J. M., and Pihl, A.,** A useful new experimental metastasis model in athymic, nude mice. Characteristics of the LOX human malignant melanoma, *Int. J. Cancer,* 41, 442, 1988.
49. **Fodstad, Ø., Kjønniksen, I., Aamdal, S., Nesland, J. M., Boyd, M. R., and Pihl, A.,** Extrapulmonary, tissue-specific metastasis formation in nude mice injected with FEMX-I human melanoma cells, *Cancer Res.,* 48, 4382, 1988.
50. **Kjønniksen, I., Storeng, R., Pihl, A., McLemore, T. L., and Fodstad, Ø.,** A new human tumor metastasis model in atymic nude rats, *Cancer Res.,* 49, 5148, 1989.
51. **Kjønniksen, I., Nesland, J. M., Pihl, A., and Fodstad, Ø.,** A nude rat model for studying metastasis of human tumor cells to bone and bone marrow, *J. Natl. Cancer Inst.,* 82, 408, 1990.
52. **Fodstad, Ø., Høifødt, H. K., and Kjønniksen, I.,** Host-dependent differences in human tumor metastasis patterns in athymic, nude mice and rats — relationship to soil factors, *Proc. Am. Assoc. Cancer Res.,* 31, 68, 1990.
53. **Kerbel, R. S., Man, M. S., and Dexter, D.,** A model of human cancer metastasis: extensive spontaneous and artificial metastasis of a human pigmented melanoma and derived sublines in nude mice, *J. Natl. Cancer Inst.,* 72, 93, 1984.
54. **Fodstad, Ø., Tveit, K. M., Rofstad, E. K., and Pihl, A.,** Spontaneous alteration in growth rate of two human melanoma xenografts. Concurrent change in chemosensitivity, *Eur. J. Cancer Clin. Oncol.,* 19, 1175, 1983.
55. **Wennerberg, J.,** Changes in growth pattern of human squamous-cell carcinomas of the head and neck during serial passages in nude mice, *Int. J. Cancer,* 33, 245, 1984.
56. **Horvath, G., Alm, P., Tropé, C., and Ferno, M.,** Changes in tumor morphology in one human endometrial carcinoma heterotransplanted into nude mice, *In Vivo,* 3, 113, 1989.
57. **Donhuijsen, K., Budach, V., Van Beuningen, D., and Schmidt, U.,** Instability of xenotransplanted soft tissue sarcomas, *Cancer,* 61, 68, 1988.
58. **Reid, L. M., Holland, J., Jones, C., Wolf, B., Niwayama, G., Williams, R., Kaplin, N. O., and Sato, G.,** Some of the variables affecting the success of transplantation of human tumors into the athymic nude mouse, in *Proc. Symp. on the Use of Athymic (Nude) Mice in Cancer Res.,* Houchens, D. P. and Ovejera, A. A., Eds., Gustav Fischer, New York, 1978, 107.
59. **Povlsen, C. O., Rygaard, J., and Fogh, J.,** Long-term growth of human tumors in nude mice: evaluation of stability, in *The Nude Mouse in Experimental and Clinical Research,* Fogh, J. and Giovanella, B. C., Eds., Academic Press, New York, 1982, 79.
60. **Fodstad, Ø., Olsnes, S., and Pihl, A.,** Inhibitory effect of abrin and ricin on the growth of transplantable murine tumors and of abrin on human cancers in nude mice, *Cancer Res.,* 37, 4559, 1977.
61. **Rofstad, E. K., Fodstad, Ø., and Lindmo, T.,** Growth characteristics of human melanoma xenografts, *Cell Tissue Kinet.,* 15, 545, 1982.
62. **Houghton, J. A. and Taylor, D. M.,** Maintenance of biologic and biochemical characteristics of human colorectal tumours suring serial passaging in immune-deprived mice, *Br. J. Cancer,* 37, 199, 1978.
63. **Gazdar, A. F., Carney, D. N., Russell, E. K., Sims, H. L., Baylin, S. B., Bunn, P. A., Guccion, J. G., and Minna, J. D.,** Establishment of continuous, clonable cultures of small-cell carcinoma of the lung which have amine precursor uptake and decarboxylation cell properties, *Cancer Res.,* 40, 3502, 1980.
64. **Povlsen, C. O., Visfeldt, J., Rygaard, J., and Jensen, G.,** Growth patterns and chromosome constitutions of human malignant tumours after long-term serial transplantation in nude mice, *Acta Path. Microbiol. Scand.,* 83, 709, 1975.
65. **Spang-Thomsen, M. and Visfeldt, J.,** Growth curves and radiation-induced synchronization of a human malignant melanoma transplanted to nude mice, in *Proc. 2nd Int. Workshop on Nude Mice,* Nomura, T., Ohsawa, N., Tamaoki, N., and Fujiwara, K., Eds., University of Tokyo Press, Tokyo, 1977, 327.
66. **Fiebig, H. H., Schuchhardt, C., Henss, H., Fiedler, L., and Löhr, G. W.,** Comparison of tumor response in nude mice and in the patients, *Behring Inst. Mitt.,* 74, 343, 1984.
67. **Visfeldt, J., Povlsen, C. O., and Rygaard, J.,** Chromosome analyses of human tumours following heterotransplantation to the mouse mutant nude, *Acta Path. Microbiol. Scand.,* 80, 169, 1972.

68. **Mach, J. P., Carrel, S., Merenda, C., Sordat, B., and Cerottini, J. C.,** *In vivo* localisation of radiolabelled antibodies to carcinoembryonic antigen in human colon carcinoma grafted into nude mice, *Nature (London),* 248, 704, 1974.
69. **Hwang, K., Fodstad, Ø., Morgan, A. C., and Oldham, R. K.,** Radiolocalization of xenografted human malignant melanoma by a monoclonal antibody (9.2.27) to a melanoma-associated antigen in nude mice, *Cancer Res.,* 45, 4150, 1985.
70. **Zalutsky, M. R., Garg, P. K., Friedman, H. S., and Bigner, D. D.,** Labeling monoclonal antibodies and $F(ab')_2$ fragments with the α-particle-emitting nuclide astatine-211: preservation of immunoreactivity and *in vivo* localizing capacity, *Proc. Natl. Acad. Sci. U.S.A.,* 86, 7149, 1989.
71. **Colapinto, E. V., Zalutsky, M. R., Archer, G. E., Noska, M. A., Friedman, H. S. Carrel, S., and Bigner, D. D.,** Radioimmunotherapy of intracerebral human glioma xenografts with ^{131}I-labeled $F(ab')_2$ fragments of monoclonal antibody MEL-14, *Cancer Res.,* 50, 1822, 1990.
72. **Godal, A., Bruland, Ø., Haug, E., Aas, M., and Fodstad, Ø.,** Unexpected expression of the 250 kD melanoma-associated antigen in human sarcoma cells, *Br. J. Cancer,* 53, 839, 1986.
73. **Bruland, Ø., Fodstad, Ø., Skretting, A., and Pihl, A.,** Selective radiolocalization of two radiolabelled anti-sarcoma monoclonal antibodies in human osteosarcoma xenografts, *Br. J. Cancer,* 56, 21, 1987.
74. **Kamel-Reid, S. and Dick, J. E.,** Engraftment of immune-deficient mice with human hematopoietic stem cells, *Science,* 242, 1706, 1988.
75. **Kamel-Reid, S., Letarte, M., Sirard, C., Doedens, M., Grunberger, T., Fulop, G., Freedman, M. H., Phillips, R. A., and Dick, J. E.,** A model of human acute lymphoblastic leukemia in immune-deficient SCID mice, *Science,* 246, 1597, 1989.
76. **Reichman-Fried, M., Hardy, R. R., and Bosma, M. J.,** Development of B-lineage cells in the bone marrow of scid/scid mice following the introduction of functionally rearranged immunoglobulin transgenes, *Proc. Natl. Acad. Sci. U.S.A.,* 87, 2730, 1990.
77. **Stanbridge, E. J. and Ceredig, R.,** Growth-regulatory control of human cell hybrids in nude mice, *Cancer Res.,* 41, 573, 1981.
78. **Fischinger, P. J. and DeVita, V. T.,** Governance of Science at the National Cancer Institute: perceptions and opportunities in oncogene research, *Cancer Res.,* 44, 4693, 1984.
79. **Bos, J. L.,** *Ras* oncogenes in human cancer: a review, *Cancer Res.,* 49, 4689, 1989.
80. **Meeting Report:** Gene regulation and oncogenes, Sharp, P. A., *Cancer Res.,* 49, 2188, 1989.
81. **Chauvin, C., Jacrot, M., Riondel, J., Brambilla, E., Foote, A. M., Brambilla, C., and Benabid, A. L.,** Amplification of oncogenes in lung carcinoma grafted in nude mice, *Anticancer Res.,* 9, 449, 1989.
82. **Sporn, M. B. and Roberts, A. B.,** Autocrine growth factors and cancer, *Nature (London),* 313, 745, 1985.
83. **Hansen, M. F., Koufos, A., Gallie, B. L., Philips, R. A., Fodstad, Ø., Brøgger, A., Gedde-Dahl, T., and Cavenee, W. K.,** Osteosarcoma and retinoblastoma: a shared chromosomal mechanism revealing recessive predisposition, *Proc. Natl. Acad. Sci. U.S.A.,* 82, 6216, 1985.
84. **Fearon, E. R. and Vogelstein, B.,** A genetic model for colorectal tumorigenesis, *Cell,* 61, 759, 1990.
85. **Weinberg, R. A.,** The retinoblastoma gene and cell growth control, *Trends Biochem. Sci.,* 15, 299, 1990.
86. **Shew, J. Y., Ling, N., Yang, X., Fodstad, Ø., and Lee, W. H.,** Antibodies to the retinoblastoma gene product detect specific loss of $pp110^{RB}$ in osteosarcomas and synovial sarcomas, *Oncogene Res.,* 1, 205, 1989.
87. **Reissmann, P. T., Simon, M. A., Lee, W. H., and Slamon, D. J.,** Studies of the retinoblastoma gene in human sarcomas, *Oncogene,* 4, 839, 1989.
88. **Harbour, J. W., Lai, S. L., Whang-Peng, J., Gazdar, A. P., Minna, J. D., and Kaye, F. J.,** Abnormalities in structure and expression of the human retinoblastoma gene in small cell lung cancer, *Science,* 241, 353, 1988.
89. **Lee, E. Y. H., To, H., Shew, J. Y., Bookstein, R., Scully, P., and Lee, W.-H.,** Inactivation of the retinoblastoma susceptibility gene in human breast cancers, *Science,* 241, 218, 1988.
90. **Huang, H. J. S., Yee, J. K., Shew, J. Y., Chen, P. L., Bookstein, R., Freidmann, T., Lee, E. Y. H., and Lee, W. H.,** Suppression of the neoblastic phenotype by replacement of the RB gene in human cancer cells, *Science,* 242, 1563, 1988.
91. **Bookstein, R., Shew, J. Y., Chen, P. L., Scully, P., and Lee, W. H.,** Suppression of tumorigenicity of human prostate carcinoma cells by replacing a mutated RB gene, *Science,* 247, 712, 1990.
92. **Steeg, P. S., Bevilacqua, G., and Kopper, L.,** Evidence for a novel gene associated with low tumor metastatic potential, *J. Natl. Cancer Inst.,* 80, 200, 1988.
93. **Ebralidze, A., Tulchinsky, E., and Grigorian, M.,** Isolation and characterization of a gene specifically expressed in different metastatic cells and whose deduced gene product has a high degree of homology to a Ca^{2+} binding protein family, *Genes Dev.,* 3, 1086, 1989.

Chapter 21

LIMITATIONS FOR TREATMENT STUDIES

Boudewijn J. M. Braakhuis and Guus A. M. S. van Dongen

TABLE OF CONTENTS

I. INTRODUCTION

The nude mouse-human tumor model is presently considered to be a powerful tool in the development of new cancer therapies. This model has gained an increasing amount of interest since it was felt that murine tumor systems gave unsatisfactory results.[1] A major advantage is thought to be its capacity to select a new or to improve an existing therapy for a certain tumor type.[2] For this purpose, a panel of hum an t mor xenografts derived from a particular tumor type is being used so as to mimic a group of patients with that type of tumor. The optimism about the value of the model is based on the notion that treatment response obtained in panels of xenografts representing various tumor types is comparable to that in clinical practice.[2,3] However, taking into account that relevant characteristics such as cell cycle parameters, growth rate, metastatic spread, invasive properties, origin of stroma, pharmacokinetics, and metabolism in the tumor-bearing host are different from the situation in the cancer patient,[3] it is to be expected that some experimental results cannot be translated directly to man. Some investigators have indeed reported a lack of correlation between xenograft and patient response.[4-6] The aim of this chapter is to analyze the limitations of the nude mouse xenograft model for therapy studies. The deficiencies of the model should become the variables that define either more appropriate models or set the boundary conditions of the existing model.

II. PREREQUISITES FOR TREATMENT STUDIES

A. TUMOR GROWTH

Reliable experiments in nude mice have to be performed in a clean environment under controlled atmospheric circumstances. Special sterile procedures have to be applied to protect the animals against infections. Under suboptimal conditions and in case of infections, tumors may fail to grow[7] or the activated residual immune system may negatively influence the therapeutic results. Therefore, the financial problem should first be overcome to ensure high-quality data from optimal housing and handling conditions.

For treatment studies using s.c. grown xenografts in the nude mouse, a sufficient number of well-growing tumor lines has to be available. A xenograft established from a single patient's tumor should be in serial transfer over more than four passages to be defined as a tumor line. The "take rate" in the first (man to mouse) passage varies considerably between tumor types. As examples, colorectal and ovarian cancers have a high take rate, while breast and prostate cancers are much more difficult to grow.[3] In general, only half of all transplants growing in the first passage become serially transplantable lines.[3] One still has to make certain that the number of xenografts in each passage is high enough to allow therapy experiments. Sometimes, a tumor line loses its growth potential.[8] Whether this is the consequence of the emergence of a cell population with new characteristics or a change in host factors is not known.

B. SELECTION OF TUMOR SUBTYPES

With the purpose to establish a panel of tumor lines reflecting various characteristics of the corresponding patients' tumors, one has to realize that certain tumor subtypes may have a growth advantage and, consequently, will dominate the panel. Tumor characteristics generally associated with a relatively high malignant behavior have been recognized to be related with a better transplantability (Table 1). In order to avoid a bias in the composition of the tumor panel, it is almost obligatory to investigate some of the characteristics as mentioned in Table 1. Depending on the tumor type involved, the investigator may replace tumor lines to restore the desired balance.

TABLE 1
Tumor Characteristics Related to Transplantability

Favorable	Unfavorable	Ref.
Lymph node metastasis	Primary tumor	9–11
Aneuploid cells	Euploid cells	12, 13
Growth factor autonomy	Growth factor dependency	14
Vascular invasion	Lack of vascular invasion	15
Many mitotic figures	Few mitotic figures	16
Poorly differentiated	Well differentiated	11, 17, 18
Rapidly growing	Slowly growing	19

III. INSTABILITY OF TUMOR LINES

A. BIOLOGICAL INSTABILITY

Tumor heterogeneity has become an accepted explanation for the failure of cancer therapy.[20] This heterogeneity within an individual tumor refers to the coexistence of multiple clonal neoplastic subpopulations which are capable of maintaining their genetic identity. Another aspect in heterogeneity may be at least as important, the so-called "secular heterogeneity", which is related to the variability in structure and organization of cancer tissue.[20] Heterogeneity has been found in a number of cell characteristics, such as the ability to metastasize, sensitivity to therapy, and the propensity to develop further genetic variation. Its existence has been described not only in experimental tumors, but also in human tumors.[21] The sequential appearance of subpopulations that genetically alter with increasing extent is thought to be the underlying mechanism of tumor progression, defined as the tendency of a tumor to become more aggressive during its lifetime.[22] Also, human tumor xenografts consist of various tumor cell populations[21] and they are likely to be subject to the increasing inclination to escape growth regulation. Tumor progression can have consequences for the use of a tumor panel for treatment studies in the sense that response can become unstable during the xenograft lifetime. In addition, the secular heterogeneity, referring to differences in tissue structure and organization between man and mouse (e.g., vascularization) may have significant consequences for therapy. This form of heterogeneity will be discussed in Section IV.

Changes in tumor growth rate during serial transplantation may influence the results of therapy. Clinical and experimental data indicate that an increase in growth rate is associated with higher sensitivity to anticancer drugs.[23,24] Human tumor xenografts commonly show a decrease in their volume-doubling time during the first passages.[23,25,26] Thereafter, the growth rate frequently stabilizes and each tumor line has its individual characteristic volume-doubling time, which normally ranges between a few days and 2 weeks.[27] However, sudden changes in growth rate can occur in later passages, which may have an implication for the degree of response. In 2 melanoma lines, the growth rate was reported to increase between passage 11 and 17, and 6 and 10, respectively; in one of these lines, a concomitant change in response to drugs could be demonstrated.[28] Baisch et al.[29] showed a similar growth acceleration in 2 out of 30 renal cell cancer lines, and this was observed between passages 12 and 44.

The histological appearance of xenografts appears to be an unstable phenotypic marker. Many authors have described a change in the degree of differentiation, where tumors became less differentiated as the passage number increased (Table 2). Instability is most prominent in the early passages, but changes in later passages have also been reported.[30] Whether an unstable histological appearance will affect the therapeutic outcome is not known, although in one report a coincidental change in response has been mentioned.[30]

TABLE 2
Histological Changes in Human Tumor Xenografts during Serial Transplantation

Tumor (line)	Frequency[1]	Change	Passage number[2]	Ref.
Various	9/52	Decrease differentiation	Between 2 and 6	18
Various	11/52	Increase differentiation	Between 2 and 6	18
Various	"Most"	Increase amount of stroma	Between 2 and 6	18
Various	9/52	Increase amount of necrosis	Between 2 and 6	18
Sarcoma (SAS)	1/1	Decrease differentiation	Between 16 and 26	30
Sarcoma (KUSHI)	1/1	Biphasic phenotype	Between 1 and 2	30
Astrocytoma	2/7	Decrease differentiation	"Early"	31
Sarcoma	6/21	Decrease differentiation	"Early"	32
Sarcoma	"Most"	Increase amount stroma	"Early'	32
Sarcoma	"Most"	Increase amount necrosis	"Early"	32
Lung cancer	"Most"	Decrease differentiation	From 2 on	33
Childhood	1/8	Increase amount stroma	Between 1 and 10	34
Childhood	8/8	Decrease differentiation	Between 1 and 10	34
Glioblastoma	1/1	New phenotype	"Early"	35
Breast cancer[3]	"Some"	Decrease differentiation	"Early"	36
Sarcoma	2/12	Decrease differentiation	"Early"	37

[1] Number of cases with changes/number analyzed.
[2] Indicates in which passage(s) the change has occurred.
[3] Xenografts transplanted in immunesuppressed mice.

DNA flow cytometry has been performed to assess the genetic stability of xenograft lines. In a minority of cases, a change in the DNA content per cell could be detected (Table 3). These genetically altered populations often had more chromosomal material, which is considered to be a manifestation of tumor progression.[22]

Variability in the expression of tumor markers may influence the results of hormone or specific antibody therapy. In later passages, a loss of androgen receptors may occur,[43] as well as the disappearance of oncofetal markers (i.e., AFP,[44] HCG,[45] and CEA[46,47]) from tissue and serum.

An ultimate change in a tumor line in serial tranfer is the possible overgrowth of human tumor cells by mouse cells. This phenomenon has, of course, serious consequences for further therapeutic experiments (see also Chapters 8 and 20).

With respect to the data available on the biological instability of xenografts, it can be concluded that these tissues, like patients' tumors, consist of different cell populations with ultimate consequences for their predicitive capacity in the development of new therapies. Upon continuous propagation, a xenograft will be more prone to selective pressure than a tumor in a patient. First, transplantation by itself may be a selective process, as an investigator will be inclined to choose the best-growing tumor for further transplantation. Second, after implantation of tumor tissue, only a few cells (possibly a subpopulation) will survive, which are capable of proliferating and forming the subsequent xenograft.[48] Because of the relative instability in the first three to five passages, it is advisable to use tumors from later passages for treatment experiments. Again, relevant changes in characteristics may still occur in later passages, making it necessary to monitor these lines for defined stability criteria, including their chemosensitivity profile.

B. RESPONSE INSTABILITY

With respect to the usefulness as a drug screening system, the maintenance of the sensitivity profile of a particular human tumor line should be guaranteed. A very limited number of studies have addressed the problem of response stability for a xenograft in serial transfer. In the reports

TABLE 3
Clonal Evolution during Serial Transplantation of Human Tumor Xenografts

Tumor type (name of line)	Frequency	Changes	Passage number	Ref.
Melanoma	0/1	No changes	Between 11 and 17	28
Renal cell cancer	4/30	Increase DNA-index	Between 12 and 44	29
Sarcoma (SAS)	1/1	Decrease number chromosomes	Between 16 and 26	30
Sarcoma (KUSHI)	1/1	Decrease number chromosomes	Between 14 and 18	30
Sarcoma	5/19	New sublines	"Early"	32
Colon cancer	1/1	DNA-index from 1.3 to 2.6	56	38
Osteosarcoma	1/6	DNA-index from 2.8 to 5.6	3	39
Skin	1/2	Hypo- to tetraploid	20	40
Endometrial cancer	1/1	Extra subline	5	41
Ovarian cancer	1/12	DNA-index from 1.7 to 2.8	Not indicated	42
Total	16/74			

TABLE 4
Comparison of Drug Response in Xenografts in Different Passages

Tumor type (name of line)	Response change[1]	Passage number	Drug[2]	Test system	Ref.
Melanoma (HMX-GE)	yes (– to +)	3 and 31	CCNU, DTIC	*In vivo*	28
Sarcoma (SAS)	yes (+ to –)	16 and 26	VCR, DOX, BLM	*In vitro*	30
Sarcoma (KUSHI)	yes (– to +)	3 and 18	Ara-C, MMC, DOX	*In vitro*	30
Rat adenocarcinoma (R3230AC)	yes (+ to –)	6 and 20	l-PAM	*In vivo*	49
Mesothelioma (AP)	yes (– to +)	14 and 18	MTX	*In vitro*	50
HNSCC[3] (HEp-2)	yes (+ to –)	2 and 8	CDDP	*In vitro*	51
Melanoma (HMX-GE)	no	3 and 31	VBL, Abrin	*In vivo*	28
Sarcoma (SAS)	no	16 and 26	MMC, Ara-C	*In vitro*	30
Sarcoma (KUSHI)	no	3 and 18	VCR, BLM,	*In vitro*	30
Ovarian cancer[4]	no	n.m.	CDDP, Melphalan	*In vivo*	42
Rat adenocarcinoma (13762)	no	6 and 20	l-PAM	*In vivo*	49
Mesothelioma (AP)	no	14 and 18	4HPCY, 5-FU, VBL	*In vitro*	50
HNSCC[3] (HEp-2)	no	8 and 16	DOX, BLM, 5-FU, VCR	*In vitro*	51

1 A change in response is recorded: + = sensitive, – = not sensitive.
2 Abbreviations: Ara-C, cytosine arabinoside; BLM, bleomycin; CDDP, cisplatin; DOX, doxorubicin; l-PAM, l-phenylalanine mustard; MMC, mitomycin C; MTX, methotrexate; VBL, vinblastine; VCR, vincristine; 4HPCY, 4-hydroperoxycyclophosphamide; 5-FU, 5-fluorouracil.
3 HNSCC, head and neck squamous cell carcinoma.
4 Results of five individual ovarian cancer lines; n.m.: not mentioned.

available, most investigators have analyzed variability in drug response *in vitro,* comparing different xenograft passages (Table 4). In 11 cases out of a total of 35 drug/xenograft combinations studied, either *in vitro* or *in vivo,* a variation in response was observed. There was no tendency indicating that the tumor became more sensitive or resistant in later passages. In two experiments, the change in response was reported to coincide with a change in DNA-pattern;[30,50] but in another case, gross modification of the chromosomal content could not be detected.[28] Because response instability seems to occur relatively frequently, this phenomenon should be acknowledged promptly by the reestablishment of such a tumor line from material frozen in early passages.

TABLE 5
Lack of Correlation in Drug Response Between Patient Tumor and Xenograft

Tumor type	Line	Drug	Xenograft response: Nature	Xenograft response: Criterion[1]	Patient Response[2]	Ref.
HNSCC[3]	HNX-TI	Methotrexate	Insensitive	SGD < 1	CR	4
HNSCC	HNX-W	Methotrexate	Insensitive	SGD < 1	PR	4
Lung	158	5-Fluorouracil	Insensitive	P	NC	5
Liver	264	5-Fluorouracil	Insensitive	P	PR	5
Ovary	PP	Cisplatin	Insensitive	SGD < 1	CR	28
Ovary	JC	Cisplatin	Sensitive	SGD > 2	P	28
Ovary	JN	Cisplatin	Sensitive	SGD > 2	P	28
Melanoma	JA I	Cisplatin	Insensitive	SGD < 2	S	52
Melanoma	MO	Cisplatin	Sensitive	SGD > 2	P	52
Melanoma	NI	Dacarbazine	Sensitive	SGD > 2	P	52
Melanoma	DA I	Dacarbazine	Sensitive	SGD > 2	P	52
Lung[4]	HX72	Mitomycin C	Sensitive	SGD > 2	P	53
Lung[4]	HX95	Mitomycin C	Sensitive	SGD > 2	P	53

[1] SGD stands for specific growth delay; i.e., the number of doubling times of the control group that is gained by the treatment. A tumor line may be considered sensitive if SGD exceeds 2. P, progression: the product of two tumor diameters is more than 125% of initial value.

[2] Patient tumor response as used in References 4, 5, 28, and 53: CR, complete remission; PR, partial remission; NC, no change; P, progression. Tumor response as defined in Reference 52: S (= sensitive) is defined as NC or better.

[3] HNSCC, head and neck squamous cell carcinoma.

[4] These xenografts were maintained in immunosuppressed mice.

IV. XENOGRAFT VS. PATIENT TUMOR RESPONSE

A. LACK OF CORRELATION

Positive correlations between drug responses in the xenografts and those in the original patients' tumors have brought about much enthusiasm on the use of the nude mouse-human tumor model to study new cancer therapies. Recently, Mattern et al.[3] have reviewed all cases reported in the literature. Applying strict criteria for activity (see Table 5), they concluded that, in the majority of comparisons (125 out of 138), the response in xenografts and that in patients appeared to correlate well. It must be stressed, however, that in two-thirds of the cases, the positive correlation was based on the absence of a response, both in man and mouse. Upon analysis of the 40 comparisons where sensitivity was observed, the responses were actually different in 13 cases. Table 5 shows the details on these 13 comparisons and the data do not indicate that the model would either select false positive or false negative drugs. In fact, there appears to be a tendency for the nude mouse model to underestimate the activity of antimetabolites, contrary to the overestimation of effects from agents causing DNA damage.

With respect to tumor heterogeneity, a xenograft may not represent all cell populations in the original tumor tissue and therefore not reflect the clinical response. To circumvent this problem, another approach may be used which is based on the comparison of the efficacy of a drug in a particular tumor type in patients with the response of the same drug in a panel of xenografts from that tumor type. When drugs are ranked according to their activity, the model gives a fair reflection of the clinical results.[3] However, the number of lines with sensitivity is higher than would be expected from clinical data, and, more importantly, the model can dramatically overestimate the activity of certain drugs.[3] As an example, a too-high activity has been found for the nitrosourea derivatives in xenografts of colon and small cell lung cancer. Furthermore, the activity of 5-fluorouracil seems too low in colon cancer lines.[3] These negative correlations in responses shown for individual, as well as panel comparisons, justify an analysis of the variables that define the xenograft and patient responses. Recognition of the mechanisms underlying the differences in responses is essential for a more rational use of the model.

B. VARIABLES OF INFLUENCE ON RESPONSE

1. Methodological Aspects

Differences in response criteria are important variables in the assessment of drug effects in xenografts and patients. Generally, the definition of a response in the patient is based on the scoring of the percentage of tumor size reduction. In xenografts, a substantial delay in growth is often considered to be a good parameter of activity.[4,52] Investigators have not yet agreed upon the best definition of a true response in xenografts.[5,52] A large-scale comparative study of drugs tested in panels of xenografts, as well as in phase II trials in patients, is needed to elucidate methodological problems. Moreover, biological aspects with specificity for a particular tumor type should be taken into consideration to define response criteria in the nude mouse model.

2. Tumor Aspects

Xenografts usually show a faster growth than patients' tumors,[33] which can be attributed to an increase in growth fraction and/or a coinciding decrease in cell loss.[54] At the time of treatment, the size of the xenografts is often small and in log-phase of growth. In patients, treatment is usually started at the time tumor lesions are larger than 1 cm and poorly vascularized. As a result of rapid growth in xenografts, the model is likely to overestimate the activity of cell cycle-specific drugs. In this respect, the high efficacy of nitrosoureas in this model may be, at least partially, explained.[3]

Tumor blood flow is an important variable for oxygen supply, transport of nutrients, and drugs, and may thus influence response to therapy. Only a few comparative studies have been performed on tumor vascularization in man and nude mice. Kallinowski et al.[55] have shown that blood flow characteristics in human tumor xenografts in nude rats are comparable to the values determined in malignant and normal tissues in patients. However, it has been demonstrated that the median distance between blood vessels and tumor cells was much longer in human tumors than in xenografts.[56] Obviously, this latter effect may have consequences for the efficacy of chemotherapy or radiotherapy. More data on tumor blood flow have to be gathered to allow for conclusions.

3. Pharmacological Aspects

The choice of the host may be of influence on the antitumor response. Strain, sex, age, and nutritional status of the mice have a pronounced effect on the pharmacokinetic profile of doxorubicin and cyclophosphamide.[57] Furthermore, the route of administration,[58] the presence of a tumor, and even the tumor type involved may interfere with drug pharmacokinetics.[57] The extent of drug toxicity is also determined by variables, such as the presence of a tumor,[4] the strain of mice,[58] and the route of administration.[59] The therapeutic efficacy can even vary with the hour of injection,[60] which has been shown for 5-fluorouracil by Peters et al.[61] Many data showing variability in normal mice may be of relevance in nudes. Slight, inconsistent differences in enzyme activity between normal and nude mice have been reported.[62,63] Nude mice are able to tolerate at least similar or higher doses of certain antitumor agents when compared to normal mice,[64] suggesting a higher drug clearance or tolerance in nude mice.

Schabel et al.[65] have already pointed out that many differences exist between man and animals with respect to drug pharmacokinetics, metabolism, and elimination and, thus, antitumor activity and host toxicity. At this moment, the best pharmacological parameter which correlates with response (and toxicity) is thought to be the area under the concentration vs. time curve (CxT).[66] Exceptions to this rule have been reported which suggest that the peak or a certain threshold concentration may also be of importance.[66] Relationships between man and mouse

TABLE 6
Dose Ratio (mg/m² Basis) and CxT Ratio for Human MTD to Mouse LD_{10}

Drug	Dose ratio[1]	CxT ratio[1]	Mouse	Ref.
Doxorubicin	5.0	0.8	CDF1	67
PALA	2.8	3.3	CDF1	67
F-ara-AMP	0.03–0.1	0.1	CDF1	67
Diaziquone	1.0	1.0	CDF1	67
AMSA	0.9	1.3	CDF1	67
Pentostatin	0.7	1.1	CDF1	67
JM-40	4.0	1.3	CDF1	68
Carboplatin	1.1	1.0	CDF1	68
Methotrexate	6.6	>5[2]	Balb/c, nude	6
5-Fluorouracil	0.4	>1[2]	Balb/c, nude	6
Mitomycin C	0.3	<1[2]	Balb/c, nude	69
Vincristine	0.3	<1[2]	Balb/c, nude	69
Cyclophosphamide	1.0	<1[2]	Balb/c, nude	69
ACNU	0.7	<1[2]	Balb/c, nude	69

[1] Man to mouse ratio.
[2] Estimated data from corresponding report.

concerning the treatment dose and their CxT-values have been analyzed with the aim to improve the efficiency of phase I clinical trials.[67,68] For most drugs, it appeared that CxT values are quite similar in man and mouse, when the drug is administered at the maximum tolerable dose (MTD) in man, and at the LD_{10} in mice, both expressed as milligrams per square meter (mg/m²) (Table 6). As examples, for doxorubicin and JM-40, man to mouse dose ratios are relatively high, whereas the CxT values are quite similar. For both drugs, it was shown that in man the peak plasma concentrations are higher and their elimination more rapid.[67,68] Also, flavone acetic acid is eliminated more rapidly in man than in the mouse.[70] Other drugs, like acivicin[71] and methotrexate,[6] disappear faster from the blood in mice than in man. These latter two drugs follow the standard rule saying that the rate at which a drug disappears from the blood is inversely related to the size of a mammal.[72] Table 6 further shows a high CxT ratio for PALA and a low value for fludarabine phosphate (F-ara-AMP), both drugs being antimetabolites. These differences may be explained by assuming that PALA is more and F-ara-AMP is less toxic to murine than to human tissue. Here, pharmacokinetic profiles may be related to unrevealed intracellular detoxifications mechanisms, which appear to be species-specific.

To improve the predictability of the nude mouse-human tumor model for clinical effectiveness of a particular drug, Inaba et al.[6] have stressed the importance of choosing the "rational" dose in the mouse. These investigators have attempted to find a dose equivalent to the CxT value calculated in patients treated with that drug at the MTD.[6,69] In the nude mouse, the CxT values at LD_{10} appeared to be much higher for mitomycin C, ACNU, cyclophosphamide, and vincristine than those observed in patients at MTD. On the contrary, the CxT values of methotrexate and 5-flurouracil were considerably lower in the nude mouse than in man and specifically so for methotrexate.[6] These findings are a possible explanation for the lack of correlation between patient and xenograft response; that is, the model estimates the activity of antimetabolites too low and of alkylating agents too high (Table 5, and Reference 3). Based on their results, Inaba et al.[6] concluded that not the LD_{10}, but a dose with pharmacokinetic equivalence to the clinical dose may eventually result in a better predictability of the nude mouse-human tumor model.

An alternative route of drug metabolism in nude mice may be another reason for a difference in therapeutic outcome. For some agents, a species-dependent variability in drug metabolism has been reported, although this did not necessarily interfere with therapeutic efficacy. So, in mouse plasma, only two of the six metabolites of epidoxorubicin were detectable that are present in human plasma.[73] A flavone acetic acid metabolite was found in mouse plasma that is undetectable in human plasma, whereas two other metabolites could only be revealed in man.[70]

Another source of error causing difficulties in the translation of therapy results from mouse to man is the concentration of naturally occurring metabolites in plasma. These metabolites may prevent cell death by interfering with the antitumor activity of drugs such as antimetabolites. As an example, the concentration of folates in the plasma may inhibit the therapeutic activity of antifolate drugs such as methotrexate. In C57BL6 euthymic mice, folate (mainly methyltetrahydrofolate) levels of 7.8 m*M* are reported;[74] whereas, in man, levels of 2 to 30 n*M* are considered to be in the normal range.[75] In case folate levels are also high in nude mice, it is not yet clarified why these animals tolerate only a small dose of methotrexate relative to that in patients (Table 6). On a milligram per square meter basis, man can be treated with a dose six times higher than Balb/c nude mice and, accordingly, the CxT is much higher in man.[6] It can be speculated that murine cells have a lower capacity to be rescued by the circulating folates as compared to human cells. This could be an alternative explanation for the lack of effect of methotrexate in human head and neck cancer xenografts.[4]

Besides high concentrations of circulating metabolites, tissue levels of certain naturally occurring metabolites can also be a source of error. Large reservoirs of uridine were described for some organs of C57BL6 euthymic mice.[76] These reservoirs are thought to be important in preventing the antitumor activity of antimetabolites like PALA and pyrazofurin, but could also form the basis for tissue-specific 5-fluorouracil toxicity. In this respect, tissues with high pools of uridine (such as liver, kidney, and spleen) are less susceptible to 5-fluorouracil-induced toxicity than tissues with low uridine pools such as gut and tumor.[76] Recently, preliminary data in man demonstrated plasma uridine levels to be similar to those in mice; the concentration of uridine in human liver, however, appeared to be twofold lower.[77] These data illustrate that plasma and tissue levels of normally occurring metabolites may differ between mouse and man and therefore cause difficulties in the translation of drug responses.

4. Immunological Aspects

A generally accepted theory is the role of the immune system in the development and growth of a tumor. The immune system per se can reject a tumor or can be stimulated to do so, and it also has the ability to interfere with cancer therapies. The nude mouse may express a significant immune response to implanted tumors,[78] and this response may have an additional positive effect on the treatment results, as has been shown for radiotherapy.[79] If drugs affecting the immune system are being evaluated in nude mice, translation of results to man might prove to be difficult. In this respect, agents like cisplatin[80] and cyclophosphamide[81] have been reported to influence the immune system.

For studies on immunotherapy in the nude mice, one has to realize that the immune system of this animal is devoid of a certain part (the T cell compartment). In general, the immune system in patients is complete, although a part may be affected in far-advanced disease or hematological malignancies. If drugs such as cytokines are investigated with the aim to induce an immunological response in the nude mouse, it is obviously difficult to translate the results directly to patients. If the sole aim for a cytokine is to assess its antitumor effect caused by activation of the immune system, it has to be certain that this product does not directly affect tumor cells (as can be shown for murine interferon-γ[82]). For most cytokines, it will be impossible to discriminate between a host-mediated and a direct anti-proliferative effect.

In case of the investigation of mouse monoclonal antibodies in the nude mouse model to detect or treat human tumor xenografts, a mouse anti-mouse antibody response is usually not elicited. With respect to patients, the development of human anti-mouse antibodies is a common phenomenon upon repeated monoclonal antibody rejections and interfering with the outcome of therapy results.

With respect to immunotherapy, and especially for cytokines, it is clear that several mechanisms are still unrecognized, which cause difficulties in the direct translation of mouse therapy results to man. Therefore, additional information on the value of the nude mouse-human tumor model for screening of the activity of biological products has to be acquired in the near future (see also Chapter 18).

V. CONCLUSIONS

For the use of the nude mouse-human tumor model in the developmental better anticancer therapies, the best approach is the set up of a tumor panel of a particular tumor type with characteristics representative for a group of patients with that tumor type. Selective pressure of a specific characteristic ultimately dominating the panel should be recognized and avoided. Once a tumor line is established, it should be monitored continuously for retention of the defined stability criteria. Instability of xenografts has been reported with respect to histology, tumor markers, DNA-ploidy, genetic composition, and sensitivity to antitumor agents.

A direct translation of therapy results from nude mouse to man should be carried out with precaution. Dissimilarities in response to some drugs have been reported in approximately 10% of the individual patient/xenograft comparisons. In general, it appears that the model tends to estimate the activity of antimetabolites too low and that of agents inducing DNA damage too high.

Variables of influence on response caused by differences between the xenograft model and patients originate from aspects related to the methodology to determine the response, tumor growth and blood supply, pharmacology of the drugs studied, and the immunological status of the host. Knowledge of these variables will define the limitations of the nude mouse-human tumor model for its use in the development of new cancer therapies.

REFERENCES

1. **Staquet, M. J., Byar, D. P., Green, S. B., and Rozencweig, M.,** Clinical predictivity of transplantable tumor systems in the selection of new drugs for solid tumors: rationale for a three-stage strategy, *Cancer Treat. Rep.*, 67, 753, 1983.
2. **Winograd, B., Boven, E., Lobbezoo, M. W., and Pinedo, H. M.,** Human tumor xenografts in the nude mouse and their value as test models in anticancer drug development (review), *In Vivo*, 1, 1, 1987.
3. **Mattern, J., Bak, M., Hahn, E. W., and Volm, M.,** Human tumor xenografts as model for drug testing, *Cancer Metast. Rev.*, 7, 263, 1988.
4. **Braakhuis, B. J. M., Schoevers, E. J., Heinerman, E. C. M., Sneeuwloper, G., and Snow, B.,** Chemotherapy of human head and neck cancer xenografts with three clinically active drugs: *cis*-platinum, bleomycin and methotrexate, *Br. J. Cancer*, 48, 711, 1983.
5. **Fiebig, H. H., Schuchhardt, C., Henss, H., Fiedler, L., and Löhr, G. W.,** Comparison of tumor response in nude mice and the patients, *Behring Inst. Mitt.*, 74, 343, 1984.
6. **Inaba, M., Kobayashi, T., Tashiro, T., and Sakurai, Y.,** Pharmacokinetic approach to rational therapeutic doses for human tumor-bearing nude mice, *Jpn. J. Cancer Res. (Gann)*, 79, 509, 1988.
7. **Kyriazis, A. P., Dipersio, L., Michael, J. G., and Pesce, A. J.,** Influence of mouse hepatitis virus (MHV) infection on the growth of human tumors in the athymic mouse, *Int. J. Cancer*, 23, 402, 1979.
8. **Braakhuis, B. J. M., Sneeuwloper, G., and Snow, G. B.,** The potential of the nude mouse model for the study of head and neck cancer, *Arch. Otorhinolaryngol.*, 239, 69, 1984.
9. **Fogh, J., Orteo, J., Sharkey, F. E., Fogh, J. M., and Daniels, W. P.,** Twenty-three new human tumour lines established in nude mice, *Exp. Cell Biol.*, 48, 229, 1980.
10. **Giovanella, B. C., Stehlin, J. S., Fogh, J., and Sharkey, F. E.,** Serial transplantation of human malignant tumors in nude mice and their use in experimental chemotherapy, in *Proc. Symp. Use of Athymic (Nude) Mice in Cancer Res.*, Houchens, D. P. and Ovejera, A. A., Eds., Fischer, Stuttgart, 1978, 163.

11. **Braakhuis, B. J. M., Laarman, D. W., Nauta, J. J. P., van Walsum, M., van Dongen, G. A. M. S., and Snow, G. B.,** Transplantability of human head and neck squamous cell carcinomas in athymic nude mice, *Arch. Otolaryngol. Head Neck Surg.,* 115, 1076, 1989.
12. **Gershwin, M. E., Lentz, D., and Owens, R. B.,** Relationship between karyotype of tissue culture lines and tumorigenicity in nude mice, *Exp. Cell Biol.,* 52, 361, 1984.
13. **Wennerberg, J.,** Changes in growth pattern of human squamous cell carcinomas of the head and neck during serial passages in nude mice, *Int. J. Cancer,* 33, 245, 1984.
14. **Renwick, D. E., Franchi, A. J., Pouysesgur, J. M., and Lagarde, A. E.,** Differential tumorigenicity of growth factor dependent versus independent CCL39 lung fibroblasts in nude mice, *J. Natl. Cancer Inst.,* 77, 105, 1986.
15. **Wennerberg, J., Willen, R., Björklund, A., and Tropé, C.,** Histopathological characteristics predictive for growth of squamous cell carcinomas of the head and neck heterotransplanted to nude mice, *Anticancer Res.,* 6, 1165, 1986.
16. **Boven, E.,** Biology of human ovarian cancer xenografts, in *Human Tumour Xenografts in Anticancer Drug Development,* Winograd, B., Peckham, M. J., Pinedo, H. M., Eds., Springer, Berlin, 1988, 11.
17. **Giovanella, B. C. and Fogh, J.,** The nude mouse in cancer research, in *Advances in Cancer Research,* Klein, G. and Weinhouse, S., Eds., Academic Press, Orlando, FL, 44, 1985, 77.
18. **Sharkey, F. E., Spicer, J. H., and Fogh, J.,** Changes in histological differentiation of human tumors transplanted to athymic nude mice: a morphometric study, *Exp. Cell Biol.,* 53, 100, 1985.
19. **Mattern, J., Haag, D., Wayss, K., and Volm, M.,** Significance of proliferation for the growth of xenografted human tumours in nude mice, *Anticancer Res.,* 1, 15, 1981.
20. **Heppner, G. H. and Miller, B. E.,** Therapeutic implications of tumor heterogeneity, *Semin. Oncol.,* 16, 91, 1989.
21. **Spang-Thomsen, M., Brünner, N., Engelholm, S. A., and Vindeløv, L. L.,** Therapy studies in nude mice, in *Immune-Deficient Animals in Biomedical Research,* Rygaard, J., Brünner, R., Græm, N., and Spang-Thomsen, M., Eds., Karger, Basel, 1987, 316.
22. **Nowell, P.C.,** Mechanisms of tumor progression, *Cancer Res.,* 46, 2203, 1986.
23. **Mattern, J., Wayss, K., and Volm, M.,** Effect of five antineoplastic agents on tumor xenografts with different growth rates, *J. Natl. Cancer Inst.,* 72, 1335, 1984.
24. **Shackney, S. E., McCormack, G. W., and Cuchural, G. J.,** Growth rate patterns of solid tumors an their relation to responsiveness to therapy, an analytical review, *Ann. Int. Med.,* 89, 107, 1978.
25. **Houghton, J. A. and Taylor, D. M.,** Growth characteristic of human colorectal tumours during serial passage in immune-deprived mice, *Br. J. Cancer,* 37, 213, 1978.
26. **Shapiro, W. R., Basler, G. A., Chernik, N. L., and Posner, J. B.,** Human brain tumor transplantation into nude mice, *J. Natl. Cancer Inst.,* 62, 447, 1979.
27. **Steel, G. G. and Peckham, M. J.,** Human tumour xenografts: a critical appraisal, *Br. J. Cancer,* 41 (Suppl. IV), 133, 1980.
28. **Fodstad, Ø., Rofstad, E. K., Tveit, K. M., and Pihl, A.,** Spontaneous alteration in growth rates of two human melanoma xenografts. Concurrent changes in chemosensitivity, *Eur. J. Cancer Clin. Oncol.,* 19, 1175, 1983.
29. **Baisch, H., Otto, U., and Klöppel, G.,** Long-term serial transplantation of 30 different human renal cell carcinomas into NMRI (nu/nu) mice: flow cytometric, histologic and growth studies, *J. Natl. Cancer Inst.,* 76, 269, 1986.
30. **Ishii, S., Yamawaki, S., Saki, T., Usui, M., Ubayama, Y., Minami, A., Yagi, T., Isu, K., and Kobayashi, M.,** Characteristics of human soft tissue sarcomas in xenografts and *in vitro Clin. Orthop.,* 173, 251, 1983.
31. **Mandybur, T. I., Sawaya, R., and Ormsby, I.,** The morphology and biologic behavior of human glioblastoma growing in nude mice, *Cancer,* 58, 1061, 1986.
32. **Donhuijsen, K., Budach, V., van Beuningen, D., and Schmidt, U.,** Instability of xenotransplanted soft tissue sarcomas, *Cancer,* 61, 68, 1988.
33. **Mattern, J., Haag, D., Wayss, K., and Volm, M.,** Growth kinetics of human lung tumours in nude mice, *Exp. Cell Biol.,* 49, 34, 1981.
34. **Rousseau-Merck, M. F., Cottreau, D., and Kahn, A.,** Isozyme pattern in serially xenotransplanted childhood tumors, *Cancer Res.,* 44, 1163, 1984.
35. **Jacobsen, P. F., Parker, T. J., and Papadimitriou, J. M.,** Ependymal differentiation of a glioblastoma multiforme after heterotransplantation into nude mice, *J. Natl. Cancer. Inst.,* 79, 771, 1987.
36. **Bailey, M. J., Ormerod, M. G., Imrie, S. F., Humphreys, J., Roberts, J. D. B., Gazet, J. C., and Neville, A. M.,** Comparative functional histopathology of human breast carcinoma xenografts, *Br. J. Cancer,* 43, 125, 1981.
37. **Roholl, P. J. M., Rutgers, D. H., Rademakers, L. H. P. M., de Weger, R. A., Elbers, J. R. J., and Unnik, J. A. M.,** Characterization of human soft tissue sarcomas in nude mice. Evidence for histogenic properties of malignant fibrous histiocytomas, *Am. J. Pathol.,* 131, 559, 1988.
38. **Vindeløv, L. L., Spang-Thomsen, M., Visfeldt, J., Povlsen, C. O., Jensen, G., and Nissen, N. L.,** Clonal evaluation demonstrated by flow cytometric DNA analysis of a human colonic carcinoma grown in nude mice, *Exp. Cell Biol.,* 50, 216, 1982.

39. **Bauer, H. C. F., Brosjö, O., Brostrom, L. A., Nilsson, O. S., Reinholt, F. P., and Tribukait, B.,** Growth and ploidy of human osteosarcoma xenografts in serial passage in nude mice, *Eur. J. Cancer Clin. Oncol.*, 22, 821, 1986.
40. **Tilgen, W., Boukamp, P., Breitkreutz, D., Dzarlieva, R. T., Engstner, M., Haag, D., and Fusenig, N. E.,** Preservation of morphological functional, and karyotypic traits during long-term culture and *in vivo* passage of two human skin squamous cell carcinomas, *Cancer Res.*, 43, 5995, 1983.
41. **Horvath, G., Alm, P., Tropé, C., and Ferno, M.,** Changes in tumor morphology in one human endometrial carcinosarcoma heterotransplanted into nude mice, *In Vivo,* 3, 113, 1989.
42. **Friedlander, M. L., Russell, P., Taylor, I. W., and Tattersall, M. H. N.,** Ovarian tumour xenografts in the study of bioloty of human epithelial ovarian cancer, *Br. J. Cancer,* 51, 319, 1985.
43. **Stebbings, W. S. L., Vinson, G. P., Farthing, M. J. G., Balkwill, F., and Wood, R. M. F.,** Effect of steroid hormones on human colorectal adenocarcinoma xenografts, of known steroid-receptor status, in nude mice, *J. Cancer Res. Clin. Oncol.,* 115, 439, 1989.
44. **Burg-Kurland, C. L., Purnell, D. M., and Combs, J. W.,** Immunocytochemical evaluation of primary human esophageal carcinomas and their xenografts for keratin, beta-chorionic gonadotropin, placental lactogen, alpha-fetoprotein, carcino-embryonic antigen, and nonspecific cross-reacting antigen, *Cancer Res.,* 46, 5730, 1986.
45. **Selby, P. J., Heyderman, E., Gibbs, J., and Peckham, M. J.,** A human testicular teratoma serially transplanted in immune-deprived mice, *Br. J. Cancer* , 41, 52, 1980.
46. **Stark, M. and Schlipköter, H. W.,** Xenotransplants of human carcinomas. Change of properties in short-term and long-term passages, in *Immune Deficient Animals,* Sordat, B., Ed., Karger, Basel, 1984, 250.
47. **Russel, P. J., Wills, E. J., Philips, J., Wass, J., Jelbart, M., Gregory, P., and Raghavan, D.,** Features of squamous and adenocarcinoma in the same cell in a xenografted human transitional cell carcinoma: evidence of a common histogenesis?, *Urol. Res.,* 16, 79, 1988.
48. **Barkla, D. H. and Tutton P. J. M.,** Temporal morphologic changes in human colorectal carcinomas following xenografting, *Am. J. Pathol.,* 110, 315, 1983.
49. **Bogden, A. E., Kelton, D. E., Cobb, W. R., Gulkin, T. A., and Johnson, R. K.,** Effect of serial passage in nude athymic nude mice on the growth characteristics and chemotherapy responsiveness of 13762 and R3230 AC mammary tumor xenografts, *Cancer Res.,* 38, 59, 1978.
50. **Nissen, E., Arnold, W., Weiss, H., Naundorf, H., and Tanneberger, St.,** Biological characterization of a mesothelioma line in the nude mice. II. Some characteristics of cells cultivated *in vitro* prior to and after transplantation in nude mice, *Arch. Geschwulstforsch.,* 49, 544, 1979.
51. **Heinerman, E. C. M.,** Clonogenic Growth of Head and Neck Cancer Cells, Ph.D. thesis, Free University, Amsterdam, Optimax, Wi jk bi j Duurstede, The Netherlands, 1987, Chap. 5.
52. **Osieka, R.,** Human malignant melanoma: preclinical and clinical drug evaluation, *Behring Inst. Mitt.,* 74, 353, 184.
53. **Shorthouse, A. J., Smyth, J. F., Steel, G. G., Ellison, M., Mills, J., and Peckham, M. J.,** The human tumour xenograft — a valid model in experimental chemotherapy?, *Br. J. Surg.,* 67, 715, 1980.
54. **Kubota, T., Nakada, M., Tsuyuki, K., Inada, T., Asunama, F., Ishibiki, K., and Abe, O.,** Cell kinetics and chemosensitivity of human carcinomas serially transplanted into nude mice, *Jpn. J. Cancer Res. (Gann),* 77, 502, 1986.
55. **Kallinowski, F., Schlenger, K. H., Runkel, S., Kloes, M., Stohrer, M., Okunieff, P., and Vaupel, P.,** Blood flow, metabolism, cellular microenvironment, and growth rate of human tumor xenografts, *Cancer Res.,* 49, 3759, 1989.
56. **Lauk, S., Zietman, A., Skates, S., Fabian, R., and Suit, H. D.,** Comparative morphometric study of tumor vasculature in human squamous cell carcinomas and their xenotransplants in athymic nude mice, *Cancer Res.,* 49, 4557, 1989.
57. **Donelli, M. G., D'Incalci, M., and Garattini, S.,** Pharmacokinetic studies of anticancer drugs in tumor-bearing animals, *Cancer Treat. Rep.,* 68, 381, 1984.
58. **Moore, J. V.,** Dose response curves after *in vivo* experimental chemotherapy: influence of route of administration on biological outcomes, *Cancer Chemother. Pharmacol.,* 15, 91, 1985.
59. **Guarino, A. M., Rozencweig, M., Kline, I., Penta, J. S., Venditti, J. M., Lloyd, R. A., Holzworth, R. A., and Muggia, F. M.,** Adequacies and inadequacies in assessing murine toxicity data with antineoplastic agents, *Cancer Res.,* 39, 2204, 1979.
60. **Burns, F. R.,** Circadian rhythmicity in DNA syntheses in untreated and saline-treated mice as a basis for improved chronochemotherapy, *Cancer Res.,* 41, 2795, 1981.
61. **Peters, G. J., van Dijk, J., Nadal, J. C., van Groeningen, C. J., Lankelma, J., and Pinedo, H. M.,** Diurnal variation in the therapeutic efficacy of 5-fluorouracil against murine colon cancer, *In Vivo,* 1, 113, 1987.
62. **Litterst, C. L., Sikic, B. I., Mimnaugh, E. G., Guarino, A. M., and Gram, T. E.,** *In vitro* drug metabolism in male and female athymic, nude mice, *Life Sci.,* 22, 1723, 1978.

63. **Freudenthal, R. I., Leber, A. P., Emmerling, D. C., Kerchner, G. A., and Ovejera, A. A.,** Comparison of the drug metabolizing enzymes in the liver and kidneys from homozygous nude swiss, heterozygous normals swiss, homozygous normal swiss and DRA/2 mice, *Res. Commun. Chem. Pathol. Pharmacol.*, 15, 1976.
64. **Houchens, D. P,. Johnson, R. K., Gaston, M. R., Goldin, A., and Marks, T.,** Toxicity of cancer chemotherapeutic agents in athymic (nude) mice, *Cancer Treat. Rep.*, 61, 103, 1977.
65. **Schabel, F. M., Griswold, D. P., and Corbett, T. H.,** Testing therapeutic hypotheses in mice and man: observations on the therapeutic activity against solid tumors of mice treated with anticancer drugs that have demonstrated potential clinical activity for treatment of advanced solid tumors of man, *Meth. Cancer Res.*, 17, 3, 1979.
66. **Powis, G.,** Anticancer drug pharmacodynamics, *Cancer Chemother. Pharmacol.*, 14, 177, 1985.
67. **Collins, J. M., Zaharko, D. S., Sedrick, R. L., and Chabner, B. A.,** Potential roles for preclinical pharmacology in phase I clinical trials, *Cancer Treat. Rep.*, 70, 73, 1986.
68. **Van Hennik, M. B., van der Vijgh, W. J. F., Klein, I., Elferink, F., Vermorken, J. B., Winograd, and Pinedo, H. M.,** Comparative pharmacokinetics of cisplatin and three analogues in mice and humans, *Cancer Res.*, 47, 6297, 1987.
69. **Ueyama, Y.,** Utilization of nude mice in research on human cancer screening system of anticancer agents using human tumor xenografts in nude mice, *Prog. Clin. Biol. Res.*, 229, 287, 1987.
70. **Cummings, J., Double, J. A., Bibby, M. C., Famer, P., Evans, S., Kerr, D. J., Kaye, S. B., and Smyth, J. F.,** Characterization of the major metabolites of flavone acetic acid and comparison of their disposition in humans and mice, *Cancer Res.*, 49, 3587, 1989.
71. **McGovern, J. P., Williams, M. G., and Stewart, J. C.,** Interspecies comparison of acivicin pharmacokinetics, *Drug Metab. Dispos.*, 16, 18, 1988.
72. **Mordenti, J.,** Man versus beast; pharmacokinetic scaling in mammals, *J. Pharm. Sci.*, 75, 1028, 1986.
73. **Maessen, P. A., Mross, K. B., Pinedo, H. M., and van der Vijgh, W. J. F.,** Metabolism of epidoxorubicin in animals: absence of glucuronidation, *Cancer Chemother. Pharmacol.*, 20, 85, 1987.
74. **Mullin, R. J., Keith, B. R., and Duch, D. S.,** Distribution and metabolism of calcium leucovorin in normal and tumor tissue, in *The Expanding Role of Folates and Fluoropyrimidines in Cancer Chemotherapy*, Rustum, Y. and McGuire, J. J., Eds., Plenum, New York, 1988, 25.
75. *Wissenschaftliche Tabellen Geigy, Teilband Hämatologie und Humangenetik*, Ciba Geigy, Basle, 1979, 69.
76. **Darnowski, J. W. and Handschumacher, R. E.,** Tissue uridine pools: evidence *in vivo* of a concentrative mechanism for uridine uptake, *Cancer Res.*, 46, 3490, 1986.
77. **Rusello, O., Gamba, E., Bruzzo, C., Maioli, D., Civalleri, D., Rosso, K., Nicolin, A., and Sobrero, A.,** Pharmacokinetics of ^{3}H-uridine: evidence for large tissue pools of this nucleoside in man, in *Proc. 6th NCI-EORTC Symp. New Drugs in Cancer Therapy*, Amsterdam, 1989, no. 133.
78. **Tompkins, M. B., Rama Rao, G. V., and Tompkins, W. A. F.,** Increased tumorigenicity and resistance to antibody lysis of human colon tumor cells xenografted in congenitally athymic mice, *Cancer Res.*, 39, 2160, 1979.
79. **Rofstad, E. K.,** Local tumor control following single dose irradiation of human melanoma xenografts: relationship to cellular radiosensitivity and influence of an immune response by the athymic mouse, *Cancer Res.*, 49, 3163, 1989.
80. **Kleinerman, E. S. and Zwelling, L. A.,** The effect of *cis*-diamminedichloroplatinum (II) on immune function *in vitro* and *in vivo*, *Cancer Immunol. Immunother.*, 12, 191, 1982.
81. **Berd, D., Mastrangelo, M. J., Engstrom, P. D., Paul, A., and Maguire, H.,** Augmentation of the human immune response by cyclophosphamide, *Cancer Res.*, 42, 4862, 1982.
82. **Morikawa, K., Fan, D., Denkins, Y. M., Levin, B., Gutterman, J. U., Walker, S. M., and Fidler, I. J.,** Mechanisms of combined effects of γ-interferon and 5-fluorouracil on human colon cancers implanted into nude mice, *Cancer Res.*, 49, 799, 1989.

Chapter 22

NEW DRUG DEVELOPMENT

Benjamin Winograd

TABLE OF CONTENTS

I. INTRODUCTION

In spite of all its limitations and high expenses, the nude mouse-human tumor model has become more frequently integrated in the evaluation of, and screening for, new compounds and therapies.

New Drug Development is a program of preclinical research and conducted with the objective of selecting new drugs for the treatment of diseases, for example, cancer. A variety of disciplines are involved and contribute to such a comprehensive program. At first, new compounds are discovered for the treatment of, for example, cancer, and through a cascade of evaluation steps (Table 1) and selection procedures, one of these compounds may finally be accepted for clinical use.

The lack of adequate knowledge of carcinogenesis and of the specific differences between normal and cancer cells makes research into new drugs for cancer therapy particularly difficult. It is this lack of knowledge which makes preclinical test methods poor models for human cancer, with limited predictive value for clinical efficacy.

The nude mouse-human tumor model bears the potentiality of an improved predictiveness, as exemplified in several of the previous chapters of this book. However, one still needs to await a broad prospective validation of this model. For these reasons, adequate clinical evaluation of as many as possible new drugs or therapies is still essential in order to find new treatments for cancer.

In order to understand the unique situation of New Drug Development in oncology, it may be helpful to compare it with other therapeutic areas. Cancer in approximately 50% of patients has a fatal outcome for which adequate or curative therapy is not available. For this reason, promising anticancer drugs with a tolerable toxicity profile are readily accepted for clinical evaluation. In other therapeutic, noncancer areas, one will insist on new drugs being evaluated not only to be "active", but also "safe". This explains why the emphasis in the development of new anticancer drugs is put on screening for active compounds. Even a narrow therapeutic index (activity/toxicity) is sufficient to initiate clinical evaluation. The acceptance of clinical toxicities is the reason that phase I studies can only be conducted in cancer patients.

In this chapter, issues, methodologies, and strategies in New Anticancer Drug Development will be exemplified for chemotherapeutic or cytotoxic drugs as the largest class of drugs available for anticancer therapy.

II. DISCOVERY

The two processes which determine the types of compounds being discovered as new "lead compounds" are those of acquisition and screening. In 1955, the U. S. National Cancer Institute (NCI) initiated a large-scale screening program for the discovery of new anticancer drugs. This program relied heavily on two murine leukemias used as primary screens.[1] More recently, other institutions and several pharmaceutical companies have become involved in anticancer drug discovery, using a variety of different primary screens.

In the 1980s, it became apparent that changes in the strategy of anticancer drug discovery were necessary because

1. The clinical activity of drugs discovered by the NCI screen was mainly confined to the hematological malignancies.[2]
2. Drugs inactive in the NCI screen showed clinical activity.
3. Only very few drugs for the effective treatment of common solid tumors had been identified.
4. Advances in research opened up new approaches to cancer therapy.

TABLE 1
Steps in Anticancer Drug Development

Stages/phases	Functions/steps
Discovery	Acquisition
	Screening
Preclinical — Development	Formulation
	Toxicology
Clinical — Development	Phase I studies
	Phase II studies
	Phase III studies
Registration	Compilation of preclinical and clinical data

A. ACQUISITION

Acquisition of new compounds can proceed in two ways: the rational or the random way. Main sources for a random acquisition are synthetic chemicals and natural sources. Any chemical, even those synthesized for a completely different purpose, might bear the potential of an anticancer drug. Compounds from natural sources, originating from (extracts of) yeasts, plants, soils, etc., represent at least 50% of the total pool of clinically available drugs. In addition, natural products often seem to be more potent than chemicals as antitumor agents. Reports on the NCI experience illustrate how large numbers of compounds can be acquired and selected for screening of cytostatic/cytotoxic activity.[3,4]

In the past, rational drug acquisition was limited almost solely to the design of analogs of existing drugs. The term "structure-activity relationship" (SAR) is commonly used to attribute improved activity to a particular chemical substructure of the series of analogs which are being evaluated. This approach has resulted in significant improvements of some clinically active drugs.

As anticancer research progresses, innovative ideas and new targets for selective chemotherapeutic and immunological destruction of malignant cells are being put forward.[5] Given these new targets, rational drug discovery will assume a greater significance in the future.

B. SCREENING

There is a general misunderstanding about the difference between "testing" and "screening" for antitumor activity. Whereas research laboratories worldwide perform antitumor testing in certain models in a more or less predefined manner, there are very few institutes performing antitumor screening. This is mainly due to the fact that "screening" involves large numbers of compounds being evaluated in preclinical models with predefined methodology, endpoints, and selection criteria which require extensive resources.

Screening can be defined as a combination of testing and selecting a few compounds from a large variety for further evaluation in clinical studies. It is obvious that the result of any screen will depend equally on the choice of the model, the prior definition of evaluation and selection parameters, and the proper validation of the screen prior to general use. The most important question for any screen is: how well is it predictive for clinical activity?

Thus, drug discovery is, at least partially, a number game and the function of screening is to systematically select active drugs from a large pool of products. After completion of the screening step, compounds can be distinguished as those that meet the activity criteria of the screen (positives) and compounds that do not meet these criteria (negatives). The real value of a screen can only be estimated in retrospect and after the clinical evaluation of a drug, giving rise to the following categories of compounds:

1. True positives: drugs active in the screening for which clinical activity is proven
2. False positives: drugs active in the screen which do not have any clinical activity
3. True negatives: drugs inactive in the screen which do not have any clinical activity
4. False negatives: drugs that are clinically active and retrospectively proven to be inactive in the screen

In a Drug Development Program, screening is usually done by applying sequential screens: the "primary screen" should ideally be sensitive, simple and selective, and cost and time efficient.[6] Using the above categories, the primary screen should deselect as many true negatives as possible and keep the number of false negatives as small as possible. It should be realized, however, that the choice of the primary screen is critical for the outcome of any drug development program because this first filter might impose a strong bias on the selection of candidate drugs for further evaluation. It is generally accepted that a "secondary screen" may be more sophisticated, labor intensive, and expensive. At this step, compounds are thoroughly evaluated and those with the most promising profile are selected for preclinical development. Such a procedure of drug screening seems conceptually correct; but in a therapeutic area where neither the etiology nor the best targets for therapy are well defined, such a sequential approach might need reconsideration.

1. The Tumor Panel Screen of the NCI

The major example of a sequential screen in anticancer drug development is the tumor panel screen of the NCI. In the period 1955 to 1985, the principal approach to drug discovery used by the NCI was empirical screening in transplantable tumor models.[3,6] More than 400,000 synthetic compounds and natural products were screened for antitumor activity during this period using a sequential methodology (see also Chapter 11, Table 1). Although a number of experimental models were employed, the vast majority of compounds were tested in murine leukemias (L1210 and/or P388) only, as these two models served as primary screen. L1210 was selected at the time, as it predicted well for clinical activity;[1] whereas P388 was preferred as prescreen later on because it was sensitive to most classes of clinically effective drugs and sufficiently restrictive to avoid overloading the (secondary) panel screen.[7]

Over the years, the NCI has contributed to the discovery of 50% of all clinically useful anticancer drugs with use of the tumor panel screen.[3] The results of animal-human correlations, which distinguished the two murine leukemias as models that predicted well for clinical antitumor activity, were constantly open to criticism. In 1985, it was decided that the sensitive leukemia-based strategy for drug discovery had passed an adequate evaluation period and that there were a number of reasons to change, including

1. Agents inactive against P388 were found active in other preclinical tumor models.[8,9]
2. Agents active against P388 are often inactive against preclinical (solid) tumor models.
3. Clinically active drugs are inactive in P388; for example, hexamethylmelamine, busulfan, *o,p′*-DDD and bleomycin.[10]
4. Many drugs have been discovered that are useful and broadly active against human leukemias and lymphomas, but hardly any drug has been found with broad activity against the common human solid tumors.

The decision of the NCI to change their screening system stimulated institutes worldwide to start various ways for the discovery of potential anticancer drugs.

2. Models for Anticancer Drug Testing

A variety of test models is being used in preclinical drug evaluation, which can be divided into *in vitro* test systems and *in vivo* test systems, using tumor material of murine or human origin.

In vitro test systems are characterized by their different endpoints serving for evaluation of drug effects. Typical endpoints have been described as: changes in cell metabolism,[11] in cell membrane function,[12] DNA metabolism,[13] and cellular reproductive capacity.[14] Under *in vitro* test conditions, cells are removed from their normal environment. Therefore, the test can be relatively simple and the results may depend on a variety of factors involved in the experiment: for example, test factors, tumor factors, and drug factors. The ultimate value of *in vitro* chemosensitivity assays will depend on the demonstration that the response of cells *in vitro* accurately translates into antitumor activity *in vivo*. Several studies comparing preclinical *in vitro* and *in vivo* data, as well as comparing *in vitro* and clinical results, have shown that *in vitro* test systems may predict relatively well for drug resistance (±90%), but less for drug sensitivity (60 to 90%). The reason for these differences are complex, but probably originate from inadequate representation of the *in vitro* situation for the environmental factors experienced by tumor cells within tissue *in vivo*.[15]

At present, there is a general disenchantment with murine tumor models for drug screening because of their lessened value in the discovery of drugs which are active in the treatment of solid tumors. The perceived failure should be put into the correct context, with particular reference to the difference in criteria used for evaluating drug sensitivity in murine tumor models vs. clinical trials[16] and the high sensitivity of tumor models used in the past.[17]

For reasons of efficiency and cost, murine tumors have been widely used in the i.p.-i.p. assay, commonly referred to as "the test tube in the mouse assay",[18] in which the tumor and the test drug are both injected i.p. The evaluation is rather simple; that is, calculating the increase in lifespan (ILS) effected by the drug treatment on the basis of a daily count of the number of surviving mice. With inclusion of physiological barriers into this i.p.-i.p. assay, the tumor model becomes more realistic and more difficult to treat. This can be done by implantation of the tumor i.p. as "brei", followed by a period of growth prior to i. p. treatment, or by s.c. transplantation of tumor tissue with subsequent i.p. or i.v. drug treatment. It is well known that there are ample murine solid tumor models (tumor implantation s.c.) that cannot be treated successfully or even cured by presently known anticancer drugs.

A relatively new *in vivo* model for anticancer drug testing incorporates human tumor xenografts grown in nude mice. Since the first transplantation of a human tumor into nude mice,[19] this model has caught the attention of many experimental oncologists. The labor- and cost-intensiveness, together with the long duration of the assay, have limited its widespread use in screening of anticancer drugs. At present, a thorough validation of this model is ongoing, to which this book is trying to make its contribution. Lengthy studies have provided evidence that transplanted human tumors retain the most important biological characteristics of the tumor of origin (see Chapter 8), including a biological instability as is known to occur in patients' tumors (see Chapter 20). In addition, several investigators have reported good correlations between drug effects in the human tumor xenograft and clinical results in the donating patient's tumor.[20] The overall impression is that chemoresponsive human tumor types give rise to tumor lines that also respond well to chemotherapy.[21] Today, the human tumor xenograft model is probably the most valuable test system in the selection of analogs (see also Chapter 14).

3. New Drug Screening Approaches

During the past five years, three *in vitro* test systems have been introduced as primary screen for large-scale screening. Shortly after the initial reports of Hamburger and Salmon (in 1977 and 1978) on the selective cloning of human tumors in primary culture for drug sensitivity testing, it was suggested to use this assay for drug screening.[22] Positive data on comparative studies with series of analogs and on *in vitro* phase II trials, as well as good predictions of the individual patient's response to chemotherapy, supported the application of the Human Tumor Colony Forming Assay (HTCFA).

As a potential alternative to the screening panel, the NCI initiated a multi-institutional program to evaluate the HTCFA for *de novo* drug screening.[9,23] Using fresh samples of tumors which are most readily available (breast, colon, kidney, lung, melanoma, and ovary), 14 out of 79 compounds previously evaluated as inactive in the murine leukemia P388 were shown to be active. Subsequently, 2 of the 14 drugs were selected for clinical evaluation (dihydrolenperone and chloroquinoxaline sulfonamide). Thus, the ability of the HTCFA to identify different and potentially disease-specific drug candidates has been demonstrated. However, due to technical limitations inherent to the HTCFA methodology, namely its requirement of fresh tumor samples and difficulties in representative colony formation, screening cannot be carried out on a large scale.

The NCI disease-oriented *in vitro* anticancer drug discovery project has just been implemented for routine use. More than 60 human tumor cell lines derived from 7 cancer types have been selected. These meet certain minimum quality assurance criteria and have reproducible profiles for growth and drug sensitivity.[24] Cell survival is determined in a dose-response fashion in microtiter plates by a protein-binding dye assay (SRB).[25] It is conceivable that, in addition to identifying new agents with biological activity profiles analogous to currently known antitumor drugs, the new screen will provide the opportunity to pursue "disease-oriented" screening. This means that a new drug could demonstrate differential sensitivity in the *in vitro* screen in either "solid tumors" vs. "leukemias" or a certain tumor type in particular. In fact, the first few compounds with such features have already been identified.[26]

A third assay, which has recently been developed at Wayne State University, is presently being used as primary screen: the two tumor disk-diffusion soft-agar assay.[17] This assay has been developed to uncover compounds with preferential cytotoxicity for the solid-tumor cell type over the leukemic cell type. Two different tumors are co-cultivated on a soft-agar disk: a leukemia (L1210) and a solid (drug insensitive) tumor. The leukemia can be regarded useful to determine cytotoxic activities, as known from conventional antitumor agents. The solid tumor cells, on the other hand, have the potential for indicating activities of undiscovered, possibly solid-tumor-specific antitumor agents. After co-cultivating the two tumor cell lines, the test agent is placed on a filter-paper disk on top of the soft agar. Four different results are possible: (1) the agent is not cytotoxic at the concentration tested; (2) cells of the leukemia and the solid tumor are equally inhibited in growth; (3) the leukemia cells are inhibited to a greater extent than the solid tumor cells; and (4) the reverse situation may also occur.

With regard to potential new *in vivo* approaches for drug screening, two models seem worthwhile of being routinely evaluated: murine solid tumors and human tumor xenografts. In principle, there are two options to integrate one of these models in drug screening: using them as secondary screen or using them upfront. Despite the amount of labor and expense, and considering the slow overall progress of effective cancer therapy and the high costs of clinical drug evaluation, there are enough data supporting their use upfront.

With the "wise"[27] use of murine solid tumors, that is, appropriate model selection, schedule and dose of treatment, and strict evaluation criteria, these tumors should be able to uncover new leads in cancer therapy. As for the human tumor xenografts, there are enough data presented in several chapters of this book to justify their upfront use for drug screening. *One* tumor line cannot be representative for human solid tumors and the selection of the test conditions must be chosen adequately. In addition, specific characteristics of human tumor lines, such as hormone dependency, drug resistance, and antigenic profile, can be used to screen structure-activity relationships of particular compounds or to open new avenues for rational drug design.

III. PRECLINICAL DEVELOPMENT

Once a drug is selected as a candidate for clinical evaluation, two important and rather critical steps have to be completed prior to the initiation of clinical studies: pharmaceutical formulation and animal toxicology.

TABLE 2
Safe Starting Dose for Clinical Studies Derived from Different Animal Toxiciology Endpoints

1/3 TDL[1] dog	1/3 LD_{10}[2] mouse	1/10 LD_{10} mouse	Ref.
22/28[3]	27/29	29/29	29
12/12	12/12	12/12	30
20/21	16/21	21/21	31
6/7	—	6/7	32

[1] TDL, toxic dose low.
[2] LD_{10}, lethal dose for 10% of the animals.
[3] Safe starting dose in x/y different drugs.

A. FORMULATION

For anticancer agents, the development of an i.v. formulation is generally preferred. Since these drugs are to be administered at a maximum tolerated dose, a variation in the bioavialability from an oral dosage form would complicate the first clinical evaluation.

In comparison with other drug classes, there are a few factors which make the development of a dosage form specifically difficult: anticancer drugs are often very potent, not readily (or even hardly) soluble in aqueous solutions, and unstable in solutions. Limited solubility of a drug candidate is regularly a reason to discontinue its overall development. Various approaches and experimental methodologies for solubilization are being used to enable clinical evaluation of an anticancer drug. The solution of the drug might add to its toxic effects, for example, local tissue damage from alcoholic solvents or lung toxicity from microparticles. In addition, the antitumor efficacy might be altered, and it is therefore mandatory to confirm the biological activity of the formulated drug. As many of these drugs have a narrow therapeutic index and any component which needs to be added to the formulation might contribute or modify the toxicity of the drug, animal toxicology should be performed with the final dosage form.[28]

B. TOXICOLOGY

Anticancer drugs are powerful pharmacological tools with an extremely narrow therapeutic index: biologically inactive doses may be within one log of lethal doses.[29] The goals of preclinical toxicologists and pharmacologists involved in anticancer drug development are fourfold:

1. The establishment of a reliable method for determination of a safe, though not homeopathic, starting dose for clinical trials
2. The prediction of the qualitative toxic effects in man based on animal experiments
3. The minimization of the number of animals and species to achieve the above goals
4. The development of methods to limit the number of patients required for assessment of the maximum tolerated dose of an investigational drug in phase I studies

Before a new agent is considered for initial clinical trials, knowledge of its quantitive and qualitative toxic potential must be developed by systematic studies in laboratory animals. For a new anticancer drug, this knowledge is of particular relevance because, for all presently available drugs, efficacy is closely related to the degree of toxicity produced.

Several retrospective evaluations of comparative qualitative interspecies toxicity have been performed in the past. The recommendations for a safe starting dose for clinical studies have changed with time. Table 2 provides an overview of these evaluations, in search for a safe starting dose.

At present, there are mainly two procedures which are advocated to determine a safe starting dose: (1) the lowest value of 1/10 of the mouse lethal dose for 10% of the animals (LD_{10}) and 1/3 of the toxic dose low (TDL) in the dog (recommendation of the U.S. Food and Drug Administration); and (2) 1/10 LD_{10} in the mouse, provided this dose is safe in the rat (Joint Steering Committee of the European Organization for Research on Treatment of Cancer and Cancer Research Campaign, 1990).

The major objectives of qualitative toxicology studies in animals are to determine the target organ(s) of toxicity and to investigate if drug-induced side effects are reversible, delayed, cumulative, manageable, or predictable. Although preclinical toxicological studies conducted in several species have served well to reveal the specific test parameters that act as indicators of toxicity in man, it is to be expected that unique or unusual toxic effects may remain undetected because of limitations in observations or inherent species differences. Once a toxic effect in an organ system is observed during an animal study, it should be viewed with a realistic understanding of the following limitations of animal toxicological data:

1. The toxic effect may develop in man in an organ system predicted to be susceptible from the animal data, but may be expressed in a different specific clinical, morphological, or chemical parameter.
2. The adverse reaction may appear in man at a greater or lesser dose level.
3. The toxic effect may follow a different order of appearance in relation to the total spectrum of qualitative toxic effects inherent to any compound.
4. The toxic effect may not develop in man.[33]

C. PHARMACOKINETICS

Defining a safe starting dose is only the first step in clinical testing of anticancer drugs. After the starting dose has been evaluated, subsequent doses are escalated until the maximum tolerated dose (MTD) is reached. This process of dose escalation is governed by a fundamental conflict: on the one hand, there is a need to go slowly in order to avoid a sudden jump from "no observable toxicity" to a "lethal dose"; on the other hand, there is a need to go rapidly, so that only a few patients are treated at ineffective doses.[34]

During the past decade, the most widely accepted approach to solve this fundamental conflict was the use of the "modified Fibonacci series" of dose escalation. Consequently, this method for determination of the highest possible, albeit safe, starting dose was the main focus in this step of anticancer drug development.

More recently, the concept of the area under the concentration-vs.-time curve (C x T) as determinant of drug effect/toxicity and the application of this $C \times T$ concept to dose escalation was studied.[34] In a retrospective study, it was shown that 20 to 50% savings in the total number of dose escalations were possible by consequent determination of the preclinical pharmacokinetics prior to the start of the clinical study, and by applying the $C \times T$ concept. This pharmacokinetically guided dose escalation is presently applied in a prospective way during the performance of phase I studies at several institutes. First reports demonstrate the successful application of this concept as well as its drawbacks.

IV. CLINICAL DEVELOPMENT

The ultimate goal of each drug development program is to determine the clinical effectiveness of a new drug. An intrinsic part of such a program is the assumption that data from animals will be predictive in man. For many anticancer drugs, a dose-response relationship has been established in preclinical models and this fact is being integrated into the design of clinical (phase I) studies. On the other hand, in the sections on screening and toxicology, it has been discussed that the predictiveness of animal models for antitumor activity and toxic effects in patients often has its limitations. Due to this dilemma for the clinical evaluation of anticancer drugs, it is of utmost importance that adequate effort and resources are expended in the design and conduct of this rate-limiting step. Effective cancer chemotherapy often consists of more than one drug regimen. Before a new drug is being added to and evaluated in a drug combination, there should be proof that in single-agent (phase II) studies the drug has some activity of its own in a given tumor type.

A. PHASE I STUDIES

In phase I studies, a new drug is administered to man for the first time. Once this critical step is taken, all information about the preclinical profile of the new drug should be available to the clinical investigator. Ideally, this should include the clinical pharmacology in animals and the methodology to perform clinical pharmacokinetics in man.[35]

The goals of a phase I study are:

1. To establish the MTD that can safely be delivered to patients in a given schedule and route of administration
2. To establish the toxicity pattern of this drug and evaluate whether toxicities are tolerable, predictable, cumulative, and reversible
3. To determine the pharmacokinetics of distribution, metabolism, and elimination at different dose levels, ideally linked to the pharmacodynamics of the drug

The determination of antitumor activity is often regarded as a secondary goal in phase I due to the design of these studies. Patients in phase I studies often have advanced or refractory tumors, and in addition, non-measurable or evaluable disease. The methodology of phase I studies implies escalating doses to be administered in order to determine the MTD; thus, a significant number of patients receive a conceptually ineffective drug dose. This ethical dilemma of phase I studies was recently quantified:[36] in phase I studies with 54 different drugs, 60% of the patients were treated initially with lower drug doses than recommended for phase II evaluation. However, it is of interest to note that there is no statistically significant difference in response rates in phase I studies between drugs not active in later evaluation (2.7%) vs. those drugs that ultimately show to be effective (4.3%).

Whereas the goals and concepts of a phase I trial appear simple, there are, in fact, several major issues in the design and conduct of a phase I trial requiring clinical experience of the investigator, these being (1) patient selection factors; (2) drug starting dose, escalation, and schedules of administration; (3) patient requirements; and (4) determination of the MTD.[37]

When terminating phase I studies and recommending a safe dose for phase II evaluation, one has to make certain that the patient selection factors at the end stage of phase I be similar to the selection factors applied for phase II studies; that is, minimal or no prior chemotherapy.

B. PHASE II STUDIES

Phase II studies are conducted to determine whether a new drug has antitumor activity of value for further clinical evaluation. In initial phase II studies, the new drug is administered in a schedule considered to be (pharmacologically) optimal for the drug and at a dose close to the MTD. Due to the inadequate prediction of clinical activity by the presently used preclinical models, the main classes of human tumors should be evaluated in phase II studies. Validated new preclinical models might reduce the large number of phase II studies with negative outcome.

The presently used design of the phase II study is disease oriented. The critical factor in this approach is that prognostic variables which may affect response are taken into consideration in the study design.[35] One of the most important variables affecting response is (the extent of) prior therapy.[2] Phase II studies with new drugs are preferably conducted in previously untreated patients. Once potential therapeutic activity has been uncovered, the next step is to focus on the role of the new drug in a particular disease. The new drug may be compared to standard therapy as a single agent or it may be integrated into a new combination chemotherapy regimen. In the latter instance, it would essentially be "recycled" to a phase I setting where the appropriate dose ratios, schedule sequences of administration, as well as variables of toxicity, can be established for the new combination.[35]

C. PHASE III STUDIES

Phase III studies are conducted to determine the role of the new drug in the treatment of a particular disease. This might be as a single agent or in combination chemotherapy. This is done most adequately by controlled, randomized studies comparing the new treatment against standard therapy. Endpoints used in phase III studies are not only relative activity (e.g., response rate), but also response duration, survival, toxicity, and quality of life. Influenced by the success of combination chemotherapy in the treatment of leukemias and lymphomas, the literature in oncology witnesses the many attempts to design new regimens for a solid cancer type. In many cases, too much time is spent in uncontrolled, exploratory studies without a suitable control group. The *sine qua non* of a controlled study is the comparability of patient groups assigned to each treatment so that the only reasonable explanation of an observed difference is directly attributable to the treatment program.[35]

The analysis of randomized phase III studies focuses on the absolute efficacy and toxicity of each treatment arm and the next process of the comparison of the (two) arms. Significant differences need to be present which, in the clinical sense, means that what is observed is notable, or important; and in the statistical sense means that in comparing two groups, the mean results differ from each other by more than twice the standard error. This implies that the smaller the difference, the larger the groups have to be to show a significant difference. Phase III is a challenge. It is much more than an attempt to confirm the findings of phase I and II; it is an attempt at the total identification of a drug.[35]

V. CONCLUSIONS

Human tumor xenografts grown s.c. in nude mice have gained a place in the discovery of new anticancer agents. The labor- and cost-intensiveness, together with the long duration of the assay, have limited their widespread use in screening of new compounds. Nevertheless, considering the slow overall progress of effective cancer therapy and the high costs of clinical drug evaluation, there are enough data to incorporate the nude mouse-human tumor model in an early stage of a Drug Development Program. *One* tumor line cannot be representative for a particular human solid tumor type and the selection of the test conditions must be chosen adequately. Additional specific characteristics of human tumor lines, such as hormone dependency, drug resistance, and antigenic profile, can be of relevance to screen structure-activity relationships of particular compounds, and open new avenues for rational drug design. At the present time, the human tumor xenograft model is probably the most valuable test system in the selection of analogs.

REFERENCES

1. **Goldin, A., Schepartz, S. A., and Venditti, J. M.,** Historical development and current strategy of the National Cancer Institute drug development program, in *Methods in Cancer Research, XVI,* DeVita, V. T. and Busch, H., Eds., Academic Press, New York, 1979, 165.
2. **Marsoni, S., Hoth, D., Simon, R., Leyland-Jones, B., De Rosa, M., and Wittes, R. E.,** Clinical drug development: an analysis of phase II trials, 1970-1985, *Cancer Treat. Rep.,* 71, 71, 1987.
3. **Driscoll, J. S.,** The preclinical new drug research program of the National Cancer Institute, *Cancer Treat. Rep.,* 68, 63, 1984.
4. **Narayanan, V. L.,** Strategy for the discovery and development of novel anticancer agents, in *Structure-Activity Relationships of Anti-Tumour Agents,* Reinhoudt, D. N., Connors, T. A., Pinedo, H. M., and van de Poll, K. W., Eds., Martinus Nijhoff Publishers, Boston, 1983, 5.
5. **Burns, C. P.,** Membranes and cancer chemotherapy, *Cancer Invest.,* 6, 439, 1988.
6. **Venditti, J. M.,** Preclinical Drug Development: rationale and method, *Semin. Oncol.,* 8, 349, 1981.
7. **Venditti, J. M.,** Drug evaluation branch program: report to screening contractors, *Cancer Chemother. Rep.,* (Part 2) 5, 1, 1975.
8. **Staquet, M. J., Byar, D. P., Green, S. B., and Rozencweig, M.,** Clinical predictivity of transplantable tumor systems in the selection of new drugs for solid tumors: rationale for a three-stage strategy, *Cancer Treat. Rep.,* 67, 753, 1983.
9. **Shoemaker, R. H., Wolpert-DeFilippes, M. K., Kern, D. H., Lieber, M. M., Makuch, R. W., Melnick, N. R., Miller, W. T., Salmon, S. E., Simon, R. M., Venditti, J. M., and Von Hoff, D. D.,** Application of a human tumor colony-forming assay to new drug screening, *Cancer Res.,* 45, 2145, 1985.
10. **Goldin, A., Venditti, J. M., MacDonald, J. S., Muggia, F. M., Henney, J. E., and DeVita, V. T.,** Current results of the screening program at the division of cancer treatment, National Cancer Institute, *Eur. J. Cancer,* 17, 129, 1981.
11. **Gosalvez, M., Canero, R. G., and Blaco, M.,** A screening for selective anticancer agents among plant respiratory inhibitors, *Eur. J. Cancer,* 12, 1003, 1976.
12. **Weisenthal, L. M. and Lippman, M. E.,** Clonogenic and nonclonogenic *in vitro* chemosensitivity assays, *Cancer Treat. Rep.,* 69, 615, 1985.
13. **Weisenthal, L. M.,** *In vitro* assays in preclinical antineoplastic drug screening, *Semin. Oncol.,* 8, 362, 1981.
14. **Salmon, S. E.,** Human tumor colony assay and chemosensitivity testing, *Cancer Treat. Rep.,* 68, 117, 1984.
15. **Phillips, R. M., Bibby, M. C., and Double, J. A.,** A critical appraisal of the predictive value of *in vitro* chemosensitivity assays, *J. Natl. Cancer Inst.,* 82, 1457, 1990.
16. **Martin, D. S., Balis, M. E., Fisher, B., Frei, E., Freireich, E. J., Heppner, G. H., Holland, J. F., Houghton, J. A., Houghton, P. J., Johnson, R. K., Mittelman, A., Rustum, Y., Sawyer, R. C., Schmid, F. A., Stolfi, R. L., and Young, C. W.,** Role of murine tumor models in cancer treatment research, *Cancer Res.,* 46, 2189, 1986.
17. **Corbett, T. H. and Valeriote, F. A.,** Rodent models in experimental chemotherapy, in *Rodent Tumor Models in Experimental Cancer Therapy,* Kallman, R. F., Ed., Pergamon Press, Oxford, 1987, 233.
18. **Corbett, T. H., Valeriote, F. A., and Baker, L. H.,** Is the P388 murine tumor no longer adequate as a drug discovery model?, *Invest. New Drugs,* 5, 3, 1987.
19. **Rygaard, J. and Povlsen, C. O.,** Heterotransplantation of a human malignant tumor to "nude" mice, *Acta Path. Microbiol. Scand.,* 77, 758, 1969.
20. **Fiebig, H. H.,** Comparison of tumor response in nude mice and in patients, in *Human Tumour Xenografts in Anticancer Drug Development,* Winograd, B., Peckham, M. J., and Pinedo, H. M., Eds., E.S.O. Mongraphs, Springer, Heidelberg, 1988, 25.
21. **Steel, G. G., Courtenay, V. D., and Peckham, M. J.,** The response to chemotherapy of a variety of human tumor xenografts, *Br. J. Cancer,* 47, 1, 1983.
22. **Salmon, S. E.,** Applications of the human stem cell assay to new drug evaluation and screening, *Prog. Clin. Biol. Res.,* 48, 291, 1980.
23. **Shoemaker, R. H.,** New approaches to antitumor drug screening: the human tumor colony-forming assay, *Cancer Treat. Rep.,* 70, 9, 1986.
24. **Alley, M. C., Scudiero, D. A., Monks, A., Hursey, M. L., Czerwinski, M. J., Fine, D. L., Abbott, B. J., Mayo, J. G., Shoemaker, R. H., and Boyd, M. R.,** Feasibility of drug screening with panels of human tumor cell lines using a microculture tetrazolium assay, *Cancer Res.,* 48, 589, 1988.
25. **Rubinstein, L. V., Shoemaker, R. R., Paull, K. D., Simon, R. M., Tosini, S., Skehan, P., Scudiero, D. A., Monks, A., and Boyd, M. R.,** Comparison of *in-vitro* anticancer-drug-screening. Data generated with a tetrazolium assay versus a protein assay against a diverse panel of human tumor cell lines, *J. Natl. Cancer Inst.,* 82, 1113, 1990.
26. **Boyd, M. R.,** Status of the NCI preclinical antitumor drug discovery screen, *Princ. Practice Oncol. Updates,* 3, 1, 1989.

27. **Skipper, H. E.,** Laboratory models: some historical perspective, *Cancer Treat. Rep.,* 70, 3, 1986.
28. **Davignon, J. P. and Cradock, J. C.,** Pharmaceutical aspects of antitumour agents, *Pharmacol. Weekbl.,* 119, 1144, 1984.
29. **Goldsmith, M. A., Slavik, M., and Carter, S. K.,** Quantitative prediction of drug toxicity in humans from toxicology in small and large animals, *Cancer Res.,* 35, 1354, 1975.
30. **Penta, J. S., Rozencweig, M., Guarino, A. M., and Muggia, F. M.,** Mouse and large-animal toxicology studies of twelve antitumor agents: relevance to starting dose for Phase I clinical trials, *Cancer Chemother. Pharmacol.,* 3, 97, 1979.
31. **Rozencweig, M., Von Hoff, D. D., Staquet, M. J., Schein, P. S., Penta, J. S., Goldin, A., Muggia, F. M., Freireich, E. J., and DeVita, V. T.,** Animal toxicology for early clinical trials with anticancer agents, *Cancer Clin. Trials,* 4, 21, 1981.
32. **Grieshaber, C. K. and Marsoni, S.,** Relation of preclinical toxicology to findings in early clinical trials, *Cancer Treat. Rep.,* 70, 65, 1986.
33. **Prieur, D. J., Young, D. M., Davis, R. D., Cooney, D. A., Homan, E. R., Dixen, R. L., and Guarino, A. M.,** Procedures of preclinical toxicologic evaluation of cancer chemotherapy agents: protocols of the laboratory of toxicology, *Cancer Chemother. Rep.,* 4, 1, 1973.
34. **Collins, J. M., Zaharko, D. S., Dedrick, R. L., and Chabner, B. A.,** Potential roles for preclinical pharmacology in Phase I clinical trials, *Cancer Treat. Rep.,* 70, 73, 1986.
35. **Carter, S. K.,** Clinical trials in cancer chemotherapy, *Cancer,* 40, 544, 1977.
36. **Estey, E., Hoth, D., Simon, R., Marsoni, S., Leyland-Jones, B., and Wittes, R.,** Therapeutic response in Phase I trials of antineoplastic agents, *Cancer Treat. Rep.,* 70, 1105, 1986.
37. **Von Hoff, D. D., Kuhn, J., and Clark, G. M.,** Design and conduct of Phase I trials, in *Cancer Clinical Trials. Methods and Practice,* Buyse, M. E., Staquet, M. J., and Sylvester, R. J., Eds., Oxford University Press, Oxford, 1984, 210.

Chapter 23

PRECLINICAL PHASE II TRIALS

H. H. Fiebig and D. P. Berger

TABLE OF CONTENTS

I. INTRODUCTION

Most of the antitumor agents used in patients were initially selected for their activity in a variety of animal tumor models. The broadest screening program has been set up by the U.S. National Cancer Institute (NCI). Since 1955, the mouse leukemias Ll210 and P388 have been employed as the initial testing models. From 1975 to 1985, compounds with activity against P388 leukemia were tested in a second stage against a panel of solid tumors including four murine models and three human tumor xenografts.[1,2] Unfortunately, for the past 10 years, these tumor models have not been able to identify novel structures with activity against solid human malignancies.[3] Therefore, other strategies for primary screening have been proposed. Recently, the NCI has developed a tumor-oriented screening project in which more than 10,000 compounds per year are tested against a panel of human tumor cell lines *in vitro*.[4] However, the distribution, metabolism, and excretion of a given drug, as well as the effect on normal tissues, cannot be taken into account in *in vitro* assays. Furthermore, culture conditions may cause artificial changes in tumor biology of the *in vitro* systems, limiting the predictive value for the *in vivo* situation.

The usefulness of human tumor xenografts grown in nude mice for studies on tumor biology and cancer therapy has been demonstrated in this book. Pinedo has proposed to use four to five human tumor xenografts for the initial screening of new compounds.[5] However, such an approach requires large resources. Due to financial restraints, human tumor xenografts will rather be of value in the secondary stage of new drug development.

In this chapter, an important question concerning the xenograft model will be addressed; namely, the comparison of tumor response to chemotherapy in the nude mouse system and in the patient. Furthermore, some aspects of preclinical phase II trials will be described.

II. COMPARISON OF CHEMOSENSITIVITY

The various reports on comparisons of tumor response in nude mice and in the patient are summarized in Table 1. Only subcutaneous implants are included, which were originally derived from the respective patients. All groups except two used immunodeficient nude mice. Nowak et al.[10] and Shorthouse et al.[13] performed thymectomy and whole body irradiation in normal mice. There were differences with regard to the mouse strains and methodology used. Some groups implanted fragments, others injected tumor cell suspensions. The dose levels, drug administration schedules, number of treatment courses, and routes of application were not standardized. Some investigators included single drug therapy only, others also used combination chemotherapy. Three different "endpoints" were used for evaluation; namely, clinical response criteria (PR + CR), tumor growth inhibition (T/C%), and specific growth delay (SGD) (Table 1). Some authors did not specify evaluation of the patient's tumor response. One-third of patients bearing sensitive tumor types were evaluated repeatedly, as they were treated with several standard antitumor agents. Despite these limitations, similar results were obtained by the different laboratories.

In a total of 113 patients, 167 comparisons have been carried out (Table 2). Some 43 tumors responded both as xenograft in the corresponding patient (26%), and 111 showed no response in both systems (66%). Overall, the xenograft model predicted correctly for clinical response in 43 of 50 comparisons (sensitivity of 86%) and in 111 of 117 for resistance (specificity of 95%). The correct prediction for resistance is of limited value as most of the comparisons were obtained in tumor types which are chemoresistant in the clinic (colorectal cancer, non-small cell lung cancer, and melanoma). In contrast, the highly correct prediction for tumor sensitivity must be stressed. In this respect, the predictive value of the nude mouse-human tumor model is significantly higher as compared to *in vitro* systems.

TABLE 1
Comparison of Tumor Response to Chemotherapy in Nude Mice[1] and in the Patient

Tumor type	Number of patients	Number of comparisons	Evaluation criteria xenograft	Xenograft/Patient Correlations				Ref.
				R/R	P/P	P/R	R/P	
Bladder	1	2	PR + CR	—	2	—	—	6
Breast	2	4	PR + CR	2	2	—	—	6
	1	1	T/C%	—	1	—	—	7
	4	7	T/C%	5	2	—	—	8
Choriocarcimona	2	2	T/C%	1	1	—	—	9
Colorectal	18	27	PR + CR	2	23	2	—	6
	4	4	SGD > 2	1	3	—	—	10[2]
	2	2	T/C%	—	2	—	—	11
	1	1	T/C%	1	—	—	—	7
Head & Neck	2	2	SGD > 2	—	—	2	—	12
Lung, NSCLC	9	12	SGD > 2	—	12	—	—	13[2]
	6	9	PR + CR	—	8	—	1	6
SCLC [1]	6	8	SGD > 2	5	2	—	1	13[2]
	2	4	PR + CR	4	—	—	—	6
Melanoma	17	37	SGD > 2	8	25	1	3	14
	7	8	PR + CR	—	7	—	1	6
Ovary	7	7	SGD > 2	2	3	2	—	15
	1	1	PR + CR	1	—	—	—	6
	1	1	T/C%	1	—	—	—	16
Penile	1	1	PR + CR	1	—	—	—	6
Renal	1	2	PR + CR	—	2	—	—	6
Sarcoma	3	4	PR + CR	—	4	—	—	6
Skin, SCC	2	2	PR + CR	—	2	—	—	6
Stomach	6	9	PR + CR	4	5	—	—	6
	2	3	T/C%	—	3	—	—	7
Testes	2	2	PR + CR	2	—	—	—	6
Thymoma	1	2	PR + CR	—	2	—	—	6
Thyroid	1	1	PR + CR	1	—	—	—	6
Wilms	1	2	PR + CR	2	—	—	—	6
Total	113	167		43	111	7	6	
%		100%		26%	56%	4%	4%	

Note: NSCLC, non-small cell lung cancer; SCLC, small cell lung cancer; SCC, squamous cell carcinoma; R, clinical response (CR + PR) or meets evaluation criteria in nude mice; P, clinical progression or does not meet evaluation criteria in nude mice.

[1] After subcutaneous implantation.
[2] Thymectomized mice were used.

III. PREDICTION OF RESPONSE FOR THE INDIVIDUAL PATIENT

The response rates of the different human tumor types to single agents and combination chemotherapy are well known today. However, we cannot predict whether an individual patient will benefit from treatment. It might be possible to individualize drug regimens by assessing a patient's tumor response to a wide range of antitumor agents prior to the start of treatment. For this purpose, several *in vitro* "predictive assays" have been employed. They include long-term culture and evaluation of morphological changes,[17] biochemical assays to measure inhibition of

TABLE 2
Correlation of Tumor Response to Chemotherapy in the Xenografts and in Patients

Tumor response			Total	
Mouse / Patient		Prediction	n	%
Response	/ Response	True positive (TP)	43	26%
No Response	/ No Response	True negative (TN)	111	66%
Response	/ No Response	False positive (FP)	6	4%
No Response	/ Response	False negative (FN)	7	4%
Sensitivity [TP/(TP + FN)]			86%	
Specificity [TN/(TN + FP)]			95%	
Positive predictive value [TP/(TP + FP)]			88%	
Negative predictive value [TN/(TN + FN)]			94%	

DNA or RNA synthesis,[18] the *in vitro* colony formation,[19-22] and the reduction of tumor cells as measured by cytofluorometry.[23] These *in vitro* systems yielded highly correct prediction rates for tumor resistance. However, none of these assays could obtain clinical significance in selecting drugs for individual patients.

Since the human tumor xenograft model has demonstrated a highly correct prediction for tumor sensitivity and resistance, it might be used as a model for identifying drugs active against an individual patient's tumor. The potential of this approach has been mentioned on several occasions in the literature. However, data have been presented only from our own group.[6,24]

During a period of 5 years, we have concentrated on the use of the xenograft model as a predictive assay, and obtained a large number of treatment comparisons. More than 500 tumors have been transplanted and the patients were followed closely. Three major drawbacks of the xenograft model have become evident. First of all, only approximately 50% of the tumors of selected tumor types (colorectal cancer, lung cancer, and melanoma) can be grown in serial passage. Second, drug testing can usually be performed after two to six passages in nude mice, when tumor growth becomes regular. Therefore, chemosensitivity data become available at the earliest after 4 months in rapidly growing tumors and after 12 to 18 months in slowly growing xenografts. The third disadvantage lies within the high costs of nude mice and their complex maintenance requirements.

The nude mouse-human tumor model has only a limited practical role for individual patients, which lies in the determination of a second-line therapy for patients with metastatic disease and in pretherapeutic sensitivity testing for patients with localized disease who do not need immediate chemotherapy. However, the probability of identifying an effective second-line therapy is low and not all patients, whose tumors are growing in nude mice, will develop metastases for which they require chemotherapy.

Two-thirds of our human tumor lines were established from material obtained at surgery for the primary tumor with no immediate need to treat the patient with cytostatic agents. In 28 of 80 cases (35%), the test result was available in the xenograft model before the patient started chemotherapy; in 18 cases (23%), the patient's tumor was treated simultaneously; and in 34 cases (43%), the patients were treated first. Effective drugs were mainly found in tumor types which demonstrate high response rates in the clinic, namely testicular, breast and stomach cancers, as well as small cell lung cancers.

The results from our studies demonstrate that the xenograft model has no practical significance as a predictive assay for the individual patient. Due to the highly correct prediction for tumor sensitivity and resistance, the nude mouse-human tumor model will have considerable value for the secondary screening of new compounds.[21,25]

IV. PRECLINICAL PHASE II TRIALS

The screening program of the NCI has utilized three human tumor models for the secondary screening from 1975 to 1985. In the s.c. assay, the colon cancer CX-1 and the lung cancer LX-1 were resistant against 0 of 20 and 1 of 20 standard agents, respectively. The breast cancer MX-1 was sensitive to cisplatin, a drug without a role in breast cancer patients. In contrast, MX-1 appeared resistant against doxorubicin, 5-fluorouracil, and methotrexate, drugs with efficacy in this tumor type.[1,26] Due to the known heterogeneity of human malignancies, a panel of tumors of the same tumor type will better reflect the constitutive variation in terms of chemosensitivity, histology, expression of markers, receptors and oncogenes, as well as possible genetic changes. From a statistical point of view, 14 patients are usually required for the first stage of a clinical phase II study[27] when the required rate of tumor response is set as 20%. If not a single response is observed in a total of 14 patients, the statistical probability to falsely reject this compound is less than 5% The outcome of clinical studies, however, is markedly influenced by prognostic factors which have been determined for most tumor types in man.

If human tumor xenografts are used as a representation of human malignancy, the outcome will depend on the various characteristics and the number of the selected tumor lines included. They should have properties known to occur in a particular tumor type. A critical question is the number of lines necessary per tumor entity. Nude mouse studies are very expensive. The animals require special facilities for breeding and maintenance. Experiments using s.c. tumor implants are long-lasting and cover a period between 1.5 and 4 months. Furthermore, a selection of fast-growing tumors which are easier to handle must be avoided. Therefore, broad studies in 14 different xenografts representing various characteristics of an individual tumor type will not be a feasible approach.

The number of xenografts necessary for a particular tumor type depends mainly on the individual pattern of tumor response. This question cannot be studied systematically in patients, but is a valid option in the xenograft model. Does each tumor line have its individual response profile? Tables 3 through 5 show the response profiles of xenografts derived from colorectal cancer, gastric cancer, and small cell lung cancer towards a number of standard cytostatic agents. The response of 20 colon cancer xenografts was rather uniform. Only two tumors responded, mainly against alkylating agents. Tumors were resistant against all other standard drugs. Out of five small cell lung cancer xenografts studied, three were very sensitive, mainly against the same compounds. The two resistant xenografts were heavily pretreated in the patient, and chemoresistance of the lines in nude mice correlated with the clinical behavior of the tumors. The variability in the chemosensitivity pattern of colon and small cell lung cancer xenografts appeared rather low. For these tumor types, testing of a smaller number of lines (e.g., 6 xenografts per tumor type) may be adequate. The response of 13 gastric cancer xenografts showed a high diversity. All responsive tumors displayed an individual response profile. Only four xenografts were resistant against all standard drugs. This individual behavior suggests that more than six human tumor lines will be necessary for tumor types with a heterogenous drug response.

V. MULTICENTER COLLABORATION IN DRUG DEVELOPMENT

Within one institution, therapeutic studies in human tumor xenografts are feasible only in a limited number of tumor lines. In Europe, several laboratories have gained experience in the establishment and characterization of human tumor xenografts as well as in the design of treatment schedules. Most laboratories concentrate on a few tumor types. Following a suggestion of the New Drug Development Office of the EORTC, a cooperative program was initiated

TABLE 3
***In Vivo* Responsiveness of Human Colorectal Cancer Xenografts**

Tumor desig- nation	Doubling time, days (1xw)	Drugs[1]														Respon- sive[2]/ total
		DOX	BLEO	CCNU	CY	DTIC	FU	IFO	MITO	MTX	PLAT	VBL	VCR	VIND	VP	
CXF 90	12				–[3]		–								–	0/ 3
CXF 110	15				–		–		–							0/ 3
CXF 158	10	–		++	–	+++	–	–	–		–			–		2/ 9
CXF 164	12				–		–		–		–					0/ 4
CXF 167	16						–		–							0/ 2
CXF 243	10	–		–	–	–	–	–	–		–	–		–	–	0/11
CXF 260	6						–		–							0/ 2
CXF 264	9						–		–							0/ 2
CXF 269	10						–		–							0/ 2
CXF 270	15						–		–							0/ 2
CXF 273	18						–		–							0/ 2
CXF 280	6	–	–	+++	++	++	–	++	+++	–	+			–	–	6/12
CXF 288	17						–		±							0/ 2
CXF 355	10						–		–							0/ 2
CXF 432	14						–		–							0/ 2
CXF 457	15						–		–							0/ 2
CXF 479	23						–		–							0/ 2
CXF 886	16	–				–					–	–		–		0/ 5
CXF 975	9	–	–		–	–	–	–	–		–	–	–	–	–	0/12
CXF 1103	4	–	–	–	–	–	–	–	–		–	–	–	–	–	0/13

[1] DOX, Doxorubicin; BLEO, Bleomycin; CCNU, Cyclohexyl-chloroethyl-nitrosourea; CY, Cyclophosphamide; DTIC, Dacarbazine; FU, 5-Fluorouracil; IFO, Ifosfamide; MITO, Mitomycin C; MTX, Methotrexate; PLAT, Cisplatin; VBL, Vinblastine; VCR, Vincristine; VIND, Vindesine; VP, VP16, Etoposide.

[2] Responsive (+/++/+++).

[3] –, Progression, ≥ 125% of initial tumor size; ±, No change, 76 – 125%; +, Minor regression, 51 – 75%; ++, Partial remission, 20 – 50%; +++, Complete remission, <20%.

TABLE 4
In Vivo Responsiveness of Human Gastric Cancer Xenografts

Tumor Desig- nation	Doubling time, days (1xw)	Drugs[1] DOX	BLEO	CCNU	CY	DTIC	FU	IFO	MITO	MTX	PLAT	VBL	VCR	VIND	VP	Respon- sive[2]/ total
GXF 90	9—12	–[3]	–	++	+	–	–	++	+++	–	+++	+	–	++	–	7/14
GXF 145	5	–		–	–		–		–		–			–	–	0/ 8
GXF 180	6	–		–		–	–		±		–			–	–	0/ 8
GXF 198	14	–		–			–		–		–			–	–	0/ 7
GXF 209	17	+		+	–		–		+++		++			±	–	4/ 8
GXF 214	9	–		–			–		–		–			–	–	0/ 7
GXF 218	8	–		–			–		++		–			+	–	2/ 7
GXF 236	10	–		–			–		–		–			++	–	1/ 7
GXF 251	9—13	–	–	–	–	–	–	–	–		–	–	–	–	–	0/13
GXF 281	13—18	–	–	+++	–	±	–		+++	–	++			–	–	3/11
GXF 324	11—13	–	–	–	–	–	–	–	–		–	–	–	–	–	0/13
GXF 350	7	–		–			–		–		–			±	–	0/ 7
GXF 602	6	–		++			–		–		–			+	–	2/ 7

1 DOX, Doxorubicin; BLEO, Bleomycin; CCNU, Cyclohexyl-chloroethyl-nitrosourea; CY, Cyclophosphamide; DTIC, Dacarbazine; FU, 5-Fluorouracil; IFO, Ifosfamide; MITO, Mitomycin C; MTX, Methotrexate; PLAT, Cisplatin; VBL, Vinblastine; VCR, Vincristine; VIND, Vindesine; VP, VP16, Etoposide.

2 Responsive (+/++/+++).

3 –, Progression, ≥ 125% of initial tumor size; ±, No change, 76 – 125%; +, Minor regression, 51 – 75%; ++, Partial remission, 20 – 50%; +++, Complete remission, < 20%.

TABLE 5
***In Vivo* Responsiveness of Human Small Cell Lung Cancer Xenografts**

Tumor Desig- nation	Doubling time, days (1xw)	Drugs[1] DOX	CCNU	CY	DTIC	FU	IFO	MITO	MTX	NOVA	PLAT	VCR	VIND	VP	Respon- sive[2]/ total
LXFS 387	11,12	++[3]	+	+	+++		+	–	–		++		++	+++	8/10
LXFS 428	18	–		–							–				0/ 3
LXFS 538	8—15	±	++	++	±		++	++	–	±	±	–	+++	+	6/12
LXFS 573	11	–		+				+			–		++	++	4/ 6
LXFS 605	10,11	+	++	++	++		+++	+++		–	++	++	++	++	10/11
LXFS 650	6—12	++	–	±	–	–	±	–	–	–	–	–	++	–	2/13

[1] DOX, Doxorubicin; BLEO, Bleomycin; CCNU, Cyclohexyl-chloroethyl-nitrosourea; CY, Cyclophosphamide; DTIC, Dacarbazine; FU, 5-Fluorouracil; IFO, Ifosfamide; MITO, Mitomycin C; MTX, Methotrexate; PLAT, Cisplatin; VBL, Vinblastine; VCR, Vincristine; VIND, Vindesine; VP, VP16, Etoposide.

[2] Responsive (+/++/+++).

[3] –, Progression, ≥ 125% of initial tumor size; ±, No change, 76 – 125%; +, Minor regression, 51 – 75%; ++, Partial remission, 20 – 50%; +++, Complete remission, < 20%.

in 1986.[5] As a result, the "Xenograft Group" has been founded, in which tumors have been shared, methods were standardized, and a central evaluation has been set up. At the present time, six different xenografts per tumor type (cancer of the large bowel, breast, ovary, head and neck, small cell and non-small cell lung cancer, as well as melanomas) are being used. In the first study performed by this group, a clinically active compound (doxorubicin) and an inactive compound (*m*-AMSA, not active in solid tumors) were compared with two investigational drugs, brequinar sodium (DUP 785) and the ellipticine datelliptium.[28] Doxorubicin was shown to be active in 8 of 32 tumor models (25%), whereas *m*-AMSA was inactive.[29]

In a second study, the effect of cisplatin was compared with diaziquone (AZQ) and the novel ellipticines pazelliptine and retelliptine. Cisplatin was identified as active in several human tumor lines, whereas AZQ produced a very short-lasting tumor inhibition, which is in contrast to the clinic. The two new compounds were only marginally active.[30] Overall, these studies showed that multicenter collaboration studies using human tumor xenografts are feasible. Hopefully, they will help to reduce the number of clinical phase II trials with negative results.

ACKNOWLEDGEMENTS

We wish to thank our co-workers E. Wallbrecher, U. Winterhalter, C. Berg, and V. Haberstroh for their excellent assistance in the establishment and characterization of human tumor xenografts in nude mice. K. H. Widmer is thanked for software development and data evaluation.

REFERENCES

1. **Goldin, A., Venditti, J. M., MacDonald, J. S., Muggia, F. M., Henney, J. E., and DeVita, Jr., V. T.,** Current results of the screening program at the division of cancer treatment, National Cancer Institute, *Eur. J. Cancer*, 17, 129, 1981.
2. **Driscoll, J. S.,** The preclinical new drug research program of the National Cancer Institute, *Cancer Treat. Rep.*, 68, 63, 1984.
3. **Marsoni, S. and Wittes, R.,** Clinical development of anticancer agents — a National Cancer Institute perspective: a commentary, *Cancer Treat. Rep.*, 68, 77, 1984.
4. **Boyd, M. R.,** Status of the NCI preclinical antitumor drug discovery scree, *Principles Practice Oncol. Updates*, 3, 1, 1989.
5. **Pinedo, H. M.,** Development of new anti-cancer drugs, *Med. Oncol. Tumor Pharmacother.*, 3, 63, 1986.
6. **Fiebig, H. H., Winterhalter, B. R., and Scholz, C. C.,** Limited potential of the human tumor — nude mouse system for pretherapeutic drug testing, *ZAC Zeitschrift antimikr. antineopl. Chemotherapie*, 6, 17, 1988.
7. **Fujita, M., Hayata, S., and Taguchi, T.,** Relationship of chemotherapy on human cancer xenografts in nude mice to clinical response in the donor patient, *J. Surg. Oncol.*, 15, 211, 1981.
8. **Giovanella, B. C., Stehlin, J. S., and Shepard, R. C.,** Experimental chemotherapy of human breast carcinomas heterotransplanted in nude mice, in *Proc. 2nd Int. Workshop on Nude Mice*, Nomura, T., Ohsawa, N., Tamaoki, N., and Fujiwara, K., Eds., Gustav Fischer Verlag, Stuttgart, 1977, 475.
9. **Hayashi, H., Kameya, T., Shimosato, Y., and Mukojima, T.,** Chemotherapy of human choriocarcinoma transplanted to nude mice, *Am. J. Obstet. Gynecol.*, 131, 548, 1978.
10. **Nowak, K., Peckham, M. J., and Steel, G. G.,** Variation in response of xenografts of colorectal carcinoma to chemotherapy, *Br. J. Cancer*, 37, 376, 1978.
11. **Kubota, T., Shimosato, Y., and Nagai, K.,** Experimental chemotherapy of carcinoma of the human stomach and colon serially transplanted in nude mice, *Gann*, 69, 299, 1978.
12. **Braakhuis, B. J. M., Schoevers, E. J., Heinerman, E. C. M., Sneeuwloper, G., and Snow, G. B.,** Chemotherapy of human head and neck cancer xenografts with three clinically active drugs: *cis*-platinum, bleomycin and methotrexate, *Br. J. Cancer*, 48, 711, 1983.

13. **Shorthouse, A. J., Smyth, J. F., Steel, G. G., Ellison, M., Mills, J., and Peckham, M. J.,** The human tumor xenograft — a valid model in experimental chemotherapy?, *Br. J. Surg.*, 67, 715, 1980.
14. **Osieka, R.,** Studies on drug resistance in a human melanoma xenograft system, *Cancer Treat. Rev.*, 11 (Suppl. A), 85, 1984.
15. **Friedlander, M. L., Russell, P., Taylor, I. W., and Tattersall, M. H. N.,** Ovarian tumor xenografts in the study of the biology of human epithelial ovarian cancer, *Br. J. Cancer*, 51, 319, 1985.
16. **Sawada, M., Matsui, Y., and Okudaira, Y.,** Establishment of a new ovarian tumor line in nude mice and its application to treatment of the donor patient, *Gynec. Oncol.*, 21, 320, 1985.
17. **Limburg, H. and Krahe, M.,** Züchtung von menschlichem Krebsgewebe in der Gewebekultur und seine Sensibilitätstestung gegen neuere Zytostatika, *Dtsch. Med. Wschr.*, 89, 1983, 1964.
18. **Volm, M., Wayss, K., Mattern, J., Kleckow, M., and Vogt-Moykopf, I.,** Resistenztestung und Chemotherapieergebnis bei Bronchialtumoren, *Dtsch. Med. Wschr.*, 103, 1266, 1978.
19. **Salmon, S. E., Alberts, D. S., Meyskens, F. L., Durie, B. G. M., Jones, S. E., Soehnlein, B., Young, L., Chen, G. H., and Moon, T. E.,** Clinical correlation of *in-vitro* drug sensitivity, in *Cloning of Human Tumor Stem Cells*, Salmon, S. E., Ed., Alan R. Liss, New York, 1980, 223.
20. **Von Hoff, D. D., Casper, J., Bradley, E., Sandbach, J., Jones, D., and Makuch, R.,** Association between human tumor colony-forming assay results and response of an individual patient's tumor to chemotherapy, *Am. J. Med.*, 70, 1027, 1981.
21. **Fiebig, H. H., Schmid, J. R., Bieser, W., Henss, H., and Löhr, G. W.,** Colony assay with human tumor xenografts, murine tumors and human bone marrow. Potential for anticancer drug development, *Eur. J. Cancer Clin. Oncol.*, 7, 937, 1987.
22. **Berger, D. P., Henβ, H., Winterhalter, B. R., and Fiebig, H. H.,** The clonogenic assay with human tumor xenografts: evaluation, predictive value and application for drug screening, *Ann. Oncol.*, 1, 333, 1990.
23. **Valet, G., Warnecke, H. H., and Kahle, H.,** New possibilities of cytostatic drug testing on patient tumor cells by flow cytometry, *Blut*, 49, 37, 1984.
24. **Fiebig, H. H.,** Comparison of tumor response in nude mice and in patients, in *Human Tumour Xenografts in Anticancer Drug Development*, Winograd, B., Peckham, M. J., and Pinedo, H. M., Eds., Springer-Verlag, Berlin, 1988, 25.
25. **Fiebig, H. H.,** Feasibility and usefulness of human tumor xenografts in secondary screening, *Investigational New Drugs*, 5, 78, 1987.
26. **Ovejera, A. A. and Houchens, D. P.,** Human tumor xenografts in athymic nude mice as a preclinical screen for anticancer agents, *Semin. Oncol.*, 8, 386, 1981.
27. **Gehan, E. A.,** The determination of the number of patients required in a preliminary and follow-up trial of a new chemotherapeutic agent, *J. Chronic Dis.*, 13, 346, 1961.
28. **Boven, E., Winograd, B., Fodstad, O., Lobezzoo, M. W., and Pinedo, H. M.,** Preclinical phase II studies in human tumor lines: a European multicenter study, *Eur. J. Cancer*, 24, 567, 1988.
29. **Fiebig, H. H., Atassi, G., Boven, E., Braakhuis, B. J. M., Fodstad, O., Langdon, S., Winograd, B., Lobbezoo, M. W., and Pinedo, H. M.,** Preclinical phase II studies in human tumor xenografts; validation and first drug testing in a multicenter study, *Proc. 6th EORTC-NCI Symp. on New Drugs in Cancer Therapy*, Amsterdam, 1989, 39.
30. **Langdon, S. P., Hendriks, H. R., Berger, D., Boven, E., Braakhuis, B. J. M., Fiebig, H. H., Fodstad, O., Pratesi, G., Smyth, J. F., and Winograd, B.,** Preclinical phase II screening in human tumor xenografts: a European multicenter follow-up study, *Proc. Am. Assoc. Cancer Res.*, 32, 380, 1991.

INDEX

A

B

D

E

H

I

J

K

L

M

N

O

P

Q

R

S

T